Biomathematics

Volume 5

Edited by

K. Krickeberg · R. C. Lewontin · J. Neyman
M. Schreiber

Albert Jacquard

The Genetic Structure
of Populations

With 92 Figures

Springer-Verlag New York · Heidelberg · Berlin 1974

Professor Albert Jacquard

Institut National d'Etudes Démographiques Paris, France

Translated by

D. and B. Charlesworth

Department of Genetics, The University of Liverpool, England

Title of the French Edition:

Structures Génétiques des Populations

© Masson & Cie, Editeurs, Paris, 1970

AMS Subject Classifications (1970)

92-A-10

ISBN 0-387-06329-3 Springer-Verlag New York · Heidelberg · Berlin

ISBN 3-540-06329-3 Springer-Verlag Berlin · Heidelberg · New York

Preface to the English Edition

It is part of the ideology of science that it is an international enterprise, carried out by a community that knows no barriers of nation or culture. But the reality is somewhat different. Despite the best intentions of scientists to form a single community, unseparated by differences of national and political viewpoint, they are, in fact, separated by language. Scientific literature in German is not generally assimilated by French workers, nor that appearing in French by those whose native language is English. The problem appears to have become more severe since the last war, because the ascendance of the United States as the preeminent economic power led, in a time of big and expensive science, to a predominance of American scientific production and a growing tendency (at least among English-speakers) to regard English as the international language of science. International congresses and journals of world circulation have come more and more to take English as their standard or official language. As a result, students and scientific workers in the English speaking world have become more linguistically parochial than ever before and have been cut off from a considerable scientific literature.

Population genetics has been no exception to the rule. The elegant and extremely innovative theoretical work of Malécot, for example, is only now being properly assimilated by population biologists outside France. It was therefore with some sense of frustration that I read Prof. Jacquard's "Structures Génétiques des Populations", for I realized that this superb treatment of the theory of evolutionary genetics would be unavailable to me as a teacher because it was inaccessible to my students as readers. What I found so attractive in Jacquard's book was the lucidity and elegance of its presentation, but most especially the fusion of demographic and genetical concepts with abundant examples from human populations. The fusion of genetics and demography has been a slow process since Fisher first briefly considered the problems of gene frequency change and population growth in the "Genetical Theory of Natural Selection" in 1929. It is still far from complete, but the point of view represented by Prof. Jacquard, a geneticist and a demographer, is slowly gaining. I was thus delighted at the prospect of an English translation of "Structures Génétiques des Populations" so that this point of view could be exposed to the widest possible audience.

Prof. Jacquard has been extremely fortunate in his choice of translators. The Charlesworths have combined linguistic skill with scientific contributions in demography and genetics to make "The Genetic Structure of Populations" not merely a translation, but a new work of even greater virtue than the original.

R. C. Lewontin

Author's Preface to the English Edition

This book was originally written for French students; its style of reasoning and method of presentation were chosen to conform with current usage in French universities. There is therefore a risk in publishing an English edition. Professor Lewontin, however, feels that we should take this risk; I should like to thank him for his favourable opinion of my book.

My hesitation in seeing my book exposed to a wider public has been lessened by the help I have received from my translators, who have criticised the book as well as performing the heavy work of the translation itself; these criticisms have helped me to fill in some of the gaps that were left in the first edition, and to correct some errors. I am very grateful to them.

A. Jacquard

Translators' Preface

This book is a translation of "Structures Génétiques des Populations", which was published in 1970. There are a number of changes in the present edition. In particular, there are three new chapters at the end of the book; also, the original treatment of populations with overlapping generations has been replaced by a new version written by B. Charlesworth (Chapter 7, and Section 3 of Chapter 10). There is a new Appendix on difference equations, and an additional Section (2.2.3) in Chapter 12, by B. Charlesworth, and a number of smaller alterations throughout the book, some by Professor Jacquard and some by the translators.

We would like to thank Professor Jacquard for his cooperation with us throughout the work of translating his book, and especially for his tolerance and patience when we have made criticisms.

D. Charlesworth
B. Charlesworth

Table of Contents

Part 1
Basic Facts and Concepts

Part 2
A Reference Model: Absence of Evolutionary Factors

Part 3
The Causes of Evolutionary Changes in Populations

Part 4
The Study of Human Population Structure

Introduction

The Individual

From the very first moment of the creation of a new human being at fertilisation, the single cell which is the new individual is already endowed with its full complement of hereditary information. This single cell will divide and form millions of new cells, which will each be adapted for specific functions; millions of chemical compounds will be synthesised, which will be used in the cells themselves, or for communications between cells; mechanisms for the precise regulation of all kinds of processes will develop. All these events take place in a determined sequence: embryonic development, growth, senescence, then death.

All the information which will ensure that these events take place is contained in the original single cell; half of the information comes from the egg, and half from the spermatozoon which fertilises it. The chromosomes of the cell carry all the instructions for the development of an individual, for example a human being, or more exactly, *this particular* human being.

The genetic information which the new individual receives is the basis of his individuality. It distinguishes him from all other individuals; no man now or in the past has had precisely the same genes as another. Of course, the development of the individual, his final size and whether he grows up to be strong or weak are greatly affected by his environment. This includes the state of health of his mother before he was born, the nutrition he receives, the temperatures and irradiations to which he is subjected, and many other factors. But the effects which environmental factors will have are themselves determined by the genetic information, since this determines the internal processes which can, in fact, take place. The end result of the individual's developmental process depends on random events in his environment, but the effects of these events obey probability distributions that depend on the genetic information brought together at his conception.

At each cell division, the new cell has a structure which corresponds to the function it will carry out in the organism (e.g. nerve cell, bone cell, blood cell, etc.). Whatever the function of the cell, however, it will receive a full copy of the genetic information of the individual of which it is a member. Each cell carries all the information which determines the whole individual, including all the special characteristics that are unique to this individual. Each single cell of a particular individual is, above all, a cell of *this* individual, and not just a cell with a certain function, such as a nerve cell, or a bone cell.

There is one exception to this. The reproductive cells, eggs or spermatozoa, carry only half of the genetic information of the individual they come from. Nearly all these cells are destined to die. A tiny proportion of them will find another reproductive cell of the opposite sort, and fuse with it. In the new cell which results, there is once again a complete set of the genetic instructions which are necessary for a new human being to develop.

The Population

We have seen that each individual passes only half of his genes to any one of his offspring. It is a matter of chance which of the two genes for a character the offspring receives.

In a sufficiently large population, the genetic heritage is maintained through the generations, despite the fact that, in each generation, the genes are segregated into the gametes, and then come together in new combinations in the new individuals. A common heritage is therefore concealed behind the diversity of the individuals of a population.

If the chromosomes of an individual carry all the information ("genotype") necessary for the organism to synthesise a certain metabolic product, then he may, in fact, synthesise it, and exhibit the characteristic "phenotype", which could be advantageous or disadvantageous. Next generation, it may happen that none of his offspring receives all the information for this synthesis, and so the corresponding phenotype has disappeared. However, the information for the synthesis of this product is not lost; in a later generation, it could once again come together into a single individual, and this individual would have the same phenotype as his ancestor.

At the population level, the fundamental reality is the information that is carried by the chromosomes coiled up in the reproductive cells, not the characters which the individuals manifest. The units of information carried in the chromosomes are handed down through the generations, often unexpressed and hidden, alternately separated and paired up together according to the chances of fertilisation, but always preserved unchanged.

Above and beyond the particular group of individuals which exist at any one time, the fundamental property of a population is its common genetic heritage.

Each individual can have a small effect on this common heritage; the genetic information which he bears will form a larger or smaller proportion of the total in the next generation, depending on whether he has many or few offspring. In this way, the mean frequency of any particular unit of information can change from one generation to the next.

Population genetics is chiefly concerned with the study of these changes. Its aim is to answer the question: "What factors affect the genetic heritage of populations, and what are their effects?"

General Bibliography

Burdette, W.: Methodology in human genetics. San Francisco: Holden-Day 1961.

Cavalli-Sforza, L.L., Bodmer, W.F.: The genetics of human populations. San Francisco: W.H. Freeman 1971.

Crow, J.F., Kimura, M.: An introduction to population genetics theory. New York: Harper and Row 1970.

Elandt-Johnson, R.C.: Probability models and statistical methods in genetics. New York: Wiley 1971.

Ewens, W.J.: Population genetics. London: Methuen 1969.

Falconer, D.S.: Introduction to quantitative genetics. Edinburgh: Oliver & Boyd 1960.

Fisher, R.A.: The genetical theory of natural selection. Oxford: Clarendon Press 1930, (Reprinted. New York: Dover Publications 1958).

Fisher, R.A.: The theory of inbreeding. Edinburgh: Oliver & Boyd 1949.

Kempthorne, O.: An introduction to genetic statistics. New York: Wiley 1957.

Kimura, M.: Diffusion models in population genetics. London: Methuen 1964.

Kimura, M., Ohta, T.: Theoretical aspects of population genetics. Princeton, New Jersey: Princeton University Press 1971.

Li, C.C.: Population genetics. Chicago, Illinois: Univ. of Chicago Press 1955.

Malécot, G.: Les mathématiques de l'heredité. Paris: Masson 1948.

Malécot, G.: Probabilités et heredité. Paris: Presses Universitaires de France 1966.

Malécot, G.: The mathematics of heredity, (Translation and revised version of "Les mathématiques de l'heredité"). San Francisco: W.H. Freeman 1969.

McKusick, V.A.: Human genetics. Englewood Cliffs, New Jersey: Prentice Hall 1964.

Moran, P.A.P.: The statistical processes of evolutionary theory. Oxford: Clarendon Press 1962.

Stern, C.: Principles of human genetics 3rd ed. San Francisco: W.H. Freeman 1973.

Wright, S.: Evolution and the genetics of populations. Vol. I (1968): Genetic and biometric foundations. Vol. II (1969): The theory of gene frequencies. Chicago, Illinois: Univ. of Chicago Press.

PART 1

Basic Facts and Concepts

The aim of population genetics is to study changes in the genetic heritage, at the population level. It may be useful to the reader to be reminded of certain biological facts at the level of the individual, or even of the single cell, which relate to the transmission of this heritage from one generation to the next.

In this section, we shall also give some definitions of the terms which will be used in this book, and introduce some fundamental ideas and notation that will be used frequently.

Chapter 1

The Foundations of Genetics

1. The Mendelian Theory of Inheritance

All our ideas about the inheritance of particular characters, and also about changes in the characteristics of populations, make use of the concepts introduced a century ago by Gregor Mendel (1822–1884).

Working in a small[1] experimental garden in his monastery in Brno, in Moravia, and using peas as his experimental material, Mendel succeeded in formulating a hypothesis which explained the results of his crosses in a simple fashion.

His hypothesis was contrary to the accepted ideas of his time, and his paper published in 1865[2] remained unknown. In 1900, sixteen years after Mendel's death, his work was rediscovered, first by the Dutchman Hugo de Vries, then by the German Correns and the Austrian von Tschermak. From then onwards, the science of genetics could develop. The school of Morgan, in the U.S.A., was especially important in this. But thirty-five precious years had been lost.

1.1. Mendel's First Law. The Law of Segregation

When he crossed peas with yellow cotyledons with peas with green cotyledons, Mendel found that all the hybrids had yellow cotyledons. When he crossed the hybrids among themselves (or, rather, selfed them), he obtained two types of peas again; the character green cotyledons, which was not expressed in the first generation hybrids, reappeared in the second generation: "segregation" of the characters had occurred. Mendel also found that, in the second generation, he always obtained proportions close to 3 yellow to 1 green. These experiments were done with six other characters (round or wrinkled seeds, axial or terminal inflorescences, etc.), and similar results were obtained.

Mendel's hypothesis to explain these results can be expressed as follows (using a different terminology from Mendel's):

---

[1] 250 m² in area.

[2] "Versuche über Pflanzenhybriden" was published in the Verhandlungen des naturforschenden Vereines in Brünn **4** (1865).

1. *Each character of an individual is controlled by two "factors", the "genes", one of which the individual receives from its male parent, and one from its female parent.*

2. *When an individual carries two different genes for a particular character, one of them will be expressed ("dominant"), while the effect of the other may not be apparent ("recessive").*

3. *A reproductive cell produced by an individual bears, for each character, one and only one of the two genes which the individual carries.*

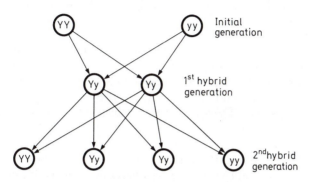

Fig. 1.1. Mendel's fundamental experiment

In the example given above, plants from the line with yellow cotyledons carry only "yellow" genes, and the genetic constitution of these plants can be written as (Y/Y). Plants from the line with green cotyledons have the genetic constitution (y/y). (In what follows, genetic constitution will be designated by pairs of letters, and will be enclosed in brackets. Lower-case letters will stand for recessive genes.)

In the cross between the yellow and green lines, individuals of the first hybrid generation receive one gene from a parent whose genetic constitution is (Y/Y), and the other from a (y/y) parent. The genetic constitution of these individuals is therefore (Y/y). The Y gene is dominant to the y gene, so that these individuals are all yellow.

The second generation is the result of crossing the hybrids, and this cross can be written as:

$$(Y/y) \times (Y/y).$$

Each parent will produce gametes of which half carry the gene Y, and half the gene y. The offspring can therefore be of three sorts:

one quarter will be   (Y/Y)
two quarters will be   (Y/y) or (y/Y)
one quarter will be   (y/y).

Because Y is dominant over y, the (Y/y) or (y/Y) peas are yellow, so we obtain the ratio 3 yellow peas to 1 green, which was the observed ratio.

Mendel's fundamental insight was the realisation that the character which was not manifest in the first hybrid generation (the character "green cotyledons", in this example) must somehow remain present in this generation, since it reappears in the next generation.

In order to explain this paradox, we have to accept the duplication of the information controlling characters. This duplication exists in all organism which reproduce sexually, although it may be only a transitory phase of the life-cycle as in many lower plants.

Mendel's fundamental hypothesis is that the hereditary material has the properties of discontinuity and stability. The genes controlling a character separate from one another, and come together in pairs in new individuals. They co-exist within individuals, in whom the action of one gene can mask the action of another, but they are themselves unaltered by this co-existence. They remain indivisible, and do not exchange parts with one another.

In 1865, such a "quantum" concept of a biological phenomenon was unacceptable. Even in 1900, Karl Pearson, in his studies of hereditary phenomena, was still using Galton's hypothesis of the fusion (and not the co-existence) of the maternal and paternal hereditary contributions. It is therefore not surprising that Mendel's hypothesis was ignored when it was first proposed.

1.2. Mendel's Second Law. Independent Assortment

Mendel also studied crosses between lines which differed for two characters: one line had yellow cotyledons and round seeds, while the other line had green cotyledons and wrinkled seeds. The first hybrid generation seeds were all yellow and round, but the second generation consisted of four types, showing the four possible combinations of the characters. The proportions were: 9/16 yellow, round, 1/16 green, wrinkled, 3/16 yellow, wrinkled, and 3/16 green, round. Mendel obtained similar results with seven different characters altogether.

Assuming that the first law is valid, these results can be explained by the single further hypothesis:

Genes controlling different characters segregate independently. We can write the genetic constitutions of the original lines as (Y/Y R/R) and (y/y r/r), and the first hybrid generation as (Y/y R/r). The gene Y is dominant to y, and R is dominant to r, so the hybrid peas are yellow and round. The hybrid plants produce four types of gametes: (YR), (yR), (Yr) and (yr). Since the genes controlling cotyledon colour are assumed to segregate independently of the genes controlling the form of the seed,

Table 1.1. Independent segregation of two characters

Gametes	Y R	Y r	y R	y r
Y R	(Y/Y R/R) Yellow Round	(Y/Y R/r) Yellow Round	(Y/y R/R) Yellow Round	(Y/y R/r) Yellow Round
Y r	(Y/Y R/r) Yellow Round	(Y/Y r/r) Yellow Wrinkled	(Y/y R/r) Yellow Round	(Y/y r/r) Yellow Wrinkled
y R	(Y/y R/R) Yellow Round	(Y/y R/r) Yellow Round	(y/y R/R) Green Round	(y/y R/r) Green Round
y r	(Y/y R/r) Yellow Round	(Y/y r/r) Yellow Wrinkled	(y/y R/r) Green Round	(y/y r/r) Green Wrinkled

we expect the frequencies of all four types of gametes to be the same — $\frac{1}{4}$ of the total. Table 1.1 shows the genotypes obtained by fusion of the four types of male ($\male$) gametes with each of the four types of female ($\female$) gametes. The genotypes are shown in brackets, and the phenotypes, which are obtained using the dominance relations of the genes, are given below them.

Remembering that the gene Y is dominant to y, and R to r, we have:

1. Four genotypes correspond to yellow, round seeds. They are: (Y/Y R/R), (Y/Y R/r), (Y/y R/R) and (Y/y R/r). These are formed by nine out of the 16 possible combinations of gamete types.

2. Two genotypes — (Y/Y r/r) and (Y/y r/r) — give yellow, wrinkled seeds. These correspond to three of the combinations of gamete types.

3. Two genotypes — (y/y R/R) and (y/y R/r) — give green, round seeds. These correspond to three of the combinations of gamete types.

4. Finally, the green, wrinkled seeds must be of the genotype (y/y r/r), which can be produced by only one combination of gamete types.

Since fertilisation occurs at random, the frequencies of all the combinations of gametes in Table 1.1 are equal, so the frequencies of the four types of pea will be 9/16, 3/16, 3/16, and 1/16.

1.3. Restriction of Mendel's Second Law. Linkage

Mendel's second law is often found not to hold, unlike the first law, which is very generally valid. Morgan, using the fruit fly, *Drosophila melanogaster*, discovered that independent segregation of characters is by no means a general rule. He found that it was possible to group the characters of an organism into a number of "linkage groups"; two genes are considered to belong to the same linkage group if there is a tendency for the parental associations of the characters to carry over into the offspring; a gene is considered not to belong to a linkage group if it segregates independently of all the genes of that group. The degree of linkage between genes can be measured by the strength of the tendency for the parental associations of the characters to carry over into the offspring; the proportion of the offspring that show the new character combinations is used as a measure of linkage.

Consider another cross between lines of peas: if plants from a line with purple flowers and long pollen grains are crossed with plants from a line with red flowers and short pollen, the first hybrid generation plants have purple flowers and long pollen. The genetic formulae for the parents are (P/P L/L) and (p/p l/l), and that for the hybrids is (P/p L/l). P is dominant to p, and L to l, so the hybrids have purple flowers and long pollen. In the second generation, four types appear, as expected: the two types showing the parental combinations of the characters, purple, long and red, short, and the two types with new combinations of the characters purple, short and red, long. But the two types with the parental associations of the characters are much more frequent than Mendel's second law would predict. Instead of the expected frequencies $9/16 = 55\%$, $1/16 = 7\%$, $3/16 = 19\%$ and $3/16 = 19\%$, the observed frequencies are 70%, 20%, 5% and 5%.

Table 1.2 shows how we can explain these results by supposing that the gametes formed by the first generation hybrid plants (P/p L/l) are more frequently types (P L) and (p l), like the parental gametes, than recombinant types (P l) and (p L). We can obtain quantitative agreement

Table 1.2. Segregation of two linked genes

Gametes		P L	P l	p L	p l
	Frequency	10/22	1/22	1/22	10/22
P L	10/22	0.207 Purple Long	0.021 Purple Long	0.021 Purple Long	0.207 Purple Long
P l	1/22	0.021 Purple Long	0.002 Purple Short	0.002 Purple Long	0.021 Purple Short
p L	1/22	0.021 Purple Long	0.002 Purple Long	0.002 Red Long	0.021 Red Long
p l	10/22	0.207 Purple Long	0.021 Purple Short	0.021 Red Long	0.207 Red Short

with the experimental data if, of every 11 gametes which carry the gene P, 10 carry the gene L, and only one carries the gene l; and if, of every 11 gametes carrying p, one carries L and 10 l.

The two characters, purple or red flower colour, and long or short length of the pollen grains, therefore belong to the same linkage group. The ratio 10:1 is very different from the 1:1 ratio which would be given by independent segregation of the genes. We therefore say that these genes are tightly linked.

By a large number of studies of the simultaneous segregation of genes, it has been possible to establish the existence of four linkage groups in *Drosophila melanogaster*, seven in peas, and ten in maize.

Mendel studied seven characters in peas. By a remarkable piece of luck, each of these seven genes belonged to a separate linkage group (which has an *a priori* probability of only one in 163, approximately); these genes therefore segregated independently. This unlucky chance meant that Mendel did not discover linkage, though it did, of course, make his results easier to interpret.

1.4. Some Definitions

Genetics, like every independent field of study, has developed a terminology, which is sometimes imprecise and sometimes over-precise and which can vary from author to author. This can confuse the novice, who may not realise that he is being confused by ambiguous use of words, rather than by the difficulty of the arguments themselves. It therefore seems important to define the "key-words" which will be used in this book.

A "*gene*" is a unit of information concerning a unit character, which is transmitted by a parent to his offspring. In sexually-reproducing organisms, individuals carry two genes for each unit character, one from each parent.

The set of genes carried by an individual is called his "*genome*".

Genes which act on the same unit character are said to be "at the same *locus*". One could equally well say that genes at the same locus are homologous (Gillois, 1964).

At each locus, an individual has two genes, one from his mother and one from his father. He transmits a copy of one of the two genes to each of his offspring. For the purposes of population genetics, the locus can be considered as the basic, indivisible unit of hereditary transmission.

The set of genes of one locus, i.e. that act on the same unit character, is called a set of "*alleles*".

For a given locus, the number of alleles is the same as the number of modes of action on the character. So far, we have only considered loci with two alleles (green or yellow cotyledons, round or wrinkled seeds, etc.), but large numbers of alleles are possible; loci with more than ten known alleles are not uncommon. In what follows, alleles will be designated by the symbols $A_1, \ldots, A_i, \ldots, A_n$.

The two genes which an individual carries at a given locus can be the same allele, or different; these two possible states are called the "*homozygous*" and "*heterozygous*" states of the locus. The genotypic formula of a homozygote is written $(A_i A_i)$, and a heterozygote is written $(A_i A_j)$.

If there are n alleles at a locus, there are n possible homozygotes: $(A_1 A_1) \ldots (A_i A_i) \ldots (A_n A_n)$. Also, $\dfrac{n(n-1)}{2}$ different heterozygotes are possible: $(A_1 A_2) \ldots (A_1 A_n)(A_2 A_3) \ldots (A_{n-1} A_n)$. The set of all these homozygous and heterozygous types constitutes the "*genotypes*" that are possible at the locus. There are $n + \dfrac{n(n-1)}{2} = \dfrac{n(n+1)}{2}$ of them.

The characteristic which an individual manifests, with respect to a unit character, is called his "phenotype". It is observable, and sometimes measurable, unlike the genotype.

The phenotypic effects of different genotypes may be the same, because the alleles at a locus can show dominance-recessivity relations. Allele A_i is said to be dominant to A_j (or, equivalently, A_j is said to be recessive to A_i), when the action of A_i, but not that of A_j, is manifest in the $(A_i A_j)$ heterozygote.

When the heterozygous genotype $(A_i A_j)$ has the same phenotype as the homozygous genotype $(A_i A_i)$, for the character concerned, dominance is said to be total.

Dominance is said to be incomplete when the $(A_i A_j)$ genotype is closer in phenotype to $(A_i A_i)$ than to $(A_j A_j)$.

Clearly, the presence of the recessive A_j gene is never entirely without effect on the organism; dominance is, in reality, merely a function of the aspect of the phenotype that is observed, and the precision of the observations.

For a unit character controlled by a locus with n alleles, the number of phenotypes is determined by the dominance relations of the alleles; there will be n phenotypes if the alleles can be ordered in sequence, with each one fully dominant to the lower ones in the sequence; if there is no dominance, there will be as many phenotypes as genotypes — $\dfrac{n(n+1)}{2}$; in other cases, there will be an intermediate number of phenotypes.

A *unit character* is one which is inherited according to Mendel's first law. In other words, it is a character controlled by the alleles of a single locus.

This definition exposes the underlying tautology in the theory as so far described. If it were not for the discovery that the words "gene", "locus" (and also "linkage group") correspond to concrete biological entities, and not only to theoretical concepts that we use to explain the facts of hereditary transmission, the whole of the above theory would remain a purely abstract, logical model.

As often happens, a hypothetical model which would explain the observed phenomena of heredity was developed before the physical entities concerned were known. Subsequent cytological and biochemical studies have shown that the Mendelian model is not merely a construct which agrees with the facts it was designed to explain, but also corresponds to the behaviour of the actual hereditary mechanism.

2. The Physical Basis of Mendelian Inheritance. The Chromosomes

In the decades following the publication of Mendel's paper, cytologists established the fact that the nuclei of cells contain filamentous structures which can be stained at the time of cell division – the chromosomes. Living cells contain a set of $2n$ chromosomes, where n is a number characteristic of the species: four in *Drosophila melanogaster*, seven in peas, ten in maize.

The regular behaviour of the chromosomes during cell division suggested to cytologists that they might be involved in hereditary transmission. At the end of the 19th century, cytologists had formulated the rules of inheritance that would apply if the chromosomes were the bearers of the hereditary determinants. When Mendel's work was rediscovered, the correspondence with the cytologists' expectations was apparent.

Without going into details in this rapidly changing field, we shall now give a summary of the physical basis of inheritance, on which population genetics is founded.

2.1. The Behaviour of the Chromosomes. Mitosis and Meiosis

Each of the millions of cell divisions which occur during the development of an individual consists of a complex series of events which chiefly involve the chromosomes. The chromosomes are invisible in the non-dividing nucleus, and they become visible as the cell prepares to divide, in the form of a set of pairs of rod-shaped objects (Fig. 1.2, stage 1).

At a later stage, each rod is seen to be divided into two identical "chromatids", which remain attached at one point, the "centromere". Meanwhile, the nuclear membrane disappears (stage 2).

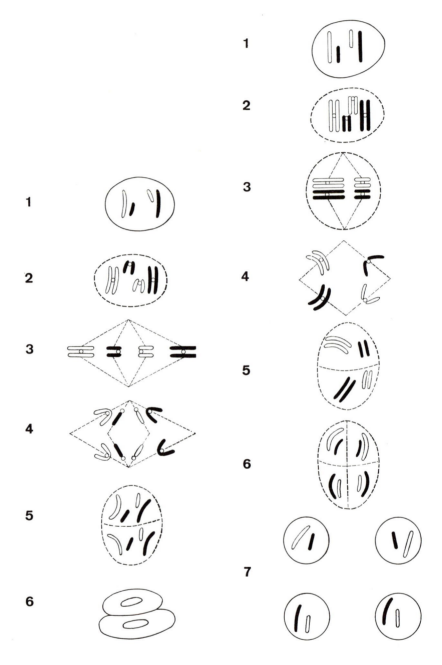

Fig. 1.2. Mitosis

Fig. 1.3. Meiosis

The double chromosomes next come to lie on the equator of the cell spindle, which forms at this time. The centromeres appear to be responsible for the movement of the chromosomes to the equatorial plane. Then the centromeres double and move apart towards the two poles of the spindle, pulling the sister chromatids with them (stage 4). A membrane forms around each of the groups of chromosomes, which both contain one chromosome of each of the original types (stages 5 and 6); finally, the chromosomes gradually lose their property of staining. This beautifully precise mechanism gives each nucleus of the two daughter cells a full and perfect copy of each of the chromosomes of the original cell.

A variant of this set of manoeuvres, meiosis, occurs during gamete formation. The sequence begins just like mitosis: the chromosomes become visible (stage 1 of Fig. 1.3), double to form a pair of chromatids joined by a centromere (stage 2), and move to the equatorial plane. But during this stage the two homologous chromosomes of each pair come together and form groups of four chromatids (stage 3).

A first division now takes place; the centromeres, each with a pair of chromatids attached, move to the poles, in such a way that each group receives only one (doubled) member of each pair of chromosomes (stages 4 and 5). The two temporary cells at the poles now divide again, and in this division the sister chromatids separate (stages 6 and 7). The final result of meiosis is four reproductive cells, each of which contains only one member of each of the pairs of chromosomes that were present in the original cell.

2.2. Consequences of Chromosome Behaviour for Hereditary Transmission of Characters

The regularity of chromosome behaviour in meiosis supports the following model: the transmission of characters from parent to offspring is mediated by elements borne on the chromosomes, and for each character there correspond two such elements, one on each of the pair of homologous chromosomes.

Each individual has two corresponding series of n chromosomes, one set of which came from his father, and one from his mother. If he has received two different elements controlling a given character, the action of one may mask that of the other. When the individual reproduces, the homologous chromosomes, maternal and paternal, separate and the descendants may again manifest the character that was masked in their ancestor: this is *segregation*.

Furthermore, the migration of the chromosomes to the two poles of the spindle, in stages 3, 4 and 5 of meiosis, occurs independently of their origin (paternal or maternal). The simple example shown in Fig. 1.3, where

$n=2$, shows how the two temporary cells (stage 6) can receive either copies of the two maternal chromosomes, or of the two paternal chromosomes, or one paternal and one maternal chromosome. This is *independent segregation* of chromosomes.

The two ideas of segregation and independence, which we have just demonstrated for the chromosomes, are exactly analogous to Mendel's two laws. Mendel's purely formal explanation of the facts of heredity is thus in perfect agreement with what we know about the physical mechanism of hereditary transmission.

Mendel's "genes", which he proposed as the "factors" mediating heredity, can therefore be equated with real "elements" carried on the chromosomes.

The only other phenomenon which remains for us to explain is linkage.

2.3. Linkage and Crossing Over

In the above description of the behaviour of chromosomes in meiosis, and its implications for genetics, we have been supposing that two characters will segregate independently if they are controlled by genes carried on different chromosomes, and will be completely linked if the genes are on the same chromosome.

In reality, exchanges can take place between the chromatids of a pair of homologous chromosomes during the 3rd stage of meiosis, when the chromosomes are aligned with one another.

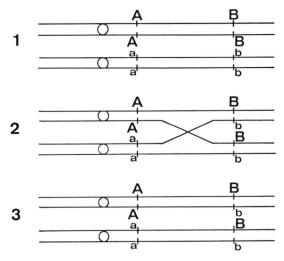

Fig. 1.4. Crossing-over

Fig. 1.4 shows an exchange between a paternal chromosome carrying genes A and B, and a maternal chromosome carrying a and b. After the exchange of segments between chromatids ("crossing-over"), some of the chromosomes which migrate to the poles will still have the parental associations of genes (AB and ab) but others will be "recombinants" (Ab and aB).

The greater the distance between the loci of the two genes on the chromosome, the greater the chance that crossing-over will occur between them[3], and thus the greater the proportion of recombinant gametes. The degrees of linkage measured by segregation of characters can therefore serve to estimate distance between points on the chromosomes. Chromosome maps have been made in several organisms, in which distances are measured, not in any of the ordinary units of distance, but by the percentages of recombinants given in crossing experiments.

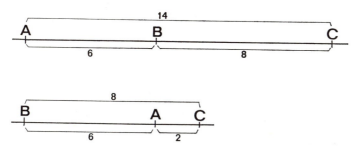

Fig. 1.5. Part of a chromosome map

In order to use recombination fractions as measures of distance, these fractions must clearly be additive. If, for example, the distance between A and B is 6%, and that between B and C is 8%, the distance between A and C must be either 14% or 2%, depending on the order of these three genes on the chromosome (Fig. 1.5). This requirement is met in practice, to a good approximation.

2.4. Human Chromosomes

The number n of chromosomes is constant for each species, and is therefore a fundamental fact about the species. The human chromosome number was not, however, correctly established until 1956.

Until this time it was thought that $n = 24$. The lack of a good enough technique, and probably also excessive respect for established opinions, had the result that the number $n = 24$ was accepted without dispute until

---

[3] We are ignoring the possibility of multiple cross-overs here.

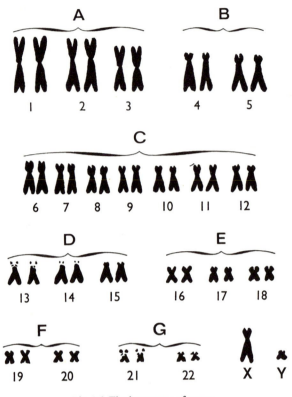

Fig. 1.6. The karyotype of a man

Tjio and Levan (1956), using an improved method, showed that n was really only 23. Man therefore has 46 chromosomes.

The 23 pairs have sufficiently distinct sizes and shapes to be classified according to an international convention. In this classification, the sex chromosome pair XX or XY, is distinguished from the 22 autosomal pairs. The autosomes are numbered according to size and the position of the centromere, starting with 1 for the largest chromosomes, which have a median centromere, and going up to 22 for the smallest chromosomes, whose centromere is near the end of the chromosome. The whole set of an individual's chromosomes is called his "karyotype".

Variations of the normal human karyotype are known, and some well-known abnormalities are associated with such variations. In 1959, Lejeune, Turpin and Gautier showed that mongolism is due to the presence of an extra chromosome 21 ("trisomy 21", i.e. 3 examples of chromosome 21, instead of the normal 2). Many other chromosome abnormalities are now known. In particular, certain types of cancer and leukaemia are associated with abnormal karyotypes.

2.5. The Sex Chromosomes

The sex chromosomes distinguish the male karyotype from the female one. In women, the 23rd chromosome pair is represented by two identical chromosomes, called the X chromosomes, which are about the same length as chromosome 6. In men, the 23rd pair consists of two chromosomes of very different size: one of them is just the same as the X chromosome of females, but the other, called Y, is small, like a chromosome 21 or 22.

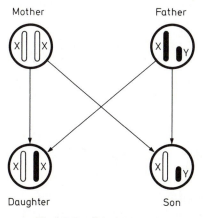

Fig. 1.7. Sex-linked inheritance

The X and Y chromosomes carry genes affecting all sorts of characters, not merely genes concerning sexual differences. Because of the difference in the sex chromosomes in men and women, characters controlled by genes on these chromosomes will be inherited in a different way from those controlled by autosomal genes: a son must inherit his Y chromosome from his father, and his X chromosome must come from his mother. Therefore a gene carried on his father's X chromosome cannot be transmitted to him. A daughter, on the other hand, receives an X from her father and one from her mother, and cannot inherit a gene carried on the Y chromosome of her father[4].

This asymmetrical type of inheritance means that the genes of the X and Y chromosomes, or "sex-linked", genes, have to be treated separately from the others.

---

[4] Many genes on the X-chromosomes are known (e. g. haemophilia, colour-blindness), but very few have been found to be carried on the Y chromosome.

2.6. Chromosome Structure. DNA

During the last twenty years there has been great progress in understanding how the structure of chromosomes is related to their two basic functions:

1. To reproduce themselves at cell division, and

2. To control biochemical reactions.

Watson and Crick, in 1953, showed that the chromosome consists of two long molecules, coiled together in a double helix (DNA).

Each of the two strands is a series of nucleotides, of which there are four types, depending on which of the four bases (adenine A, guanine G, thymine T or cytosine C) the nucleotide contains. A strand of DNA therefore has a definite sequence, e.g. TCGAGCAAGCC...

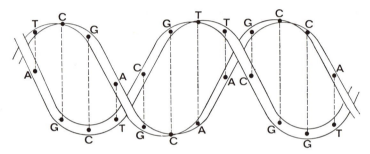

Fig. 1.8. The DNA double-helix

The nucleotides are like four "letters" of the alphabet, in which the message of the DNA is "written".

Furthermore, the two strands of the DNA double helix are complementary: an A in one corresponds to a T in the other, and a G corresponds to a C. The complementary sequence to the one written above would be AGCTCGTTCGG...

Auto-replication of DNA. In the earliest stages of mitosis or meiosis, the two strands of the DNA separate. Then a complementary strand to each of the two is synthesised, using the cell's pool of nucleotides; this synthesis is catalysed by an enzyme. The final result is a pair of DNA double helices, each identical with the original one.

Control of biochemical reactions. Proteins, which are the basis of all living matter, are extremely large molecules, with molecular weights of up to several million. Proteins consist of one or more long polypeptide chains, which are compactly folded up in a way which is determined by the amino acid sequence of the chains. Twenty types of amino acid enter into the composition of proteins.

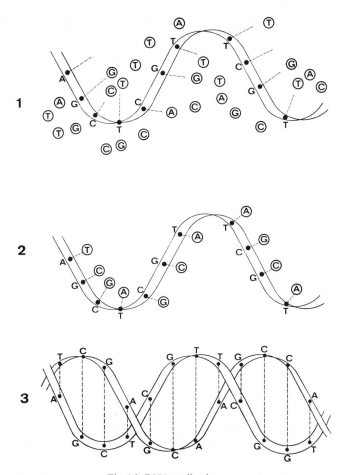

Fig. 1.9. DNA replication

Just as a strand of DNA is characterised by its base sequence, so a polypeptide chain is characterised by its amino acid sequence. The amino acids are therefore like the 20 "letters" of an alphabet in which the message of the proteins is "written".

The two sequences, the DNA sequence and the protein sequence, are related to one another by the "genetic code". An amino acid corresponds to a sequence of three nucleotides.

The mechanism by which the chromosome governs protein synthesis has become understood in recent years. The main stages are as follows:

1. A molecule of RNA is synthesised alongside a DNA helix, by a process which must be like DNA synthesis itself. The molecule of RNA is

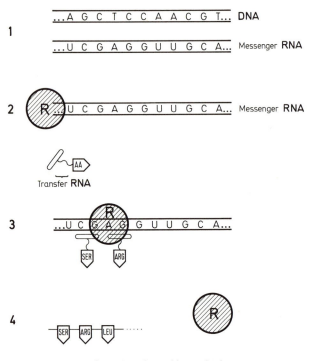

Fig. 1.10. Polypeptide synthesis

like a single strand of DNA, but contains uracil bases instead of thymine. The sequence AGCTCGAA ... in DNA therefore gives rise to the RNA sequence UCGAGCUU ...

2. This complementary RNA molecule, which is called messenger RNA, goes out of the nucleus into the cytoplasm of the cell. Ribosomes become attached to one end of the messenger RNA, and then they move along the messenger.

3. A pool of the different amino acids, attached to molecules of "transfer RNA", is present in the cell. The transfer RNA molecules are shorter than messenger RNA, and each type is specific both for the amino acid that can be attached to it, and also for a sequence of 3 bases in messenger RNA. For example, one transfer RNA type can have phenyl-alanine attached to it, and also recognises the sequence UUU in messenger RNA; a different type is specific for serine and for the sequence UCA. The transfer RNA molecules thus form the mechanism whereby the genetic code is translated – the 20 amino acids are brought into corre-spondence with the $4^3 = 64$ nucleotide triplets.

4. As a ribosome moves along the messenger RNA, each triplet in turn is recognised by a transfer RNA molecule, which has its amino acid attached to it, and, one by one, the amino acids are added to the growing polypeptide chain. The sequence of amino acids in the chain is thus determined by the base sequence of the messenger RNA.

The genetic code. It is now known that the genetic code is "degenerate": several of the 20 amino acids are coded for by more than one base triplet, e.g. UUG and CUC both code for leucine. However, three triplets do not code for amino acids, but are "nonsense" triplets, which serve as "punctuation marks"; these triplets are known to signal termination of the polypeptide chain.

It is possible that the degeneracy of the genetic code is such as to minimise the effects of mutations on the amino acids which are most important for protein structure and function.

2.7. Mutation

Chromosomes are not perfectly stable entities; changes occur in them with a small, but not negligible frequency. The nucleotide sequence of DNA can be changed by irradiation with ultra-violet light or X-rays, by certain chemicals or viruses, or simply by an error in the replication process. The types of changes produced are the replacement of one nucleotide by another, and the deletion or insertion of one or more nucleotides.

The descendants of a mutated cell will all carry the same change in their DNA. If the mutation occurred in a somatic cell, the individual in which it happened will be a "mosaic", with some of his cells genetically different from the others (this seems to be the case in some cancers); if the mutation occurs in a reproductive cell, the offspring who receives the mutant gene will differ genetically from both parents, and will transmit the mutation to his descendants.

A change of a single nucleotide of the DNA can have important consequences for the whole organism. A change in the DNA sequence which codes for a protein can abolish the synthesis of a functional protein. If the changed protein is an enzyme, for example, the biochemical process it is involved in will be abolished.

However, each cell contains two copies of the gene controlling each function, one on a paternal and one on a maternal chromosome. This greatly decreases the effects of mutations on the organism. If one of the two genes is altered and its action abolished, this is usually not apparent, since the other gene contains the information needed for the normal functioning of the cell. Therefore this heterozygosity has slight consequences, or none at all. This explains the phenomenon of recessivity and dominance; a mutant gene which cannot give rise to the synthesis of a protein, or codes for an inactive protein, will be recessive to a gene which has these capacities. If a single functional gene gives rise to enough protein for normal cellular function, the recessivity is total. If not, the non-functional gene is partially recessive, and the phenotype of the heterozygote differs from that of the dominant homozygote.

In a later generation, however, a gamete carrying the mutant gene may unite with another gamete which also carries this mutation, and so produce a homozygote for the mutation, who will manifest its effect.

2.8. Individual Diversity

Mutations are not necessarily harmful to the organism. They are random changes, which sometimes can be beneficial. Mutations are the source of individual diversity.

To understand the extent of this diversity, consider a population whose members can differ for N characters, each of which can have one of two states. The total number of possible types will be 2^N.

At present, the total human population of the earth is about 3×10^9, while the total number of human beings who have ever lived must probably be less than 10^{11}. So if humans differ for only 40 characters, every individual who has ever lived could be different from every other.

The number of spermatozoa that have been formed by all the men who have ever lived is of the order of 10^{22}. A man who is heterozygous for 75 genes, which is not a particularly large number, would produce $2^{75} = 10^{23}$ genetically different types of spermatozoa.

How is this diversity maintained? How do the proportions of the different genes change in relation to environmental conditions and the behaviour of the individual population members? These are the questions that population genetics tries to answer.

Chapter 2

Basic Concepts and Notation.
Genetic Structure of Populations and of Individuals

1. Probability

When the members of the pairs of chromosomes separate from one another during meiosis, the distribution of paternal and maternal chromosomes to the reproductive cells that are formed takes place "at random". Out of tens of millions of spermatozoa emitted, which one succeeds in fertilising the egg is the result of "chance".

The techniques which have been developed for handling processes that involve "random", or "chance" events (i.e. probability theory) are therefore frequently used in studying hereditary processes.

Of course, there is always a deterministic reason for the particular outcome of the "random" event that does, in fact, occur, but these causes are inaccessible to us, and we can only study them by observing repeated outcomes of the random event, and attaching probabilities to the various possibilities. We cannot examine a spermatozoon from a man who is heterozygous at a particular locus, and see if it has gene A_1 or A_2, but if we find individuals carrying A_1 and individuals carrying A_2 among his descendants, we would then know that his spermatozoa had a probability of $\frac{1}{2}$ of being A_1, and $\frac{1}{2}$ of being A_2. This would be all that we could say from information about his progeny. If we later examined the child produced by the spermatozoon, and find that its genotype is $(A_2 A_2)$, we would know that the spermatozoon carried the A_2 gene. A certainty would replace a probability.

Probabilistic reasoning allows us to make use of incomplete information. A probability has a meaning only in relation to a definite set of information, and it is changed if the information at our disposal is changed.

Genetics is probably the field of study in which probabilistic models correspond most closely to the reality of the phenomena being studied. Paradoxically, the early work in theoretical population genetics did not make much use of probabilistic reasoning. Since 1948, however, Malécot has re-studied a number of topics using probability theory, and this has deepened our understanding of these areas.

In the next sections, we shall recall some basic elements of probability theory.

1.1. Definition of Probability

Consider an event E, which occurs or does not occur, depending on the result of a "trial", for example, a lottery. Suppose that the trial has not yet taken place, or that its result is not yet known.

Taking account of our knowledge or opinions about the nature of the trial, we can assign to each of the different possible results e_i a number p_i which bears a relation to the strength of our belief that e_i will take place. By convention, we choose the p_i to be positive, and such that

$$\sum_i p_i = 1. \tag{1}$$

If an event E takes place if the result of the trial is e_k, or e_l, or e_m, the probability of the event is defined as:

$$P(E) = p_k + p_l + p_m$$

or, more generally:

$$P(E) = \sum_i p_i \tag{2}$$

where i refers to the set of subscripts belonging to the results that lead to event E.

If E takes place whatever the result of the trial, in other words if it is certain to take place, we have, from (1):

$$P(E) = 1.$$

If E never takes place, whatever the result of the trial (in other words E is impossible), we have:

$$P(E) = 0.$$

Example. Consider the "trial" of drawing a card at random from a pack. There are 52 possible results e_i, and we consider that they are equally probable. Therefore:

$$p_1 = p_2 = \cdots = p_{52} = \tfrac{1}{52}.$$

Consider the event E_1: "the card drawn is a diamond". This happens in 13 of the 52 possible cases, so that the probability of E_1 is:

$$P(E_1) = 13 \times \tfrac{1}{52} = \tfrac{1}{4}.$$

Consider the event E_2: "the card drawn is a king". This happens in 4 out of the 52 possible cases, therefore:

$$P(E_2) = 4 \times \tfrac{1}{52} = \tfrac{1}{13}.$$

1.2. Principle of Addition of Probabilities

Two events E_1 and E_2 are said to be incompatible when the occurrence of each depends on different results of the trial. For example, if E_1 occurs when the result of the trial is e_3 or e_5, but E_2 occurs when the result is e_4 or e_8, then E_1 and E_2 cannot both occur simultaneously: they are incompatible.

An event E_u is called the *union* of the two events E_1 and E_2, if it occurs when either E_1 or E_2 or both E_1 and E_2 occur. (This third case is impossible if E_1 and E_2 are incompatible.) We can thus write:

$$E_u = E_1 \cup E_2.$$

It is easy to see, from (2), that if two events are incompatible, the probability of their union is equal to the sum of their probabilities. In general: *if the events E_i are incompatible, two by two,*

$$P\left(\bigcup_i E_i\right) = \sum_i P(E_i). \tag{3}$$

Example. In the trial of drawing a card from a pack, the events E_1 "the card is a diamond", and E_3 "the card is a heart" are incompatible. The probability of E_4 "the card is red" is therefore

$$P(E_4) = P(E_1) + P(E_3) = \tfrac{1}{4} + \tfrac{1}{4} = \tfrac{1}{2}.$$

1.3. Principle of Multiplication of Probabilities

An event E_I is said to be the *intersection* of two events E_1 and E_2 if it occurs when both E_1 and E_2 take place. We write:

$$E_I = E_1 \cap E_2.$$

(In the case when E_1 and E_2 are incompatible, E_I cannot occur.)

The *conditional probability* of the event E_1, knowing that E_2 has occurred is defined as the ratio of the probability of the intersection of E_1 and E_2 to the probability of E_2. The conditional probability is designated by $P(E_1|E_2)$, so we can write:

$$P(E_1|E_2) = \frac{P(E_1 \cap E_2)}{P(E_2)} \tag{4}$$

which is valid when $P(E_2) \neq 0$.

Using this definition, we can write the following equalities, which express the "principle of multiplication of probabilities":

$$P(E_1 \cap E_2) = P(E_1) \times P(E_2 | E_1) = P(E_2) \times P(E_1 | E_2) \tag{5}$$

which is valid even when $P(E_1)$ or $P(E_2)$ is zero (in which case all the terms of the equalities (5) are zero, even though $P(E_2|E_1)$ or $P(E_1|E_2)$ are not defined).

Independent events. An event E_1 is said to be independent of event E_2 if the probability of E_1 is not affected by information about whether E_2 has or has not occurred. Another way of putting this is that:

$$P(E_1 | E_2) = P(E_1).$$

It then follows from (5) that:

$$P(E_1 \cap E_2) = P(E_1) \times P(E_2)$$

or: *if two events are independent, the probability of their intersection is equal to the product of their probabilities.*

Example. In the trial of drawing cards from a pack, consider the events:

E_1: "the card is a diamond"
E_2: "the card is a king"
E_5: "the card is a court card of the diamond suit (i.e. ace, king, queen or knave of diamonds)".

We have

$$P(E_1) = \tfrac{13}{52} = \tfrac{1}{4}, \qquad P(E_2) = \tfrac{4}{52} = \tfrac{1}{13}, \qquad P(E_5) = \tfrac{4}{52} = \tfrac{1}{13}.$$

It is easy to see that the probabilities of the intersections of these events are

$$P(E_1 \cap E_2) = \tfrac{1}{52}, \qquad P(E_2 \cap E_5) = \tfrac{1}{52}, \qquad P(E_1 \cap E_5) = \tfrac{1}{13}.$$

Using the definition (4), we get the 6 conditional probabilities:

$$P(E_1 | E_2) = \tfrac{1}{4} = P(E_1) \qquad\qquad P(E_2 | E_1) = \tfrac{1}{13} = P(E_2)$$

$$P(E_2 | E_5) = \tfrac{1}{4} \qquad\qquad\qquad P(E_5 | E_2) = \tfrac{1}{4}$$

$$P(E_1 | E_5) = 1 \qquad\qquad\qquad\; P(E_5 | E_1) = \tfrac{4}{13}.$$

The events E_1 and E_2 are independent, but events E_1 and E_2 are not independent of E_5.

1.4. Bayes' Theorem

Consider an event R which can occur as the result of a number of events $C_1, \ldots, C_n$, that are incompatible two by two, and which we shall call the "causes" of R. If we understand the process, we can write the probability, $P(R|C_i)$, of the event R given that one of the causes, C_i, has occurred. We also assume that we can attach *a priori* probabilities, $P(C_i)$, to the different causes.

Suppose we do an experiment and obtain the result R. How can we use this information to modify the *a priori* probabilities $P(C_i)$? Or, equivalently, what are the probabilities, $P(C_i|R)$, of C_i, knowing that R has occurred?

From (5), we can write

$$P(C_1 \cap R) = P(C_1) \times P(R|C_1) = P(R) \times P(C_1|R).$$

But R can only take place if one of the C_i has occurred, so that:

$$P(R) = P(R \cap C_1) + P(R \cap C_2) + \cdots + P(R \cap C_n) = \sum_i P(C_i) P(R|C_i)$$

therefore

$$P(C_1|R) = \frac{P(C_1) P(R|C_1)}{\sum_i P(C_i) P(R|C_i)}. \tag{6}$$

This formula expresses Bayes' theorem. It allows us to use the result of an experiment to modify our opinion about the probability of an event.

Example. Suppose we have two packs of cards. One of them, J_1, is normal. In the other, J_2, two knaves have been substituted for two of the aces. The packs are mixed, a card is drawn and it is an ace. What is the probability that the pack it was drawn from was J_1?

The *a priori* probability of the card coming from one or the other pack is:

$$P(J_1) = P(J_2) = \tfrac{52}{104} = \tfrac{1}{2}.$$

The probabilities of drawing an ace in the two packs are:

$$P(A|J_1) = \tfrac{4}{52} = \tfrac{1}{13}, \qquad P(A|J_2) = \tfrac{2}{52} = \tfrac{1}{26}.$$

We want to find $P(J_1|A)$. Eq. (6) gives:

$$P(J_1|A) = \frac{P(J_1) P(A|J_1)}{P(J_1) P(A|J_1) + P(J_2) P(A|J_2)}$$

$$= \frac{\tfrac{1}{2} \times \tfrac{1}{3}}{\tfrac{1}{2} \times \tfrac{1}{13} + \tfrac{1}{2} \times \tfrac{1}{26}} = \tfrac{2}{3}.$$

Note. This theory is based on the initial hypothesis of the existence of a trial whose results are random. We can only use probability theory in cases when the "mechanism" of the basic trial is such that probabilities can be assigned to the different results.

In the examples that have been given in the previous sections, the "mechanism" was drawing a card. We assumed that this was done in a way which made every result equiprobable, but other hypotheses would have been possible.

1.5. Random Variables

We can assign a real number x_i to each of the results e_i of a trial. We know the probabilities p_i of the results, so that we can now calculate the probability P_k that the number corresponding to the result will be $X = x_k$. This defines a "random variable X associated with the trial".

The term "probability distribution of X" is used to mean the set of probabilities P_i of the events $X = x_i$, where the x_i are the various values that X can take.

Example. The trial consists of drawing a card from a pack. The value of X is 1 for the knaves, 2 for the queens, 3 for the kings, 4 for the aces, and 0 for other cards. The values x_i are 0, 1, 2, 3 and 4, and the probabilities corresponding to the values of X are:

$$P_0 = P(X=0) = \tfrac{9}{13}, \qquad P_1 = P_2 = P_3 = P_4 = \tfrac{1}{13}.$$

These five P values constitute the "probability distribution of X".

1.6. The Expectation and Variance of a Random Variable

The expectation (or expected value) of a random variable is defined as the mean of the different values of the variable, weighted by the probabilities of these values:

$$E(X) = \sum_{i=1}^{k} P_i x_i. \tag{7}$$

In the example of the "value of a card" given above, the expectation of the value is

$$E(X) = 0 \times \tfrac{9}{13} + 1 \times \tfrac{1}{13} + 2 \times \tfrac{1}{13} + 3 \times \tfrac{1}{13} + 4 \times \tfrac{1}{13} = \tfrac{10}{13} = 0.769.$$

This example shows that the expected value of the variable need not be equal to any of the x_i values.

The expectation, being a mean, is a central value of the random variable, and is a parameter which can be used to characterise its probability distribution. The variance is another such parameter, and it is a measure of the dispersion of the x_i values about the mean.

The variance of a random variable is defined as the mean of the squared values of the deviations of the variable from its mean, weighted by the probabilities:

$$V(X) = \sum_{i=1}^{k} P_i[x_i - E(X)]^2. \qquad (8)$$

The variance is a positive number. The square root of the variance is called the "standard deviation", and is denoted by the letter σ: $\sigma_x^2 = V(X)$.

The variance can only be zero if the variable X can take just one value. The variance increases as the x_i values become more and more dispersed.

In the example of the "value of a card", the variance is:

$$V(X) = \tfrac{9}{13}(\tfrac{10}{13})^2 + \tfrac{1}{13}(1-\tfrac{10}{13})^2 + \tfrac{1}{13}(2-\tfrac{10}{13})^2 + \tfrac{1}{13}(3-\tfrac{10}{13})^2 + \tfrac{1}{13}(4-\tfrac{10}{13})^2 = 1.72.$$

Note. In practice, variances are calculated using the following formula, which is equivalent to the one given above:

$$V(X) = \sum_{i} P_i x_i^2 - [E(X)]^2.$$

1.7. Examples of Random Variables

1.7.1. The binomial distribution. Suppose that an experiment gives the result R with probability p, and a result other than R with probability $q = 1 - p$. Suppose that n independent repetitions are done, and write X (which is a number between 0 and n) for the number of times result R is obtained in these n tests.

Since we are assuming that the repetitions are independent of one another, we can calculate:

1. The probability, P_0, that R does not occur at all: $P_0 = (1-p)^n = q^n$, and
2. The probability, P_n, that R occurs in each of the n tests: $P_n = p^n$.

It is easy to show that the probability $P(X = r)$ that R occurs r times in the series of n tests is:

$$P_r = \binom{n}{r} p^r q^{n-r} \qquad (9)$$

where $\binom{n}{r} = \dfrac{n!}{r!(n-r)!}$ is the number of combinations of n things, taken r at a time (or the number of different samples of size r that can be chosen from n objects).

Eq. (9) is the definition of the binomial probability distribution. The name comes from the fact that the values of P_r are the terms of the expansion of the binomial $(p+q)^n$.

It is simple to show that the binomial distribution with parameters n and p has expectation:

$$E(X) = np$$

and variance:

$$V(X)=np(1-p).$$

The precise form of the probability distribution is obviously a function of n and p, but whatever their values, the probabilities P_r increase as r increases from zero up to a number m which is the largest whole number less than $(n+1)p$. After this maximum, the probabilities decrease as r increases from m to n.

Example. For $n=10$ and $p=\frac{1}{2}$, the highest value of P is attained when r is the largest whole number less than $(10+1)\times0.5=5.5$, i.e. $m=5$. The value of P is then

$$P_m=\frac{10!}{5!\,5!}\left(\frac{1}{2}\right)^{10}=0.246.$$

For $n=10$ and $p=0.2$, m is 2 and

$$P_m=\frac{10!}{2!\,8!}(0.2)^2\times(0.8)^8=0.302.$$

1.7.2 The Poisson distribution. Consider an event such that:

1. Its probability of occurrence in a short time-interval dt is proportional to the length of the interval, independent of when the observation is made, i.e.:

$$P(E \text{ occurs between } t \text{ and } t+dt)=\lambda\,dt.$$

2. The probability that it will occur twice during this interval is negligible.

If X is the number of times that E occurs in a unit of time, it is possible to show that the probability distribution of the random variable X is:

$$P(X=r)=e^{-\lambda}\frac{\lambda^r}{r!} \tag{10}$$

where λ is some positive number.

This equation shows that X can be any positive whole number. The probability $P(X=0)$ that the event E does not occur at all in a unit of time is $P_0=e^{-\lambda}$. It can be shown that P_r has a maximum when r is the largest whole number less than λ ($\lambda-1$, if λ is itself a whole number); as r increases above this value, P_r decreases. In the case where $\lambda\leq1$, the maximum value is P_0, and P_r is a decreasing function of r.

One interesting property of the Poisson distribution is that the expectation is equal to the variance

$$E(X)=V(X)=\lambda.$$

Note. If X is a binomial variable with parameters n and p, it can be shown that if n is increased indefinitely while p decreases towards zero, such that the product np tends to a limit λ, then the probability distribution of X tends towards a Poisson distribution with parameter λ.

2. Genetic Structures

At each locus, an individual has two genes, one of which he received from his father, and one from his mother. If the two genes carry the same information, in other words if the two alleles at this locus are the same, the individual is said to be homozygous. If they are different, he is heterozygous.

In a population, the number of different alleles present at a locus can be very high. In order to simplify things, geneticists often work with models which assume only two alleles at a locus, but this is unlike the real situation: loci with more than ten alleles are not at all uncommon.

In what follows, we shall generally be treating models with an arbitrary number, n, of alleles: $A_1, ..., A_i, ..., A_n$.

2.1. The Definition of Genic and Genotypic Structures

In order to describe the genetic composition of a population for a given locus, we must give the frequencies $p_1, ..., p_i, ..., p_n$ of the different alleles. This set of frequencies will be called the "*genic structure*" of the population, and will be designated by the letter s. s is a vector whose i-th element is the frequency of the i-th allele:

$$s = (p_1 \cdots p_i \cdots p_n).$$

There are $\dfrac{n(n+1)}{2}$ two by two combinations of the n alleles; each combination represents a genotype. There are:

n homozygous genotypes, and

$\dfrac{n(n-1)}{2}$ heterozygous genotypes.

The set of the genotype frequencies is called the "*genotypic structure*" of the population, and we will use the letter S to refer to it.

This set of frequencies includes two different kinds of frequency: the frequencies P_{ii} of the homozygous genotypes, and those of the heterozygous genotypes, P_{ij}. We cannot represent the $\dfrac{n(n+1)}{2}$ frequencies as a vector, since they cannot be ordered; we could represent them by a $n \times n$

matrix, but it would still be necessary to adopt a convention about the arrangement of the $\dfrac{n(n+1)}{2}$ elements in the n^2 positions of the matrix.

A better representation, which has a clear relation to the reality of the situation, is to write the set of frequencies in the form of a triangle with n rows (with just one number in the first row, and n in the n-th row) and n columns (with n numbers in the first column, and only one in the n-th column); the element corresponding to the homozygous genotype $(A_i A_i)$ would be written on the hypotenuse of the triangle, in the i-th row and i-th column, while the element corresponding to the heterozygous genotype $(A_i A_j)$ would be placed at the intersection of the i-th row and the j-th column. Thus:

$$
S \equiv \left|
\begin{array}{l}
P_{11} \\
\cdots\cdots\cdots \\
P_{i1} \cdots P_{ij} \cdots P_{ii} \\
\cdots\cdots\cdots\cdots\cdots \\
P_{n1} \cdots P_{nj} \cdots P_{ni} \cdots P_{nn}
\end{array}
\right.
$$

Such a triangular array will be called a "trimat".

2.2. The Relation between Genic and Genotypic Structures

When we know the genotypic structure of a population, we can obviously deduce its genic structure. The frequency of an allele A_i is given by:

$$
p_i = P_{ii} + \tfrac{1}{2} \sum_{j \neq i} P_{ij}. \tag{11}
$$

We shall use the word "*contraction*" for the operation of finding a set of gene frequencies from a set of genotype frequencies by this formula. Thus:

$$
s = \mathrm{cont}\, S
$$

is equivalent to

$$
p_i = P_{ii} + \tfrac{1}{2} \sum_{j \neq i} P_{ij}.
$$

On the other hand, knowledge of a population's genic structure does not enable us to deduce its genotypic structure. It will be useful, for reasons which will appear shortly, to define a single operation which relates two genic structures s and s' to a genotypic structure S.

We will call the genotypic structure S the "genetic product" of the genic structure s and s', when the elements P_{ij} of S are related to the elements p_i and p_i' of s and s' by:

$$
P_{ii} = p_i \times p_i', \qquad P_{ij} = p_i p_j' + p_i' p_j.
$$

This definition implies that s and s' must have the same number of elements. We will use the following symbolism for a genetic product:

$$S = s \times s'.$$

In the particular case when s and s' are the same, their product is called the "genetic square of s", and the symbolism:

$$S = \check{s}^2$$

is equivalent to

$$P_{ii} = p_i^2, \qquad P_{ij} = 2 p_i p_j.$$

Finally, notice that according to our definition of the operation of "contraction" we have:

$$\text{cont}\,(s \times s') = \frac{s + s'}{2}$$

and, in particular:

$$\text{cont}\,(\check{s}^2) = s.$$

2.3. The Probability Structures of Populations

The genic structures s, and the genotypic structures S, which we have been considering up to now, referred to the situation when enumeration of the various genes and genotypes is possible, in other words, when their exact frequencies could be known. Often, however, we want to consider cases when the population is inaccessible in space (so that we can study only a sample from it), or in time (for example, if we want to consider the future state of a population).

In order to characterise a population about which we do not have full information, we attach a probability Π_{ij} to each of the different possible genotypes $A_i A_j$. We then make the following definitions:

The "genotypic probability structure" s of a population, at present or in the future, is the set of probabilities assigned, in accordance with the information we possess about the population, to the various genotypes that an individual taken at random from the population could have.

Similarly, the "genic probability structure", $\mathscr{S}$, of a population is the set of probabilities π_i attached to the various genes which could be carried at a particular locus by a given chromosome taken at random from an individual member of the population.

It is obvious that for probability structures, just as for actual structures, we have

$$s = \text{cont}\,\mathscr{S}.$$

2.4. Probability Structures of Individuals

The definitions given above in relation to populations can easily be applied to the case of an individual.

If an individual is chosen at random from a population with genotypic structure S, the probability *a priori* that he will have genotype $(A_i A_j)$ is equal to P_{ij}, the frequency of this genotype in the population. $\mathscr{S}$ represents the information we would have if all we knew was that the individual was a member of the population. The set of these probabilities, for the various genotypes, will be called the "genotypic probability structure" of the individual.

If we choose two individuals at random from the population, they each have the same genotypic probability structure, $\mathscr{S}$. The two structures are independent of one another, and knowledge of the genotype of one of the individuals does not tell us anything about the genotype of the other. However, if we choose two individuals who are known to be identical twins, the genotype of the second will be the same as that of the first individual, so that their genotypic structures are fully dependent on one another.

Between these extremes of complete independence and full dependence, there can be any degree of dependence, according to the degree of relatedness of the two individuals. This dependence can be expressed in terms of the genotypic probability structure of one of the individuals (B), given the genotype of the other (A), which we can write $\mathscr{S}_{B|A}$.

This shows that an individual's genotypic probability structure depends on the information which we possess about him: for example, what population he comes from, which members of the population he is related to, and with what degrees of relatedness. The genotypic probability structure is modified whenever we get new information about the individual.

Finally, we define the genotypic probability structure of an individual A as the set of probabilities assigned, in accordance with the information we have about him, to the various genotypes which the individual might have. This is the trimat:

$$\mathscr{S}_A \equiv \begin{array}{|l}
\Pi_{11} \\
\cdots\cdots\cdots\cdots \\
\Pi_{i1} \dots \Pi_{ii} \\
\cdots\cdots\cdots\cdots\cdots \\
\Pi_{n1} \dots \Pi_{ni} \dots \Pi_{nn}
\end{array}$$

And, similarly, the genic probability structure of an individual A is the set of probabilities π_i that a gamete derived from the individual will carry the gene A_i. This is the vector:

$$s_A \equiv (\pi_1, \ldots, \pi_i, \ldots, \pi_n).$$

Note. In this last definition, it would be equivalent to speak of the set of probabilities π_i that a gene chosen at random from the two carried by the individual is A_i. The process of gamete formation in the first definition corresponds exactly to the random choice in this definition.

For individuals, just as for populations, we have

$$s_A = \text{cont } \mathscr{S}_A.$$

When we know that the genotype of an individual is, for example, $(A_i A_j)$, his genotypic probability structure is a set with all its elements equal to zero, except for the element Π_{ij}, which is equal to 1. His genic (or gametic) probability structure is a vector whose elements are all zero except for two:

$$\pi_i = \pi_j = \tfrac{1}{2}$$

which we can write as:

$$s_A = \tfrac{1}{2} i + \tfrac{1}{2} j$$

where i denotes the vector whose elements are all zero except for $\pi_i = 1$.

This shows that knowledge of the genotype of an individual gives us knowledge only of the probability structure of his gametes. The reappearance of a random factor at the stage of transition from an individual to his descendants is one of the fundamental properties of sexual reproduction.

It is important to distinguish clearly between actual genetic structures, and probability structures, and to understand the exact meaning of the probability structures; π_i is the probability that A_i will be carried by a gamete emitted by the individual, and not the probability that his genome includes this gene; in the example given above, we know that A_i is present in the genome, but π_i is $\tfrac{1}{2}$, not 1.

However, if the individual is homozygous $(A_i A_i)$, all his gametes will carry the gene A_i, and $\pi_i = 1$. In homozygotes there is no random element in hereditary transmission.

3. Sexual Reproduction

3.1. Genic Structures of Parents and Offspring

Sexual reproduction involves equal contributions from the male and female parents to the heredity of their offspring (at least if we confine our attention to autosomal loci, and ignore sex-linkage). This symmetry leads to the property that:

*an individual's genic probability structure is equal to the arithmetic
mean of the genic probability structures of his father and mother.* This can
be expressed symbolically by:

$$s_I = \tfrac{1}{2}(s_F + s_M).\tag{12}$$

This can be demonstrated as follows. An A_i gene can be present in a
gamete emitted by I if either:

1. the gamete carries the gene which was transmitted to I by his
father F (which has probability $\tfrac{1}{2}$) and it is of type A_i (which has the
probability $_F\pi_i$), or

2. the gamete carries the gene which was transmitted to I by his
mother, and is of type A_i (which has the probability $\tfrac{1}{2}\,_M\pi_i$). Therefore,
for all values of i:

$$_I\pi_i = \tfrac{1}{2}\,_F\pi_i + \tfrac{1}{2}\,_M\pi_i$$

which is equivalent to the condensed form given in (12).

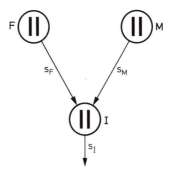

Fig. 2.1. Sexual reproduction

This relationship is of the same form as the model of "blending
inheritance", which was studied at the end of the 19th century, by the
biometricians. In this model, if X is a hereditary character, the expectation
of X in the offspring of a couple (F and M) is:

$$X_I = \tfrac{1}{2}(X_F + X_M).\tag{13}$$

If this formula were obeyed precisely, then in a random-mating population
the variance of X would be reduced by half each generation[1], and this
does not happen. Mendelian inheritance, however, has as a corollary the
maintenance of genetic variance.

---

[1] V_X would be equal to $\tfrac{1}{4}(V_{X_F} + V_{X_M})$, which gives, designating the generation by g:

$$V_g = \tfrac{1}{4}(V_{g-1} + V_{g-1}) = \tfrac{1}{2}V_{g-1}.$$

In Mendelian inheritance, as in blending inheritance, the child "corresponds to the arithmetic mean of his parents", but this statement relates to the probabilities of having the genes governing the character, and not to the measure of the character.

3.2. Genotypic Structure of Parents and Offspring

There is no simple relation between the genotypic probability structure $\mathcal{S}_I$, of an individual I, and those of his parents, without making further assumptions. We know that, for I to have the genotype $A_i A_j$, either:

1. he received an A_i gene from his father and A_j from his mother (event E_1), or

2. he received A_i from his mother and A_j from his father (event E_2).

These two events are incompatible, so that:

$$_I\Pi_{ij} = \text{prob}(E_1) + \text{prob}(E_2).$$

But E_1 and E_2 are the intersections of pairs of elementary events. Probability theory therefore requires that we know the conditional probabilities that M will transmit an A_j gene, knowing that F has transmitted an A_i gene, and *vice versa*.

These probabilities depend on the relationship between M and F. The simplest case is when their probability structures are independent. In this case, the conditional probability that M will transmit an A_j gene, given that F has transmitted an A_i gene, is simply the *a priori* probability that M will emit an A_j gene. So we can write:

$$_I\Pi_{ij} = {_F}\pi_i \times {_M}\pi_j + {_F}\pi_j \times {_M}\pi_i. \tag{14}$$

Similarly, we find, for the probability that I has the homozygous genotype $A_i A_i$, that:

$$_I\Pi_{ii} = {_F}\pi_i \times {_M}\pi_i. \tag{15}$$

Remembering the definition of the genetic product of two vectors, we can combine (14) and (15) and write:

$$\mathcal{S}_I = s_F \times s_M. \tag{16}$$

This relation is valid only for the case when information about one of the parents does not give information about the other; it is not generally valid.

We shall see later how it is possible to broaden the application of (16) by introducing parameters which measure the degree of relationship between the father and mother of an individual.

PART 2

A Reference Model:
Absence of Evolutionary Factors

Population genetics is the study of changes in the genetic composition of populations, from generation to generation. Through the gametes which they produce, and which go to form the members of the next generation, individuals can affect the future genetic composition of the population of which they are members. Although in human populations an individual can influence the population's conditions of existence (for example its social structure, or its intellectual level), he can only influence the genetic future of the population by means of the gametes he transmits.

Many factors can affect this transmission, and change the genetic structure of a population. Such changes constitute the evolution of the population. Before we study the effects of the various factors that can change the genetic composition of a population, it is, however, useful to consider the situation when no evolutionary force is acting, other than any forces for change that are imposed by the hereditary mechanism itself. This is obviously never the case for any real population. However, the concept is useful as a model which we can use for reference when we come to study the effects of the factors which we have assumed to be absent in such a population: mutation, selection, migration, non-random mating, overlapping generations and stochastic factors due to limited population size.

Chapter 3

The Hardy-Weinberg Equilibrium for one Locus

In 1908, a few years after the rediscovery of Mendel's laws, the English mathematician Hardy and the German geneticist Weinberg independently demonstrated that populations which fulfil a number of conditions, to be described in this chapter, are in a state of equilibrium with certain simple properties.

Before defining the conditions for Hardy-Weinberg equilibrium, it will be necessary to state precisely what is meant by the term "population", which has been used very loosely up to now.

1. Populations

So far, a population has been considered to be simply a group of N interbreeding individuals. However, in a real population whose future genetic state we want to study, matters are not quite so simple. For example:

1. Only some of the members of the population are old enough, but not too old, for reproduction.

2. Immigrants, who were born outside the population, will contribute to the next generation.

3. Due to births and deaths, the population size N is constantly varying.

It is a classical technique in science to start by studying the simplest possible case, and so the first case which we shall study is the ideal, simplified case in which the factors listed above are neglected. Later, the effects of these disturbing factors will be investigated.

Therefore, unless the contrary is explicitly stated, the type of "population" which we shall be considering is a group of individuals belonging to the same generation, whose members breed only with one another. We are thus assuming two conditions to hold:

① Migration does not occur;

② Individuals from different generations do not breed together.

Such closed populations are sometimes called "isolates", but this word has been used for many different ideas, especially by ethnologists. Henry (1969) has used the French word "*cercle*" for a population which satisfies the two above conditions. There is no suitable English equivalent to this word, so the word "isolate" will be used in what follows. It is therefore important to remember that "isolate" is being given the very specific definition: a population which satisfies conditions ① and ② above; in other words, a population which is cut off from other populations both in space and time.

2. The Hardy-Weinberg Principle

We shall first study the genotypic structure of a locus with n alleles $(A_1, ..., A_n)$, in an "isolate" defined as explained above.

2.1. Stability of the Genic Structure

The information carried in the chromosomes is not perfectly stable: mutations occur at a certain, usually very low, frequency; an individual of genotype $(A_i A_j)$ can therefore produce an A_k gamete.

Continuing our policy of starting by eliminating all complicating factors, we now assume that the population satisfies a further condition:

③ Mutations do not occur.

Furthermore, in a real population, different individuals may have greatly different numbers of offspring, either simply as a result of chance, or because of differing reproductive potentials associated with their different genetic constitutions. For example, if individuals who have gene A_i have, on average, more offspring than individuals with A_j, then in the next generation A_i will spread in the population, at the expense of A_j, and will have a higher frequency. The tendency of individuals of one genotype to have more offspring than individuals of another genotype is called "selection". Selection leads to changes in gene frequencies, and the rates of change are related to the intensity of the selection. Once again, we shall start by considering populations in which this cause of gene frequency change does not act, i.e. which satisfy the condition:

④ There is no selection. (In other words, the number of offspring an individual has is independent of his genotype at the locus in question.)

Now consider an isolate with known genotypic structure S. The members of the isolate will produce many gametes, of which some will enter into the formation of the new generation.

Suppose we choose one of these "successful" gametes at random. The probability π_i that it will carry the gene A_i is, assuming condition ③ (no mutation) to hold:

1, if it comes from an $(A_i A_i)$ individual,
$\frac{1}{2}$, if it comes from an $(A_i A_j)$ individual, where $i \neq j$, and
0, if it comes from an $(A_j A_k)$ individual, where $i \neq j$ and $i \neq k$.

But the probability that the individual which produced the gamete was of genotype $(A_i A_j)$ is, assuming condition ④ (no selection), equal to the frequency P_{ij} of this genotype in the population. Therefore:

$$\pi_i = P_{ii} + \tfrac{1}{2}(P_{i1} + \cdots + P_{in}). \tag{1}$$

By the definition of the operation of contraction, this equality shows that the vector representing the genic probability structure of the new generation is equal to the contracted vector of the actual genotypic structure of the preceding generation:

$$s_g = \mathrm{cont}\,(S_{g-1}) = s_{g-1}. \tag{2}$$

If the genotypic probability structure of the preceding generation is known, rather than the actual genotypic structure, a similar argument leads to the result:

$$s_g = s_{g-1}. \tag{3}$$

By Eq. (2), $s_1 = s_0$, so that by considering successive generations we obtain:

$$s_g = s_0 \tag{4}$$

which can be expressed in words as follows: in a population subjected to the four conditions which have been listed above, the genic probability structure of future generations is equal to the initial genic structure.

2.2. Genotypic Structure

Eq. (4) tells us the composition of the set of gametes which will form the next generation. However, we also want to know the genotypes of the individuals of this generation. An individual of this generation receives two genes from the preceding generation, one from his mother and one from his father. In order to have:

1. the genotype $(A_i A_i)$, the genes he receives from his mother and father must both be A_i;

2. the genotype $(A_i A_j)$, he must either have received A_i from his father and A_j from his mother, or the other way round.

Applying the algebra of probabilities given in Chapter 2, we find that the probability Π_{ii} of the first case is equal to the product of:

1. the probability of transmission of an A_i gene by the father, and

2. the probability of transmission of an A_i gene by the mother, *given that the gene the father has transmitted is* A_i. Thus:

$$\Pi_{ii} = {}_F\pi_i \times {}_M\pi_{i|i}. \tag{5}$$

Similarly, for the second case, the probability Π_{ij} is equal to the sum of:

1. the product of the probability of transmission of an A_i gene by the father and the probability of transmission of an A_j gene by the mother, given that the father has transmitted the A_i gene, and

2. the product of the probability of transmission of an A_j gene by the father and the probability of transmission of the A_i gene by the mother, given that the father has transmitted A_j. Thus:

$$\Pi_{ij} = {}_F\pi_i \cdot {}_M\pi_{j|i} + {}_F\pi_j \cdot {}_M\pi_{i|j}. \tag{6}$$

There are two situations to be considered:

a) Suppose we know the actual genic structure s_{g-1} of the generation $g-1$, i.e. we know the frequencies $p_i^{(g-1)}$ of the alleles in this generation.

For a member of generation g taken at random, the probability that one of his parents, for example his father, transmits an A_i allele to him, is equal to the frequency of this allele in the parental generation:

$$_F\pi_i = p_i^{(g-1)}.$$

The conditional probability $_M\pi_{j|i}$ that the second parent transmits an A_j allele, given that the first parent has transmitted an A_i allele, depends on the mating behaviour of the individuals in the population. For example, if all the men carrying the A_i gene married $(A_i A_i)$ women, we should have $_M\pi_{i|i} = 1$ and $_M\pi_{j|i} = 0$.

In order to make statements about the genotypic structure of the new generation, we therefore have to adopt a hypothesis about the population's mating behaviour. The simplest hypothesis is one that allows us to leave out the qualifying phrase, "given that the first parent has transmitted a particular allele". In order for this to be possible, knowledge of the gene transmitted by the father must give us no information about the gene transmitted by the mother. In other words, mating takes place at random with respect to the genotypes at the locus in question. This fifth condition can be expressed as:

⑤ For the locus in question, mates are genetically independent of one another.

Then we can write:

$$_F\pi_i = p_i^{(g-1)}, \qquad _M\pi_{i|i} = p_i^{(g-1)}, \qquad _M\pi_{j|i} = p_j^{(g-1)}.$$

Hence:

$$\Pi_{ii}=\{p_i^{(g-1)}\}^2, \qquad \Pi_{ij}=2\,p_i^{(g-1)}\times p_j^{(g-1)},$$

which, using the definition of the "genetic square" of a vector, can be written:

$$\mathscr{S}_g=\vec{s}_{g-1}^{\,2}. \tag{7}$$

Expressed in words, this is equivalent to the statement: "In a population which is subject to the five conditions listed above, the genotypic probability structure of a generation is equal to the genetic square of the genic structure of the preceding generation."

b) If, however, we do not know the actual genic structure s_{g-1} of generation $g-1$, but only its probability structure, $\mathbf{s}_{g-1}$, the assumption expressed as condition ⑤ is still not enough to enable us to say what the genotypic structure of the new generation will be.

Without any further assumptions, we know only two things about the father F and the mother M of a randomly chosen member of generation g:

1. They are a pair of mates, and

2. they belong to the same population (or, more precisely, to the same isolate, defined by conditions ① and ②).

Condition ⑤ allows us to make use of the first item of information, but not of the second.

In a particular population, the chance events involved in the transmission of genes from one generation to the next lead to fluctuations in the frequencies of the different alleles, and even sometimes to the chance elimination of alleles. Therefore two populations that are subjected to the same conditions may become different from one another, purely as the result of chance; one population might lose the allele A_i, while another might lose A_j. In this situation, the fact that the individual F has transmitted an A_i allele indicates that A_i is present in the population, and this information relates to all the members of the population, including F's mate, M.

It is obvious that the effect of chance events in changing the gene frequencies of populations is greater the smaller the population size. In the extreme case when each generation consists of one man and one woman, the probability of non-transmission of a gene is $(\frac{1}{2})^2$, because each gene takes part in two successive "draws", in each of which its chance of not being transmitted is $\frac{1}{2}$. After a number of generations, many loci will have become homozygous. Knowledge of a gene carried by one of the two population members would then tell us what genes the other has.

Relation (4) is still valid: the *a priori* probability of a particular allele in generation g is equal to its frequency in the initial generation, but this does not give us any information about the actual genic structure in generation g.

In order to overcome this problem, we have to assume that the chance events in the genetic transmission process do not affect the population; this is true if the population size is very large.

Also, according to the law of large numbers, as population size increases, the actual frequencies of the alleles $p_i^{(g)}$ converge towards their probabilities $\pi_i^{(g)}$. We therefore impose the last condition:

⑥ The population size is so large that it can be treated as infinite.

Then $p_i^{(g)} = \pi_i^{(g)}$, and Eq. (4) becomes:

$$s_g = s_0 \tag{8}$$

which is equivalent to: "In a population subjected to the six conditions listed above, the genic structure remains constant from generation to generation."

Also, using relations (7) and (8), we can write:

$$\mathscr{S}_g = \vec{s}_0^2.$$

But, from the law of large numbers, the set of genotypic frequencies $P_{ij}^{(g)}$ converges towards the set of probabilities $\Pi_{ij}^{(g)}$ of these genotypes. In other words:

$$S_g = \mathscr{S}_g.$$

Finally, therefore, we can write:

$$S_g = \vec{s}_0^2 \tag{9}$$

or: "In a population subjected to the six conditions listed above, the genotypic structure remains constant, and is equal to the genetic square of the initial genic structure."

2.3. Panmixia and Perfect Panmixia

Condition ⑤ concerns only the particular locus which we are considering, so the results which have been derived, notably relations (7) and (8), remain valid if there is choice of mating partners, provided only that this does not entail dependence between the individuals, for the locus concerned.

In practice, we often broaden this condition to cover all loci, and assume that there is complete independence between the genetic constitutions of the two mating individuals. This condition can be expressed as follows:

(5 b) "Mating is at random", or
"There is no choice of mates", or
"The mating couples are a random sample of all possible couples".

Condition (5 b) is clearly a more restrictive condition than condition ⑤, which follows from (5 b).

Populations which satisfy condition (5 b) are often called "panmictic", It is perhaps useful to distinguish between populations for which condition ⑤ is valid and populations for which condition (5 b) applies: we can call the first of these "panmictic for a particular locus", and the second "perfectly panmictic" populations.

2.4. The Hardy-Weinberg Principle

For populations in the equilibrium state described above, for one or more loci, Eqs. (4), (7), (8) and (9) are valid. These represent the essence of the Hardy-Weinberg principle. From the way the argument in the last section was developed, this principle can be seen to consist of a number of elements:

a) If conditions ①, ②, ③, and ④ are satisfied, the probabilities of encountering the different alleles among the gametes produced in any generation are equal to their frequencies in the initial generation:

$$s_g = s_0. \tag{4}$$

b) If condition ⑤ is also satisfied, the probabilities of the different genotypes in a generation constitute a trimat of probabilities equal to the genetic square of the genic structure of the preceding generation:

$$\mathscr{S}_g = \vec{s}_{g-1}^2. \tag{7}$$

c) If condition ⑥ also holds, the gene frequencies remain constant from generation to generation:

$$s_g = s_0 \tag{8}$$

and the set of genotype frequencies is a trimat equal to the genetic square of the genic structure in the initial generation:

$$S_g = \vec{s}_0^2. \tag{9}$$

In terms of the frequencies of the genes and genotypes involved, these two results can be expressed as follows:

1. The gene frequencies at any time are equal to the initial gene frequencies $p_i^{(0)}$.

2. Starting in generation 1, the genotype frequencies will be:

$$
\begin{aligned}
P_{ii} &= (p_i^{(0)})^2 \\
P_{ij} &= 2\, p_i^{(0)} \times p_j^{(0)} \qquad (i \neq j).
\end{aligned}
\tag{10}
$$

Thus the population not only remains in stable gene frequency equilibrium, but also attains, in the first generation, its equilibrium with respect to genotype frequency.

Notes. 1. Let C_{ij} be a number associated with individuals whose genotype is $(A_i A_j)$. The mean of this number for the whole population is:

$$\bar{C} = \sum_{ij} P_{ij} C_{ij}$$

the summation being taken over all the $\dfrac{n(n+1)}{2}$ genotypes $(A_i A_j)$.

If the population is in Hardy-Weinberg equilibrium, i.e. if the frequencies P_{ij} are given by the relations (10), we can write:

$$\bar{C} = \sum_{ij} p_i p_j C_{ij} \tag{11}$$

where this summation is over the n^2 combinations of the n subscripts i and the n subscripts j; the value C_{ij} enters into the summation once when $i = j$, and twice when $i \neq j$.

This result simplifies the calculation of the mean in cases when the population can be considered to be in Hardy-Weinberg equilibrium.

2. In the derivation of the Hardy-Weinberg principle, we have implicitly assumed that the initial genotypic structure was the same for both sexes. If this assumption is not valid, genotypic equilibrium is attained at the second generation, not the first. For example, consider the simple case when the initial generation consists of males who are homozygous for a gene $(A_1 A_1)$, and females homozygous for a different allele $(A_2 A_2)$. Then all members of both sexes of the first generation will be heterozygotes $(A_1 A_2)$, and the equilibrium genotypic structure, $\frac{1}{4}(A_1 A_1), \frac{1}{2}(A_1 A_2), \frac{1}{4}(A_2 A_2)$, is attained in the second generation.

3. The Classical Treatment
of the Hardy-Weinberg Equilibrium

The reasoning which has been employed to demonstrate the Hardy-Weinberg principle has the advantages of being rigorous, and of showing clearly the successive assumptions that have been made. However, it may seem rather too abstract. The Hardy-Weinberg equilibrium is usually derived by the use of a table of mating types. In order to avoid unmanageable calculations, the case of a locus with two alleles A_1 and A_2 is usually treated. The individuals can then be of three genotypes $(A_1 A_1)$, $(A_1 A_2)$ and $(A_2 A_2)$.

We shall employ the usual notation of p and q for the frequencies of the two genes, and D, H and R for the three genotype frequencies. Clearly:

$$p+q=1, \quad D+H+R=1.$$

3.1. Establishment of the Equilibrium

So far, we have made no assumptions about the initial structure of the population; the proportions of the three genotypes are simply such that D, H and R are three numbers adding up to 1.

All the genes of the individuals of genotype $(A_1 A_1)$ are A_1, and these individuals are a proportion D of the population. Half the genes of $(A_1 A_2)$ individuals are A_1, and these individuals are a proportion H of the population. The frequency of allele A_1 is therefore:

$$p=D+\frac{H}{2} \tag{12}$$

and similarly the frequency of A_2 is:

$$q=R+\frac{H}{2}. \tag{13}$$

Note that these two equations are equivalent to the statement that the genic structure (p, q) is the contracted vector of the genotypic structure, which in this case is:

$$S = \begin{array}{|l} D \\ H \quad R \end{array}$$

If we assume that mating takes place at random, with no choice of mates at all, the frequencies of the different types of mating are as shown in Table 3.1. For example, the probability that both the mates are $(A_1 A_1)$ is equal to the probability (D) that the male is $(A_1 A_1)$ multiplied by the probability (D) that the female is $(A_1 A_1)$. If the population size is large enough, the frequency of this type of mating is equal to its probability, D^2.

Table 3.1. Random mating

♀	♂		
	$(A_1 A_1)$	$(A_1 A_2)$	$(A_2 A_2)$
$(A_1 A_1)$	D^2	DH	DR
$(A_1 A_2)$	DH	H^2	HR
$(A_2 A_2)$	RD	RH	R^2

Table 3.2. The genotypes among the offspring

Parents		Offspring		
Mating type	Frequency	$(A_1 A_1)$	$(A_1 A_2)$	$(A_2 A_2)$
$(A_1 A_1) \times (A_1 A_1)$	D^2	1	0	0
$(A_1 A_1) \times (A_1 A_2)$	$2DH$	$\frac{1}{2}$	$\frac{1}{2}$	0
$(A_1 A_1) \times (A_2 A_2)$	$2DR$	0	1	0
$(A_1 A_2) \times (A_1 A_2)$	H^2	$\frac{1}{4}$	$\frac{1}{2}$	$\frac{1}{4}$
$(A_1 A_2) \times (A_2 A_2)$	$2HR$	0	$\frac{1}{2}$	$\frac{1}{2}$
$(A_2 A_2) \times (A_2 A_2)$	R^2	0	0	1

If we ignore the sex of the mating individuals, the nine types of mating reduce to six. The frequencies of these six types are given in Table 3.2. For an $(A_1 A_1) \times (A_1 A_2)$ mating, for example, all the offspring receive A_1 from the first parent, while half the offspring receive A_1 from the second parent, and half A_2. Among the offspring of this type of mating, the frequencies of the genotypes produced will therefore be $\frac{1}{2}(A_1 A_1)$, $\frac{1}{2}(A_1 A_2)$ and $0(A_2 A_2)$.

Among the offspring from all the types of mating, the proportion of $(A_1 A_1)$ individuals will be:

$$D_1 = D^2 + \tfrac{1}{2}(2DH) + \tfrac{1}{4}H^2 = \left(D + \frac{H}{2}\right)^2 = p^2$$

and similarly:

$$H_1 = \tfrac{1}{2}(2DH) + 2DR + \tfrac{1}{2}H^2 + \tfrac{1}{2}(2HR) = 2(D + \tfrac{1}{2}H)(R + \tfrac{1}{2}H) = 2pq,$$

and

$$R_1 = \tfrac{1}{4}H^2 + \tfrac{1}{2}(2RH) + R^2 = (R + \tfrac{1}{2}H)^2 = q^2.$$

The gene frequencies in the new generation will therefore be:

$$p_1 = D_1 + \tfrac{1}{2}H_1 = p^2 + pq = p(p+q) = p$$
$$q_1 = R_1 + \tfrac{1}{2}H_1 = q^2 + pq = q(p+q) = q.$$

Therefore, the gene frequencies remain constant from generation to generation.

Since, as we have just shown, the genotype frequencies D, H and R depend only on the gene frequencies, it follows that the genotypic structure also remains constant in later generations.

Thus the population attains, in the first generation of random mating, an equilibrium structure with the proportions:

$$D = p^2, \quad H = 2pq, \quad R = q^2. \tag{14}$$

3.2. Random Union of Gametes

This somewhat laborious demonstration shows that the assumption of random mating leads to the same results for the population structure as the assumption that mates are genetically independent of one another. We get the same results if we assume that the gametes (eggs and spermatozoa) unite at random. This can be demonstrated as follows.

Assuming that the gametes are shed into a common "pool", and that zygotes are formed by the fusion of pairs of zygotes taken at random from the pool, the probabilities of the genotypes will be $D=p^2$, $H=2pq$, $R=q^2$. This is shown in Table 3.3.

Table 3.3. Random union of gametes

♀	♂	A_1	A_2
	Frequency	p	q
A_1	p	p^2	pq
A_2	q	pq	q^2

This theoretical situation is in practice approximately true for some aquatic species, for example sea-urchins, which shed their gametes into the sea, so that the union of two gametes does not involve the copulation of the individuals they came from.

Since the genotypic structure of a random-mating population is determined by the genic structure of the previous generation, the results

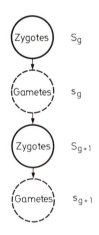

Fig. 3.1. Successive stages in the history of a population

we obtain by studying genotypes are equivalent to those obtained by considering the gametes.

The history of a population can therefore be viewed as a succession of stages in which it consists alternately of individuals (or "zygotes") and reproductive cells ("gametes") (Fig. 3.1).

In studying some types of problem, it is often easier to find the difference equation for the transition from one generation to the next in terms of the gametic stages than in terms of the zygotic stages, especially as the number of genotypes, $\dfrac{n(n+1)}{2}$, is much higher than n, the number of gamete types. We can make the transition from the functions describing changes in the gametic stage to functions for the zygotic stage and *vice versa*, by using the Hardy-Weinberg formula (result (7) of Section 2.2).

3.3. Properties of the Hardy-Weinberg Equilibrium

For a locus with two alleles, the possible states of the population, with respect to genotype frequency, can be represented on a triangular graph.

Consider an equilateral triangle ABC (Fig. 3.2). The sum of the perpendicular distances of a point inside the triangle from the sides is a constant and is equal to the altitude of the triangle. We can see this as follows. The triangles AB'C' and PA'B' (where B'C' is the parallel to BC through P, and A'P is the parallel to AC through P) are equilateral.

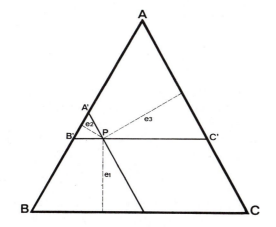

Fig. 3.2. Triangular coordinates

We therefore have:

$$\text{Altitude of } \triangle A'B'P = e_2,$$

$$\text{Altitude of } \triangle AB'C' = \text{altitude of } \triangle A'B'P + e_3$$

$$= e_2 + e_3,$$

$$\text{Altitude of } \triangle ABC = \text{altitude of } \triangle AB'C' + e_1$$

$$= e_2 + e_3 + e_1.$$

Taking the altitude of the triangle as unity, the proportions D, H and R of the three genotypes can therefore be represented by e_1, e_2 and e_3. This representation was introduced by de Finetti (1926); he represented the proportion H of heterozygotes by the distance from the base of the triangle, and this convention is usually followed.

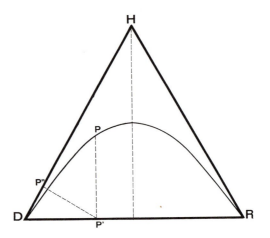

Fig. 3.3. The Hardy-Weinberg equilibrium

Thus we can represent a population by a point in a triangle; for example, a population consisting entirely of heterozygotes is represented by the apex H (Fig. 3.3).

If the population is in Hardy-Weinberg equilibrium ($D = p^2$, $H = 2pq$, $R = q^2$), these three proportions satisfy the equation:

$$H^2 = 4DR, \tag{15}$$

and the point representing the population lies on the parabola which corresponds to this equation. This parabola is called "de Finetti's parabola".

The parabola has a maximum when $D=R=\frac{1}{4}$ and $H=\frac{1}{2}$. Therefore the proportion of heterozygotes in a population in Hardy-Weinberg equilibrium cannot be greater than 50 %.

Finally, we note that the projection P' of P on the base DR of the triangle divides the base into two segments which are proportional to p and q. This can be proved as follows. Let P'' be the projection of P' on DH. Then:

$$DP'=\frac{P'P''}{\cos 30°}=\frac{R+H\cos 60°}{\cos 30°}=\frac{2}{\sqrt{3}}(R+\tfrac{1}{2}H)$$

and similarly

$$P'R=\frac{2}{\sqrt{3}}(D+\tfrac{1}{2}H).$$

Therefore

$$\frac{DP'}{RP'}=\frac{R+\tfrac{1}{2}H}{D+\tfrac{1}{2}H}=\frac{p}{q}.$$

4. The Equilibrium for Sex-Linked Genes

We have seen that women have two X chromosomes, but that men have only one. Therefore women can be homozygous or heterozygous for genes on this chromosome, which are called "sex-linked" genes; men, however, can only be "*hemizygous*".

To characterise the genetic structure of a population for a sex-linked locus, we divide it into two sub-sets, the male and female populations. In the female population, the genotypic structure $_fS_g$ in generation g is the set of the frequencies $_fP_{ij}^{(g)}$ of the different genotypes $(A_i A_j)$; the female population's gametic output has the genic structure $_fs_g$, which is the vector of the frequencies $_fp_i^{(g)}$ of the different genes A_i.

In the male population, however, the genotypic structure is identical to the genic structure, because the zygotes, like the gametes, have only one gene at these loci. A single vector, $_ms_g$, therefore characterises both the male zygotes and the spermatozoa they emit. $_ms_g$ is the vector of the frequencies $_mp_i^{(g)}$ of the various genes A_i.

4.1. Passage from One Generation to the Next

When a new generation is formed, each son receives his X chromosome from his mother. The genotypic structure of the male population in generation $g+1$ is therefore identical to the genic structure of the female population in generation g, since we are assuming that the number of children a woman has is independent of her genotype:

$$_m S_{(g+1)} = {}_f S_g .\tag{16}$$

Each daughter, however, receives one X chromosome from her father, and one from her mother. In order to be homozygous ($A_i A_i$), she must have received A_i from both her father and mother, which has the probability:

$$\Pi_{ii} = {}_m p_i^{(g)} \times {}_f p_i^{(g)}.$$

In order to be heterozygous ($A_i A_j$), she must have received A_i from her father, and A_j from her mother, or *vice versa*, and this has the probability:

$$\Pi_{ij} = {}_f p_i^{(g)} \times {}_m p_j^{(g)} + {}_f p_j^{(g)} \times {}_m p_i^{(g)}.$$

Putting the frequencies $P_{ii}^{(g+1)}$ and $P_{ij}^{(g+1)}$ of the homozygotes and heterozygotes of the female population in generation $g+1$ equal to the probabilities Π_{ii} and Π_{ij}, we can write:

$$_f S_{g+1} = {}_f S_g \times {}_m S_g \tag{17}$$

i.e. the genotypic structure of the female population in generation $g+1$, under panmixia, is the genetic product of the genic structures of the male and female populations of generation g.

The genic structure of the female population in generation $g+1$ is the contracted vector of the genotypic structure:

$$_f S_{g+1} = \mathrm{cont}\left({}_f S_g \times {}_m S_g\right) = \tfrac{1}{2}\left({}_f S_g + {}_m S_g\right)$$

or, remembering (16):

$$_f S_{(g+1)} = \tfrac{1}{2}\left({}_f S_g + {}_f S_{(g-1)}\right) \tag{18}$$

i.e. the frequency of a gene in the female population of generation $g+1$ is equal to the mean of the frequencies in the two preceding generations.

This relation enables us to determine the genic structure of the female population in any generation, given the genic structures of the female populations in the first two generations, or, equivalently, given the genic structures of the male and female populations in the initial generation.

4.2. The Equilibrium State

The overall mean genic structure of the population in generation g is:
$\bar{S}_g = \dfrac{2\,{}_f S_g + {}_m S_g}{3}$, since the females each carry two genes, and the males only one. But, from (16) and (18), we have:

$$\bar{S}_{(g+1)} = \tfrac{1}{3}\left[2\,{}_f S_{(g+1)} + {}_m S_{(g+1)}\right]$$
$$= \tfrac{1}{3}\left[\left({}_f S_g + {}_m S_g\right) + {}_f S_g\right] = \bar{S}_g$$

so that the mean genic structure remains constant:

$$\bar{s}_g = \bar{s}. \tag{19}$$

Eq. (18) can now be written in the form:

$$
\begin{aligned}
{}_f s_{(g+1)} - \bar{s} &= \tfrac{1}{2}({}_f s_g + {}_m s_g) - \tfrac{1}{3}(2\,{}_f s_g + {}_m s_g) \\
&= -\tfrac{1}{2}\left({}_f s_g - \frac{2\,{}_f s_g + {}_m s_g}{3}\right) \\
&= -\tfrac{1}{2}({}_f s_g - \bar{s}).
\end{aligned}
$$

Therefore:

$$
{}_f s_{(g+1)} - \bar{s} = \left(-\tfrac{1}{2}\right)^{(g+1)}({}_f s_0 - \bar{s}) \tag{20}
$$

which shows that the deviations of the allele frequencies in the female population from the mean allele frequencies for the whole population are halved each generation.

Fig. 3.4 illustrates the process of rapid approach to the equilibrium frequency via a series of damped oscillations, for the case of a locus with two alleles. The frequencies in the original population are:

Females: ${}_f p_0 = 75\,\%$, ${}_f q_0 = 25\,\%$,

Males: ${}_m p_0 = 15\,\%$, ${}_m q_0 = 85\,\%$.

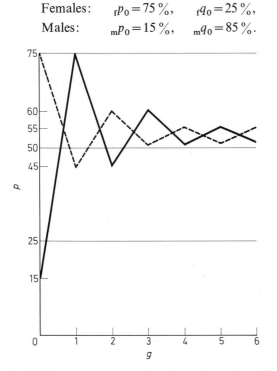

Fig. 3.4. The convergence of sex-linked genes towards equilibrium

Despite the large initial difference between the sexes, a value closer than 1 % to the equilibrium frequency:

$$\bar{p} = \frac{2 \times 75 + 15}{300} = 55\%, \quad \bar{q} = \frac{2 \times 25 + 85}{300} = 45\%$$

is reached in the fifth generation:

Generation (g)	0	1	2	3	4	5	∞
$_fp_g$	75	45	60	52.5	56.25	54.37	55
$_mp_g$	15	75	45	60	52.5	56.25	55

5. The Hardy-Weinberg Principle in Human Populations

5.1. Autosomal Loci with Two Alleles

We have seen that, for an autosomal locus with two alleles, the three genotype frequencies under random mating are:

Genotype: $A_1 A_1$ $A_1 A_2$ $A_2 A_2$

Frequency: p^2 $2pq$ q^2

where p is the frequency of allele A_1 and q is the frequency of allele A_2.
These proportions can be represented by parabolas, as in Fig. 3.5.

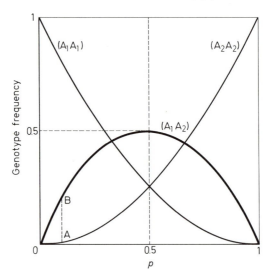

Fig. 3.5. The genotypic frequencies at equilibrium in the two-allele case

5.1.1. Rare recessive genes. If the gene A_2 is completely recessive, its effect is manifest only in $A_2 A_2$ individuals. If the frequency of these individuals in the population is R, the frequency of the A_2 gene is $q = \sqrt{R}$.

For uncommon genes, the frequency of heterozygous "carrier" individuals, who do not manifest the effect of the gene, is much higher than the frequency of homozygotes showing the trait, since the heterozygote frequency is:

$$2pq = 2(1 - \sqrt{R})\sqrt{R}$$

which is much larger than R, if R is small.

Fig. 3.5 illustrates this effect. The points A and B give the frequencies of homozygotes and heterozygotes respectively, for a rare gene. The value of B can be seen to be much greater than that of A.

Cystic fibrosis is an example of a character controlled by this type of gene. This disease is characterised by a number of symptoms, but all affected individuals have elevated sodium chloride concentrations in their sweat. Pedigrees of families which contain affected members support the hypothesis that this disease is controlled by an autosomal recessive gene.

The frequency of cystic fibrosis varies from country to country, between about 1 in 1 000 births and 1 in 3 000, with a mean of about 1 in 2 000, i.e. 5×10^{-4}.

The frequency of the gene for cystic fibrosis is therefore: $q = \sqrt{\dfrac{5}{10\,000}} = 2.2\%$, and the frequency of hererozygotes is: $2q(1-q)$ $= 4.3\%$. Thus carriers of the gene are 87 times more frequent than affected individuals. For the present size of the population of France, these percentages show that there must be more than two million unaffected carriers of the cystic fibrosis gene, while the number of affected individuals would be about 25 000 if their longevity were unaffected by the disease; their actual number is even lower than this. These calculations are based on the assumption that the population is random-mating with respect to this gene. This is probably not the case, but deviations of the actual situation from panmixia do not affect the order of magnitude of the figures we have obtained here.

5.1.2. Rare dominant genes. If a gene A_2 is fully dominant, all the individuals who carry the gene, whether they are homozygous ($A_2 A_2$) or heterozygous ($A_1 A_2$) for it, will show the trait. If the trait is rare, we can show, by a similar argument to the one for rare recessive genes, that the proportion of heterozygotes among the affected individuals is much higher than the proportion of homozygotes. For this type of gene, it is often not possible to find homozygotes at all, either because they are too rare to be found by investigators, or else because the double dose of the gene causes death of the individuals as embryos, or shortly after birth.

The gene for invagination of the upper lateral incisors is an example of this type of gene. In a study of schoolchildren in Loir-et-Cher, Sutter (1966) found that 1.4 % of the children were affected. The frequency of the corresponding gene is therefore approximately 0.7 %, and the homozygotes, if they exist, would have a frequency of 0.005 %, i.e. one in 20 000 individuals.

5.1.3. Non-deleterious genes.

For many loci, alleles are known which do not have any detectable deleterious effect on the viability or reproductive potential of the individuals who carry them. These alleles are often found in high frequencies.

An example of such a gene is the gene for the ability to taste phenyl thiocarbamide (PTC). This chemical tastes very bitter to some people, but is practically tasteless for others. The ability to taste PTC is controlled by the "taster" allele T at this locus, which is dominant to the "non-taster" allele t.

Snyder (1932) found the proportions of the two phenotypes in a sample of 3 643 white Americans to be: taster 70.2 % and non-taster 29.8 %. The allele frequencies are therefore $q = \sqrt{0.298} = 0.545$, and $p = 1 - q = 0.455$.

Another example is the ability to roll up the tongue in a lengthwise direction, which seems to be controlled by a dominant gene R, so that the individuals who cannot roll up their tongues are homozygotes (rr). Stern (1973) gives the frequency of the R gene in the United States as about $p = 40 \%$, so that the "non-roller" individuals constitute $(\frac{60}{100})^2 = 36 \%$ of the population.

5.1.4. Co-dominant genes.

When two alleles show co-dominance, there are three phenotypes corresponding to the three genotypes. In this case, we can calculate not only the frequencies of the two alleles, but also test the genotype frequencies for agreement with the values predicted by the Hardy-Weinberg principle.

The MN blood group is a well-known case of co-dominance. Red blood cells carry "antigens" on their membranes. The presence of such antigens can be detected by the use of corresponding "antibodies", which are found in certain sera. Antibody synthesis is usually the result of an immunological reaction, but in some cases this synthesis occurs autonomously. The cells carrying a particular antigen, X, are specifically agglutinated by serum containing the antibody anti-X.

Many red-blood-cell antigen-antibody systems are known, and the MN blood group system is just one of them. Every individual has either the antigen M, or the antigen N, or both, on his red blood cells; the blood of M individuals is agglutinated by serum containing anti-M, while blood from N individuals is agglutinated by anti-N serum, and blood from MN individuals is agglutinated by either type of serum.

This blood group is controlled by a locus with two alleles, A_M and A_N, which determine the synthesis of the M and N antigens, respectively. The correspondence between the genotypes and the phenotypes is as follows:

Genotype	$(A_M A_M)$	$(A_M A_N)$	$(A_N A_N)$
Phenotype (blood group)	M	MN	N

Race and Sanger (1962) give the following data for British people:

	Blood group			Total
	M	MN	N	
Number of individuals	363	634	282	1 279
Frequency	28.38%	49.57%	22.05%	100%

Remembering that A_M is represented twice in M individuals, and once in MN individuals, we can calculate the frequencies p and q of A_M and A_N in the population:

$$p = \frac{363 + \frac{1}{2} 634}{1\,279} = 53\%, \quad q = 47\%.$$

If we assume that the population has reached equilibrium, the Hardy-Weinberg principle leads to the following theoretical frequencies:

$$p^2 = 0.281, \quad 2pq = 0.498, \quad q^2 = 0.221.$$

The agreement with the observed frequencies is so close that it is unnecessary to test it statistically.

5.2. Autosomal Loci with Three Alleles

The AB0 blood group is an example of a character controlled by a locus with three alleles[1]. We can designate the alleles as A_A, which leads to the synthesis of the A antigen, A_B, which leads to the formation of the B antigen, and A_0 which does not lead to the synthesis of either

---

[1] In reality, this locus has more than three alleles; the account given here is only an outline.

antigen. The alleles A_A and A_B are co-dominant with respect to each other, but both are dominant to A_0. The six possible genotypes therefore correspond to four phenotypes, as follows:

Genotype	Phenotype (blood group)
$(A_A A_A)$ and $(A_A A_0)$	A
$(A_B A_B)$ and $(A_B A_0)$	B
$(A_A A_B)$	AB
$(A_0 A_0)$	0

If p, q and r are the frequencies of the three alleles in a panmictic population, then, by the Hardy-Weinberg principle, the frequencies of the various phenotypes are:

Phenotype	Frequency under panmixia
A	$p^2 + 2pr$
B	$q^2 + 2qr$
AB	$2pq$
0	r^2

Race and Sanger (1962) give the following observed numbers and frequencies of these phenotypes, in Great Britain.

Phenotype	Observed number	Observed frequency
A	79 334	0.417
B	16 280	0.086
AB	5 781	0.030
0	88 789	0.467

In order to calculate the gene frequencies p, q and r, we note that:

$$r = \sqrt{f_0}, \quad p = 1 - \sqrt{f_0 + f_B}, \quad q = 1 - \sqrt{f_0 + f_A}$$

where f_0, f_A, f_B are the frequencies of individuals with blood group 0, A and B. We thus obtain:

$$r = \sqrt{\tfrac{467}{1000}} = 68.3\ \%$$

$$p = 1 - \sqrt{\tfrac{553}{1000}} = 25.7\ \%$$

$$q = 1 - \sqrt{\tfrac{884}{1000}} = 6.0\ \%.$$

The theoretical phenotype frequencies corresponding to these frequencies are:

$$0.417 \qquad 0.085 \qquad 0.031 \qquad 0.467.$$

Clearly the agreement between the observed and theoretical frequencies is partly due to the method of calculating p, q and r, but the agreement for all phenotypes shows that the assumption of Hardy-Weinberg equilibrium is acceptable. This could be tested by a χ^2 test, deducting appropriate number of degrees of freedom for the parameters which have been estimated.

5.3. Sex-Linked Genes

For genes carried on the X chromosome, the phenotypes that are possible are different in men and women. For a locus with two alleles, the genotype frequencies under panmixia are:

	Males		Females		
Genotype:	(A_1)	(A_2)	$(A_1 A_1)$	$(A_1 A_2)$	$(A_2 A_2)$
Frequency:	p	q	p^2	$2pq$	q^2

where p is the frequency of allele A_1 in the whole population, and the population is assumed to have reached equilibrium.

5.3.1. Rare recessive sex-linked genes. If a trait is controlled by a recessive sex-linked gene A_2, all men who carry A_2 will show the trait, but among women only the homozygotes $(A_2 A_2)$ will show it. The manifestation of the trait may also differ in the two sexes.

This explains how a hereditary trait can appear to "skip a generation", which was considered for a long time to be contrary to the Mendelian theory. A marriage between an affected man and a non-carrier female can be written as $(A_2) \times (A_1 A_1)$. All the sons of this marriage will receive the A_1 gene from their mother, and will have the genotype (A_1). They will therefore not show the trait. All the daughters will have the genotype $(A_1 A_2)$ and will therefore be carriers, but will not show the trait, as the A_2 gene is assumed to be recessive. Thus none of the offspring will show the trait, and this seems to contradict the idea that the character is hereditary.

In the next generation, the daughters, even if they marry normal, unaffected men, will have sons half of whom are affected: the marriage is $(A_1 A_2) \times (A_1)$, so their daughters will have the genotypes $(A_1 A_1)$ and $(A_1 A_2)$, while the sons will have the genotypes (A_1) and (A_2). The (A_2) sons will show the trait again. We therefore see that the disappearance

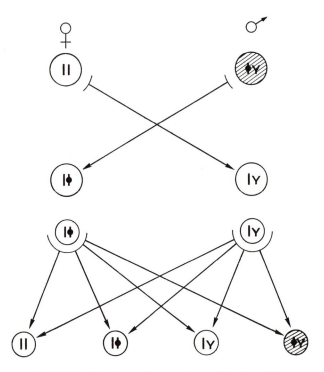

Fig. 3.6. The inheritance of a recessive sex-linked condition

of a trait for one generation confirms Mendel's laws, rather than contradicts them (Fig. 3.6).

The frequency of recessive traits controlled by sex-linked genes is
much higher in men than women. If we designate these frequencies by T_m
and T_f, we have:

$$\frac{T_f}{T_m} = \frac{q^2}{q} = q$$

so the ratio of T_f to T_m can be very small if the gene is rare.

Examples. The best-known cases of sex-linked inheritance are
haemophilia and colour-blindness. Haemophilia clearly has an effect on
the number of offspring of the various possible types of marriages, so
one of the conditions for the Hardy-Weinberg principle to hold is
violated. There is therefore no point in testing for agreement with Hardy-
Weinberg frequencies in this case.

The term "colour-blindness" covers several types of defect. The most
frequent of these, "deuteranomaly", is found in about 4% of men. If

we assume that marriages take place at random with respect to colour vision, the proportion of colour-blind women would be $(\frac{4}{100})^2 = 0.16\%$, which is 25 times less than the frequency among men. However, the frequency of unaffected carrier heterozygous women would be $2 \times 0.04 \times (1-0.04) = 8\%$, i.e. double the frequency of affected men.

The control of colour-blindness is, in reality, more complicated than this, and the frequency of colour-blind women is even lower than the figure given above. Nevertheless, this example demonstrates the different frequencies of affected individuals in the different sexes, for sex-linked traits.

5.3.2. Rare dominant sex-linked genes. When a trait is controlled by a dominant, sex-linked gene, all the carrier women will show the trait. The theoretical frequency, $T_f = q^2 + 2pq$, of carrier women is approximately equal to twice the frequency, $T_m = q$, of affected men:

$$\frac{T_f}{T_m} = \frac{q^2 + 2pq}{q} = q + 2p = 2 - q \approx 2 \quad \text{(if q is small).}$$

The pattern of transmission of such traits is very different from that for recessive sex-linked genes. The marriage of an affected man (A_2) and a normal woman $(A_1 A_1)$ gives daughters who all have the genotype $(A_1 A_2)$ and who are all affected. The sons are all (A_1) and are normal. The marriage of a normal man (A_1) and an affected woman, whose genotype is $(A_2 A_2)$ or $(A_1 A_2)$, gives in the first case all affected offspring, and in the second case half the offspring affected whatever their sex. There is therefore no "skipping" of generations, as happens in the recessive case.

Ocular albinism is an example of a condition which shows this mode of inheritance. Affected males have very little pigment in the iris and fundus of the eye, while carrier females have an abnormally translucent iris, and the fundus shows a pattern of pigmented and unpigmented spots. This gene is therefore only partially dominant.

5.4. Y-Linked Genes

For genes carried on the Y chromosome, there is obviously no female genotype, since this chromosome is present only in men.

Traits controlled by Y-linked genes have a very simple pattern of inheritance: all the sons, but none of the daughters, have the phenotype of the father.

A small number of conditions, all affecting the skin, appear (but this is disputed) to have this mode of transmission, which is sometimes called "holandry". These conditions are all very rare and are more important in medical genetics than in population genetics.

Chapter 4

The Equilibrium for Two Loci

In the last chapter, we considered the situation when individuals are classified with respect to the alleles at one locus. What is the equilibrium situation when no forces for change are active, if we classify individuals according to the genes present at two loci?

1. The Role of Individuals

In each generation, the new individuals are formed by the fusion of two gametes, and they in turn produce gametes which contain the same genes as were carried by these gametes, but reassorted as a result of the meiotic process.

Let A_i and B_j be the genes carried at two loci, by the paternal gamete, and A_k and B_l be the genes carried at these loci in the maternal gamete. The genotype of the individual formed by the fusion of these two gametes can be written as

$$\frac{A_i B_j}{A_k B_l}.$$

When this individual reproduces, he will produce four types of gametes:

$\left.\begin{array}{l} A_i B_j \\ A_k B_l \end{array}\right\}$ These combinations are the same as those in the parental gametes.

$\left.\begin{array}{l} A_i B_l \\ A_k B_j \end{array}\right\}$ These combinations are different from those in the parental gametes.

If the loci A and B are not linked, segregation of the genes at the two loci takes place independently, according to Mendel's second law. The frequencies of the four types of gametes are therefore equal, and we can write:

$$p_{ij} = p_{il} = p_{kj} = p_{kl} = \tfrac{1}{4}.$$

If, on the other hand, the loci are linked, the parental types of gametes, $A_i B_j$ and $A_k B_l$, will have higher frequencies than the "recombinant" types, $A_i B_l$ and $A_k B_j$.

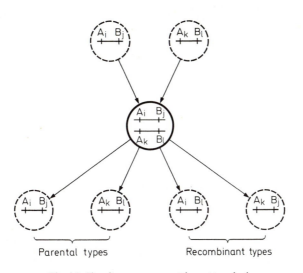

Parental types   Recombinant types

Fig. 4.1. Simultaneous segregation at two loci

Let r stand for the proportion of recombinant gametes; if linkage is total, no recombinant gametes are formed, and $r=0$; if the genes are unlinked, there are equal numbers of recombinant and non-recombinant gametes, so $r=\frac{1}{2}$; thus r lies between the limits: $0 \leq r \leq \frac{1}{2}$.

The frequencies of the four gamete types produced by an $\left(\dfrac{A_i B_j}{A_k B_l}\right)$ individual are therefore, in general:

$$p_{ij}=p_{kl}=\tfrac{1}{2}(1-r), \qquad p_{il}=p_{kj}=\tfrac{1}{2}r.$$

In a population where the matings that take place, and the numbers of offspring produced, are assumed to be independent of genotype, the role of individuals, from the point of view of the genetic changes in the population, is confined to their function as the classical "urn" of probability theory from which balls of different colours are drawn: each individual is an "urn" in which the chromosomes he received from his father and his mother are mixed, and from which the chromosomes which will constitute the next generation are drawn, after crossing-over has taken place. For each chromosome pair, which one of the pair goes into a particular gamete is a matter of chance.

2. Genic Structure

We shall first study the changes in frequency of the various associations of the genes A_i and B_j, in the gametes produced by successive generations. Let

$p_{ij}^{(g)}$ be the frequency of gametes carrying A_i and B_j among the gametes produced by the individuals of generation g,

$p_i^{(g)}$ be the frequency of the A_i gene in generation g, and

$p_j^{(g)}$ be the frequency of the B_j gene.

We have:

$$p_i^{(g)} = \sum_j p_{ij}^{(g)} \quad \text{and} \quad p_j^{(g)} = \sum_i p_{ij}^{(g)}.$$

2.1. The Recurrence Relation for the Transition from One Generation to the Next

The gametes produced in generation g are of two sorts:

1. A fraction $(1-r)$ have not undergone recombination. A non-recombinant gamete can only be of the type $A_i B_j$ if the individual that produced it was itself formed by two gametes, one of which was $A_i B_j$.

The probability of an $A_i B_j$ gamete among the gametes produced in generation g, given that there has been no recombination, is therefore equal to the frequency of this type of gamete in generation $g-1$:

$$P\{\text{gamete } A_i B_j \,|\, \text{non-recombinant}\} = p_{ij}^{(g-1)}.$$

2. A fraction r are recombinant. A recombinant gamete can be of the type $A_i B_j$ if the individual that produced it was formed from two gametes, of which one carried A_i (and any allele at the B locus), and one carried B_j (and any allele at the A locus). Since we are assuming perfect panmixia, and thus independence of the uniting gametes, the probability of an $A_i B_j$ gamete among gametes produced in the g-th generation, given that there has been recombination, is equal to the product of the frequencies of A_i and B_j in generation $g-1$:

$$P\{\text{gamete } A_i B_j \,|\, \text{recombinant}\} = p_i^{(g-1)} \times p_j^{(g-1)}.$$

Finally, the probability of an $A_i B_j$ gamete in generation g, whether recombinant or not, is:

$$P(\text{gamete } A_i B_j) = r\,P\{\text{gamete } A_i B_j \,|\, \text{recombinant}\}$$

$$+ (1-r)\,P\{\text{gamete } A_i B_j \,|\, \text{non-recombinant}\}.$$

Putting the frequencies equal to the probabilities (which is made possible by the assumption of condition ⑥: large population size) we can write:

$$p_{ij}^{(g)} = (1-r)\,p_{ij}^{(g-1)} + r\,p_i^{(g-1)} \times p_j^{(g-1)}. \tag{1}$$

2.2. The Constancy of Gene Frequencies

Eq. (1) enables us to show that the frequency of a gene remains constant from generation to generation, as follows:

$$
\begin{aligned}
p_i^{(g)} &= \sum_j p_{ij}^{(g)} \\
&= (1-r) \sum_j p_{ij}^{(g-1)} + r\, p_i^{(g-1)} \sum_j p_j^{(g-1)} \\
&= (1-r) p_i^{(g-1)} + r\, p_i^{(g-1)} \\
&= p_i^{(g-1)}.
\end{aligned}
$$

This result is obvious *a priori:* the fact that we are considering two loci simultaneously does not affect this fundamental property of equilibrium populations. From now onwards, when we are dealing with equilibrium populations, we can leave out the superscript for generation number and write p_i for $p_i^{(g)}$, and p_j for $p_j^{(g)}$.

2.3. The Approach to Equilibrium

Eq. (1) can be written in the form:

$$
p_{ij}^{(g)} - p_i p_j = (1-r) \left[p_{ij}^{(g-1)} - p_i p_j \right].
$$

We therefore have:

$$
p_{ij}^{(g)} - p_i p_j = (1-r)^g \left[p_{ij}^{(0)} - p_i p_j \right],
$$

where $p_{ij}^{(0)}$ is the value of p_{ij} in the initial generation. Writing D_{ij} for the difference between the frequency of $A_i B_j$ gametes produced in generation g and the product of the gene frequencies of A_i and B_j, we have:

$$
D_{ij}^{(g)} = (1-r)^g D_{ij}^{(0)}. \tag{2}
$$

Eq. (2) shows that p_{ij} tends towards the product of the frequencies p_i and p_j.

The speed with which the population approaches equilibrium becomes greater as $1-r$ becomes smaller, in other words, as r approaches its maximum value of $\frac{1}{2}$; the approach to equilibrium is therefore faster for unlinked than for linked genes. For completely independent genes, $r = \frac{1}{2}$, so that:

$$
p_{ij}^{(g)} = p_i p_j + (\tfrac{1}{2})^g D_{ij}^{(0)}.
$$

In this case, the equilibrium is approached very rapidly, with the same rate of convergence as for sex-linked genes.

Table 4.1. The convergence towards equilibrium in the two-locus case. The values of $D_{ij}^{(g)}$ when $D_{ij}^{(0)} = 0.25$ are given

g	r						
	0.01	0.05	0.10	0.20	0.30	0.40	0.50
0	0.25	0.25	0.25	0.25	0.25	0.25	0.25
5	0.23775	0.19345	0.14762	0.08192	0.04202	0.01944	0.00781
10	0.22610	0.14968	0.08717	0.02684	0.00706	0.00151	0.00024
15	0.21501	0.11582	0.05147	0.00880	0.00119	0.00012	0.00001
20	0.20448	0.08962	0.03039	0.00288	0.00020	0.00001	0
30	0.18493	0.05366	0.01060	0.00031	0.00001	0	0
40	0.16724	0.03213	0.00370	0.00003	0	0	0
50	0.15125	0.01924	0.00129	0	0	0	0

However, if the loci A and B are tightly linked, $(1-r)^g$ tends towards zero extremely slowly, and equilibrium is reached only after many generations. Table 4.1 shows some examples of the calculation of $D_{ij}^{(g)}$ starting with a value of 0.25 for $D_{ij}^{(0)}$. When the recombination frequency is 0.30, an initial value of D_{ij} of 0.25 is reduced to 0.00001 in 30 generations, but if the recombination frequency is 0.01, an initial D_{ij} of 0.25 is reduced only to 0.15 in 50 generations.

3. Genotypic Structure

So far, we have studied only the gametes produced by successive generations. What changes take place in the genotypic structure of the individuals in successive generations?

The genotypic structure is the set of frequencies $P_{ik, jl}$ of individuals which have alleles A_i and A_k at the A locus, and B_j and B_l at the B locus. If the A locus has m alleles, and the B locus n, the number of different genotypes $(A_i A_k B_j B_l)$ is:

$$\frac{m(m+1)}{2} \times \frac{n(n+1)}{2}.$$

The relation between the frequencies of the different types of gametes and the frequency of a genotype among the zygotes produced depends on whether the genotype is homozygous for both loci, or for just one locus, or heterozygous for both loci.

The double homozygote $(A_i A_i B_j B_j)$ is the result of the union of two $A_i B_j$ gametes. Its frequency is therefore:

$$P_{ii, jj}^{(g+1)} = \{p_{ij}^{(g)}\}^2. \tag{3}$$

The single homozygote of genotype $(A_i A_i B_j B_l)$ is the result of the union of an $A_i B_j$ gamete and an $A_i B_l$ gamete. Hence:

$$P_{ii,\,jl}^{(g+1)} = 2\,p_{ij}^{(g)} \times p_{il}^{(g)}. \tag{4}$$

Finally, the double heterozygote $(A_i A_k B_j B_l)$ is the result of either:
1. the union of an $A_i B_j$ gamete and an $A_k B_l$ gamete, or
2. the union of an $A_i B_l$ gamete and an $A_k B_j$ gamete. Therefore:

$$P_{ik,\,jl}^{(g+1)} = 2\,\{p_{ij}^{(g)} \times p_{kl}^{(g)} + p_{il}^{(g)} \times p_{kj}^{(g)}\}. \tag{5}$$

From Eq. (2), we know the frequencies $p_{ij}^{(g)}$ of the different types of gametes. Eqs. (3), (4) and (5) enable us to calculate the frequencies $P_{ik,\,jl}^{(g)}$ of the genotypes, in terms of the gene frequencies p_i and p_j for the two loci, the recombination frequency r, and the initial coefficients $D_{ij}^{(0)}$.

At equilibrium, when the D_{ij} coefficients have become zero, we find:

$$\begin{aligned}
\hat{P}_{ii,\,jj} &= (p_i\,p_j)^2 \\
\hat{P}_{ii,\,jl} &= 2\,(p_i)^2\,p_j\,p_l \\
\hat{P}_{ik,\,jl} &= 4\,p_i\,p_j\,p_k\,p_l.
\end{aligned} \tag{6}$$

The equilibrium structure is therefore independent of:
1. the original genotypic structure, and
2. the tightness of the linkage between the two loci. It depends solely on the gene frequencies.

This result is important; for a population that is close to its equilibrium state (which seems a reasonable assumption for most human populations), the finding of an association between two characters is not evidence that the gene controlling the characters are linked. Only the study of pedigrees can give information about linkage.

4. Two Loci, Each with Two Alleles

4.1. Gamete Frequencies

The case when each locus has two alleles has been much studied. In this section, the frequencies of the alleles A_1 and A_2 at the first locus will be designated p and q, and the frequencies of the alleles B_1 and B_2 at the second locus will be written as u and v.

The gametes are of four types: $A_1 B_1$, $A_2 B_1$, $A_1 B_2$, and $A_2 B_2$. Their frequencies will be written as p_{11}, p_{21}, p_{12}, and p_{22}. There are nine genotypes:

$(A_1 A_1 B_1 B_1)$ $(A_1 A_2 B_1 B_1)$ $(A_2 A_2 B_1 B_1)$

$(A_1 A_1 B_1 B_2)$ $(A_1 A_2 B_1 B_2)$ $(A_2 A_2 B_1 B_2)$

$(A_1 A_1 B_2 B_2)$ $(A_1 A_2 B_2 B_2)$ $(A_2 A_2 B_2 B_2)$

Writing a superscript g to indicate the generation, Eq. (2) becomes:

$$p_{11}^{(g)} = p\,u + (1-r)^g\,(p_{11}^{(0)} - p\,u)$$
$$p_{21}^{(g)} = q\,u + (1-r)^g\,(p_{21}^{(0)} - q\,u)$$
$$p_{12}^{(g)} = p\,v + (1-r)^g\,(p_{12}^{(0)} - p\,v)$$
$$p_{22}^{(g)} = q\,v + (1-r)^g\,(p_{22}^{(0)} - q\,v).$$

The four initial difference terms $(p_{11}^{(0)} - p\,u)$ etc. are not independent: if we write D_0 for the difference term for the $A_1 B_1$ gametes, i.e. $D_0 = (p_{11}^{(0)} - p\,u)$, then, since $p_{11}^{(0)} + p_{21}^{(0)} = u$, we have:

$$p_{21}^{(0)} = u - p_{11}^{(0)} = u - (D_0 + p\,u) = u\,q - D_0.$$

Hence:

$$p_{21}^{(0)} - u\,q = -D_0.$$

Similarly we find that:

$$p_{12}^{(0)} - p\,v = -D_0 \quad \text{and} \quad p_{22}^{(0)} - q\,v = D_0.$$

So the set of four equations above becomes:

$$p_{11}^{(g)} = p\,u + (1-r)^g\,D_0$$
$$p_{21}^{(g)} = q\,u - (1-r)^g\,D_0$$
$$p_{12}^{(g)} = p\,v - (1-r)^g\,D_0 \tag{7}$$
$$p_{22}^{(g)} = q\,v + (1-r)^g\,D_0$$

where D_0 can be written:

$$D_0 = p_{11}^{(0)} - (p_{11}^{(0)} + p_{12}^{(0)})(p_{11}^{(0)} + p_{21}^{(0)})$$
$$= p_{11}^{(0)}(1 - p_{11}^{(0)} - p_{12}^{(0)} - p_{21}^{(0)}) - p_{12}^{(0)}\,p_{21}^{(0)}$$
$$= p_{11}^{(0)}\,p_{22}^{(0)} - p_{12}^{(0)}\,p_{21}^{(0)}.$$

Clearly, a D coefficient can be calculated for any generation g. D_g is often called the "coefficient of linkage disequilibrium", for the generation in question. It is easy to see from Eqs. (7) that $D_g = (1-r)^g\,D_0$.

4.2. Fusion of Two Populations

To illustrate the approach of a population to equilibrium, consider the case of the fusion of two populations of equal size, one of which consists of individuals of the genotype $(A_1 A_1 B_1 B_1)$, and the other of $(A_2 A_2 B_2 B_2)$ individuals.

The initial state of the population formed by the mixing of these two originally distinct populations can be represented by the following table of frequencies:

	$A_1 A_1$	$A_1 A_2$	$A_2 A_2$
$B_1 B_1$	$\frac{1}{2}$	0	0
$B_1 B_2$	0	0	0
$B_2 B_2$	0	0	$\frac{1}{2}$

This population produces gametes in the following proportions:

	A_1	A_2
B_1	$\frac{1}{2}$	0
B_2	0	$\frac{1}{2}$

The gene frequencies are:

$$p=q=\tfrac{1}{2}, \quad u=v=\tfrac{1}{2}.$$

We know that these frequencies will remain constant from generation to generation. Knowing the initial gamete frequencies, Eqs. (3), (4) and (5) enable us to determine the genotype frequencies in the first generation after the populations fused. These are:

$$\begin{bmatrix} \frac{1}{4} & 0 & 0 \\ 0 & \frac{1}{2} & 0 \\ 0 & 0 & \frac{1}{4} \end{bmatrix}.$$

The individuals of this generation will produce the four types of gametes in the proportions:

$$\begin{bmatrix} \frac{1}{4}+\frac{1}{4}(1-r) & \frac{1}{4}r \\ \frac{1}{4}r & \frac{1}{4}+\frac{1}{4}(1-r) \end{bmatrix}.$$

Clearly, it soon becomes very laborious to calculate each generation from the preceding one, in this way. It is easier to use the results which we derived in Section 4.1, giving

1. The equilibrium gamete frequencies:

$$\begin{bmatrix} p\,u & p\,v \\ q\,u & q\,v \end{bmatrix} = \begin{bmatrix} \frac{1}{4} & \frac{1}{4} \\ \frac{1}{4} & \frac{1}{4} \end{bmatrix}.$$

2. The initial coefficient of linkage disequilibrium:

$$D_0 = \tfrac{1}{4}.$$

3. The gamete frequencies in any generation g. Substituting the equilibrium gamete frequencies and the initial coefficient of linkage disequilibrium into Eqs. (7), these are:

$$\begin{bmatrix} \frac{1}{4} & \frac{1}{4} \\ \frac{1}{4} & \frac{1}{4} \end{bmatrix} + (1-r)^g \begin{bmatrix} \frac{1}{4} & -\frac{1}{4} \\ -\frac{1}{4} & \frac{1}{4} \end{bmatrix} = \frac{1}{4} \begin{bmatrix} 1+(1-r)^g & 1-(1-r)^g \\ 1-(1-r)^g & 1+(1-r)^g \end{bmatrix}.$$

4. The genotype frequencies in any generation g. From Eqs. (3), (4) and (5), these are:

$$\frac{1}{16} \begin{bmatrix} \{1+(1-r)^g\}^2 & 2\{1-(1-r)^{2g}\} & \{1-(1-r)^g\}^2 \\ 2\{1-(1-r)^{2g}\} & 4\{1-(1-r)^{2g}\} & 2\{1-(1-r)^{2g}\} \\ \{1-(1-r)^g\}^2 & 2\{1-(1-r)^{2g}\} & \{1+(1-r)^g\}^2 \end{bmatrix}.$$

5. The equilibrium genotype frequencies are found either by applying Eqs. (6), or by setting $(1-r)^g = 0$ in the above formula, giving:

$$\frac{1}{16} \begin{bmatrix} 1 & 2 & 1 \\ 2 & 4 & 2 \\ 1 & 2 & 1 \end{bmatrix}.$$

This example shows how the calculations are simplified by first finding recurrence relations for the gametic structures in successive generations and then using these to derive the genotypic structures.

Using the figures given in Table 4.1 (p. 70), we find that, for loci with recombination frequencies greater than 0.2, equilibrium is almost reached in five generations.

Notes. The genotype frequencies at equilibrium are independent of the linkage between the loci. However, the rate of approach to equilibrium when two populations that were originally distinct are suddenly mixed, is related to the recombination fraction between the loci for which the populations differed.

Strictly, in order to use this type of data as evidence for linkage, the new population formed by the mixture of the two original populations must be a panmictic population, which is unlikely to be the case.

The divergences between the frequencies of genotypes and their equilibrium frequencies can give information about the history of populations, and the extent to which they have been subject to immigration from other populations.

4.3. Instantaneous Attainment of Equilibrium

In order for equilibrium for two loci to be attained after only a single generation of panmixia, all the initial difference coefficients

$$D_{ij}^{(0)} = p_{ij}^{(0)} - p_i p_j$$

must be zero.

In the case of two loci with two alleles, we can write, in the notation of Section 4.1:

$$D_0 = p_{11}^{(0)} p_{22}^{(0)} - p_{21}^{(0)} p_{12}^{(0)} = 0. \tag{8}$$

This condition is met in several different situations, for example:

1. The initial population consists entirely of double heterozygotes $(A_1 A_2 B_1 B_2)$. The initial genotype matrix is therefore:

$$\begin{bmatrix} 0 & 0 & 0 \\ 0 & 1 & 0 \\ 0 & 0 & 0 \end{bmatrix}.$$

The following proportions of gametes are produced:

$$\begin{bmatrix} \frac{1}{4} & \frac{1}{4} \\ \frac{1}{4} & \frac{1}{4} \end{bmatrix}.$$

This satisfies Eq. (8), therefore the population attains the following equilibrium structure in the first generation:

$$\begin{bmatrix} \frac{1}{16} & \frac{1}{8} & \frac{1}{16} \\ \frac{1}{8} & \frac{1}{4} & \frac{1}{8} \\ \frac{1}{16} & \frac{1}{8} & \frac{1}{16} \end{bmatrix}.$$

2. The initial population is the result of the fusion of two populations which are both in equilibrium with respect to the A locus, while one is homozygous $B_1 B_1$ and the other is homozygous $B_2 B_2$, so that its initial genotype matrix is:

$$\alpha \begin{bmatrix} p_1^2 & 2 p_1 q_1 & q_1^2 \\ 0 & 0 & 0 \\ 0 & 0 & 0 \end{bmatrix} + \beta \begin{bmatrix} 0 & 0 & 0 \\ 0 & 0 & 0 \\ p_2^2 & 2 p_2 q_2 & q_2^2 \end{bmatrix}.$$

where α and β are proportional to the sizes of the two populations, and p_1 and p_2 are the frequencies of A_1 in the populations. The frequencies of the gametes produced will be:

$$\begin{bmatrix} p_1 & q_1 \\ p_2 & q_2 \end{bmatrix}.$$

Condition (8) is equivalent, in this notation, to:

$$(p_1 q_2 - p_2 q_1) = 0$$

which leads to:

$$p_1 = p_2.$$

Therefore equilibrium is attained in the first generation if the frequency of A_1 is the same in both the populations.

3. The initial population is the result of the fusion of two populations, each of which is in equilibrium for both the loci, so that the initial genotype matrix is:

$$\alpha \begin{bmatrix} p_1^2 u_1^2 & 2 p_1 q_1 u_1^2 & q_1^2 u_1^2 \\ 2 p_1^2 u_1 v_1 & 4 p_1 q_1 u_1 v_1 & 2 q_1^2 u_1 v_1 \\ p_1^2 v_1^2 & 2 p_1 q_1 v_1^2 & q_1^2 v_1^2 \end{bmatrix}$$

$$+ \beta \begin{bmatrix} p_2^2 u_2^2 & 2 p_2 q_2 u_2^2 & q_2^2 u_2^2 \\ 2 p_2 u_2 v_2 & 4 p_2 q_2 u_2 v_2 & 2 q_2^2 u_2 v_2 \\ p_2^2 v_2^2 & 2 p_2 q_2 v_2 & q_2^2 v_2^2 \end{bmatrix}$$

where α and β are proportional to the sizes of the two populations, and p_1, q_1, u_1, v_1 are the frequencies of the genes A_1, A_2, B_1 and B_2 in the first population, and p_2, q_2, u_2 and v_2 are their frequencies in the second population.

The gamete frequencies of the initial generation are:

$$\begin{bmatrix} \alpha p_1 u_1 + \beta p_2 u_2 & \alpha q_1 u_1 + \beta q_2 u_2 \\ \alpha p_1 v_1 + \beta p_2 v_2 & \alpha q_1 v_1 + \beta q_2 v_2 \end{bmatrix}$$

and condition (8) becomes:

$$(p_1 - p_2)(u_1 - u_2) = 0.$$

Equilibrium is therefore attained instantaneously if the frequencies of the genes are the same in both populations, for one of the two loci. (Note that case (2) becomes the same as this one if $u_1 = 1$ and $u_2 = 0$.)

5. The Detection and Measurement of Linkage

We have seen that the existence of linkage between two loci cannot be inferred from an association between characters in a population. At equilibrium, the associations depend only on the frequencies of the genes, and not on the tightness of the linkage between them. To study linkage, it is therefore essential to do family studies.

In order to detect recombination between two loci, it is obvious that the individual which produces the gametes must be heterozygous at both the loci. It is easy to make such double heterozygotes in most plant or animal species, by crossing pure lines, but if we want to study man, we cannot do this, and have to use indirect methods. A number of methods of analysing human data so as to extract as much information as possible from the pedigrees that are available have been proposed. They are mostly rather complex. We shall now study three such methods: the "sib pair" method of Penrose (1935, 1946, 1953), which permits the detection of linkage, Morton's "sequential analysis" method (Morton, 1955, 1956, 1957) and Smith's (1953, 1959) "Bayesian" method. These two last give estimates of recombination frequencies, as well as allowing linkage to be detected.

5.1. Detection of Linkage – Penrose's Method

The method proposed by Penrose is based on the classification of pairs of brothers and sisters, or "sib pairs", according to whether they are alike or unlike for the two characters in question. The pairs fall into four categories according to whether the members of the pair are the same (S_1) or different (D_1) for the first character, and similarly for the second character (S_2) and (D_2). The numbers of sib pairs in the four categories are denoted by n_1, n_2, n_3 and n_4, as shown in Table 4.2.

We suppose that the first character is controlled by a locus with two alleles, of which one, A, is dominant to the other, a. Similarly, the second character is supposed to be controlled by a locus with two alleles, B and b, where B is dominant to b.

Table 4.2. The classification of sib-pairs

		First character			
		Same		Different	
Second character	Same	$S_1 S_2$	n_1	$D_1 S_2$	n_2
	Different	$S_1 D_2$	n_3	$D_1 D_2$	n_4

If the loci are independent, it is clear that:

$$n_1 n_4 = n_2 n_3$$

except for deviations due to sampling.

However, if the loci are linked, some marriages will produce mainly unlike sib pairs, while others will produce mainly alike pairs. Then:

$$n_1 n_4 > n_2 n_3 .$$

The reasoning is as follows:

A parent can have any of the nine possible genotypes, so there are 81 possible mating types, if we take account of the sexes. It would be tedious to work out the nature of the offspring for all these cases. We can see, however, that the distribution of the offspring into the four categories of Table 4.2 does not depend on the linkage of the loci except for the marriages which involve one double heterozygote ($Aa\,Bb$) and when the other marriage partner is either:

1. also a double heterozygote ($Aa\,Bb$), or
2. homozygous for the recessive gene at one of the loci, and heterozygous at the other ($Aa\,bb$) or ($aa\,Bb$), or
3. homozygous recessive at both loci ($aa\,bb$).

We shall examine only the first of these cases in detail. First of all, suppose that, in the parents, the dominant genes A and B are carried on the same chromosome, and the recessive genes, a and b, on the other (this situation is known as "coupling"). The gametes produced by individuals with this constitution have the following probabilities:

$$AB: \frac{1-r}{2}, \quad Ab: \frac{r}{2}, \quad aB: \frac{r}{2}, \quad ab: \frac{1-r}{2}$$

where r is the recombination frequency between the loci.

By making a table like Table 1.2, we find the following probabilities of the genotypes of the offspring:

1. Dominant gene at both loci: $\dfrac{2+(1-r)^2}{4}$.

2. Dominant gene at the first, and homozygous recessive at the second: $\dfrac{1-(1-r)^2}{4}$.

3. Homozygous recessive at the first locus, dominant gene at the second: $\dfrac{1-(1-r)^2}{4}$.

4. Homozygous recessive at both loci: $\dfrac{(1-r)^2}{4}$.

From these probabilities, it is easy to derive the probabilities for two of the offspring to be:

1. The same for both characters: $\frac{1}{8}\{3+2(1-r)^4\}$.
2. Different for both characters: $\frac{1}{8}\{1+2(1-r)^4\}$.
3. Alike for the first character, but unlike for the second: $\frac{1}{4}\{1-(1-r)^4\}$.
4. Unlike for the first character, and alike for the second: $\frac{1}{4}\{1-(1-r)^4\}$.

If the number of *independent* sib-pairs examined is N, the expectations of the numbers of pairs in the four categories are not such that:

$$n_1\,n_4 = n_2\,n_3,$$

since:

$$E(n_1\,n_4) = \tfrac{1}{64}\{3+8(1-r)^4+4(1-r)^8\}\,N^2$$

$$E(n_2\,n_3) = \tfrac{1}{16}\{1-(1-r)^4\}^2 \times N^2,$$

so that the difference between the two expectations is:

$$E(n_1\,n_4 - n_2\,n_3) = \frac{16(1-r)^4-1}{64}\,N^2.$$

This difference is equal to zero only if $r=\frac{1}{2}$, i.e. when the two loci are independent.

Similar results are obtained for the case when the dominant genes are on different chromosomes ("repulsion"). It is also possible, though tedious, to show that, for the three other types of marriage, the probabilities of the four categories of sib-pairs are such that $n_1\,n_4 > n_2\,n_3$, when the genes are linked.

We have seen that, for marriages of other than the four types mentioned, $n_1\,n_4 = n_2\,n_3$. So linkage between the loci will lead to an excess of pairs falling into categories 1 and 4, provided that enough families are examined for at least some of the marriages to be of the informative types.

Although the explanation of this method is complicated, its application to the analysis of data is simple. We classify the pairs of sibs into the four categories, and then test the table, by a χ^2 test, against the null hypothesis $n_1\,n_4 = n_2\,n_3$, which is the situation we would expect to obtain in the absence of linkage.

In addition to its simplicity, this method has the advantage that it only requires information about one generation.

However, it is necessary to study rather large numbers of families, in order to obtain statistically significant results (notice also that only one pair of sibs from each family can be used, since the method is based on the assumption that the sib pairs are independent, and thus must come from different families).

This method allows us to detect linkage, but not to measure it. It also has the disadvantage that it ignores any information which we may have about the phenotypes of the parents. Morton's method, which we shall study next, allows this information to be used fully.

5.2. Estimation of Recombination Fractions – Morton's Method

It is possible to calculate the probability of a given pedigree which concerns two characters, in terms of the frequencies of the genes controlling the characters. If at least one of the parents of a family is heterozygous for the two loci, the probability p of the family depends on the recombination frequency r, i.e.:

$$p = f(r).$$

If we study a set of pedigrees $1, 2, \ldots, n$ which satisfy this condition, the probability of this sample of families is:

$$P = f_1(r) \times f_2(r) \times \cdots \times f_n(r).$$

We can then compare the two hypothesis:

H_0: the two loci are independent, i.e.:

$$P\left(\tfrac{1}{2}\right) = f_1\left(\tfrac{1}{2}\right) \times f_2\left(\tfrac{1}{2}\right) \times \cdots \times f_n\left(\tfrac{1}{2}\right).$$

H_1: the loci are linked, with recombination frequency ρ i.e.:

$$P(\rho) = f_1(\rho) \times f_2(\rho) \times \cdots \times f_n(\rho).$$

We represent the logarithm of the ratio of these two probabilities by Z:

$$Z = \log \frac{P(\rho)}{P\left(\tfrac{1}{2}\right)} = \sum_i \log \frac{f_i(\rho)}{f_i\left(\tfrac{1}{2}\right)} = \sum_i z_i(\rho). \qquad (9)$$

If, for some values of ρ, $P(\rho)$ is distinctly higher than $P\left(\tfrac{1}{2}\right)$, i.e. if Z is greater than zero, this suggests that the genes are linked. The convention is therefore to take as the estimate of recombination fraction r the value of ρ which corresponds to the maximum value of Z.

We need to be able to test whether this maximum value of Z is significantly different from zero. If α is the risk of rejecting H_0 when H_0 is, in fact, true (this is called a type I error), the deviation of Z from zero will be considered significant when:

$$Z > \log \frac{1}{\alpha}.$$

With $\alpha = \tfrac{1}{100}$, Z must be greater than 2. With $\alpha = \tfrac{1}{1000}$, Z must exceed 3.

The great advantage of this method is that information can be accumulated gradually and used to give provisional conclusions, which are progressively modified as new information is obtained.

Example of the calculation of the probability of a pedigree. Consider two loci, the first with two alleles A and a, where a is recessive to A, and the second with two co-dominant alleles, B_1 and B_2. In a family with five members, suppose the parents' genotypes are:

$$\female \ (Aa \, B_2 B_2), \qquad \male \ (Aa \, B_1 B_2)$$

and the phenotypes of the offspring are:

$$aB_1 B_2, \ AB_2 \ \text{and} \ AB_1 B_2.$$

What is the probability of finding such a pedigree?

We do not know if, in the father, the genes A and B_1 are on the same chromosome, or the genes a and B_1. If the population has reached its equilibrium for these two loci (which we could find out by studying the genetic structure of the population), we can assume that these two possibilities are equally probable.

In the first case, the probabilities of the various gametes which the parents can produce are:

$$\male \begin{cases} (AB_1): \tfrac{1}{2}(1-\rho) & (aB_2): \tfrac{1}{2}(1-\rho) \\ (AB_2): \tfrac{1}{2}\rho & (aB_1): \tfrac{1}{2}\rho \end{cases}$$

$$\female \ (AB_2): \tfrac{1}{2} \qquad (aB_2): \tfrac{1}{2}.$$

Therefore the probabilities of the different possible phenotypes among the offspring are:

$$AB_1 B_2: \tfrac{1}{4}(2-\rho) \qquad AB_2: \tfrac{1}{4}(1+\rho)$$

$$aB_1 B_2: \tfrac{1}{4}\rho \qquad aB_2: \tfrac{1}{4}\rho.$$

The probability for these parents to have three children with the observed genotypes is therefore:

$$P_1(\rho) \propto \rho(1+\rho)(2-\rho).$$

In the case of the other possible genetic constitution of the father, a similar argument (with ρ replaced by $1-\rho$) gives:

$$P_2(\rho) \propto (1-\rho)(2-\rho)(1+\rho).$$

So the ratio $\dfrac{P(\rho)}{P(\tfrac{1}{2})}$ for this pedigree is:

$$\frac{P(\rho)}{P(\tfrac{1}{2})} = \frac{\rho(1+\rho)(2-\rho)+(1-\rho)(2-\rho)(1+\rho)}{\tfrac{1}{2}\times\tfrac{3}{2}\times\tfrac{3}{2}+\tfrac{1}{2}\times\tfrac{3}{2}\times\tfrac{3}{2}}$$

$$= \tfrac{4}{9}(1+\rho)(2-\rho)$$

and $Z=\log 4-\log 9+\log(1+\rho)+\log(2-\rho)$ from which it is easy to calculate a series of values of Z corresponding to a series of values of ρ.

5.3. Smith's "Bayesian" Method of Estimating Recombination Fractions

In the maximum likelihood method of estimating recombination fractions, we take as our estimate of r the value of the parameter ρ which maximises the probability $P(\rho)$ of the observations obtained. We can raise the objection to this method that it does not take into account any *a priori* opinions about the system in question.

If we choose two loci at random, we can have *a priori* opinions about the probability that they are located on the same chromosome, and also about the most probable distance between them. In other words, we have opinions about whether they are likely to be linked, and how close linkage is likely to be. The reasoning behind this is as follows.

We know that the autosomal genes in man are located on 22 pairs of chromosomes, whose lengths are unequal. We can write the lengths as $l_1, l_2, ..., l_{22}$, with $L=\sum_i l_i$. Assuming that genes are uniformly distributed along the length of the chromosomes, the probability that a gene is located on the i-th chromosome is equal to $\dfrac{l_i}{L}$, and the probability that two genes are both located on this chromosome is $\left(\dfrac{l_i}{L}\right)^2$. Therefore the probability that two loci are located on the same chromosome is:

$$P=\sum_i \left(\frac{l_i}{L}\right)^2$$

where the summation is over all 22 chromosomes.

Now suppose that the two genes are both on the i-th chromosome. The distance between them must be between 0 and l_i. The set of pairs of loci on the same chromosome can be represented by the points M in a two-dimensional diagram (Fig. 4.2); the distance of the first locus from one end of the chromosome is represented by the abcissa, and the distance of the other locus from the same end is represented by the ordinate. Thus the point M lies within a square OABC of side l_i. If M lies on the diagonal OB, the two loci are located at the same point, and the distance between them is zero. If M is not on the diagonal, the loci are distinct, and the distance between them is given by the length of the line M'M from M, parallel with one of the axes, to the diagonal.

Assuming that the genes are distributed uniformly along the chromosome, the point M is a random variable with a uniform probability

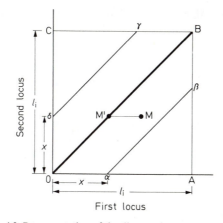

Fig. 4.2. Representation of the distance between two genes

density inside the square OABC. It follows that the probability that the distance between two loci on the i-th chromosome is less than x is given by:

$$\text{Prob}_i(d<x)=\frac{\text{Area of surface }(O\alpha\beta B\gamma\delta)}{\text{Area of surface }(OABC)}$$

$$=\frac{l_i^2-(l_i-x)^2}{l_i^2}$$

$$=\frac{2l_i x-x^2}{l_i^2}$$

where α, β, γ and δ are points defined by:

$$O\alpha=O\delta=B\beta=B\gamma=x.$$

The probability density of the random variable d for the i-th chromosome is therefore given by:

$$\psi_i(x)=\frac{d}{dx}[\text{Prob}(d<x)]=\frac{2(l_i-x)}{l_i^2}.$$

If we arrange chromosomes in order of decreasing size $l_1 \ldots l_{22}$, the probability that the distance between a pair of loci will be less than x, is thus:

$$\text{Prob}(d<x)=\sum_{i\in C}\frac{l_i^2}{L^2}+\sum_{i\in \bar{C}}\frac{2l_i x-x^2}{L^2}$$

where the first summation is over the set C of chromosomes shorter than x and the second summation is over the set of chromosomes which are equal to or longer than the length x.

The probability density of the length x is therefore given by:

$$\psi(x) = \frac{2}{L^2} \sum_{i \in \bar{C}} (l_i - x). \tag{10}$$

The graph of this function thus consists of a series of straight-line segments (see Renwick, 1971).

If we assume some relationship between the distance x between two loci and the recombination frequency r between the loci, the probability distribution of r, which we can consider as a random variable taking values between 0 and $\frac{1}{2}$, can be found from Eq. (10). The relation between x and r that we choose must take account of the fact that multiple cross-overs may occur, if the loci are sufficiently far apart. If the map distances are measured in "centiMorgans", i.e. in terms of percentage frequencies of recombinants, which are assumed to be additive (in this system of units, the largest chromosome, number 1, has a length of 300 centiMorgans), we can assume that $r = x$ when x is less than 30 centiMorgans.

Suppose that $\psi(r)$, where $0 \leq r < \frac{1}{2}$, is the *a priori* density function for r which we derive in this way, and that $P(\frac{1}{2})$ is the *a priori* probability that the two loci are unlinked $(r = \frac{1}{2})$.

If we have data on a number of pedigrees, we can calculate the probability $P(0|r)$ of the set of observations. Applying Bayes' theorem, we can then obtain the *a posteriori* distribution of r as:

$$\psi(r|0) = \frac{\psi(r) P(0|r)}{P(0)}$$

$$= \frac{\psi(r) P(0|r)}{\int_0^{\frac{1}{2}} \psi(r) P(0|r) \, dr + P(\frac{1}{2}) P(0|\frac{1}{2})} \tag{11}$$

$$P(\tfrac{1}{2}|0) = P(\tfrac{1}{2}) \frac{P(0|\frac{1}{2})}{P(0)}.$$

Following Smith (1953, 1959) we write:

$$\varLambda = \int_0^{\frac{1}{2}} \frac{\psi(r) P(0|r) \, dr}{P(0|\frac{1}{2})}.$$

$\varLambda$ is thus the mean value of the ratio of the probability density assuming linkage to the probability density assuming no linkage between the genes. In practice, this value is calculated by summation rather than by integration; we choose the successive values of r in the interval $0-\frac{1}{2}$ to be sufficiently close that this is a good approximation. For example:

$$\varLambda = \frac{1}{P(0|\frac{1}{2})} [\psi(0.00) P(0|0.00) + \psi(0.01) P(0|0.01) + \cdots + \psi(0.49) P(0|0.49)].$$

Eqs. (11) then become:

$$\psi(r|0)=\psi(r)\frac{P(0|r)}{[\varLambda+P(\tfrac{1}{2})]\,P(0|\tfrac{1}{2})}$$

$$P(\tfrac{1}{2}|0)=P(\tfrac{1}{2})\frac{1}{\varLambda+P(\tfrac{1}{2})}. \tag{12}$$

Finally, we obtain, not a single estimated recombination fraction, but a set of probabilities of the different values that r can take. This is obviously a more informative result than those given by the two methods we described previously. It also has the advantage of using our *a priori* knowledge of the human karyotype.

5.4. The Linkage Map of Man

It is not surprising that the human linkage map is fragmentary at present, if we consider the difficulty of accumulating a sufficient amount of useable data. In *Drosophila melanogaster*, well over 100 genes have been located on each of the major chromosomes (as a result of an enormous number of experimental crosses done in laboratories all over the world). In man, however, we have only located a handful of genes.

Recently, much progress has been made in mapping human genes, and it has even proved possible, by studying chromosomal elimination in inter-species cell hybrids, and by studies of human chromosomal variants, to specify the chromosomes on which certain genes are located. Renwick (1971) has reviewed present knowledge of linkage in man.

It is much easier to map the sex-linked genes, since we already know that all such genes are on the X-chromosome, and can estimate the recombination frequency without first having to get evidence for linkage. However, most of the sex-linked genes which we know at present cause serious diseases, and linkage studies can only be done using pedigrees which contain members who have two such diseases; these, happily, are rare.

Cavalli-Sforza and Bodmer (1971) review the present state of our knowledge of the map of the X-chromosome. The genes which have so far been mapped appear to fall into two groups with a large distance between the groups. One group contains the genes for the Xg blood group, ichthyosis, ocular albinism and angiokeratoma while the other group contains the genes for the Xm blood group, deuteranomaly, G6PD, protanomaly and haemophila A.

Chapter 5

The Inheritance of Quantitative Characters

Up to now, we have studied qualitative characters which could manifest themselves in a limited number of different states: for example, the blood groups (an individual can be A, B, AB or 0), or sensitivity to PTC (an individual can be a taster or a non-taster). But there are many characters for which individuals cannot be classified into a definite number of types, but can only be characterised by measurement. For example, height, or head circumference or body-weight, are characters of this sort.

Continuously-varying characters seem to contradict Mendelian theory, whose essence is the idea of discontinuity. If the character "height" were controlled by a locus with a finite number of alleles, then we should observe a finite number of heights among individuals. Clearly, this is not the case, and yet height is a hereditary character.

Two ideas contribute to the explanation of this seeming paradox. First, a character can be affected by more than one locus. Mendel discovered a character whose inheritance could be explained in this way. From a cross of peas with white flowers and peas with purple flowers, he obtained a second hybrid generation consisting of plants whose flowers varied continuously in colour from white to purple. Mendel formulated the hyptothesis, which was later confirmed, that this character is controlled by more than one pair of genes; the number of possible combinations of the different alleles is thus so high that their effects seem to be continuous.

Second, we know that individuals with exactly the same genes, for example a pair of monozygotic twins, can show differences in the expression of a character; the expression of a character therefore depends on environmental conditions, such as nutrition, irradiation etc., and not only on the genes which the individual received at conception. Of course, the effect which environmental influences will have on the development of an individual will depend on his genotype: certain conditions may favour one genotype, but be harmful for another.

To a particular genotype, therefore, there corresponds not a single measure, but a set of measures of a character, which has a certain distribution of values about a mean; the genotype defines a frequency distribution of phenotypic values:

$$\text{Prob}\,(Q = x) = f\,(x, g, e)$$

where Q is a quantitative character which can have the measure x, g is the genotype, and e is the environment.

Obviously we cannot define the function $f(x, g, e)$ fully. We have to approximate, as in all quantitative sciences.

In practice, we do not attempt to study the frequency distribution of a quantitative character, but confine our attention to the mean $\bar{x}$ and the variance V_x of the character.

1. The Mean

Suppose we have a group of individuals of genotype g living in an environment e. We can calculate the mean $\bar{x}$ of the individual measures of Q, and write:

$$\bar{x} = f\,(g, e)$$

to show that $\bar{x}$ is a function of the genotype and of the environment. If two groups differ in either g or e, their values of $\bar{x}$ may not be the same.

The function f is in general not simple. The effect of the genotype cannot in general be considered to be independent of the environment. However, as a first approximation, it seems reasonable to assume that for certain characters, over a limited range of environmental conditions, the effect of interactions is negligible compared with the genotypic and environmental effects themselves. With this assumption, we can write:

$$\bar{x} = G + E$$

to express the important idea that the mean measure of the character is equal to a value corresponding to the genotype, plus a correction due to the environment.

1.1. Definition of Additive Effects and Dominance Deviations

We shall first study a character controlled by a single locus with n alleles $A_1, \ldots, A_i, \ldots, A_n$.

Let $p_1, \ldots, p_i, \ldots, p_n$ be the frequencies of these alleles in the population. We further assume (and this is essential for the whole of the following argument) that the population is in Hardy-Weinberg equilibrium. In other words, we assume that the genotype frequencies are given by:

$$P_{ii} = p_i^2, \qquad P_{ij} = 2 p_i p_j.$$

We saw earlier (Chapter 3, Section 2.4) that, in such a population, one can calculate the mean of a character, weighted by the genotype frequencies, by the formula

$$\bar{C} = \sum_{ij} p_i p_j C_{ij}.$$

This formula will be used later on.

The value of a character in an individual can be considered to be a random variable; we shall assume that it has a mean μ_{ij}, which depends on the genotype $(A_i A_j)$ of the individual, and variance V_E, which is independent of genotype. A constant variance implies that there is no interaction between the effects of the genotypes and the environment, as we have already assumed. With these assumptions, we can write the measure of the character as:

$$C_{ij} = \mu_{ij} + \varepsilon$$

where ε is a random variable with a certain probability distribution with mean 0, and variance V_E.

If the origin for the measures is taken in such a way that the mean of the C_{ij} values (and thus of the μ_{ij} values) is zero, we have:

$$\sum_{ij} p_i p_j \mu_{ij} = 0. \tag{1}$$

At the locus in question, each individual has two genes. If the effects of different alleles on the character are strictly additive, with no dominance, we can write:

$$\mu = \alpha^* + \alpha^{**}$$

where α^* and α^{**} represent the effects of genes contributed by the two parents.

In general, however, the genes are not perfectly additive, and we cannot write this equation. In the more general case, we can partition μ into an "additive" contribution, and a contribution due to the effect of "dominance":

$$\mu_{ij} = \alpha_i + \alpha_j + \delta_{ij}. \tag{2}$$

We need to define the n parameters α_i, and the $\dfrac{n(n+1)}{2}$ parameters δ_{ij}.

To do this, we have to make another assumption. Fisher (1918, 1941) proposed the idea of partitioning μ so as to minimise the sum of the squared values of the δ's for the set of the genotypes of a random-mating population. This is equivalent to minimising the expression:

$$\sum_{ij} p_i p_j \delta_{ij}^2 = \sum_{ij} p_i p_j (\mu_{ij} - \alpha_i - \alpha_j)^2. \tag{3}$$

This is done by finding the n partial derivatives of the expression with respect to α_i and setting them equal to zero:

$$\frac{\partial}{\partial \alpha_i} \sum_{ij} p_i p_j (\mu_{ij} - \alpha_i - \alpha_j)^2 = 0. \tag{4}$$

In this sum, the square of α_i occurs:

1. in the term $p_i^2 (\mu_{ii} - 2\alpha_i)^2$, and
2. in the $n-1$ terms $2 p_i p_j (\mu_{ij} - \alpha_i - \alpha_j)^2$ where $i \neq j$.

The partial derivative of expression (3) with respect to α_i is therefore:

$$-4 p_i^2 (\mu_{ii} - 2\alpha_i) - 4 p_i \sum_{j \neq i} p_j (\mu_{ij} - \alpha_i - \alpha_j) = -4 p_i \sum_j p_j \delta_{ij}$$

and the Eqs. (4) become:

$$\sum_j p_j \delta_{ij} = 0. \tag{5}$$

These n equations, together with the $\dfrac{n(n+1)}{2}$ equations (2), allow us to calculate the α_i and δ_{ij} values, from knowledge of the μ_{ij}.

1.2. Determination of the Additive Effects and Dominance Deviations

Multiplying the n equations (5) by p_i, and summing them, we have:

$$\sum_{ij} p_i p_j \delta_{ij} = 0 \tag{6}$$

which is equivalent to the statement that: "the mean of the dominance deviations of the genotypes in a panmictic population is zero".

But Eq. (6) can be written:

$$\sum_{ij} p_i p_j \mu_{ij} - \sum_{ij} p_i p_j \alpha_i - \sum_{ij} p_i p_j \alpha_j = 0.$$

From (1), since $\sum_i p_i = 1$, we have:

$$\sum_i p_i \alpha_i + \sum_j p_j \alpha_j = 0.$$

Therefore

$$\sum_i p_i \alpha_i = 0 \tag{7}$$

which is equivalent to the statement that: "the mean of the additive effects of the genes is zero".

Returning to the Eqs. (5) we see that they can be written in the form:

$$\sum_j p_j (\mu_{ij} - \alpha_i - \alpha_j) = 0.$$

Hence:

$$\alpha_i = \sum_j p_j \mu_{ij}, \tag{8}$$

since

$$\sum_j p_j = 1 \quad \text{and} \quad \sum_j p_j \alpha_j = 0.$$

This is equivalent to: "the additive effect of a gene is equal to the mean of the values of the genotypes which contain the gene, weighted by the frequencies of the other genes in these genotypes".

The dominance effects are found from the Eqs. (2), which can be written in the form:

$$\delta_{ij} = \mu_{ij} - \sum_k p_k \mu_{ik} - \sum_k p_k \mu_{jk}. \tag{9}$$

Also, we note that the covariance of the additive and dominance effects in the population is zero:

$$\begin{aligned}
\text{Cov}(\alpha, \delta) &= \sum_{ij} p_i p_j (\alpha_i + \alpha_j) \delta_{ij} \\
&= 2 \sum_i p_i \alpha_i \sum_j p_j \delta_{ij} \\
&= 0, \quad \text{from Eq. (5)}.
\end{aligned}$$

To summarise, we can partition the mean μ_{ij} of the character in the genotype $(A_i A_j)$ into the additive effects α_i and α_j, and the dominance deviation δ_{ij}; the sum $\alpha_i + \alpha_j$ is sometimes called the "breeding value" of the genotype.

It is important to remember that this partition is an arbitrary one; in order to do it, we had to assume that the dominance effects could be found by minimising the sum of squares of their values. This has the advantage of giving additive and dominance effects which have zero covariance, but other methods of partition could equally well have been used.

The values α_i of the additive effects of the genes are clearly not related simply to the physiological effects of the genes, since they depend on the frequencies of the other genes. The α_i values can therefore be different in different populations, if the populations have different gene frequencies p_i, even if the values μ_{ij} of the genotypes are the same.

Furthermore, since characters are manifested only by zygotes in which there are two genes for each unit character, it is nonsense to talk about the "effect" of a single gene. Additive effects are simply a useful concept in certain types of argument, and have no other biological significance.

The concept of additive effects is, nevertheless, very useful; this can be illustrated by the answer to the question. "What is the effect of a change in the genetic composition of a population, on a quantitative character?".

1.3. The Effect of a Small Change in Gene Frequency

Suppose that, for some reason (for example because of immigration), the frequency of a gene A_i undergoes a slight change Δp_i, which is balanced by a change in the frequency of another allele A_j: $\Delta p_j = -\Delta p_i$.

The mean of the character in the population,

$$M = \sum_{ij} p_i p_j \mu_{ij}$$

undergoes a change, ΔM, which is approximately given by:

$$\Delta M = \Delta p_i \frac{\partial M}{\partial p_i} + \Delta p_j \frac{\partial M}{\partial p_j}$$

$$= \Delta p_i \left\{ \frac{\partial M}{\partial p_i} - \frac{\partial M}{\partial p_j} \right\}.$$

But

$$\frac{\partial M}{\partial p_i} = \frac{\partial}{\partial p_i} \sum_{ij} p_i p_j \mu_{ij} = 2 \sum_j p_j \mu_{ij}$$

since p_i appears in p_i^2 and in terms of the type:

$$p_i p_k \ (i \neq k) \quad \text{and} \quad p_l p_i \ (l \neq i).$$

But, by (8) $\sum_j p_j \mu_{ij} = \alpha_i$, hence:

$$\Delta M = 2 \Delta p_i (\alpha_i - \alpha_j). \tag{10}$$

This can be expressed in words as follows: if the frequency of one gene in a population increases at the expense of another, the mean of the corresponding character is changed by an amount equal to twice the product of the change in the gene frequency and the difference between the additive effects of the two genes.

1.4. The Case of a Single Locus with Two Alleles

When the locus in question has only two alleles, A_1 and A_2, the calculations are considerably simplified.

Let the frequencies of A_1 and A_2 in the population be p and q, and let the mean values of the character for individuals of the genotypes $(A_1 A_1)$, $(A_1 A_2)$ and $(A_2 A_2)$ be i, j and k, respectively.

We assume that the population is in Hardy-Weinberg equilibrium, so that the frequencies of these three genotypes are p^2, $2pq$ and q^2.

Condition (1), which expresses the fact that the measures are "normalised", becomes:

$$p^2 i + 2pqj + q^2 k = 0. \tag{11}$$

To find the "additive" and "dominance" effects, we assign additive effects α_1 to allele A_1 and α_2 to A_2, and write:

Genotype $(A_1 A_1)$: $i = 2\alpha_1 + \delta_{11}$
Genotype $(A_1 A_2)$: $j = \alpha_1 + \alpha_2 + \delta_{12}$
Genotype $(A_2 A_2)$: $k = 2\alpha_2 + \delta_{22}.$

We have to find values for α_1 and α_2 such that the population mean of the squares of the dominance effects is minimised. This mean is:

$$p^2 \delta_{11}^2 + 2pq\, \delta_{12}^2 + q^2\, \delta_{22}^2 = p^2 (i - 2\alpha_1)^2 + 2pq(j - \alpha_1 - \alpha_2) + q^2 (k - 2\alpha_2)^2.$$

This expression is a minimum when the two partial derivatives with respect to α_1 and α_2 are both zero, i.e.:

$$-4p^2 (i - 2\alpha_1) - 4pq(j - \alpha_1 - \alpha_2) = 0$$

$$-4pq(j - \alpha_1 - \alpha_2) - 4q^2 (k - \alpha_2) = 0$$

or

$$pi + qj = \alpha_1 + \alpha_1 p + q\alpha_2 \tag{12}$$

$$pj + qk = \alpha_2 + \alpha_1 p + q\alpha_2.$$

Multiplying the first equation by p, and the second by q, and adding, we obtain:

$$p^2 i + 2pqj + q^2 k = (\alpha_1 p + q\alpha_2)(1 + p + q).$$

But, by (11), the left hand side of this equality is zero, hence:

$$\alpha_1 p + \alpha_2 q = 0 \tag{13}$$

which is the two-allele version of Eq. (7). Taking account of (13), the Eqs. (12) become:

$$\alpha_1 = pi + qj$$
$$\alpha_2 = pj + qk \tag{14}$$

which corresponds to (8).

If we write $\alpha_1 - \alpha_2 = \alpha$, it is easy to show that:

$$\alpha_1 = \alpha q$$
$$\alpha_2 = -\alpha p \tag{15}$$

(Eq. (13) becomes $\alpha_1 p + q(\alpha_1 - \alpha) = 0$, therefore $\alpha_1 - \alpha q = 0$).

The parameter α is often called the "average effect of the gene substitution", and we find that, if the frequency p undergoes a small change Δp, the population mean value for the character changes, according to Eq. (10), by:

$$\Delta M = 2\Delta p(\alpha_1 - \alpha_2) = 2\alpha\Delta p.$$

Notice that the dominance effect of genotype $(A_1 A_1)$ is:

$$\delta_{11} = i - 2\alpha_1 = i - 2pi - 2qj$$
$$= i - 2pi - 2qj + (p^2 i + 2pqj + q^2 k)$$

since the term in brackets is equal to zero. Hence:

$$\delta_{11} = i(1 - 2p + p^2) - 2q(1 - p)j + q^2 k$$
$$= q^2(i - 2j + k).$$

If we write d for the difference between the value j of the heterozygote and the mean of the values i and k of the homozygotes: $d = j - \dfrac{i+k}{2}$, we see that δ_{11} can be written:

$$\delta_{11} = -2q^2 d$$

and, similarly

$$\delta_{12} = 2pqd \qquad\qquad (16)$$

$$\delta_{22} = -2p^2 d.$$

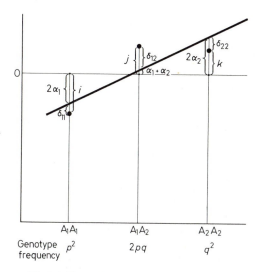

$A_1 A_1$	$A_1 A_2$	$A_2 A_2$
Genotype frequency p^2	$2pq$	q^2

Fig. 5.1. The analysis of a quantitative character

As we found for the general case, the covariance between the additive and dominance effects is zero. In the present notation, this can be shown as follows:

$$\mathrm{Cov}\,(\alpha,\delta)=p^2\times 2\alpha_1\times\delta_{11}+2pq(\alpha_1+\alpha_2)\,\delta_{12}+q^2\times 2\alpha_2\times\delta_{22}$$
$$=p^2(2q)(-2q^2\,d)+2pq(q-p)\,\alpha(2pq\,d)+q^2(-2\alpha p)(-2p^2\,d)$$
$$=0.$$

Finally, we see that the genotypic values can be analysed into two parts:

1. The additive effects of the genes, which are functions of the gene frequencies p and q, and of the average effect of the substitution, α.

2. The dominance deviations, which are functions of p and q, and of the difference d between the value of the heterozygote, and the mean values of the homozygotes.

These results are summarised in Table 5.1.

Fig. 5.1 illustrates the analysis.

Table 5.1. The analysis of a quantitative character

Genotype	(A_1A_1)	(A_1A_2)	(A_2A_2)
Frequency	p^2	$2pq$	q^2
Genotypic value	i	j	k
Breeding value	$\begin{cases}2\alpha_1\\2(pi+qj)\\2q\alpha\end{cases}$	$\begin{aligned}&\alpha_1+\alpha_2\\&pi+j+qk\\&(q-p)\,\alpha\end{aligned}$	$\begin{aligned}&2\alpha_2\\&2(pj+qk)\\&-2p\alpha\end{aligned}$
Dominance effect	$-2q^2\,d$	$2pq\,d$	$-2p^2\,d$

$$\alpha=\alpha_1-\alpha_2=p(i-j)+q(j-k),\quad d=j-\frac{i+k}{2}$$

1.4.1. Special cases. a) *No dominance.* If there is no dominance, the value of the heterozygote is equal to the mean of the values of the two homozygotes, i.e. $j=\dfrac{i+k}{2}$, $d=0$. Hence $\delta_{11}=\delta_{12}=\delta_{22}=0$.

Also, $\alpha_1=q\left(\dfrac{i-k}{2}\right)=i/2$, since Eq. (11) becomes:

$$p^2\,i+pq(i+k)+q^2\,k=0,\quad\text{i.e.}\quad pi+qk=0,\quad\text{so that}\quad qk=-pi.$$

Similarly, we find that $\alpha_2=p\left(\dfrac{k-i}{2}\right)=k/2$. Therefore $\alpha=\dfrac{i-k}{2}$.

Thus, in the case of no dominance, the additive effects, α_1 and α_2, are independent of the gene frequencies.

b) *Complete dominance.* If gene A_2 is fully dominant, $j=k$, so that:

$$\alpha=p(i-k), \qquad \alpha_1=pq(i-k), \qquad \alpha_2=p^2(i-k)$$
$$\delta_{11}=q^2(i-k), \qquad \delta_{12}=-pq(i-k), \qquad \delta_{22}=p^2(i-k).$$

Thus all the parameters into which the genotypic values are analysed, are functions of the difference $(i-k)$ between the dominant and recessive phenotypes.

1.5. Characters Controlled by Several Loci

If a character is controlled by more than one locus, the effects of the different loci may or may not be additive. If they are additive, we can write:

$$\bar{G}=\sum_l \bar{G}_l$$

with the summation being taken over all loci.

If the effects are not additive, we have a supplementary effect of their interaction, over and above the effects of the individual loci; we can thus write:

$$\bar{G}=\sum_l \bar{G}_l+\sum_{lk} I_{lk}$$

where I_{lk} is the interaction effect for loci l and k.

These interactions are sometimes called "epistasis": two loci are epistatic if their joint effect is not the sum of their individual effects, but there is an interaction.

When discussing the additivity of genes, it is therefore important to remember that two different situations may be important; if we are considering genes at the same locus, interactions are called "*dominance*", but interactions between genes at different loci are called "*epistasis*".

The analysis of a character which is controlled by genes at many loci is extremely difficult. In practice, we have to make a number of simplifying assumptions in order to construct models which can explain the observations, to a good approximation, and allow us to predict the effects of changes in the system. It would, however, be illusory to hope that such studies will enable us to accept or reject any of the models.

Nevertheless, it is one of the greatest successes of Mendelian genetics that it can explain the facts of the inheritance of apparently continuous variation, using simple models based on discontinuous factors.

A good example of this is the study of the inheritance of skin colour in American negroes.

1.6. An Example of a Character Controlled by Several Genes:
Skin Colour

The early work of Davenport (1910, 1913) on the descendants of marriages between negroes and white people, led to the hypothesis that skin colour is controlled by two loci, each with two alleles: whites would have the genotype $(A_1 A_1 B_1 B_1)$ and negroes would be $(A_2 A_2 B_2 B_2)$; the alleles were assumed to show no dominance or epistasis, and the darkness of the skin was assumed to depend only on the number of A_2 and B_2 genes.

More recent research in the United States has shown that a four-, five-, or six-locus model fits the observed distribution of skin-colour better than a two-locus model, when all the loci are assumed to affect skin colour equally, and to show no dominance and no epistasis (Stern, 1953). This hypothesis does not claim to represent reality. It is highly improbable that all the genes contribute equally to the colour of the skin, but we adopt this model because of its simplicity.

The negro population of the United States is a hybrid population, whose ancestry is about 70 % African negro, and about 30 % white European.

If we consider a 5-locus model, each individual carries ten genes, each of which can be of type 2, which darkens the skin, or of type 1, which does not cause pigmentation of the skin. Type 2 genes are derived from the African ancestors, and type 1 from the European ancestors. There are thus 11 possible phenotypes, corresponding to the number (0 to 10) of type 2 genes in the genotype.

At each locus, the probability that a gene is of type 2 is 0.7, so the random variable x "number of type 2 genes in an individual" is distributed according to the binomial distribution with $p = 0.7$ and $N = 10$. The probability of having r type 2 genes is therefore:

$$P(r) = \binom{10}{r} 0.7^r \times 0.3^{10-r}.$$

This corresponds to the following distribution of type 2 genes in the population:

Number of type 2 genes	Frequency (per thousand)	Number of type 2 genes	Frequency (per thousand)
0	–	6	201
1	0.1	7	267
2	2	8	232
3	9	9	121
4	37	10	28
5	103		

Of course, the colour of the skin varies continuously, and there are not eleven distinct groups in the population. However, one can set *a priori* levels of coloration, and divide the members of the population into classes according to whether they fall above or below these levels. The proportions of individuals in the various classes agree very well with the frequencies in the table above. Our model is therefore adequate to explain the results.

We can now ask: if the populations of American white and black people were to mix and form a single panmictic population (in other words, if the choice of marriage partner became independent of skin colour), what would be the resulting distribution of skin colour?

At present, the black community is about 10 % of the total population of the United States. Type 2 alleles are therefore about $\frac{1}{10} \times 0.7 = 7 \%$ of the genes, and type 1 alleles about 93 % of the total. When the new Hardy-Weinberg equilibrium became established, the probability of having r type 2 genes would be:

$$P'(r) = \binom{10}{r} 0.07^r \times 0.93^{10-r},$$

which corresponds to the distribution (from Stern, 1953):

Number of type 2 genes	Frequency (per thousand)	Number of type 2 genes	Frequency (per thousand)
0	484	6	0.02
1	364	7	—
2	123	8	—
3	25	9	—
4	3	10	—
5	0.3		

The lightest-skinned class, 0, which at present forms 90 % of the population, would form only 48 % of the mixed population, but categories 1 and 2, which are very slightly coloured, would increase in frequency, so that individuals of classes 0, 1 and 2 would be 97 % of the total population, compared with 90 % at present. These two classes can hardly be distinguished from class 0. Only 3 % of individuals would fall into classes 3 or higher, and be truly distinct from the "whites", compared with 10 % at present. Classes 9 and 10, the darkest of all, would practically disappear; in a population of 200 million people, the number of people in class 9 would be 80, and the number in class 10 would be 0.01, which is nearly zero; the respective numbers in these two classes are 2.4 and 0.5 million at present.

The fusion of two populations thus results in the disappearance of the extreme phenotypes of the smaller sized population.

2. The Variance

The variance of a character in a population is the mean of the squared deviations of the individuals' values from the mean:

$$V = \frac{1}{N} \sum_i (x_i - \bar{x})^2.$$

The variance measures the dispersion of the measures around the mean. We have seen that the measure of the phenotype can be analysed into components corresponding to the effects of the genotype and of the environment, and that the effect of the genotype can be further analysed into the additive effects of genes, the dominance deviations for genes at the same loci, and the interaction effects between genes at different loci.

In an analogous way, we can analyse the variance of the phenotype, V_P, into:

$$V_P = V_G + V_E + V_{GE} \tag{17}$$

where V_G is the variance due to genotypic differences, V_E is the variance due to environmental differences, and V_{GE} is due to the interaction between genotypes and environments.

2.1. Environmental Variance

In order to measure the variance due to the environment, we have to study individuals which have the same genotype. This can be done in some plant and animal species in which we can produce inbred lines by repeated consanguineous mating.

In man, we can estimate the effect of the environment by studying pairs of twins. Galton was the first to study twins to try and find out the relative important of "nature and nurture" for the expression of a character. A large number of twin studies have since been done.

Twin births constitute about 0.8 % (in South Africa) to 1.8 % (in Belgium) of all births. Twins can be "monozygotic", resulting from a single fertilised egg, or "dizygotic", resulting from two separate fertilised eggs. Monozygotic twins are sometimes called "identical" twins. They are completely identical genetically, while dizygotic twins are no more similar genetically than an ordinary pair of sibs.

Studies of monozygotic twins reared apart can be used to partition the variance of a quantitative character in man. From this type of data we can obtain two types of variance: V_B, the variance due to differences between the twin pairs, and V_W, the variance due to differences between members of the same twin pair.

Differences between members of the same twin pair must be due to environmental influences. If we assume that there is no tendency for the two members of a twin pair to be adopted by similar types of families, and thus to be brought up in similar environments, and if we also assume that there are no genotype-environment interactions, it can be shown that V_W is an estimate of V_E, and that V_B is an estimate of $2V_G + V_E$. Thus, data of this kind provide estimates of both V_E and V_G.

An example of this type of study is provided by a study (Shields, 1962) of the I.Q. of 19 pairs of monozygotic twins who were separated shortly after birth and brought up by different adoptive parents. In this data, V_W was 9.28, and V_B was 61.46. Hence our estimate of V_E is equal to 9.28, and our estimate of V_G is:

$$V_G = \frac{V_B - V_E}{2} = \frac{V_B - V_W}{2} = 26.09.$$

2.1.1. The diagnosis of monozygosity. To use the method just described, we need to know which twins are monozygotic and which are dizygotic.

One method for the diagnosis of monozygosity is to make skin grafts. If the donor and recipient of the grafts are genetically different, the graft will degenerate. However, a pair of monozygotic twins have the same genotype and grafts between twins will remain healthy.

More often, the diagnosis of monozygosity is made on the basis of similarity between the twins for a number of traits which are easy to determine, e.g. their sex, the blood groups AB0, MN, Rh, etc. Identical twins are necessarily the same for all these characters, but dizygotic twins may or may not be the same, with certain probabilities. Penrose and Smith (1955) have published tables of these probabilities; they also give the probabilities that identical and non-identical twins have given differences in certain characteristics of their fingerprints and in the palm patterns.

The tables enable one to calculate the probability of wrongly classifying a pair of twins as monozygotic, when they are the same for a number of characters; the probability is usually small, provided that several characters are examined.

About one third of twin pairs are monozygotic. This proportion can be independently estimated from the numbers of twin pairs whose members are of different sexes. Such twins cannot be monozygotic, and they should represent about half of all dizygotic pairs. Thus:

$$P\text{(different-sex twins)} = \frac{1 - P\text{(identical twins)}}{2}.$$

Hence:

$$P\text{(identical twins)} = 1 - 2P\text{(different-sex twins)}.$$

The proportion of different-sex twin pairs is usually about $\frac{1}{3}$, so the proportion of identical twins is also approximately $\frac{1}{3}$.

2.2. Genotypic Variance

The measure of the effect of the genes on a character can be analysed into additive, dominance and epistatic effects:

$$G = A + D + I.$$

In an analogous way, we can analyse the variance into:

$$V_G = V_A + V_D + V_I + 2\,\text{Cov}(A, D) + 2\,\text{Cov}(D, I) + 2\,\text{Cov}(I, A).$$

The analysis soon becomes extremely complex if several loci with epistatic effects are involved (this subject is discussed by Kempthorne, 1957). We shall therefore study the case when these effects do not exist (or rather when it is not too unrealistic to assume that they do not exist). We can then write:

$$V_G = V_A + V_D + 2\,\text{Cov}(A, D).$$

But, if we analyse G into additive effects and dominance deviations by the method described earlier, we know that the covariance of A and D is zero, so that:

$$V_G = V_A + V_D. \tag{18}$$

V_A corresponds to the variance due to the additive effects of gene substitutions. The ratio V_A/V_P is called "*heritability*", and is denoted by h^2. It is an important component of the correlations between relatives for quantitative characters, as we shall see in Chapter 6.

In the case of a character controlled by a single locus, we can write, in the notation used earlier in this chapter:

$$V_A = \sum_{ij} p_i p_j (\alpha_i + \alpha_j)^2$$

$$= 2 \sum_{ij} p_i p_j \alpha_i^2 + 2 \sum_{ij} p_i p_j \alpha_i \alpha_j.$$

But, by (7),

$$\sum_{ij} p_i p_j \alpha_i \alpha_j = \left(\sum_i p_i \alpha_i\right)\left(\sum_j p_j \alpha_j\right) = 0.$$

Now we have:

$$\sum_{ij} p_i p_j \alpha_i^2 = \sum_j p_j \sum_i p_i \alpha_i^2 = \sum_i p_i \alpha_i^2.$$

Hence:

$$V_A = 2 \sum_i p_i \alpha_i^2. \tag{19}$$

The additive variance is equal to twice the weighted mean of the squared additive effects.

2.3. The Case of a Locus with Two Alleles

In the notation of Section 1.4, we have:

$$V_A = p^2 (2\alpha_1)^2 + 2pq(\alpha_1 + \alpha_2)^2 + q^2 (2\alpha_2)^2$$
$$= 2p\alpha_1^2 + 2q\alpha_2^2 + 2[p^2 \alpha_1^2 + 2pq\alpha_1\alpha_2 + q^2 \alpha_2^2].$$

But the term in brackets is equal to $(p\alpha_1 + q\alpha_2)^2$, and this is zero, from (13). Therefore:

$$V_A = 2(p\alpha_1^2 + q\alpha_2^2)$$

which is the two-allele equivalent of Eq. (19).

But we have seen that $\alpha_1 = \alpha q$, and $\alpha_2 = -\alpha p$, so that:

$$V_A = 2pq\alpha^2. \tag{20}$$

The dominance variance can be written:

$$V_D = p^2 \delta_{11}^2 + 2pq \delta_{12}^2 + q^2 \delta_{22}^2$$
$$= p^2 (-2q^2 d)^2 + 2pq(2pqd)^2 + q^2 (-2p^2 d)^2 \tag{21}$$
$$= 4p^2 q^2 d^2.$$

where $d = j - \dfrac{i+k}{2}$ is the difference between the value of the heterozygote and the mean of the two homozygotes.

Chapter 6

Genetic Relationships between Relatives

The fact that related individuals resemble one another was no doubt the starting point of all genetical thought. The most casual observation shows that children resemble their parents, or sometimes are strikingly like a more distant relative, and that sibs resemble one another. It is also very noticeable that the extent of resemblance differs for different characters.

The primary cause of this resemblance is obviously the fact that both parents transmit genes to each of their offspring. The role of both father and mother in heredity has, however, been accepted for only about 150 years. Until the early 19th century, the "ovists" considered that the ovum alone is the origin of the child, while the "spermists" believed the opposite[1].

The measurement of the degree of genetic relationship between related individuals is used not only to satisfy our curiosity about the rôle of heredity in the manifestation of a character, but also in medicine, where it can play a part in the diagnosis of certain hereditary conditions.

The information that is taken into account when making a diagnosis includes the results of observations on the individual in question (e.g. his blood pressure, heart rhythm, temperature, etc.), but can also include information obtained from his relatives. In this chapter we shall consider the question: "What information about an individual A can we gain from our knowledge (certain or probabilistic) about another individual, B, who is related in a given way to A?".

The statement that two individuals are related is equivalent to saying that one or more of the ancestors of one individual are also ancestors of the other. Two individuals are related when they have parts of their pedigrees in common.

At first sight, the number and complexity of the possible relationships between individuals seems too great for it to be possible to classify them, let alone devise a measure of relatedness. However, using the concept

---

[1] In Latin, the adjective "consanguineus" is used only for relationships through the male line.

of "identity" of genes, we can simplify the complexity, and can obtain results which are completely rigorous.

Before we try to define the genetic consequences of the relationship of two individuals, we must establish a measure of relatedness. In the first section of this chapter, we shall see how such a measure can be defined.

In the second section, expressions for the genetic structures of two related individuals will be derived.

Finally, in the third section, we shall see how these results, together with those from Chapter 5, can be used to analyse the "resemblance" between relatives, in terms of the correlations between related individuals.

1. The Measure of Relatedness

1.1. Identity by Descent

However complex the relationship between two individuals A and B may be, the genetic implication of this relationship is simply that there is a certain possibility that a gene in A and one in B may both be a copy of a given gene in one of their common ancestors.

Two genes are called "identical by descent"[2] if they are copies of a given ancestral gene. Two genes which are identical by descent must therefore be the same allele, since we are assuming that mutations do not happen. To indicate that two genes G_1 and G_2 are identical by

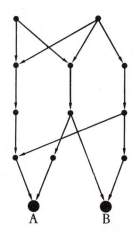

Fig. 6.1. The genetic relationship between relatives

[2] Throughout this chapter, the word "identical" will often be used instead of "identical by descent", for the sake of brevity.

descent, we shall write: $G_1 \equiv G_2$, and $G_1 \not\equiv G_2$ to indicate that they are not identical by descent.

If we know the genotype of individual A, we know the following things about his relative B:

1. B belongs to a certain population, characterised by a given genotypic structure.

2. B is related in a particular way to A, and this relationship determines the probability that a gene in B is identical by descent with a gene in A.

The relationship between A and B can therefore be characterised, from the point of view of genetics, by the probabilities of identity by descent of their genes.

1.2. The Definition of Coefficients of Identity

1.2.1. Coefficients of identity. At a given locus, two individuals A and B have a total of four genes: G_A transmitted to A by his father, G_A^* transmitted by his mother, G_B and G_B^* (we are considering autosomal genes). These genes can be identical with one another, or not, depending on the pedigrees of A and B.

In the case when G_A and G_A^* are identical, and G_A^* and G_B are identical, G_B is necessarily identical to G_A; the relation of identity is transitive: two genes which are each identical to a third gene are identical to each other.

Taking the property of transitivity into account, there are fifteen possible cases with respect to the identity of the genes of A and B. These fifteen "identity modes" are as follows (Gillois, 1964):

S1: $G_A \equiv G_A^* \equiv G_B \equiv G_B^*$

S2: $(G_A \equiv G_A^* \equiv G_B) \not\equiv G_B^*$

S3: $(G_A \equiv G_A^* \equiv G_B^*) \not\equiv G_B$

S4: $(G_A \equiv G_B \equiv G_B^*) \not\equiv G_A^*$

S5: $(G_A^* \equiv G_B \equiv G_B^*) \not\equiv G_A$

S6: $(G_A \equiv G_A^*) \not\equiv (G_B \equiv G_B^*)$

S7: $(G_A \equiv G_A^*) \not\equiv G_B \not\equiv G_B^*$

S8: $(G_B \equiv G_B^*) \not\equiv G_A \not\equiv G_A^*$

S9: $(G_A \equiv G_B) \not\equiv (G_A^* \equiv G_B^*)$

S10: $(G_A \equiv G_B) \not\equiv G_A^* \not\equiv G_B^*$

S11: $(G_A^* \equiv G_B^*) \not\equiv G_A \not\equiv G_B$

S12: $(G_A \equiv G_B^*) \not\equiv (G_A^* \equiv G_B)$

S13: $(G_A \equiv G_B^*) \not\equiv G_A^* \not\equiv G_B$

S14: $(G_A^* \equiv G_B) \not\equiv G_A \not\equiv G_B^*$

S15: $G_A \not\equiv G_B \not\equiv G_A^* \not\equiv G_B^*$

These fifteen identity modes can be represented as in Fig. 6.2. In this diagram, each group of four dots represents the four genes at the locus in question in the two individuals, in the arrangement given at the top of the diagram. Genes which are identical are joined by a line.

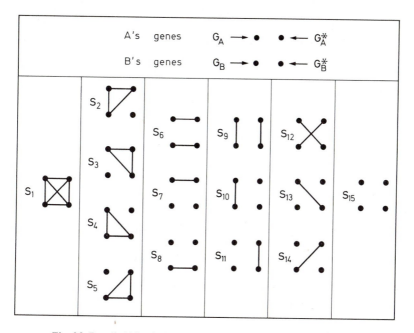

Fig. 6.2. Detailed identity modes. (Identical genes are linked by a line)

By analysis of the pedigree which contains A and B, we attach a probability δ_i to each of the fifteen possible cases S_i. These probabilities are called the "*coefficients of identity by descent*" of A and B. They contain all the information about the relationship of A and B that we require for genetic purposes.

Thus, each possible type of relationship has a set of fifteen coefficients of identity. For example, if A and B have no common ancestor, and if, for both A and B, their fathers and mothers were unrelated, none of the cases of identity is possible for any pair of genes from A and B, so we have:

$$\delta_1 = \delta_2 = \cdots = \delta_{14} = 0, \quad \delta_{15} = 1.$$

If, however, A and B are identical twins, whose parents were unrelated, each gene in A is identical with the gene in B which he received from the same parent, so that:

$$\delta_1 = \delta_2 = \cdots = \delta_8 = \delta_{10} = \cdots = \delta_{15} = 0, \quad \delta_9 = 1.$$

1.2.2. Condensed coefficients of identity. We are usually interested in the genotype of an individual, without regard to which genes came from which of his parents; the genotype $(A_i A_j)$ is the same, whether A_i

came from the father and A_j from the mother, or the other way round: A_j from the father and A_i from the mother. If we disregard the origins of the genes in this way, some of the fifteen identity modes are equivalent, and there is no point in distinguishing them. For example, the modes in which one of A's genes is identical to one of B's genes, but no other genes are identical with one another (S 10, S 11, S 13 and S 14) are equivalent, from this point of view.

If we ignore the maternal or paternal origins of the genes of A and B, we find that the fifteen identity modes reduce to nine "condensed" modes. In defining these nine modes, the following factors are considered:

1. Whether the two genes of A are identical: (I_A) or not: $(\tilde{I}_A)$.

2. Whether the two genes of B are identical or not: I_B and $\tilde{I}_B$, respectively.

3. Whether any genes in A are identical with genes in B. The number of genes that are identical between A and B can be 0, 1, 2 or 4 (the number of lines of identity which can be drawn between A's genes and B's genes, in Fig. 6.3). These numbers (N) are given in brackets alongside the identity modes in the table below.

The nine "condensed" identity modes are:

$$S'1: \quad I_A, I_B \ (N=4) \qquad\qquad S'6: \quad \tilde{I}_A, I_B \ (N=0)$$
$$S'2: \quad I_A, I_B \ (N=0) \qquad\qquad S'7: \quad \tilde{I}_A, \tilde{I}_B \ (N=2)$$
$$S'3: \quad I_A, \tilde{I}_B \ (N=2) \qquad\qquad S'8: \quad \tilde{I}_A, \tilde{I}_B \ (N=1)$$
$$S'4: \quad I_A, \tilde{I}_B \ (N=0) \qquad\qquad S'9: \quad \tilde{I}_A, \tilde{I}_B \ (N=0)$$
$$S'5: \quad \tilde{I}_A, I_B \ (N=2)$$

These possible cases can be represented as in Fig. 6.3. Identity between A's and B's genes is symbolised by a line.

Each condensed mode corresponds to one or more of our original fifteen modes of identity; for example, S'8 corresponds to S 10, S 11, S 13 and S 14.

If the pedigree containing A and B is known, we can attach a probability, Δ_i, to each of the possible condensed modes. These probabilities are called the "condensed coefficients of identity by descent". It is easy to verify that:

$$\Delta_1 = \delta_1 \qquad\qquad\qquad \Delta_6 = \delta_8$$
$$\Delta_2 = \delta_6 \qquad\qquad\qquad \Delta_7 = \delta_9 + \delta_{12}$$
$$\Delta_3 = \delta_2 + \delta_3 \qquad\qquad \Delta_8 = \delta_{10} + \delta_{11} + \delta_{13} + \delta_{14} \qquad\qquad (1)$$
$$\Delta_4 = \delta_7 \qquad\qquad\qquad \Delta_9 = \delta_{15}$$
$$\Delta_5 = \delta_4 + \delta_5$$

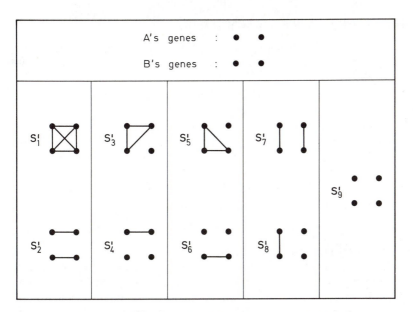

Fig. 6.3. Condensed identity modes. (Identical genes are linked by a line)

Cotterman (1940) introduced a similar method of measuring rela-
tionship. His k_0, k_1 and k_2 are the same as Δ_9, Δ_8 and Δ_7 here. Denniston
(1967) extended Cotterman's method to include the relationship between
two inbred individuals. See also Harris (1964).

1.2.3. Coefficients of kinship. Finally, we are often not concerned
with the genotypes of individuals, but only with their genic structures.
When this is the case, the sole effect of the relationship between A and B
is that there is a certain probability that a gene from A is identical by
descent with a gene from B. From this point of view, modes $S'3$, $S'5$
and $S'7$ are equivalent.

Malécot (1948) introduced the use of this probability as a measure
of relationship, and called it the "*coefficient de parenté*". Many different
translations of this are in use. We shall employ the term "coefficient of
kinship", with the definition:

"*The coefficient of kinship Φ_{AB} of two individuals, A and B, is the
probability that a gene taken at random from A is identical with a gene at
the same locus taken at random from B*".

If we know the coefficients of identity δ_i, or the condensed coefficients
Δ_i, of the pair of individuals A and B, we can find the value of Φ_{AB}.

The following relations are easy to verify:

$$\Phi_{AB}=\delta_1+\tfrac{1}{2}(\delta_2+\delta_3+\delta_4+\delta_5+\delta_9+\delta_{12})+\tfrac{1}{4}(\delta_{10}+\delta_{11}+\delta_{13}+\delta_{14})$$
$$\Phi_{AB}=\Delta_1+\tfrac{1}{2}(\Delta_3+\Delta_5+\Delta_7)+\tfrac{1}{4}\Delta_8. \tag{2}$$

Note. If A and B are identical twins whose parents are unrelated, $\Phi_{AB}=\tfrac{1}{2}$, and not 1, as one might think at first.

Also, if A and B are the same individual, Φ_{AB} represents the coefficient of kinship Φ_{II} of I with himself. Two genes taken from I can be either:

1. The same gene (probability $\tfrac{1}{2}$); in this case they must be identical, or

2. The paternal and the maternal gene (probability $\tfrac{1}{2}$); in this case, the probability that they are identical is given by the coefficient of kinship Φ_{FM} of I's parents. We therefore have:

$$\Phi_{II}=\tfrac{1}{2}(1+\Phi_{FM}).$$

If I's parents are unrelated, $\Phi_{FM}=0$, so that $\Phi_{II}=\tfrac{1}{2}$.

The coefficient of kinship, Φ, is the most condensed measure of relationship between two individuals; it contains less information than the condensed coefficients of identity, and these in turn contain less information than the full set of coefficients of identity. Φ can be used to study problems involving genic structure, but is not, in general, sufficient to solve problems involving genotypic structures.

1.2.4. Inbreeding coefficients. Consider the offspring O of two individuals, A and B. At a given locus, he has two genes, one of which is a copy of one of A's genes at this locus, and the other a copy of a gene carried by B. The probability that the two genes which O receives are identical by descent is therefore Φ_{AB}. We define the "inbreeding coefficient" as this probability:

"*The inbreeding coefficient, f, of an individual is equal to the coefficient of kinship, Φ, of his father and mother.*"

To summarise this section, the relationship between two individuals can be measured in several ways, which can be very detailed or less so, according to the purpose of the measurement.

For a complete characterisation of the relationship, we need to know the set of fifteen "coefficients of identity by descent", δ_i.

If there is no need to distinguish between the paternal and maternal origin of the genes, we can use the set of nine "condensed coefficients of identity by descent", Δ_i.

Finally, if we are not interested in the genotypes, but simply in the presence of identical genes, we can use the "coefficient of kinship", Φ.

When the coefficient of kinship is applied to a pair of breeding individuals, the inbreeding coefficient of their offspring is equal to this coefficient of kinship.

It is important to understand the distinction between coefficients of kinship and inbreeding coefficients: a coefficient of kinship applies to two individuals, and corresponds to the fact that they have a common ancestor, while an inbreeding coefficient concerns an individual, and corresponds to the fact that his parents are related.

Knowledge of the coefficients of identity δ_i or Δ_i for A and B also tells us the inbreeding coefficients of A and B. The following relations are easy to see:

$$\begin{aligned} f_A &= \delta_1 + \delta_2 + \delta_3 + \delta_6 + \delta_7 \\ &= \Delta_1 + \Delta_2 + \Delta_3 + \Delta_4, \\ f_B &= \delta_1 + \delta_4 + \delta_5 + \delta_6 + \delta_8 \\ &= \Delta_1 + \Delta_2 + \Delta_5 + \Delta_6. \end{aligned} \tag{3}$$

1.3. The Calculation of Coefficients of Identity

As we have already stated, the coefficients of identity, δ_i and Δ_i, and the coefficient of kinship, Φ, characterise a *known* relationship between two individuals, A and B.

In theory, the number of ancestors of an individual increases the farther back one goes. n generations ago, an individual has 2^n ancestors, which would be approximately a million, for $n = 20$. Since human population sizes are limited, any two individuals must therefore have a considerable number of common ancestors. In practice, it is clearly impossible to trace genealogies back for ever. We cannot, therefore, measure the relatedness of A and B in absolute terms, but only with respect to the part of the pedigree which we know; we assume that all other ancestors of A and B were neither inbred nor related to one another. In other words, we assume that there is zero probability of identity of genes in A and B by descent from one of these ancestors.

Despite this simplification, it is usually very laborious to calculate the sets of δ_i or Δ_i coefficients. Φ, however, can be calculated quite easily.

1.3.1. The calculation of coefficients of kinship. Suppose first of all that the individuals A and B, for whom we want to calculate the coefficient of kinship Φ_{AB}, have just one common ancestor, C. A gene taken at random from A and one taken at random from B can be identical only if both are derived from C, and if either:

1. both are copies of the same gene of C, which has a probability of $\frac{1}{2}$, or

2. they are copies of the two different genes of C, but these are identical because C himself is inbred; this has a probability of $\frac{1}{2} f_C$.

If there are n generations between A and C, the probability that a gene in A came from C is $(\frac{1}{2})^n$; similarly, if there are p generations between B and C, the probability that a gene in B came from C is $(\frac{1}{2})^p$. The probability that a gene in A is identical with a gene in B is therefore:

$$\Phi_{AB}=(\tfrac{1}{2})^{n+p+1}(1+f_C). \tag{4}$$

If A and B have more than one common ancestor, there are several lines of descent from the various common ancestors to A and B; two lines of descent are distinct if they differ by at least one link.

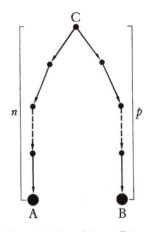

Fig. 6.4. The calculation of the coefficient of kinship

We assume that the members of the pedigree whose parents are not given in the common pedigree of A and B are all unrelated to one another. Then the transmission of a gene to A and B by one line of descent excludes the possibility of its transmission via other lines of descent. We can therefore use the principle of the addition of probabilities of incompatible events, and write:

$$\Phi_{AB}=\sum_i (\tfrac{1}{2})^{n_i+p_i+1}(1+f_{C_i}) \tag{5}$$

where the summation is over all the lines of descent.

Unless we are dealing with a particular case, it seems illogical to assume that the common ancestors are unrelated to one another, and yet that they may be inbred. If we make the further assumption that they are not inbred, Eq. (5) becomes

$$\Phi_{AB}=\sum_i (\tfrac{1}{2})^{n_i+p_i+1}. \tag{6}$$

In Chapter 8, when we deal with the changes in the mean inbreeding coefficient in a population of limited size, we shall see how this formula can be modified to take into account the "ambient inbreeding".

Some particular cases. In the following examples, we assume that the common ancestors are not related to one another, and not inbred, unless this is shown in the pedigree. It is easy to derive the following coefficients of kinship:

Parent-offspring: $\Phi_{PO} = (\frac{1}{2})^{1+0+1} = \frac{1}{4}$

Half-sibs: $\Phi_{HS} = (\frac{1}{2})^{1+1+1} = \frac{1}{8}$

Full-sibs: $\Phi_{FS} = (\frac{1}{2})^{1+1+1} + (\frac{1}{2})^{1+1+1} = \frac{1}{4}$

First cousins: $\Phi_{FC} = (\frac{1}{2})^{2+2+1} + (\frac{1}{2})^{2+2+1} = \frac{1}{16}$.

Double first cousins: there are four lines of descent between double first cousins, therefore:

$$\Phi_{DC} = (\frac{1}{2})^{2+2+1} \times 4 = \frac{1}{8}$$

Second cousins: $\Phi_{SC} = (\frac{1}{2})^{3+3+1} \times 2 = \frac{1}{64}$

Uncle-nephew: $\Phi_{UN} = (\frac{1}{2})^{2+1+1} \times 2 = \frac{1}{8}$.

Offspring of a brother-sister mating: Fig. 6.5 shows the pedigree. There are six lines of descent: ACB and ADB have two links, while ACEDB, ACFDB, ADECB and ADFCB have four links. Hence:

$$\Phi_{AB} = 2(\frac{1}{2})^{2+1} + 4(\frac{1}{2})^{4+1} = \frac{3}{8}.$$

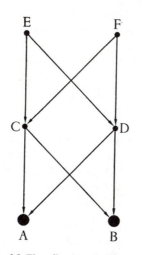

Fig. 6.5. The offspring of sib-mating

An equivalent way of deriving this result is as follows. Two genes taken at random, one from A and one from B could either:

1. have both come from C or from D (probability $\frac{1}{2}$). The probability that they are identical in either case is $\frac{1}{2}$; or

2. in half the cases, one gene came from C and one from D. Since C and D are sibs, and the coefficient of kinship of sibs is $\frac{1}{4}$, the probability in this case that the two genes are identical is $\frac{1}{4}$. Hence:

$$\Phi_{AB} = \frac{1}{2} \times \frac{1}{2} + \frac{1}{2} \times \frac{1}{4} = \frac{3}{8}.$$

In any complex relationship, it becomes extremely laborious to trace all the possible lines of descent. The calculations can be done more systematically (and are easier to program for a computer) by using the fact that the coefficient of kinship of two individuals, X and Y is related to the coefficients of kinship among their parents as follows:

$$\Phi_{XY} = \frac{1}{4}(\Phi_{FF'} + \Phi_{FM'} + \Phi_{FM} + \Phi_{MM'}).$$

This follows directly from the definition of Φ.

1.3.2. The calculation of coefficients of identity by descent. In order to find the probabilities of all the modes of identity (15 full cases, or 9 condensed cases) between the four genes of two individuals, we have to carry out the following series of operations on their pedigrees:

1. Find all the ancestors of A and B;

2. Define all the possible combinations of the origins of the four genes: G_A, G_A^*, G_B and G_B^*;

3. Determine the probability of each combination;

4. For each combination, calculate the probabilities of the various modes of identity which are possible;

5. Finally, add the probabilities for each possible mode of identity.

If the pedigree of A and B is at all extensive, the number of combinations of the possible origins of the four genes becomes very great, and the calculations can only be done by using a computer. The logic of this type of calculation has been explained by Jacquard (1966) and revised and developed by Nadot and Vayssein (1973); the method is based on a condensed form of pedigree in which we exclude all the ancestors other than those which have at least two offspring in the common pedigree of A and B. The use of this type of condensed pedigree results in a great economy of the calculations in complex cases.

Some examples. For the simplest types of relationship, the calculation of the coefficients of identity is not difficult. Table 6.1 gives some examples of condensed coefficients of identity.

Table 6.1. Condensed coefficients of identity

Relationship	Δ_1	Δ_2	Δ_3	Δ_4	Δ_5	Δ_6	Δ_7	Δ_8	Δ_9	Φ
Parent-offspring	0	0	0	0	0	0	0	1	0	$\frac{1}{4}$
Half-sibs	0	0	0	0	0	0	0	$\frac{1}{2}$	$\frac{1}{2}$	$\frac{1}{8}$
Full sibs	0	0	0	0	0	0	$\frac{1}{4}$	$\frac{1}{2}$	$\frac{1}{4}$	$\frac{1}{4}$
First-cousins	0	0	0	0	0	0	0	$\frac{1}{4}$	$\frac{3}{4}$	$\frac{1}{16}$
Double first-cousins	0	0	0	0	0	0	$\frac{1}{16}$	$\frac{6}{16}$	$\frac{9}{16}$	$\frac{1}{8}$
Second-cousins	0	0	0	0	0	0	0	$\frac{1}{16}$	$\frac{15}{16}$	$\frac{1}{64}$
Uncle-nephew	0	0	0	0	0	0	0	$\frac{1}{2}$	$\frac{1}{2}$	$\frac{1}{8}$
Offspring of sib-matings	$\frac{1}{16}$	$\frac{1}{32}$	$\frac{1}{8}$	$\frac{1}{32}$	$\frac{1}{8}$	$\frac{1}{32}$	$\frac{7}{32}$	$\frac{5}{16}$	$\frac{1}{16}$	$\frac{3}{8}$

In this table, the values of the coefficients of kinship are also shown. The reader may like to verify the relation:

$$\Phi = \Delta_1 + \tfrac{1}{2}(\Delta_3 + \Delta_5 + \Delta_7) + \tfrac{1}{4}\Delta_8.$$

These methods can be used even for complex cases. As an example we shall describe a study of the Jicaque Indians of Honduras (Chapman and Jacquard, 1971). This tribe is highly inbred, since it was founded by eight individuals who left another community and went to live in an isolated mountain area, about 100 years ago. The genealogy of the group has been worked out, back to its founders. Fig. 6.6 shows the pedigree of two brothers whose parents have so many common ancestors that it is

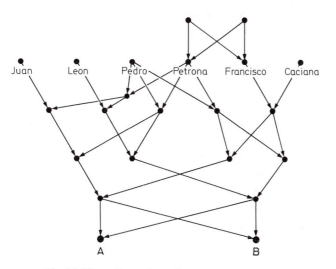

Fig. 6.6. The pedigree of two Jicaque Indian brothers

practically impossible to describe their relationship in words (their mothers were sisters, their fathers half-brothers, two of their grandfathers were brothers, and were cousins of one of the grandmothers, and uncles of a fourth grandparent, and half-brothers of a fifth, etc.).

It is almost impossible to calculate the coefficients of identity for such a complicated case by hand. The calculations took 10 seconds (a long time) on an IBM 360-75 computer. The Δ_i values obtained were as follows:

$$\Delta_1 = 0.0657 \qquad\qquad \Delta_7 = 0.2544$$
$$\Delta_2 = 0.0073 \qquad\qquad \Delta_8 = 0.3657$$
$$\Delta_3 = \Delta_5 = 0.0867 \qquad\qquad \Delta_9 = 0.0859$$
$$\Delta_4 = \Delta_6 = 0.0238$$

which gives $\Phi_{AB} = 0.3710$.

1.4. Sex-Linked Genes

We have so far considered only autosomal genes. For autosomal loci, an individual's father and mother contribute precisely the same amounts of genetic information. When we come to consider sex-linked loci, i.e. genes carried on the X chromosome, we require a completely different set of formulae from those we have derived above.

A woman has two X chromosomes, one of which she received from her mother, and one from her father. She passes on a copy of one of these chromosomes to each of her offspring, regardless of their sex. A man, however, has only one X chromosome, which he inherited from his mother (while he received a Y chromosome from his father). He transmits his X chromosome to his daughters only, and not to his sons.

In order to measure the relatedness of two individuals, A and B, with respect to sex-linked loci, we proceed in a similar way to how we studied the autosomal case, and define a "coefficient of kinship for sex-linked genes" $\Phi_X(AB)$; this is equal to the probability that two genes taken at random from A and B, at a given X-linked locus, will both be copies of the same gene in a common ancestor of A and B. Just as in the autosomal case, we can calculate this probability by listing all the possible lines of descent from A and B back to each of their common ancestors C_i. The formula analogous to Eq. (5) is as follows:

$$\Phi_X(AB) = \sum_i \lambda_i (\tfrac{1}{2})^{n_i' + p_i' + 1} (1 + F_{C_i}) \qquad (7)$$

where the summation is over all possible lines of descent, n_i' is the total number of steps minus the number of steps from a son to his mother, that occur in the line of descent from C_i to A; p'_i is the same for B; F_{C_i} is equal to 1 if C_i is a male, or to the inbreeding coefficient of C_i, if C_i is female; λ_i is equal to zero if either of the paths of descent from C_i to A or from C_i to B contains two males in succession, i.e. if there is a passage from a father to a son; if there is no such step in the lines of descent, $\lambda_i = 1$.

These rules are derived directly from the way in which the X chromosome is inherited.

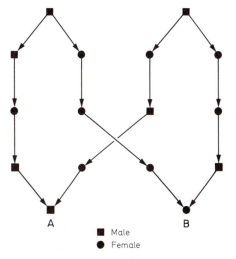

Fig. 6.7. Pedigree to demonstrate the calculation of Φ_X, the coefficient of kinship for sex-linked genes

The pedigree shown in Fig. 6.7, gives the following value of $\Phi_X(AB)$:

$$\Phi_X(AB) = 0 + (\tfrac{1}{2})^{2+3+1}(1+1) = (\tfrac{1}{2})^5.$$

The coefficient of kinship, Φ_X, for X-linked genes depends on the sexes of the ancestors who appear in the pedigree leading back to the common ancestors of the two individuals in question. Two pairs of individuals who are related in the same way (e.g. two pairs of second cousins) can therefore have different values of Φ_X for their sex-linked genes, if the sexes of corresponding members of the two pedigrees are not identical throughout. Fig. 6.8 shows the four possible ways in which a male and female first-cousin pair can be related. Cases I and II are called "parallel" cousins (in case I their fathers were sibs, and in case II their mothers were sibs). Cases III and IV are called "cross" cousins; in case III, the father of the male was a brother of the female's mother, while in case IV the male's mother was a sister of the female's father.

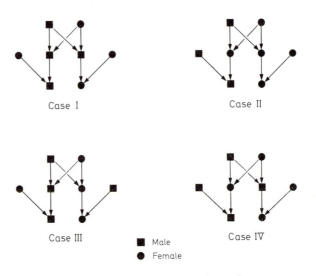

Case I

Case II

Case III

■ Male
● Female

Case IV

Fig. 6.8. The four types of cousin marriage

Using Eq. (7), we obtain:

$$\Phi_X \text{ (cousins of type I)} \ \ = 0$$
$$\Phi_X \text{ (cousins of type II)} \ = \tfrac{3}{16}$$
$$\Phi_X \text{ (cousins of type III)} = 0$$
$$\Phi_X \text{ (cousins of type IV)} = \tfrac{1}{8}.$$

2. The Genetic Structures of Related Individuals

We have seen that there are two ways of defining the genetic probability structure of an individual; his genic probability structure, s, is the vector of the probabilities that a gamete produced by him contains the different possible alleles at the locus in question; his genotypic probability structure, $\mathscr{S}$, is the set of probabilities of the various combinations of two alleles that he can have, at this locus.

We now wish to ask: "If we know the genic probability structure, s_A, or the genotypic probability structure, $\mathscr{S}_A$, of an individual A, what is the structure s_B or $\mathscr{S}_B$ of another individual B, who has a given relationship to A, when A and B are members of a population in Hardy-Weinberg equilibrium with known genic structure?".

2.1. The Relation between the Genic Structures of Related Individuals

If A and B are related, the presence of a particular gene in A increases the probability that B will have this gene, since A and B may both inherit this gene from a common ancestor. It follows that the genic probability structures of two related individuals are not independent; information about one is also information concerning the other.

Suppose we know s_A; this is probabilistic information about A's genes. If A is non-inbred (i.e. if $f_A = 0$), knowledge of the state of one of A's genes at the locus in question tells us no more about the other gene than we knew already, knowing s_A. But if A is inbred, the second gene is not independent of the first. We shall see that, in the first case, the relation between the genic structures s_A and s_B can be expressed as a function of their coefficient of kinship alone; in the second case, however, we also have to introduce the coefficients of identity, Δ_i.

In the next sections, we shall study the case when A is non-inbred, and will give the treatment of the general case at the end of Section 2, as a consequence of the relation between the genotypic structures of relatives.

First of all, we shall derive a useful property of the coefficient of kinship between two individuals.

2.1.1. Coefficients of kinship between individuals.
The coefficient of kinship Φ between an individual A, and a related individual B, is equal to the arithmetic mean of the coefficients of kinship of B with the father and mother of A:

$$\Phi_{AB} = \tfrac{1}{2}(\Phi_{FB} + \Phi_{MB}). \tag{8}$$

This can be proved as follows. A gene carried by A can have come from either his father F or his mother M. A gene taken at random from B can be identical to a gene taken at random from A if either:

1. A received the gene from his father F (probability $\tfrac{1}{2}$), and it is identical to the gene from B (probability Φ_{FB}); or

2. A received the gene from his mother (probability $\tfrac{1}{2}$), and it is identical to the gene from B (probability Φ_{MB}).

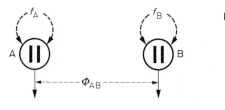

Fig. 6.9. The coefficient of kinship and the inbreeding coefficients of two individuals

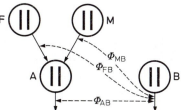

Fig. 6.10. Coefficients of kinship between two related individuals

These two cases are mutually exclusive, so we have:

$$\Phi_{AB} = \tfrac{1}{2}\Phi_{FB} + \tfrac{1}{2}\Phi_{MB}.$$

This relation enables us to write the probability structure of B, given that of A, $s_{B|A}$, as a function of Φ_{AB} alone, when A is non-inbred.

2.1.2. The relation between the genic structures of related, non-inbred individuals. We assume that the genic probability structure s_A, of an individual is known, and that A is not inbred.

A gene of B can be identical to a gene of A either because it is identical to the gene which A received from his father (which has the probability Φ_{FB}), or because it is identical to the gene A received from his mother (which has the probability Φ_{MB}). These two possibilities are mutually exclusive, since A is assumed to be non-inbred.

Suppose one of B's genes is chosen at random. There are two possible situations:

I: The gene is identical to one of A's genes. From relation (7), the probability of this is:

$$\Phi_{FB} + \Phi_{MB} = 2\Phi_{AB}.$$

If this situation occurs, the probability that this gene is allele A_i is equal to the probability that a gene of A is A_i, hence:

$$s_{B|I} = s_A.$$

II: The gene is not identical to either of A's genes. This has the probability:

$$1 - 2\Phi_{AB}.$$

If this is the situation, the probability that this gene is A_i is equal to the probability that a gene from this population is A_i, hence:

$$s_{B|II} = s$$

where s is the genic structure of the population of which B is a member.

We therefore have the result:

$$s_{B|A} = 2\Phi_{AB}\, s_A + (1 - 2\Phi_{AB})\, s. \tag{9}$$

Some special cases. This formula can readily be applied to the case when the actual genic structure of A, rather than the probability structure, is known. If A has the genotype $(A_i A_j)$, we have seen that his genic structure can be written:

$$s_A = \tfrac{1}{2}(i + j)$$

where i is the vector whose elements are all zero, except for π_i which is equal to 1.

Eq. (9) then becomes:

$$s_{B|A_i A_j} = \Phi_{AB}(i+j) + (1 - 2\Phi_{AB})s.$$

Similarly, if we know only that A has the gene A_i, his genic probability structure is:

$$s_A = \tfrac{1}{2}(i+s)$$

and Eq. (9) becomes:

$$s_{B|A_i} = \Phi_{AB}\, i + (1 - \Phi_{AB})\, s.$$

Alternative proof of Eq. (9).[3] The result of the arguments given in the two preceding sections can be proved more rigorously using only the basic axioms of probability theory concerning the additivity of the probabilities of incompatible events, and the algebra of conditional probabilities.

Let

α be a gene taken at random from A;

β be a gene taken at random from B;

A_k be the copy of allele A_k in an ancestor of A.

By definition

$$\Phi_{AB} = \mathrm{prob}\,(\alpha \equiv \beta).$$

We want to find the probability ψ_k that a gene taken at random from B is identical to A_k, and that this gene is present in A. We shall write:

$$\psi_k = \mathrm{prob}\,\{(\beta \equiv A_k) \cap (A_k \subset A\}.$$

We can write:

$$\psi_k = \mathrm{prob}\,\{(\beta \equiv A_k) \cap (A_k \subset A) \cap (\alpha \equiv A_k)\}$$
$$+ \mathrm{prob}\,\{(\beta \equiv A_k) \cap (A_k \subset A) \cap (\alpha \not\equiv A_k)\} \qquad (10)$$
$$= X + Y$$

since the events $\alpha \equiv A_k$ and $\alpha \not\equiv A_k$ are complementary.

Since α is taken from A, the event $\alpha \equiv A_k$ implies the event $A_k \subset A$. Therefore the first term, X, in Eq. (9) can be written:

$$X = \mathrm{prob}\,\{(\beta \equiv A_k) \cap (\alpha \equiv A_k)\}$$
$$= \mathrm{prob}\,\{(\beta \equiv \alpha) \cap (\alpha \equiv A_k)\}$$

since the relation of identity is transitive.

The events $(\beta \equiv \alpha)$ and $(\alpha \equiv A_k)$ are independent, hence, finally:

$$X = \mathrm{prob}\,(\beta \equiv \alpha) \times \mathrm{prob}\,(\alpha \equiv A_k)$$
$$= \Phi_{AB} \times {}_A p_k.$$

---

[3] This proof is due to Daniel Courgeau.

The second term, Y, of Eq. (9) can be written:

$$Y = \text{prob} \{(\beta \equiv A_k) \cap (A_k \subset A)\}$$

$$\times \text{prob} [(\alpha \not\equiv A_k) \text{ given that } \{(\beta \equiv A_k) \cap (A_k \subset A)\}].$$

But the probability of the event in square brackets is $\frac{1}{2}$: it is the probability of taking a gene from A and finding it not to be A_k, when A is known to have the gene A_k, and A is assumed to be non-inbred. We therefore have:

$$Y = \psi_k \times \tfrac{1}{2}.$$

Eq. (10) therefore becomes:

$$\psi_k = \Phi_{AB} \times {}_A\pi_k + \tfrac{1}{2} \psi_k.$$

Hence

$$\psi_k = 2 \Phi_{AB} \times {}_A\pi_k.$$

A gene in B can be allele A_k either because it is identical by descent with one of the genes in A, and this gene is A_k (this event has the probability $\psi_k = 2 \Phi_{AB} \times {}_A p_k$), or else because it is not identical by descent with either of A's genes (probability $1 - \Phi_{AB}$), and is A_k (probability p_k, the frequency of A_k in the population).

Thus, finally, we have:

$$s_{B|A} = 2 \Phi_{AB} \, s_A + (1 - \Phi_{AB}) \, s.$$

2.2. The Relation between the Genotypic Structures of Related Individuals

For many purposes, the genic structure of an individual is insufficient information; for example, if we are considering a trait controlled by a recessive gene A_r which gives rise to a disease when homozygous, the probability that an individual is affected is given by the probability of the combination $(A_r A_r)$, not by the probability of the presence of the gene A_r.

In this section we consider what information we obtain about the genotype of an individual B from knowledge of the genotype of a related individual, A. We want to find the probabilities of the various genotypes which are possible for B, i.e. the genotypic probability structure, $\mathscr{S}_{B|A}$.

We need to know the relations between both of A's genes, and both of B's genes. The coefficient of kinship, Φ_{AB}, of A and B is therefore not a sufficiently informative measure of the relationship between them. We have to know the values of the nine condensed coefficients of identity, Δ_i (it is not necessary to know the full coefficients, δ_i, because we are not interested in the paternal or maternal origin of the genes, but simply in the resulting genotype).

The genotype of A is assumed to be known. There are two possible situations; if A is heterozygous, we know that his two genes are not identical by descent; if A is homozygous, the genes may or may not be identical by descent. Let us start by examining the first situation.

2.2.1. The situation when A is heterozygous. We first assume that A is heterozygous $(A_i A_j)$, $i \neq j$. We therefore know that his two genes are not identical by descent. The probabilities of identity modes $S'1$, $S'2$, $S'3$ and $S'4$, given that A is heterozygous, are therefore zero. We now want to find the probabilities of modes $S'5$, $S'6$, $S'7$, $S'8$ and $S'9$; from relation (3), these probabilities, given that A is heterozygous, are, respectively:

$$\frac{\Delta_5}{\Delta_5 + \Delta_6 + \Delta_7 + \Delta_8 + \Delta_9} = \frac{\Delta_5}{1 - f_A}$$

and

$$\frac{\Delta_6}{1 - f_A}, \dots, \frac{\Delta_9}{1 - f_A},$$

where f_A is the inbreeding coefficient of A.

We shall examine each of these five modes, and find the probabilities for the different possible genotypes of B to which they give rise.

Identity mode $S'5$ [i.e., in the notation of Section 1.2.2, $\tilde{I}_A$, I_B ($N = 2$)]:
This mode has the probability $\dfrac{\Delta_5}{1 - f_A}$. In this situation, the two genes of B are identical, and also identical with one or other of A's genes. B can therefore only be $(A_i A_i)$ or $(A_j A_j)$, and these have the same probability. Writing S_{ij} for the set of frequencies whose elements are all zero, except for $P_{ij} = 1$, and $\mathscr{S}_{B|S'5}$ for B's genotypic probability structure with the mode $S'5$, we have:

$$\mathscr{S}_{B|S'5} = \tfrac{1}{2} S_{ii} + \tfrac{1}{2} S_{jj}.$$

Identity mode $S'6$ [$\tilde{I}_A$, I_B ($N = 0$)]: This situation has the probability $\dfrac{\Delta_6}{1 - f_A}$. The two genes of B are identical with one another, but not with either of A's genes. B must therefore be homozygous, and his genotype will be $(A_k A_k)$ with probability p_k, the frequency of the A_k gene in the population. B's genotypic probability structure is therefore equal to the trimat whose elements are all zero, except for the hypotenuse, which consists of the values of p_i. This trimat will be denoted by S_H. It represents the genotypic structure of a population with the same genic structure as the population we are considering, but whose members are all homozygous, and can be called the "*structure of the corresponding homozygous population*". Thus:

$$\mathscr{S}_{B|S'6} = S_H.$$

Identity mode S′7 $[\tilde{I}_A, \tilde{I}_B\ (N=2)]$: In this situation, which has probability $\dfrac{\Delta_7}{1-f_A}$, each of B's genes is identical with one of A's genes. B must therefore have the genotype $(A_i A_j)$. Hence:

$$\mathscr{S}_{B|S′7}=S_{ij}.$$

Identity mode S′8 $[\tilde{I}_A, \tilde{I}_B\ (N=1)]$: The probability of this situation is $\dfrac{\Delta_8}{1-f_A}$. One of B's genes is identical with one of A's genes, while the other is independent. B's genotype is therefore either $(A_i A_k)$ or $(A_j A_k)$, where A_k is any allele at this locus; the probability of each of these two possibilities is $\frac{1}{2}p_k$, where p_k is the frequency of A_k in the population. The trimat of the probabilities of the different possible genotypes for B is therefore equal to the genetic product of the genic structure of A with the genic structure of the population, $s_A \times s$. This genetic product also represents the genotypic probability structure of an offspring of the cross of A with an individual taken at random from the population. This will be written as:

$$S_F \equiv s_A \times s.$$

Finally, we have:

$$\mathscr{S}_{B|S′8}=S_F.$$

Identity mode S′9 $[\tilde{I}_A, \tilde{I}_B\ (N=0)]$: This situation has the probability $\dfrac{\Delta_9}{1-f_A}$. B's two genes are independent, and both are independent of both of A's genes. B can have any genotype $(A_k A_l)$, with the probabilities, p_k^2 and $2p_k p_l$, which correspond to the different genotype frequencies in a panmictic population; we shall represent this set of frequencies as S_P. Hence:

$$\mathscr{S}_{B|S′9}=S_P.$$

Gathering all the possible situations together, we have the following genotypic structure for B, given that A's genotype is $(A_i A_j)$, $i \neq j$:

$$\mathscr{S}_{B|A_i A_j}=\frac{1}{1-f_A}\left[\tfrac{1}{2}\Delta_5(S_{ii}+S_{jj})+\Delta_6 S_H+\Delta_7 S_{ij}+\Delta_8 S_F+\Delta_9 S_P\right]. \quad (11)$$

2.2.2. The situation when A is homozygous. Now suppose that A's genotype is $(A_i A_i)$. All of the nine modes of identity are possible. We shall examine each of them in turn, and find the probabilities of the different possible genotypes for B.

Identity mode S′1 $[I_A, I_B\ (N=4)]$: B's two genes are both identical to A's two. B's genotype is therefore $(A_i A_i)$, hence:

$$\mathscr{S}_{B|S′1}=S_{ii}.$$

Identity modes S′ 2 *and* S′ 6 $[I_A, I_B (N = 0)$ and $\tilde{I}_A, I_B (N = 0)]$: B's genes are identical with one another, but independent of A's genes. Therefore B is homozygous, and the probability that his genotype is $(A_k A_k)$ is p_k, hence:

$$\mathscr{S}_{B|S′2} = \mathscr{S}_{B|S′6} = S_H.$$

Identity modes S′ 3 *and* S′ 8 $[I_A, \tilde{I}_B (N = 2)$ and $\tilde{I}_A, \tilde{I}_B (N = 1)]$: One of B's genes is identical with one or both of A's genes, while the other is independent. B's genotype is therefore $(A_i A_k)$, where A_k is an allele at this locus, with probability p_k. Hence:

$$\mathscr{S}_{B|S′3} = \mathscr{S}_{B|S′8} = s_A \times s = S_F.$$

Identity modes S′ 4 *and* S′ 9 $[I_A, \tilde{I}_B (N = 0)$ and $\tilde{I}_A, \tilde{I}_B (N = 0)]$: B's genes are independent of one another, and of A's genes. B can have any genotype $(A_k A_l)$, with the frequencies p_k^2 and $2 p_k p_l$ corresponding to the frequencies in a panmictic population. Hence:

$$\mathscr{S}_{B|S′4} = \mathscr{S}_{B|S′9} = S_P.$$

Identity modes S′ 5 *and* S′ 7 $[\tilde{I}_A, I_B (N = 2)$ and $\tilde{I}_A, \tilde{I}_B (N = 2)]$: Each of B's genes is identical with one of A's genes; they are therefore both A_i. Thus:

$$\mathscr{S}_{B|S′5} = \mathscr{S}_{B|S′7} = S_{ii}.$$

Finally, gathering all the probabilities together, we have:

$$\mathscr{S}_{B|A_i A_i} = (\Delta_1 + \Delta_5 + \Delta_7) S_{ii} + (\Delta_3 + \Delta_8) S_F + (\Delta_2 + \Delta_6) S_H$$
$$+ (\Delta_4 + \Delta_9) S_P. \tag{12}$$

Eqs. (11) and (12) enable us to find the probabilities of the various possible genotypes for an individual who is a member of a panmictic population of known genic structure, knowing the genotype of an individual who has a given relationship with him. These equations appear complex, as they involve the following variables:

1. The nine condensed coefficients of identity.

2. The structures S_{ii}, S_{ij} and S_{jj} corresponding to the three genotypes which can be formed from A's two genes.

3. The two reference genotypic structures of the population. These are S_P, the structure of the corresponding panmictic population:

$$S_P \equiv \left|
\begin{array}{l}
p_1^2 \\
\dotfill \\
\dots 2 p_i p_j \dots p_i^2 \\
\dotfill \\
\dotfill p_n^2
\end{array}
\right.$$

and S_H, the structure of the corresponding homozygous population:

$$S_H \equiv \begin{vmatrix} p_1 \\ \cdots\cdots \\ 0\ldots p_i \\ \cdots\cdots\cdots \\ 0\ldots 0\ldots p_n \end{vmatrix}$$

4. The structure, $S_F = s_A \bowtie s$, of an individual resulting from the cross of A and a randomly-chosen member of the population.

In some of the simplest types of relationship, which are the ones most commonly met with in practice, some of the Δ_i coefficients become zero, and the formulae are thus simpler than in this general case.

2.2.3. Some particular cases. a) A *non-inbred.* If A's father and mother are unrelated, we have:

$$f_A = 0, \quad \text{hence} \quad \Delta_1 = \Delta_2 = \Delta_3 = \Delta_4 = 0.$$

Eq. (11) then becomes:

$$\mathscr{S}_{B|A_iA_i} = (\Delta_5 + \Delta_7)\, S_{ii} + \Delta_6\, S_H + \Delta_8\, S_F + \Delta_9\, S_P$$

which is the equivalent of Eq. (10) when $f_A = 0$, and $A_i = A_j$. We can therefore write:

$$\mathscr{S}_{B|A_iA_j} = \frac{\Delta_5}{2}\,(S_{ii} + S_{jj}) + \Delta_6\, S_H + \Delta_7\, S_{ij} + \Delta_8\, S_F + \Delta_9\, S_P$$

and this is valid both for $A_i = A_j$ and for $A_i \neq A_j$.

b) B *non-inbred.* In this case, $f_B = 0$, and $\Delta_1 = \Delta_2 = \Delta_5 = \Delta_6 = 0$. Eqs. (10) and (11) therefore become:

$$\mathscr{S}_{B|A_iA_j} = \frac{1}{1 - f_A}\,(\Delta_7\, S_{ij} + \Delta_8\, S_F + \Delta_9\, S_P)$$

$$\mathscr{S}_{B|A_iA_i} = \Delta_7\, S_{ii} + (\Delta_3 + \Delta_8)\, S_F + (\Delta_4 + \Delta_9)\, S_P.$$

Notice that these equations do not contain S_H.

c) *Both A and B non-inbred.* Eqs. (11) and (12) reduce to:

$$\mathscr{S}_{B|A_iA_j} = \Delta_7\, S_{ij} + \Delta_8\, S_F + \Delta_9\, S_P \tag{13}$$

whether $A_i = A_j$ or $A_i \neq A_j$.

In this case, Δ_7 is the probability that both of B's genes are identical to one of A's genes; Δ_8 is the probability that one of B's genes is identical with one of A's genes; Δ_9 is the probability that neither of B's genes is identical with a gene of A.

Eq. (13) is equivalent to the formula derived by Li (1955) by the use of "ITO matrices". This method is simpler than the one given here, but it is valid only for the case when A and B are non-inbred (which is, however, the most usual case, in practice).

d) A *and* B *non-inbred and at least one of the four parents of* A *and* B *is assumed to be independent of the other three.* Thus one of the four genes must be non-identical with the others, and identity mode S′ 7 is impossible, $\Delta_7 = 0$. Eq. (13) then becomes:

$$\mathscr{S}_{B|A_iA_j} = \Delta_8 S_F + \Delta_9 S_P.$$

By (2), we know that $\Delta_8 = 4\Phi_{AB}$ (since $\Delta_1 = \Delta_3 = \Delta_5 = \Delta_7 = 0$), and $\Delta_9 = 1 - 4\Phi_{AB}$, hence:

$$\mathscr{S}_{B|A_iA_j} = 4\Phi_{AB} S_F + (1 - 4\Phi_{AB}) S_P. \tag{14}$$

Of the various measures of relationship, only the coefficient of kinship enters into this equation.

These examples show how restrictive are the conditions for B's genotypic structure to be expressed solely as a function of the coefficient of kinship of A and B; A and B must both be non-inbred, but also one parent of A or B must be unrelated to the parents of the other. An important case when this condition is satisfied is when A and B are linked by only one path of descent.

Eq. (14) is therefore valid for the relations parent-offspring, uncle-nephew and first cousins, but not for sibs or double first cousins.

e) Using the values of Δ_i or Φ from Section 1.3.2, we can write out Eq. (13) or (14) as appropriate, for the various relationships examined earlier. The values of $\mathscr{S}_{B|A_iA_j}$ are as follows:

Parent-offspring:	S_F
Half-sibs:	$\frac{1}{2} S_F + \frac{1}{2} S_P$
Full sibs:	$\frac{1}{4} S_{ij} + \frac{1}{2} S_F + \frac{1}{4} S_P$
First cousins:	$\frac{1}{4} S_F + \frac{3}{4} S_P$
Double first cousins:	$\frac{1}{16} S_{ij} + \frac{6}{16} S_F + \frac{9}{16} S_P$
Second cousins:	$\frac{1}{16} S_F + \frac{15}{16} S_P.$

Offspring of brother-sister matings: in this case, A and B are both inbred, so we have to return to Eqs. (11) and (12), for the cases when A is heterozygous or homozygous, respectively:

$$\mathscr{S}_{B|A_iA_j} = \frac{4}{3}\left(\frac{1}{16} S_{ii} + \frac{1}{16} S_{jj} + \frac{5}{32} S_{ij} + \frac{1}{16} S_F + \frac{1}{32} S_H + \frac{1}{16} S_P\right),$$

$$\mathscr{S}_{B|A_iA_i} = \frac{13}{32} S_{ii} + \frac{7}{16} S_F + \frac{1}{16} S_H + \frac{3}{32} S_P.$$

2.2.4. Loci with two alleles. When there are only two alleles, only three genotypes are possible, and we can write out the trimats of probabilities in full, which may make Eqs. (11) and (12) easier to understand.

Let p and $q = 1 - p$ be the frequencies of the two alleles, A_1 and A_2, in the population. The two reference genotypic structures are:

$$S_P = \left| \begin{matrix} p^2 & \\ 2pq & q^2 \end{matrix} \right. \qquad\qquad S_H = \left| \begin{matrix} p & \\ 0 & q \end{matrix} \right.$$

The structure S_F is:

$$\left| \begin{matrix} p/2 & \\ \tfrac{1}{2} & q/2 \end{matrix} \right. \qquad \text{if A is heterozygous } (A_1 A_2)$$

and

$$\left| \begin{matrix} p & \\ q & 0 \end{matrix} \right. \qquad \text{if A is homozygous } (A_1 A_1).$$

If A is heterozygous, Eqs. (11) and (12) give:

$$\left| \begin{matrix} \varPi_{11} & \\ \varPi_{12} & \varPi_{22} \end{matrix} \right. = \frac{1}{1 - f_A} \left\{ \frac{\Delta_5}{2} \left| \begin{matrix} 1 & \\ 0 & 0 \end{matrix} \right. + \frac{\Delta_5}{2} \left| \begin{matrix} 0 & \\ 0 & 1 \end{matrix} \right. + \Delta_7 \left| \begin{matrix} 0 & \\ 1 & 0 \end{matrix} \right. \right.$$

$$\left. + \Delta_8 \left| \begin{matrix} p/2 & \\ \tfrac{1}{2} & q/2 \end{matrix} \right. + \Delta_6 \left| \begin{matrix} p & \\ 0 & q \end{matrix} \right. + \Delta_9 \left| \begin{matrix} p^2 & \\ 2pq & q^2 \end{matrix} \right. \right\} \qquad (15)$$

where $\varPi_{11}$, $\varPi_{12}$ and $\varPi_{22}$ are the probabilities that B will have the genotype $(A_1 A_1)$, $(A_1 A_2)$ and $(A_2 A_2)$, respectively. If A is homozygous $A_1 A_1$, Eqs. (11) and (12) give:

$$\left| \begin{matrix} \varPi_{11} & \\ \varPi_{12} & \varPi_{22} \end{matrix} \right. = (\Delta_1 + \Delta_5 + \Delta_7) \left| \begin{matrix} 1 & \\ 0 & 0 \end{matrix} \right. + (\Delta_3 + \Delta_8) \left| \begin{matrix} p & \\ q & 0 \end{matrix} \right.$$

$$+ (\Delta_2 + \Delta_6) \left| \begin{matrix} p & \\ 0 & q \end{matrix} \right. + (\Delta_4 + \Delta_9) \left| \begin{matrix} p^2 & \\ 2pq & q^2 \end{matrix} \right. . \qquad (16)$$

2.2.5. Example of the use of Eqs. (15) and (16) [4]. These equations enable us to answer two types of questions:

1. If an individual has a relative who manifests a certain trait, what is the probability that he himself will be affected?

2. What is the probability that such an individual will be a carrier for the trait?

---

[4] All the arguments in this section assume that we know the genotype of A, but of no other relative of B.

Let us take as an example a recessive disease, e.g. cystic fibrosis, which is controlled by a gene whose frequency q is about 2 % in European populations. The proportion of individuals with the disease is therefore approximately $q^2 = 4 \times 10^{-4}$.

Suppose an unborn child has a relative A who has cystic fibrosis. From Eq. (16), the probability that the child will be affected is:

$$\Pi_{22} = (\Delta_1 + \Delta_5 + \Delta_7) + (\Delta_3 + \Delta_8 + \Delta_2 + \Delta_6)\, q + (\Delta_4 + \Delta_9)\, q^2$$

and the probability that he is a carrier is:

$$\Pi_{12} = (\Delta_3 + \Delta_8)\, p + 2(\Delta_4 + \Delta_9)\, p\, q.$$

With the value of q given above, we have:

$$\Pi_{22} = (\Delta_1 + \Delta_5 + \Delta_7) + \tfrac{2}{100}(\Delta_2 + \Delta_3 + \Delta_6 + \Delta_8) + \tfrac{4}{10000}(\Delta_4 + \Delta_9)$$

$$\Pi_{12} = \tfrac{98}{100}(\Delta_3 + \Delta_8) + \tfrac{392}{10000}(\Delta_4 + \Delta_9).$$

For example, if A is a first cousin of the child in question,

$$\Pi_{22} = \tfrac{1}{800} + \tfrac{15}{40000} = 16 \times 10^{-4}$$

so the child has four times the risk of being affected, compared with a child with no affected relatives.

$$\Pi_{12} = \frac{98}{16 \times 100} + \frac{392 \times 15}{16 \times 10000} = 9.8\,\%$$

so that his risk of being a carrier is increased by a factor of nearly 2.5.

As another example, if A is a double first cousin of the unborn child:

$$\Pi_{22} = \frac{1}{16} + \frac{3}{400} + \frac{9}{4 \times 10000} = 7.0\,\%$$

$$\Pi_{12} = \frac{6 \times 98}{16 \times 100} + \frac{392 \times 9}{16 \times 10000} = 38.9\,\%.$$

In this case, the risk of being affected is multiplied by 175, and the risk of being a carrier is multiplied by nearly 10.

If A is the brother or sister of the unborn child, and their parents are also brother and sister (this situation is obviously most likely to be encountered in animal populations):

$$\Pi(A_1 A_1) = (\tfrac{1}{16} + \tfrac{1}{8} + \tfrac{7}{32}) + \tfrac{2}{100}(\tfrac{1}{32} + \tfrac{1}{8} + \tfrac{1}{32} + \tfrac{5}{16}) + \tfrac{4}{10000}(\tfrac{1}{32} + \tfrac{1}{16}) = 41.6\,\%$$

$$\Pi(A_1 A_2) = \tfrac{98}{100}(\tfrac{1}{8} + \tfrac{5}{16}) + \tfrac{392}{10000}(\tfrac{1}{32} + \tfrac{1}{16}) = 43.2\,\%.$$

In this case, the risk of being affected is multiplied by 1000, and the risk of being a carrier by just over 10.

Finally, suppose that A and B are the two Jicaque Indians whose pedigree is shown in Fig. 6.6, and whose Δ_i coefficients we calculated in Section 1.3.2, and that A has cystic fibrosis. If the frequency of the gene for this disease in the population is very low, the genotypic probability structure of B can be written, from Eq. (16):

$$
0.407 \begin{vmatrix} 0 \\ 0 \quad 1 \end{vmatrix} + 0.452 \begin{vmatrix} 0 \\ p \quad q \end{vmatrix} + 0.031 \begin{vmatrix} p \\ 0 \quad q \end{vmatrix} + 0.110 \begin{vmatrix} p^2 \\ 2pq \quad q^2 \end{vmatrix}
$$

$$
\approx \begin{vmatrix} 0.141 \\ 0.452 \quad 0.407 \end{vmatrix}
$$

Hence the probability that B will have cystic fibrosis is 0.407, compared with 0.250 if his parents had been unrelated; the probability that he will be a carrier, however, is reduced from the value, 0.5, if his parents were unrelated, to 0.452.

These examples show the importance of the equations we have derived for the relations between the genotypic structures of related individuals. They permit us to use information about one individual to deduce something about another. This is a useful tool in the diagnosis of diseases.

2.3. The Relations between the Genic Structures of Inbred Individuals

In Section 2.1 we saw that we can only use the coefficient of kinship of two individuals to tell us something about the genic structure of one, given the genotype of the other, when this other individual is non-inbred.

Eqs. (11) and (12), however, are valid for any related individuals, and we can use these in the general case.

If A's genotype is $(A_i A_j)$, we know, from Eq. (11), that a gene from B can be:

A_i, with probability:

$$
\frac{1}{1-f_A} \left[\frac{\Delta_5}{2} + \frac{\Delta_7}{2} + \frac{\Delta_8}{4} + \frac{\Delta_8}{2} p_i + \Delta_6 p_i + \Delta_9 p_i \right].
$$

A_j, with probability:

$$
\frac{1}{1-f_A} \left[\frac{\Delta_5}{2} + \frac{\Delta_7}{2} + \frac{\Delta_8}{4} + \frac{\Delta_8}{2} p_j + \Delta_6 p_j + \Delta_9 p_j \right]
$$

or A_k, different from A_i and A_j, with probability:

$$
\frac{1}{1-f_A} \left[\frac{\Delta_8}{2} p_k + \Delta_6 p_k + \Delta_9 p_k \right].
$$

We can therefore write:

$$s_{B|A_iA_j} = \frac{1}{1-f_A} \left[\left(\Delta_5 + \Delta_7 + \frac{\Delta_8}{2} \right) s_A + \left(\frac{\Delta_8}{2} + \Delta_6 + \Delta_9 \right) s \right]. \quad (17)$$

Similarly, if A is homozygous $(A_i A_i)$, Eq. (12) shows that a gene from B can be:

A_i, with probability:

$$(\Delta_1 + \Delta_5 + \Delta_7) + \tfrac{1}{2}(\Delta_3 + \Delta_8)(1 + p_i) + (\Delta_2 + \Delta_6 + \Delta_4 + \Delta_9) p_i$$

or A_k, different from A_i, with probability:

$$\tfrac{1}{2}(\Delta_3 + \Delta_8) p_k + (\Delta_2 + \Delta_6 + \Delta_4 + \Delta_9) p_k.$$

Hence:

$$s_{B|A_iA_i} = \left(\Delta_1 + \Delta_5 + \Delta_7 + \frac{\Delta_3}{2} + \frac{\Delta_8}{2} \right) s_A$$

$$+ \left(\frac{\Delta_3}{2} + \frac{\Delta_8}{2} + \Delta_2 + \Delta_6 + \Delta_4 + \Delta_9 \right) s. \quad (18)$$

These two equations, (17) and (18), are valid in the general case, unlike Eq. (9), which is valid only when A is non-inbred. To verify that this is the case, set $f_A = 0$, whence $\Delta_1 = \Delta_2 = \Delta_3 = \Delta_4 = 0$. Then Eq. (17) becomes:

$$s_{B|A_iA_j} = \left(\Delta_5 + \Delta_7 + \frac{\Delta_8}{2} \right) s_A + \left(\frac{\Delta_8}{2} + \Delta_6 + \Delta_9 \right) s$$

$$= 2\Phi_{AB} s_A + (1 - 2\Phi_{AB}) s.$$

Eq. (18) also reduces to this equation.

It is important to notice that Eqs. (17) and (18) can only be used when we know whether A is homozygous or heterozygous.

2.4. Other Points

2.4.1. Limitations of the coefficient of kinship. Section 2.3 shows the extent to which the measurement of the relationship between two individuals becomes complicated if one or other of them is inbred.

Although the coefficient of kinship does not at first sight appear to have any limitations, we find that it is not, in general, a sufficiently informative measure to enable us to describe the relation between the genic structures of two individuals; when the individual whose genic structure we know is inbred, we cannot find the probability of the presence of an allele in a related individual, given that the allele is present (or given its probability of being present) in the first, in terms of the coefficient of kinship alone.

The advantages of thinking in terms of the coefficients of identity are illustrated by comparing this method with Sewall Wright's (1921) method of "path coefficients". This method gives results which are equivalent to the results of the methods used in Section 2.2 of this chapter, but is much more difficult to follow, especially for relationships of any complexity.

2.4.2. The method of "path coefficients". This method has been the classical way of studying the genetic consequences of relationships between individuals. A very lucid account of the method is given by Li (1955) in his book "*Population Genetics*" (pp. 144–187).

The method of path coefficients is based on the analysis of the chain of "causality" between one or more initial events, or "causes", and a final event, or "effect". The strength of the relation between events is measured by the correlation coefficients between them.

This method enables one to measure the correlation between a zygote and a gamete which it produces, or between a gamete and the zygote to which it will contribute. Using these fundamental steps, the correlation between any two zygotes or gametes in a pedigree can be found. This is simple to do for simple types of relationship, but becomes extremely laborious when a pedigree contains many common ancestors.

Also, the method using coefficients of identity is more closely and explicitly related to the fundamental events which give rise to the correlations. It emphasises the two stages of the analysis: first we find the effects of the relationship on the genetic structures of the individuals we are considering, and then we deduce the resulting correlations between the relatives. This second stage of the analysis will be dealt with in Section 3 of this chapter.

2.4.3. The "reversibility" property of the Hardy-Weinberg equilibrium. Consider a population in Hardy-Weinberg equilibrium with genic structure s. The genic probability structure of a member P of this population is s_P.

The genic structure of the offspring, O, of P is given by:

$$s_{O|P} = \tfrac{1}{2}(s_P + s_M) = \tfrac{1}{2} s_P + \tfrac{1}{2} s \tag{19}$$

since all we know about O's other parent, M, is that he is a member of this population.

Now suppose that we know the genic structure, s_O, of the offspring, rather than that of his parent. A gene transmitted by P to O has a probability of $\tfrac{1}{2}$ of being found in one of O's gametes. Therefore:

$$s_{P|O} = \tfrac{1}{2} s_O + \tfrac{1}{2} s. \tag{20}$$

Eqs. (19) and (20) are particular cases of Eq. (9) for the case of parent and offspring, where $\Phi = \frac{1}{4}$. These equations demonstrate the perfect symmetry between the parent-offspring and offspring-parent relationships, in a population in Hardy-Weinberg equilibrium; these relationships are both characterised by the fact that the individuals concerned have a gene in common at each locus, without regard to which individual is the parent and which the offspring.

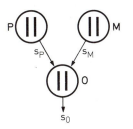

Fig. 6.11. The genetic relationship between parents and offspring

This shows that the hypothesis of Hardy-Weinberg equilibrium is analogous to the concept of "reversibility" in physics. It corresponds to the ideal case when the direction of the determinism of events can be reversed.

3. Resemblance between Relatives

In this section, we shall study the resemblance between related individuals in their phenotype for a quantitative character (e.g. height or head circumference). The closeness of the resemblance between two individuals who are related in a particular way will be measured by the correlation coefficient of the character corresponding to this relationship.

3.1. The Determination of the Covariance between Relatives

Let P be the measure of a character controlled by a locus with n alleles $(A_1, \ldots, A_n)$. We assume that the population is in Hardy-Weinberg equilibrium, with gene frequencies $p_1, \ldots, p_n$, at this locus. We saw in Chapter 5 that, under certain assumptions, P can be analysed into a genetic component, μ_{ij}, which depends on the genotype $(A_i A_j)$, and an environmental component, ε, which is independent of the genotype. Since ε and μ_{ij} are independent, the covariances in terms of P are the same as the covariances in terms of μ.

In Chapter 5, we also showed that μ can be analysed into several elements, according to the equation:

$$\mu_{ij} = \alpha_i + \alpha_j + \delta_{ij}$$

where μ_{ij} is the mean of the measure of the character in $(A_i A_j)$ individuals, α_i and α_j are the additive effects of the A_i and A_j genes, and δ_{ij} is the effect of interaction (dominance deviation) of these two genes.

This analysis is performed arbitrarily, in such a way as to minimise the mean value of δ_{ij}^2 in a population in Hardy-Weinberg equilibrium. If the origin of the measures is chosen so as to make their mean equal to zero, we have:

$$\sum_{ij} p_i p_j \mu_{ij} = 0, \quad \sum_i p_i \alpha_i = 0, \quad \sum_i p_i \delta_{ij} = 0,$$

$$\alpha_i = \sum_j p_j \mu_{ij}, \quad \text{Cov}(\alpha, \delta) = 0. \tag{21}$$

The expectation of the value of an $(A_i A_j)$ individual A is:

$$\mu_A = \alpha_i + \alpha_j + \delta_{ij}.$$

Now suppose that an individual B is related to A, and that the nine condensed coefficients of identity Δ_i are known. Eqs. (11) and (12) give the probabilities of the different genotypes which B can have; each genotype has a corresponding expected value of the character.

We can therefore calculate the expectation of the product $\mu_A \times \mu_B$, and hence the covariance of the μ values between individuals who are related in the way given by the Δ_i coefficients. The calculations are somewhat laborious, and fall into the following stages:

a) A's two genes are identical by descent with probability:

$$f_A = \Delta_1 + \Delta_2 + \Delta_3 + \Delta_4.$$

In this situation, the only identity modes with non-zero probabilities are $S' 1$, $S' 2$, $S' 3$ and $S' 4$. Hence Eq. (12) becomes:

$$\mathscr{S}_{B|A_i \equiv A_i} = \frac{1}{f_A} (\Delta_1 S_{ii} + \Delta_2 S_H + \Delta_3 S_F + \Delta_4 S_P).$$

Still assuming that A's genes are identical, A's value for the measure of the character is:

$$\mu_A^* = 2\alpha_i + \delta_{ii}, \quad \text{with probability } p_i.$$

The expectation of the measure of the character for B, given that A's genotype is $(A_i A_i)$ is:

$$\mu_B^* = \frac{1}{f_A} [\Delta_1 (2\alpha_i + \delta_{ii}) + \Delta_3 \alpha_i + \Delta_2 \mu_H]$$

where μ_H is as defined below.

We can calculate:

μ_F: the mean of the μ values for individuals whose genotypic structure is S_F:

$$\mu_F = \sum_k p_k(\alpha_i + \alpha_k + \delta_{ik})$$

$$= \alpha_i, \quad \text{from (21)}.$$

μ_P: the mean of μ for individuals whose genotypic structure is S_P:

$$\mu_P = \sum_{kl} p_k p_l(\alpha_k + \alpha_l + \delta_{kl}) = 0.$$

μ_H: the mean of μ for individuals whose genotypic structure is S_H:

$$\mu_H = \sum_k p_k(2\alpha_k + \delta_{kk}) = \sum_k p_k \delta_{kk}.$$

b) The probability that A's two genes are not identical is $1 - f_A$. In this situation, identity modes $S'1$, $S'2$, $S'3$ and $S'4$ have probabilities of zero. Eqs. (11) and (12) for the genotypic probability structure of B, given the genotype of A, therefore both become:

$$\mathscr{S}_{B|A_i \neq A_j} = \frac{1}{1 - f_A} \left[\frac{\Delta_5}{2} (S_{ii} + S_{jj}) + \Delta_7 S_{ij} + \Delta_8 S_F + \Delta_6 S_H + \Delta_9 S_P \right].$$

The expectation of the measure of the character when A has the genotype $(A_i A_j)$ is:

$$\mu_A^{**} = \alpha_i + \alpha_j + \delta_{ij}.$$

Assuming this genotype for A, the expectation for B is:

$$\mu_B^{**} = \frac{1}{1 - f_A} \left[\frac{\Delta_5}{2} (2\alpha_i + 2\alpha_j + \delta_{ii} + \delta_{jj}) \right.$$

$$\left. + \Delta_7 (\alpha_i + \alpha_j + \delta_{ij}) + \Delta_8 \frac{(\alpha_i + \alpha_j)}{2} + \Delta_6 \mu_H \right]$$

$$= \frac{1}{1 - f_A} \left[\left(\Delta_5 + \Delta_7 + \frac{\Delta_8}{2} \right) (\alpha_i + \alpha_j) + \frac{\Delta_5}{2} (\delta_{ii} + \delta_{jj}) + \Delta_7 \delta_{ij} + \Delta_6 \mu_H \right].$$

c) We next calculate the mean measures of the character among individuals who have the same relationship as A and B.

For the A individuals, we have:

$$\bar{\mu}_A = f_A \sum_i p_i(2\alpha_i + \delta_{ii}) + (1 - f_A) \sum_{ij} p_i p_j(\alpha_i + \alpha_j + \delta_{ij})$$

$$= f_A \sum_i p_i \delta_{ii}$$

$$= f_A \mu_H.$$

For the B individuals:

$$\bar{\mu}_B = f_A \sum_i p_i \mu_B^* + (1 - f_A) \sum_{ij} p_i p_j \mu_B^{**}$$

$$= \sum_i \left[\Delta_1 p_i (2\alpha_i + \delta_{ii}) + \Delta_3 p_i \alpha_i + \Delta_2 p_i \mu_H + \frac{\Delta_5}{2} \sum_{ij} p_i p_j (\delta_{ii} + \delta_{jj}) \right.$$

$$\left. + \sum_{ij} \left(\Delta_5 + \Delta_7 + \frac{\Delta_8}{2} \right) p_i p_j (\alpha_i + \alpha_j) + \Delta_7 \sum_{ij} p_i p_j \delta_{ij} + \Delta_6 \sum_{ij} p_i p_j \mu_H \right]$$

$$= (\Delta_1 + \Delta_2 + \Delta_5 + \Delta_6) \mu_H$$

$$= f_B \mu_H.$$

d) Finally, we calculate the covariance between μ_A and μ_B for the group of individuals who have the same relationship as A and B:

$$\mathrm{Cov}(\mu_A, \mu_B) = f_A \sum_i p_i (\mu_A^* - \bar{\mu}_A)(\mu_B^* - \bar{\mu}_B) + (1 - f_A) \sum_{ij} p_i p_j (\mu_A^{**} - \bar{\mu}_A)(\mu_B^{**} - \bar{\mu}_B).$$

After making the appropriate substitutions, this becomes:

$$\mathrm{Cov}(\mu_A, \mu_B) = (4\Delta_1 + 2\Delta_3 + 2\Delta_5 + \Delta_8) \sum_i p_i \alpha_i^2$$

$$+ (4\Delta_1 + \Delta_3 + \Delta_5) \sum_i p_i \alpha_i \delta_{ii} + \Delta_1 \sum_i p_i \delta_{ii}^2 \qquad (22)$$

$$+ \Delta_7 \sum_{ij} p_i p_j \delta_{ij}^2 + (\Delta_2 - f_A f_B) \mu_H^2.$$

If we now write:

V_A for the additive genetic variance in the population:

$$V_A = 2 \sum_i p_i \alpha_i^2;$$

V_D for the dominance variance in the population:

$$V_D = \sum_{ij} p_i p_j \delta_{ij}^2;$$

V_H for the dominance variance in the corresponding homozygous population:

$$V_H = \sum_i p_i \delta_{ii}^2;$$

$\mathrm{Cov}_H(A, D)$ for the covariance of additive and dominance effects in the corresponding homozygous population:

$$\mathrm{Cov}_H(A, D) = \sum_i p_i \alpha_i \delta_{ii};$$

D_H for the mean of the dominance deviations in the corresponding homozygous population:

$$D_H = \sum_i p_i \delta_{ii};$$

Φ_3 for the probability that three genes taken at random from the four genes of A and B are identical:

$$\Phi_3 = \Delta_1 + \frac{\Delta_3}{2} + \frac{\Delta_5}{2}$$

and $\Phi_4 = \Delta_1$ for the probability that all four genes of A and B are identical, (22) becomes:

$$\mathrm{Cov}(\mu_A, \mu_B) = 2\,\Phi_{AB}\,V_A + \Delta_7\,V_D + \Phi_4\,V_H + 2(\Phi_3 + \Phi_4)\,\mathrm{Cov}_H(A, D) \\ + (\Delta_2 - f_A\,f_B)\,D_H^2. \tag{23}$$

Gillois (1964) derived this equation in a different form. It enables us to calculate the covariance for a character in the most general case, for any relationship, and whatever the dominance relations.

3.2. Some Particular Relationships

Eq. (23) is complex because it covers all cases. It becomes simpler when the relationship under consideration satisfies certain conditions.

a) A *non-inbred*. Eq. (23) becomes:

$$\mathrm{Cov}(\mu_A\,\mu_B) = 2\,\Phi_{AB}\,V_A + \Delta_7\,V_D + 2\Phi_3\,\mathrm{Cov}_H(A, D).$$

b) B *non-inbred*. The same equation results as for the case when A is non-inbred. This result is obvious from the symmetry of the ways in which A and B affect the covariance.

c) *Both* A *and* B *non-inbred*. In this case $\Phi_3 = 0$, hence:

$$\mathrm{Cov}(\mu_A, \mu_B) = 2\,\Phi_{AB}\,V_A + \Delta_7\,V_D.$$

d) A *and* B *non-inbred, and one of the four parents of* A *and* B *is unrelated to the other three*. In this case $\Delta_7 = 0$, so that:

$$\mathrm{Cov}(\mu_A, \mu_B) = 2\,\Phi_{AB}\,V_A.$$

Here again we find that the equations are considerably simplified by the presence in one of the individuals of a gene which is not identical to any of the other three genes of the two individuals. In this case, the covariance is simply a function of the coefficient of kinship, Φ_{AB}, and of the additive variance.

e) In the simple relationships for which the Δ_i coefficients have already been calculated, the covariances of genotypic value are as follows (if we assume that μ_{ij} is independent of ε, as at the beginning of Section 3.1, then these are also the phenotypic covariances between these types of relatives):

Relations	Covariance
Parent-offspring:	$\frac{1}{2} V_A$
Half-sibs:	$\frac{1}{4} V_A$
Full sibs:	$\frac{1}{2} V_A + \frac{1}{4} V_D$
First cousins:	$\frac{1}{8} V_A$
Double first cousins:	$\frac{1}{4} V_A + \frac{1}{16} V_D$
Second cousins:	$\frac{1}{32} V_A$
Uncle-nephew:	$\frac{1}{4} V_A$
Offspring of sib-matings:	$\frac{3}{4} V_A + \frac{7}{32} V_D + \frac{1}{16} V_H + \frac{1}{2} \mathrm{Cov}_H(A, D) - \frac{1}{32} D_H^2.$

3.3. The Case of a Locus with Two Alleles

In Chapter 5 (Section 1.4), we saw that, if a character is controlled by two alleles A_1 and A_2, the additive and dominance effects can be expressed in terms of:

p and q, the gene frequencies,

$\alpha = p(i-j) + q(j-k)$, the average effect of the gene substitution, and

$d = j - \dfrac{i+k}{2}$, the difference between the value of the character in the heterozygote and the mean of the values in the homozygotes.

In Section 2.3 of Chapter 5, we saw that the variances due to additive effects and dominance effects, in the two-locus case, are:

$$V_A = 2pq\alpha^2 \quad \text{and} \quad V_D = 4p^2 q^2 d^2.$$

The other parameters in Eq. (23) can be found, as follows:

$$\begin{aligned}
\mathrm{Cov}_H(A, D) &= p\,\alpha_1\,\delta_{11} + q\,\alpha_2\,\delta_{22} \\
&= p(q\alpha)(-2q^2 d) + q(-p\alpha)(-2p^2 d) \\
&= 2pq(p-q)\,\alpha d, \\
V_H &= p\,\delta_{11}^2 + q\,\delta_{22}^2 \\
&= p\,4q^4 d^2 + q\,4p^4 d^2 \\
&= 4pq(p^3 + q^3) d^2, \\
D_H &= p\,\delta_{11} + q\,\delta_{22} = -2pq^2 d - 2p^2 q d \\
&= -2pq d.
\end{aligned}$$

The last result shows that $D_H^2 = V_D$, but this is not generally true for more than two alleles.

Finally, Eq. (23) takes the form:

$$
\begin{aligned}
\mathrm{Cov}\,(\mu_A, \mu_B) &= 4\,\Phi_{AB}\,p\,q\,\alpha^2 + 4(\Delta_7 + \Delta_2 - f_A\,f_B)\,p^2\,q^2\,d^2 \\
&\quad + 4pq(p^3 + q^3)\,\Phi_4\,d^2 + 8pq(p-q)(\Phi_3 + \Phi_4)\,\alpha\,d \\
&= 4pq\,[\,\Phi_{AB}\,\alpha^2 + (\Delta_7 + \Delta_2 - f_A\,f_B)\,p\,q\,d^2 + (p^3 + q^3)\,\Phi_4\,d^2 \\
&\quad + 2(p-q)(\Phi_3 + \Phi_4)\,\alpha\,d\,]. \tag{24}
\end{aligned}
$$

Special cases

1. If there is no dominance, i.e. if $d = 0$, we have:

$$
\mathrm{Cov}\,(\mu_A, \mu_B) = 4\,p\,q\,\Phi_{AB}\,\alpha^2.
$$

In this case, the coefficient of kinship is the only measure of the relationship which enters into the covariance.

2. If there is total dominance, we have seen that $\alpha = p(i - k)$ and $d = -\dfrac{i-k}{2}$, hence:

$$
\alpha = -2\,p\,d
$$

and

$$
V_A = 8p^3\,q\,d^2, \qquad V_D = 4p^2\,q^2\,d^2, \qquad V_H = 4pq(p^3 + q^3)\,d^2,
$$

$$
\mathrm{Cov}_H(A, D) = -4p^2\,q(p-q)\,d^2.
$$

These terms of the covariance depend on the gene frequencies. For example, if $p = 0.9$, $q = 0.1$, i.e. the dominant gene is rare, we obtain:

$$
V_A = 0.583\,d^2, \qquad V_D = 0.032\,d^2, \qquad V_H = 0.263\,d^2,
$$

$$
\mathrm{Cov}_H(A, D) = -0.259\,d^2.
$$

The most important term is thus V_A, while the dominance variance is small.

If the two alleles have the same frequency, $p = 0.5$, $q = 0.5$, we obtain:

$$
V_A = 0.5\,d^2, \qquad V_D = 0.25\,d^2, \qquad V_H = 0.25\,d^2,
$$

$$
\mathrm{Cov}_H(A, D) = 0.
$$

In this case the covariance of additive and dominance effects in the corresponding homozygous population is zero, while the terms V_H and V_D are equal in value.

As a last example, suppose $p=0.1$, $q=0.9$, i.e. the dominant gene is common. Then:

$$V_A = 0.007\,d^2, \qquad V_D = 0.032\,d^2, \qquad V_H = 0.263\,d^2,$$

$$\mathrm{Cov}_H(A, D) = 0.288\,d^2.$$

In this case, the most important terms are V_H and $\mathrm{Cov}_H(A, D)$.

These examples illustrate the effect of the genic structure of the population on the covariance of a character between relatives.

3.4. The Interpretation of Observed Correlations between Relatives

For the purpose of comparing observed resemblances between relatives with their expectations on the basis of the theory we have just derived, it is convenient to express them in terms of correlation coefficients, rather than covariances. The correlation coefficient between two variates X and Y is defined as:

$$r_{XY} = \frac{\mathrm{Cov}(X, Y)}{\sqrt{V_X\,V_Y}}. \tag{25}$$

It is thus a dimensionless number, and can take values between $+1$ and -1, depending on the strength of the relation between the two variates. The correlation coefficient has the advantage that it is independent of the scale on which X and Y are measured.

When we are considering the correlation coefficient between relatives of a particular kind, e.g. parents and offspring, V_X and V_Y are obviously the same (V_P). Eq. (25) therefore simplifies to:

$$r_{AB} = \frac{\mathrm{Cov}(A, B)}{V_P}.$$

The theoretical values of the correlation coefficients for various degrees of relationship can therefore be found by dividing the covariance values shown on p. 136 by V_P.

In Chapter 5, we referred to the concept of heritability, which we defined as the proportion of the total phenotypic variance which is due to additive gene effects:

$$h^2 = \frac{V_A}{V_P}.$$

It is obvious from the table of covariances on p. 136 that h^2 is the sole contributor to the correlations between non-inbred individuals who have at least one parent who is unrelated to the three other parents (parent-offspring, half-sibs, etc.). For example, the correlation coefficient for parents and offspring is equal to $h^2/2$.

For other types of relationship, the proportion of the total variance that is due to dominance effects enters into the correlation. For example, the correlation for full sibs is equal to $(\frac{1}{2} V_A + \frac{1}{4} V_D)/V_P$.

These considerations suggest that the theory of the inheritance of quantitative characters based on discrete, Mendelian genes could be tested in man by measuring the phenotypic correlations between sets of individuals related in different ways, and comparing the correlations with those expected from the table of covariances. We should also be able to use such observed correlations to estimate the proportions of phenotypic variance for a character that are due to environmental, additive genetic, and dominance effects.

Unfortunately, there are several theoretical and practical difficulties. First, many quantitative characters in man, such as height and I.Q., show large correlations between husband and wife. Mating is therefore not at random with respect to the genes affecting these characters, and so one of the assumptions on which our analysis is based is violated. Fisher (1918) and Wright (1921) have considered this problem, and given approximate methods of correcting for assortative mating, and obtaining estimates of heritability when mating is not at random for the phenotypes in question. An excellent account of this is contained in Crow and Kimura (1970). It is, however, beyond the scope of this book.

Even more serious is the undoubted existence of environmental correlations between related individuals for many characters, such as height or I.Q., which are affected by social class and economic status. For such characters, we cannot assume that the observed phenotypic correlations between related individuals are due solely to heredity. This problem is an extremely serious one, and greatly limits the use of the theory we have derived in estimating components of variance in man.

To use the formulae we have derived, we therefore have to utilise data on quantitative characters which are not subject to correlation between mates, and which are not likely to be liable to environmental sources of correlation. Such a character is dermal ridge count. Holt (1961) found a correlation coefficient for the counts in parents and offspring of 0.48, and a correlation of 0.50 for full sibs. The correlation between mates was only 0.05, which is not significantly different from zero. These data suggest that h^2 for dermal ridge count is close to 1; there seems to be little effect of dominance or environmental variation on this character.

In experimental animals and plants, it is possible to eliminate the problems of non-random mating and environmental correlations by suitable experimental design; more reliable estimates of the components of variation can therefore be obtained. This subject is reviewed by Falconer (1960). An example of the type of analysis which can be done is provided by the work of Clayton, Morris and Robertson (1957) on abdominal bristle number in *Drosophila melanogaster*. These workers found a correlation of 0.25 between parents and offspring, 0.12 between half-sibs, and 0.26 between full sibs. This is consistent with a value of h^2 about 0.50, and no dominance variation. The degree of hereditary determination of this character is therefore much lower than for dermal ridge count in man.

Chapter 7

Overlapping Generations

One of the conditions which we had to assume in order to derive the Hardy-Weinberg equilibrium was that the members of the population belong to distinct generations, and that individuals from different generations do not breed with one another. For such a population, the set of ancestors of the population's members has no element in common with the set of the population's descendants.

In reality, many populations are not like this at all; species which do behave like this include the annual plants, which survive the winter as seeds, and many species of insects, which overwinter as immature stages, so that individuals which breed together in the next year are necessarily members of the same generation. For human populations, however, the idea of generations, though meaningful within a family, has no meaning in the population as a whole; there is a continuous range of ages, and individuals of different ages can breed together. In many animal and plant species there are annual breeding-seasons, with individuals surviving over several years to participate in more than one such season. It is therefore important to find out whether these differences from the model we have assumed up to now have any significant effects on the results we have derived. In the first section of this chapter, we shall be concerned with the basic concepts formulated by demographers to deal with the changes in the size and age-composition of populations with overlapping generations; in the second section, these concepts will be applied to the problem of the equilibrium genetic structure of populations with overlapping generations. Population size will be assumed throughout to be so large that chance fluctuations may be ignored. Much of the material discussed in this chapter will be useful in treating natural selection, in Chapter 10. The discussion will largely be in terms of human populations, but we shall refer to other species where this is appropriate.

1. The Demographic Description of a Population

When we are studying changes in the genetic structure of a population, the only part of the life-cycle which is important is the period up to the end of reproductive life; although longevity beyond this period is of importance to the individual, it has no rôle in determining the genetic make up of the population in the future. In man, the end of the reproductive period is much more clearly defined for women than for men; a woman becomes incapable of reproduction after the age of about 45–50, but there is no such sharp cut-off in men. Furthermore, a woman is able to have far fewer children than a man, so that the birth-rate of a population is determined largely by the women. Because of these two facts, we shall study changes in the demographic structure of populations largely in terms of women only, without taking the number of men into account. Specifically, we shall assume that the average number of births to women of a given age, at a particular time, is independent of the number and age-distribution of the men present in the population. It is probable that this assumption is also reasonable for many species other than man.

Let us assume that a woman is capable of reproduction only between the ages of 15 and 45. A woman's number of offspring depends on two factors:

1. Her capacity to survive up to the age of 15, and throughout the reproductive period.

2. Her capacity to conceive a child during the reproductive period. This in turn depends on whether she marries, becomes divorced or widowed, etc., and on her physiological characteristics (fertility, length of the sterile period after each birth, age of menopause etc.).

We shall now define some of the demographic parameters that can be used to characterise a population, and will give some techniques for estimating them.

1.1. Demographic Parameters

1.1.1. The life-table. Let us consider a group of 1000 women who were all born in the same year (year 0). Because of the death of some of them, the size of this "cohort" will steadily diminish as they grow older. Let us write D_x for the number of the group who die during the year x. The number of survivors at the *beginning* of the x-th year is therefore given by:

$$S_x = 1000 - (D_0 + D_1 + \cdots + D_{x-1}).$$

The ratio of the number of deaths D_x during the x-th year to the number of women S_x who survived up to the beginning of this year is called the death rate at age x:

$$d_x = \frac{D_x}{S_x}.$$

The death rate at age x is therefore the probability that a woman who has survived up to age x dies between her x-th and $x+1$th birthdays,

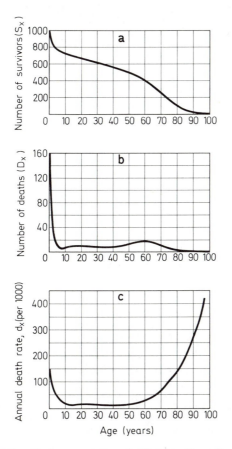

Fig. 7.1. Graphical representation of a life-table. (From Pressat, 1966)

while the D_x values are proportional to the *a priori* probabilities (calculated at birth) of dying in this interval. We shall also need to define l_x, the probability of surviving from year 0 to the *beginning* of year x:

$$l_x = \frac{S_x}{1000}.$$

The values of l_x, D_x and d_x are usually given as a table for a series of ages, starting with age 0, and going up to 100. This is called the "life table" of the group of women in question.

A smooth curve is often drawn through the points representing l_x, D_x and d_x, and the resulting graphs are usually similar to the ones shown in Fig. 7.1.

Table 7.1. A condensed life-table

Age (x)	Frenchwomen born in about 1830			Frenchwomen born in about 1900		
	S_x	$_5D_x$	$_5d_x$	S_x	$_5D_x$	$_5d_x$
0	1000	268	268	1000	180	180
5	732	37	54	820	12	15
10	695	23	35	808	12	15
15	672	27	39	796	20	25
20	645	29	44	776	20	25
25	616	29	46	756	21	27
30	587	29	48	737	21	28
35	558	30	50	716	21	29
40	528	30	54	695	22	30
45	498	31	61	673	13	20

For the purposes of genetics, a year-by-year analysis is usually too detailed, since other data are never available in so precise a form. A condensed table is therefore often used, in which the number of survivors of each five-year period are given. We shall write:

$$_5D_x = S_x - S_{x+5}$$

for the number of women who die between their x-th and $x+5$th birthdays. The death rates for these periods are written as $_5d_x$; they are the ratios of $_5D_x$ to S_x. In addition, for genetic puposes we can ignore women aged over 45, and we then obtain a condensed life table like Table 7.1, which gives the mortality parameters for two cohorts of French women; one cohort consists of women born in about 1830, and the other of women born in about 1900.

Finally, we must distinguish between "vertical" tables of this sort, which follow the changes in time of a cohort of women who are all the same age, and "horizontal" tables, which give the mortality parameters D'_x and d'_x corresponding to deaths of women at different ages, who belong to different cohorts, during a given year.

1.1.2. The fecundity table. Just as one can note the successive deaths of women who were all born at the same time, we can also follow the cohort and note the numbers of children they produce. We can thus draw up a "fecundity table", giving the number $_5b_x$ of births in each five-year period, and also the fecundity $_5F_x = \dfrac{_5b_x}{S_x}$, where S_x is the number of women who have survived up to age x. The fecundity represents the expectation of the number of offspring during the coming five years, of a woman who has reached age x.

Table 7.2. A condensed fecundity table

Age (x)	Frenchwomen born in about 1830			Frenchwomen born in about 1900		
	S_x	$_5b_x$	$_5F_x$	S_x	$_5b_x$	$_5F_x$
15	672	91	135	796	56	70
20	645	464	720	776	508	655
25	616	589	955	756	499	660
30	587	475	810	737	313	425
35	558	328	565	716	165	230
40	528	153	290	695	49	70
Total		2 100	3 475		1 590	2 110

Table 7.2 shows the fecundity tables for the two cohorts of women which we considered before – Frenchwomen born in 1830 and 1900; Table 7.2 gives the values of $_5b_x$ and $_5F_x$, and also the total numbers of live-born children of these cohorts of women ($\sum _5b_x$) and the total numbers of children which they would have had if none of them had died, $\sum _5F_x$ (i.e. if each woman had had the same number of children as the average number for women who survived up to the age of 45).

Tables 7.1 and 7.2 show how greatly the population has changed with respect to both fecundity and mortality. For the women born in 1830, the probability of surviving up to the age of 40 was barely greater than 0.5, but was nearly 0.7 for the women born in 1900. At the same time, the mean number of children fell from approximately 2 to 1.6 per woman; the difference is even greater for women who survived to the end of the reproductive period – the mean number of children of such women fell from about 3.5 to 2.1.

Such rapid changes are by no means the rule in human populations; there must have been many periods during which a population's demographic parameters changed very slowly. In Section 2 of this chapter, we will show that a population with overlapping generations will come to an equilibrium with constant gene frequencies and Hardy-Weinberg proportions of the genotypes. This equilibrium situation will be valid whether the population's demographic parameters remain constant, or change with time, provided that there is no selection, i.e. no differences between the demographic parameters of different genotypes. However, it is much simpler to prove the result for the case of constant demographic parameters than when they are changing, so we shall assume from now onwards that the age-specific mortality rates and fecundities are constant in time.

1.2. The Future Demographic Structure of a Population

1.2.1. Age-classes, cycles and age-structure. The central problem of the science of demography is to predict the future size, age-composition and rate of growth of a population, given its present size and age-composition, and knowledge of the life- and fecundity tables. In order to tackle this problem, it is convenient to define the population's demographic parameters in terms of "age-classes" rather than years. Let age-class 1 be the ages 0–4, age-class 2, 5–9 years, etc. Let $l(x)$ be the probability of survival from age-class 1 to the *beginning* of age-class x. Similarly, we define the fecundity for age-class 1, $F(1)$, as $_5F_0$, $F(2) = _5F_5$, etc.

In organisms with discrete breeding-seasons, but which can survive for more than one breeding-season, the concept of age-classes arises naturally; in continuously-reproducing organisms, such as man, it is an approximation, but the approximation can be made extremely good by taking sufficiently short intervals to define the age-classes.

We shall now investigate the future age-structure of a population, given its life- and fecundity tables. First we note that the age-specific fecundities $F(x)$ refer to all births, regardless of the sex of the offspring, whereas we want to consider only the female population. In man, and many other organisms, the sex ratio at birth varies very little either in time or place, so that we can assume that the number of males born per 100 females is constant. In man, this number is always found to be close to 105. In other words, the proportion of female births, c, in man can be taken as 0.488 and assumed to be constant.

Changes in the female population therefore depend on the set of survival probabilities $l(1)$, $l(2)$, ..., $l(x)$ etc., and on the set of female fecundities $f(1), f(2), ..., f(x)$ etc., which are equal to $F(1) \times c, F(2) \times c$ etc.

Finally, let us assume that females start reproducing in age-class α, and cease reproducing after age-class ω. For $x < \alpha$ or $x > \omega$, $f(x) = 0$; for $\alpha \leq x \leq \omega$, $f(x) > 0$. In human populations, we may take α to be 4 and ω to be 9, corresponding to reproduction starting at age 15, and ceasing after age 45.

An interval of time equal in length to the period which we have chosen for the age-classes, will be called a "cycle". Instead of following changes in the population in terms of generations, or years, it is convenient to use cycles as the units of time.

We can characterise the state of the population in the t-th cycle by the number of females present in each of the age-classes 1, 2, ..., ω at the beginning of the cycle. Let this number be $N_f(t, x)$ for the x-th age-class. If we ignore mortality within age-classes, $N_f(t, x)$ is equal to the number of individuals who entered age-class x during cycle $t - 1$. If we write $B_f(t)$

for the number of females born into the population during cycle t, we have, for $x = 1, 2, \ldots, \omega$ and for $t \geq x$:

$$N_f(t, x) = B_f(t - x)\, l(x). \tag{1}$$

For $t \geq \omega$, we can therefore determine the size and age-structure of the population during any cycle, in terms of the numbers of females belonging to the various age-classes at the beginning of the cycle; all we need in order to do this is the set of successive values of $B_f(t)$ and the set of $l(x)$ values for $x = 1 \ldots \omega$. We may accordingly concentrate our attention on deriving an expression for $B_f(t)$ in terms of t.

Let us assume that the first cycle for which we have information is $t = 0$, and that we want to calculate the value of $B_f(t)$ for some future cycle t, such that $t \geq \omega$. The oldest women giving birth to children during cycle t are members of age-class ω; they were born during cycle $t - \omega$. The youngest are in age-class α; they were born in cycle $t - \alpha$. The number $B_f(t)$ therefore depends only on the numbers of births of females during the cycles $t - \omega, \ldots, t - x, \ldots, t - \alpha$. If we ignore variation in fecundity between individuals of different ages, but who belong to the same age-class, then the $B_f(t - x)\, l(x)$ individuals aged x at the beginning of cycle t will produce $c\, F(x)$ daughters during that cycle. The total number of female births in cycle t is therefore:

$$B_f(t) = \sum_{x = \alpha}^{\omega} c\, B_f(t - x)\, l(x)\, F(x) = \sum_{x = \alpha}^{\omega} B_f(t - x)\, l(x)\, f(x). \tag{2}$$

Referring to Appendix A, Eq. (A.16), we see that Eq. (2) is a "renewal equation". We can equate the $l(1) f(1) \ldots l(\omega) f(\omega)$ with $a_1 \ldots a_m$, and the $B_f(0) \ldots B_f(t)$ with $u_0 \ldots u_n$. It follows from Eqs. (A.11) and (A.12) that:

$$B_f(t) = \sum_{j=0}^{\omega-1} A_j\, \lambda_j^t \tag{3}$$

where the λ_j are the roots of the characteristic equation:

$$\sum_{x = \alpha}^{\omega} \lambda^{-x}\, l(x)\, f(x) = 1 \tag{4}$$

and the A_j are determined by the first ω values $B_f(0) \ldots B_f(\omega - 1)$.

From Eq. (A.15) and Eq. (4), it follows that, for large t:

$$B_f(t) = A_0\, \lambda_0^t \tag{5}$$

where λ_0 is the root of Eq. (4) which has the largest modulus.

1.2.2. The roots of the characteristic equation. Eq. (4) has as many roots as the number of age-classes, and the problem arises as to which of them has the largest modulus. This root may be found as follows. Consider all real, positive values of λ. Then λ^{-x} is a decreasing function of λ. Since $l(x)f(x) > 0$ for all x between α and ω, there must be a value of λ, $\hat{\lambda}$ say, such that Eq. (4) is satisfied. For values of λ less than $\hat{\lambda}$, the left-hand-side of Eq. (4) will be greater than 1; for values greater than $\hat{\lambda}$, it will be less than 1. $\hat{\lambda}$ is therefore the single real, positive root of Eq. (4). This root also has the largest modulus of all the roots. We can see this as follows. Let the root λ_j be written in the form:

$$\lambda_j = m_j e^{i\Theta_j} = m_j(\cos\Theta_j + i\sin\Theta_j)$$

where m_j is the modulus of λ_j.

Substituting into Eq. (4) we have:

$$\sum_{x=\alpha}^{\omega} m_j^{-x}[\cos(-x\Theta_j) + i\sin(-x\Theta_j)]\, l(x)f(x) = 1. \tag{6}$$

Equating the real parts of the two sides of this equation, we have:

$$\sum_{x=\alpha}^{\omega} m_j^{-x}[\cos(-x\Theta_j)]\, l(x)f(x) = 1. \tag{7}$$

In the case when $\lambda_j = \hat{\lambda}$, Θ_j is zero, and $m_j = \hat{\lambda}$. For the negative real roots, $\Theta_j = n\pi$, where n is an integer, and $\cos(-x\Theta_j) = (-1)^x$. For the complex roots, $|\cos(-x\Theta_j)| < 1$. Hence:

$$\left|\sum_{x=\alpha}^{\omega} m_j^{-x}\cos(-x\Theta_j)\, l(x)f(x)\right| \le \sum_{x=\alpha}^{\omega} m_j^{-x}l(x)f(x). \tag{8}$$

Clearly, this is an equality only when $\lambda_j = \hat{\lambda}$. In this case, the right-hand-side of Eq. (8) is equal to 1. For the other roots, we must therefore have $\sum_{x=\alpha}^{\omega} m_j^{-x} l(x)f(x) > 1$. This is impossible for any root for which $m_j \ge \hat{\lambda}$.

This completes the proof that λ_0, the root of Eq. (4) which has the largest modulus, is the single real, positive root $\hat{\lambda}$.

1.2.3. The intrinsic rate of natural increase and equilibrium age-structure of a population. Eq. (5) shows that when t is sufficiently large the number of births during cycle t is such that:

$$B_f(t) = \lambda_0 B_f(t-1) = \lambda_0^n B_f(t-n). \tag{9}$$

The population thus settles down to a state of constant growth at a rate $\hat{r}$, such that:

$$\frac{B_f(t+1) - B_f(t)}{B_f(t)} = \hat{r}$$

where $\hat{r} = \lambda_0 - 1$, and $\hat{r}$ is the real root > -1 of the equation:

$$\sum_{x=\alpha}^{\omega} (1+r)^{-x} l(x) f(x) = 1. \tag{10}$$

The parameter $\hat{r}$ is called the "intrinsic rate of natural increase" of the population.

When the population has attained its equilibrium rate of growth, $\hat{r}$, it is clear that its age-composition must be constant, since, from Eqs. (1) and (9) we have:

$$N_f(t, x) = B_f(t-x) l(x) = \hat{\lambda}^{-x} B_f(t) l(x)$$

or

$$\frac{N_f(t, x)}{B_f(t)} = (1+\hat{r})^{-x} l(x). \tag{11}$$

The relative proportions of females in the different age-classes will therefore remain constant once the equilibrium rate of growth has been attained.

In summary, we see that *a population whose age-specific mortality and fecundity parameters remain constant tends towards a "stable age-structure", such that the sizes $N_f(x)$ of the age-classes are proportional to $(1+\hat{r})^{-x} l(x)$ where $\hat{r}$, the intrinsic rate of natural increase, is the rate of change in the population's size when it has reached its equilibrium age-structure.*

This fundamental result was derived by Lotka (1922). It is illustrated in Table 7.3, which shows that the population moves quite rapidly

Table 7.3. The approach towards stable age-structure, with the demographic parameters of the two cohorts of Frenchwomen

Age-class	Initial age structure	1830 cohort				1900 cohort			
		1st cycle	10th cycle	20th cycle	30th cycle	1st cycle	10th cycle	20th cycle	30th cycle
1	150	170	167	168	168	107	123	118	120
2	150	111	122	122	122	128	98	103	102
3	100	143	116	116	115	154	99	107	105
4	100	97	114	111	111	104	108	110	107
5	100	97	109	106	107	102	119	110	109
6	100	96	103	101	102	102	128	108	112
7	100	96	97	97	97	101	120	109	114
8	100	95	88	92	92	101	103	114	115
9	100	95	86	87	87	101	102	121	115
Total size	1000	995.1	998.7	1029.6	1053.5	958.8	688.5	463.0	306.1

towards its stable state. The initial structure was purposely chosen to be very different from the stable age-structure. The resulting total population sizes at the stable state are also shown in Table 7.3. We see that the two cohorts have different equilibrium age-structures and population sizes.

1. *Stable age-structure:* The "1830" population leads to a stable population with a higher proportion of young individuals than the "1900" population; the proportion of individuals aged less than 15 is 40.5 % in the first case, compared with 32.7 % in the second.

2. *Total size:* The "1830" population slowly increases in size, by 5.4 % in 30 cycles (i.e. 150 years), whereas the "1900" population has decreased to 30.6 % of its initial size after 30 cycles. The first population is thus expanding slowly, but the second is rapidly declining in size.

The change in the total population size during the course of a five-year cycle will itself change as the age-structure of the population changes. For example, the "1830" population suffers an initial decline in size (due to the low initial proportion of women of reproductive age) but then begins to expand. A population's growth rate at any instant is therefore determined not only by its mortality and fecundity parameters, but also by its age-structure.

1.3. The Intrinsic Rate of Natural Increase

In order for a population to remain in a state of expansion, in the long term, its mortality and fecundity parameters must be such that an initial group of 1000 females will produce at least 1000 daughters by the end of their reproductive life-span. With the notation we have used above, this will happen when R_0 is greater than one, where

$$R_0 = \sum_{x=\alpha}^{\omega} l(x) f(x). \tag{12}$$

R_0 is called the "net reproduction rate"; R_0 is clearly a function of the mortality and fecundity parameters.

For the "1830" and "1900" cohorts of women, we have:

$$R_0(1830) = 0.488 \times 2.100 = 1.025,$$

$$R_0(1900) = 0.488 \times 1.590 = 0.776.$$

But the replacement of 1000 females by 1000 new ones will take a certain length of time, which depends on the mean age of mothers at the birth of their daughters. If two populations have the same net reproduction rate R_0, but the females in one population reproduce earlier than in the other, the size of the first population will change faster than that of the second.

The mean age of mothers at the time of birth of their children is calculated from the formula:

$$T = \frac{\sum_x x\, l(x)\, f(x)}{\sum_x l(x)\, f(x)}. \tag{13}$$

Assuming that all births to women in a given age-class occur at the median age of the class, the values (in years, rather than age-class units) for our two cohorts of Frenchwomen are:

$$T(1830) = \frac{27.95}{1.025} + 2.5 = 31.01$$

and

$$T(1900) = \frac{19.82}{0.776} + 2.5 = 28.04.$$

We can use the parameters R_0 and T to obtain an approximate estimate of $\hat{r}$. If $\hat{r}$ is small, we can write:

$$(1 + \hat{r})^{-x} = 1 - x\hat{r}$$

to a good approximation. Eq. (10) then becomes:

$$\sum_x (1 - \hat{r}x)\, l(x)\, f(x) = 1$$

which gives

$$\hat{r} = \frac{\sum_x l(x)\, f(x) - 1}{\sum_x x\, l(x)\, f(x)}$$

or, from Eqs. (12 and (13),

$$\hat{r} = \frac{R_0 - 1}{R_0\, T}. \tag{14}$$

For the two cohorts of Frenchwomen, we find:

$$\hat{r}(1830) = \frac{1.025 - 1}{31.8} = +0.08\,\% \text{ per year},$$

$$\hat{r}(1900) = \frac{0.776 - 1}{21.75} = -1.3\,\% \text{ per year}.$$

1.4. The Male Population

So far, we have ignored the male population. For the purposes of genetics, it is necessary to take males into account, since they contribute just as much to the genetic make-up of the population as females. It is clear that we can construct a life-table for males just as well as for females. We can therefore determine the probability of survival, $l^*(x)$, from age-class 1 to the beginning of age-class x for a male by just the same methods as for females. However, the treatment of male fecundity is not so easy. We have assumed that the number of male and female births to the population is determined solely by the reproductive potential of females and by the sex-ratio at birth, which is assumed to be constant. As the age-composition of the population of females changes, it is clear that the expected number of offspring of a male entering a given age-class will change. We cannot, therefore, regard male fecundity as independent of time, except when the age-structure of the female population is stable.

The problem of finding an expression for male fecundity can be examined in the following way. We shall only consider a population which has reached its stable age-structure. Let us assume that the sex-ratio is constant; then the number of male births during each cycle is related to the total number of births by:

$$B_m(t) = (1 - c) B(t). \tag{15}$$

The number $N_m(x, t)$ of males belonging to age-class x at the beginning of cycle t is related to the number of male births during cycle $(t - x)$ by:

$$N_m(t, x) = B_m(t - x) l^*(x).$$

Using Eqs. (9) and (15), this gives:

$$\frac{N_m(t, x)}{B_m(t)} = (1 + \hat{r})^{-x} l^*(x). \tag{16}$$

Thus the male population, as well as the female population, tends towards a constant age-structure, which is a function only of $\hat{r}$ (which depends on the fecundity of the females) and the probabilities of survival $l^*(x)$ of the males.

Just as for females, we can also define a fertile period for males, between ages α^* and ω^*, and can assign age-specific fecundities $F^*(x)$ and $f^*(x) = (1 - c) F^*(x)$. The number of births during a cycle t is:

$$B(t) = \sum_{x = \alpha^*}^{\omega^*} N_m(t, x) F^*(x)$$

or, when stable age-structure has been attained:

$$B(t) = \sum_{x=\alpha^*}^{\omega^*} B_m(t)(1+\hat{r})^{-x} l^*(x) F^*(x)$$

which gives:

$$\sum_{x=\alpha^*}^{\omega^*} (1+\hat{r})^{-x} l^*(x) f^*(x) = 1. \tag{17}$$

The parameters α, ω and $F(x)$ for the females, and the parameters α^*, ω^* and $F^*(x)$ for the males, are not independent since the two equations, (10) for females and (17) for males, must have the same solution $\hat{r}$. Let α_m be the smaller of the two numbers α and α^*, and ω_M the larger of ω and ω^*. Then, adding Eqs. (10) and (17), we obtain:

$$\sum_{x=\alpha_m}^{\omega_M} (1+\hat{r})^{-x} [l(x) f(x) + l^*(x) f^*(x)] = 2 \tag{18}$$

where $f(x)$ is zero for those age-classes in the interval $(\alpha_m - \omega_M)$ which do not belong to the interval $(\alpha - \omega)$, and similarly for $f^*(x)$.

2. The Equilibrium Genetic Structure of a Population with Overlapping Generations

Up to now, we have been entirely concerned with the demographic characteristics of populations. We are now in a position to use the knowledge we have gained to investigate the equilibrium genetic structure of populations with overlapping generations, assuming that all the other conditions for Hardy-Weinberg equilibrium are met; now, however, instead of the condition that individuals from different generations do not breed together, we assume that matings can take place between individuals of different ages.

2.1. Genic and Genotypic Structures of Populations with Overlapping Generations

Let us consider a continuously-breeding population of the type studied in the preceding section of this chapter, and assume that, as before, we have arbitrarily broken it up into a set of age-classes $1, 2, \ldots, \omega$. In general, we may expect the genotypic structures of males and females to be different, and to vary between age-classes. The magnitude of these differences will obviously be greatly influenced by the initial state of the population. Let us consider a single, autosomal locus with n alleles $A_1, \ldots, A_n$. Then we can represent the frequencies of the various genotypes $(A_i A_j)$ among females belonging to the x-th age-class in cycle t by

$S_f(t, x)$ and the analogous frequencies among males as $S_m(t, x)$. These quantities will be referred to as the genotypic structures for age-class x at time t, for females and males respectively.

We can similarly define the genic structures of females and males for age-class x at time t as:

$$s_f(t, x) = \text{cont } S_f(t, x)$$

$$s_m(t, x) = \text{cont } S_m(t, x). \tag{19}$$

Our problem is to determine the course of evolution of these genetic structures in a random-mating population with overlapping generations. To do this, we introduce a further concept. For any cycle, t, we can define the frequencies of the various alleles among the genes derived from female parents in the new individuals who were born into the population during the cycle. If there are $B(t)$ births during cycle t, the total number of such maternally-derived genes is $B(t)$, since each individual has one gene at each locus from his female parent. If $B_i(t)$ of these maternally-derived genes are allele A_i, the desired frequency of A_i is:

$$p_i(t) = \frac{B_i(t)}{B(t)}.$$

We can represent the genic structure of maternally-derived genes entering the population in cycle t as:

$$s_f(t) = [p_1(t), \ldots, p_i(t), \ldots, p_n(t)].$$

Similarly we can consider the number of new paternally-derived copies of the i-th allele entering the population in cycle t. If we write $B_i^*(t)$ for this number, the corresponding frequency among paternally-derived genes is $p_i^*(t) = \dfrac{B_i^*(t)}{B(t)}$. The set of these frequencies is the genic structure of paternally-derived genes entering the population in cycle t:

$$s_m(t) = [p_1^*(t), \ldots, p_n^*(t)].$$

It is obvious that $s_m(t)$ and $s_f(t)$ need not be equal to one another, in any given cycle, since males and females may reproduce at different ages, so that, unless the genic structures are the same for all age-classes, the males will contribute a different array of genes to the new individuals from the array which the females contribute.

Since each individual receives one gene from his father and one from his mother, the genic structure of individuals born during cycle t is:

$$s(t) = \tfrac{1}{2} [s_f(t) + s_m(t)]. \tag{20}$$

Note that this genic structure is the same for male and female births.

Clearly, if we consider times greater than or equal to both ω and ω^*, so that no individuals of reproductive age in the cycle in question were alive at the beginning of the initial cycle, $t=0$, we have, assuming no selection:

$$s_f(t, x) = s(t - x)$$
$$s_m(t, x) = s(t - x). \tag{21}$$

We therefore need to investigate the behaviour of $s_f(t)$ and $s_m(t)$ since the values of $s_f(t, x)$ and $s_m(t, x)$ can always be deduced from these.

2.2. The Evolution of the Genetic Structure of a Population

If we restrict our attention to values of t greater than or equal to both ω and ω^*, so that all the new individuals born into the population come from parents who were born in or after cycle 0, we can proceed as follows. During cycle $t - x$, the number of female births was $B_f(t - x) = cB(t - x)$; the genic structure among these new-born individuals was $s(t - x)$. During cycle t, these females will reproduce, and will contribute the array of genes $B_f(t - x) s(t - x) l(x) F(x)$ to the new-born individuals. The total number of maternally-derived genes contributed to the population in cycle t is $B(t)$, so if we sum over all reproductive age-classes, we obtain the genic structure of maternally-derived genes entering the population:

$$s_f(t) = \sum_{x=\alpha}^{\omega} B(t - x) s(t - x) l(x) f(x) / B(t). \tag{22}$$

When the population has reached stable age-structure, we have:

$$B(t - x) = (1 + \hat{r})^{-x} B(t)$$

which gives:

$$s_f(t) = \sum_{x=\alpha}^{\omega} (1 + \hat{r})^{-x} s(t - x) l(x) f(x). \tag{23}$$

Similarly the genic structure of the genes derived from males is:

$$s_m(t) = \sum_{x=\alpha^*}^{\omega^*} B_m(t - x) s(t - x) l^*(x) F^*(x) / B(t).$$

But

$$B_m(t - x) = (1 - c) B(t - x) = (1 - c)(1 + \hat{r})^{-x} B(t).$$

Hence:

$$s_m(t) = \sum_{x=\alpha^*}^{\omega^*} (1 + \hat{r})^{-x} s(t - x) l^*(x) f^*(x).$$

Finally, therefore:

$$s(t) = \tfrac{1}{2}[s_f(t) + s_m(t)]$$

$$= \tfrac{1}{2} \sum_{x=\alpha}^{\omega} (1+\hat{r})^{-x} s(t-x)\, l(x)\, f(x) + \tfrac{1}{2} \sum_{x=\alpha^*}^{\omega^*} (1+\hat{r})^{-x} s(t-x)\, l^*(x)\, f^*(x)$$

$$= \tfrac{1}{2} \sum_{x=\alpha_m}^{\omega_M} (1+\hat{r})^{-x} s(t-x)\, [l(x)\, f(x) + l^*(x)\, f^*(x)]. \qquad (24)$$

The evolution of the genic structure of the population is thus governed by a renewal equation. Using the results given in Appendix A, we know that the solution of Eq. (24) is of the form:

$$s(t) = \sum_{j=0}^{\omega_M - 1} A_j \lambda_j^t$$

where the λ_j are the roots of the equation:

$$\tfrac{1}{2} \sum_{x=\alpha_m}^{\omega_M} \lambda^{-x} (1+\hat{r})^{-x} [l(x)\, f(x) + l^*(x)\, f^*(x)] = 1.$$

Comparing this with Eq. (18), we see that $\lambda = 1$ is a solution of this equation. By a similar argument to that in Section 1.2.2, it can be shown that this is the root which has the largest modulus. It therefore follows that, as t increase, the values of all the other terms λ_j^t tend towards zero, so that $s(t)$ becomes constant.

Thus, in a population with overlapping generations, the genic structure of new individuals entering the population during one cycle becomes the same for maternally- and paternally-derived genes, and becomes constant in time. From Eqs. (21), we see that this means that the genic structure for any age-class becomes constant, and is the same in both sexes. It should be noted that the genic structure will be independent of time, and the same for males and females, right from $t = 0$, if the initial genic structures of males and females are the same, and are the same in all age-classes.

However, when there are initial differences in the genic structures of different age-classes or sexes, or both, the genic structure of any age-class, for example age-class one, will change at first, and will only gradually approach its equilibrium state. All alleles which are represented in individuals of reproductive or pre-reproductive age in the initial population will remain in the population during the approach towards the equilibrium, and afterwards; the frequencies in the equilibrium set of allele frequencies, s, will depend on the initial composition of the popula-

Table 7.4. The approach towards equal genotype frequencies in all age-classes, for an autosomal locus with two alleles, with the demographic parameters of the 1830 cohort of Frenchwomen. The initial age composition was with 100 individuals in each age-class; the frequency of A_1 was 1 in the first five classes, and 0 in the others

Cycle	Age-class	Age-structure[a]	Genotype frequencies			Frequency of A_1
			$(A_1 A_1)$	$(A_1 A_2)$	$(A_2 A_2)$	
2	1	161.9	0.273	0.499	0.229	0.523
	5	0.573	1	0	0	1
	9	0.558	0	0	1	0
8	1	153.1	0.362	0.479	0.158	0.602
	5	0.620	0.842	0.151	0.007	0.918
	9	0.335	1	0	0	1
20	1	153.3	0.513	0.406	0.080	0.717
	5	0.650	0.519	0.403	0.078	0.721
	9	0.504	0.625	0.332	0.044	0.790
40	1	162.9	0.530	0.396	0.074	0.728
	5	0.638	0.531	0.395	0.074	0.729
	9	0.515	0.536	0.392	0.072	0.732
60	1	172.6	0.531	0.395	0.074	0.729
	5	0.637	0.531	0.395	0.074	0.729
	9	0.516	0.531	0.395	0.074	0.729
80	1	182.7	0.531	0.395	0.074	0.729
	5	0.638	0.531	0.395	0.074	0.729
	9	0.516	0.531	0.395	0.074	0.729

[a] The total number of individuals entering age-class 1 during the cycle is given for age 1, and the numbers entering age-classes 5 and 9 are given as fractions of this number.

tion, with respect to the sex-ratios and genic structures of each age-class.

The course of change in the genic structure of a population with overlapping generations is illustrated in Table 7.4.

2.3. The Evolution of the Genotypic Structure of a Population

The results we have obtained show merely that the genic structure of a population with overlapping generations eventually becomes constant. We must now consider what happens to the genotypic structures defined in Section 2.1.

As we have seen, $s_f(t)$ and $s_m(t)$ tend to a common constant value, s. If mating is at random with respect to genotype, we can apply the considerations of Chapter 3 to the equilibrium population. The genic structure of the gametes produced by each age-class is s, so that the genotypic structure of individuals born in cycle t is therefore

$$S_f(t, 1) = S_m(t, 1) = \tilde{s}^2. \tag{25}$$

2.4. Conclusions

The results derived in this chapter show that, in a population with overlapping generations, which satisfies all the other conditions for Hardy-Weinberg equilibrium discussed in Chapter 3, a Hardy-Weinberg equilibrium genotypic structure is eventually attained. Equilibrium is, however, approached only gradually, compared with its attainment after one generation when the generations do not overlap.

PART 3

The Causes of Evolutionary Changes in Populations

Using the model situation in which no evolutionary forces are acting, other than those imposed by the genetic mechanism itself, we have derived certain properties of the genetic structure of populations. Like all scientific models, this equilibrium model does not give an exact representation of the real situation. Its chief purpose is to provide a starting-point from which we can explore more complex situations. This is what we shall do in this section, when models that are closer to the real situation are introduced.

In the earlier sections of this book, we have assumed that the five principal forces for evolutionary change were absent, namely that there was no migration, no mutation, no selection, random mating, and no stochastic changes in the genetic composition of the population (in other words, infinite population size). In the next five chapters, we shall study the ways in which the equilibrium situation is altered when each in turn of these five assumptions is dropped.

Chapter 8

Finite Populations

In the arguments in Chapter 3 leading to the Hardy-Weinberg principle, we found that, in order to obtain the *actual* genic structure s_g, or genotypic structure S_g, of the population in a future generation g, we had to assume that the population was "infinitely" large.

With conditions ① to ④ (non-overlapping generations, no mutation, migration or selection) we can obtain the genic probability structure s_g; condition ⑤ (random mating) enables us to obtain the genotypic probability structure $\mathscr{S}_g$. But we can say nothing about the actual structures s_g and S_g unless we assume that condition ⑥ holds, i.e. that the population size is infinite.

If we do not make this assumption, two problems arise. First, it is invalid to assume that non-identical genes are independent. We met this problem earlier, in Section 2.2 of Chapter 3, when we saw that, without condition ⑥, we could not relate the genotypic probability structure $\mathscr{S}_g$ of one generation to the genic probability structure s_{g-1} of the preceding generation.

Second, unless the population size is very large, we cannot consider the probabilities of the genotypes to be the same as their frequencies, since this equivalence is based on the "law of large numbers".

In this chapter, we shall study these two aspects of the problem of finite population size. We shall assume that all the other conditions for the establishment of Hardy-Weinberg equilibrium are met (i.e. we assume non-overlapping generations, and no selection, mutation, migration or choice of mates).

1. Identity by Descent of Genes in Finite Populations

1.1. The Inbreeding Coefficient and Coefficient of Kinship of a Population

The inbreeding coefficient f of an individual has been defined as the probability that the two genes which he has at a locus are "identical by descent"; in other words that they are both copies of one particular gene which was carried by one of his ancestors. This probability sums up our information about the common ancestors of the individual's parents.

The concept of inbreeding coefficient can be extended to a population by the following definition:

The inbreeding coefficient α of a population is the probability that the two genes of a member of the population taken at random are identical by descent.

If individuals with inbreeding coefficient f_x constitute a proportion p_x of the population, the inbreeding coefficient α of the population is equal to the weighted mean of the inbreeding coefficients of the population members:

$$\alpha = \sum_x p_x f_x.$$

This is the same formula as for Bernstein's (1930) α coefficient.

We can similarly define a coefficient of kinship β for a population, corresponding to the coefficient of kinship Φ of two individuals:

The coefficient of kinship, β, of a population is the probability that two genes taken at random from two separate individuals are identical by descent.

In a panmictic population (i.e. when mating is at random between individuals of the same generation), the value of α_g in one generation is related to the value of β_{g-1} in the preceding generation, as follows. The two genes of an individual in generation g were taken at random from two individuals of generation $g-1$ (his father and mother); the probability of identity of his two genes is therefore equal to the probability of identity of two genes from two random individuals in the preceding generations:

$$\alpha_g = \beta_{g-1}. \tag{1}$$

However, it is clear that this is true only for random-mating populations.

1.2. Increase of the Inbreeding Coefficient in a Finite Population

Consider a population consisting of $N_m^{(g)}$ males and $N_f^{(g)}$ females in generation g.

A member A of the population in generation g has two genes at a given locus, one of which is a copy of a gene in his father F, and the other is a copy of one of his mother's (M) genes. The paternal gene is a copy of one of the four genes in the paternal grandfather and grandmother, FF and FM, of the individual (Fig. 8.1); the maternal gene is a copy of one of the four genes of the maternal grandfather and grandmother, MF and MM.

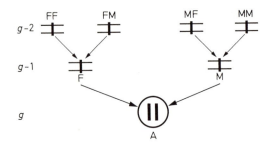

Fig. 8.1. The transmission of genes through two generations

The possible origins of A's genes are therefore:
1. Both from his grandfathers: probability $\frac{1}{4}$.
2. Both from his grandmothers: probability $\frac{1}{4}$.
3. One from a grandfather and one from a grandmother: probability $\frac{1}{2}$.

If mating is at random, without regard to relationships between individuals, the two grandfathers may be the same individual. The probability that this is the case is:

$$\frac{1}{N_m^{(g-2)}}.$$

This follows from the assumption of random mating, which implies that, if the first grandfather is known, the second is an individual chosen at random from among the $N_m^{(g-2)}$ male members of generation $g-2$. Each of the male members has the same chance of being the second grandfather, so that the probability that both grandfathers are the same individual is $\frac{1}{N_m^{(g-2)}}$.

Similarly, the probability that the two grandmothers are the same individual is $\frac{1}{N_f^{(g-2)}}$.

Finally, the probability that A received both his genes from the same individual in generation $g-2$ is:

$$\frac{1}{4}\frac{1}{N_m^{(g-2)}} + \frac{1}{4}\frac{1}{N_f^{(g-2)}}$$

which we can designate as $\frac{1}{N_e^{(g-2)}}$ where N_e is the harmonic mean of $2N_m$ and $2N_f$. N_e is called the "*effective population size*" (Wright, 1931).

The different possible origins of A's two genes can be divided into two groups:

1. The two genes may come from one ancestor in generation $g-2$, with probability $\frac{1}{N_e^{(g-2)}}$. If this is the case, the genes can be identical as a result of either of two situations:

a) They may both be copies of the same gene of this ancestor; this has probability $\frac{1}{2}$.

b) They are copies of the two genes of this ancestor (probability $\frac{1}{2}$), and these are identical by descent (which has probability α_{g-2}, from the definition of α in the previous section).

2. The two genes come from two ancestors in generation $g-2$, with probability $1-\frac{1}{N_e^{(g-2)}}$. In this case, the probability that the genes are identical is equal to β_{g-2}.

Thus the probability that A's two genes are identical is:

$$\alpha_g = \frac{1}{N_e^{(g-2)}} \left\{ \tfrac{1}{2} + \tfrac{1}{2}\alpha_{g-2} \right\} + \left\{ 1 - \frac{1}{N_e^{(g-2)}} \right\} \beta_{g-2}.$$

Substituting from Eq. (1), we have:

$$\alpha_g = \frac{1}{2\,N_e^{(g-2)}} + \left\{ 1 - \frac{1}{N_e^{(g-2)}} \right\} \alpha_{g-1} + \frac{1}{N_e^{(g-2)}}\alpha_{g-2}. \qquad (2)$$

If we know the values of α for the two initial generations, and the series of effective population sizes $N_e^{(g)}$, we can use this equation to calculate the values of α_g for any generation.

This equation shows that the course of change of the mean inbreeding coefficient of a population is determined by the numbers of males and females in the population.

1.3. Constant Effective Population Size

If N_e remains constant from generation to generation, Eq. (2) is a linear difference equation. It can be solved easily, by noting that it can be written in the homogeneous form:

$$1 - \alpha_g = \left(1 - \frac{1}{N_e}\right)(1-\alpha_{g-1}) + \frac{1}{2N_e}(1-\alpha_{g-2}) \qquad (3)$$

which has the solution (see Appendix A):

$$1 - \alpha_g = a\,\lambda_1^g + b\,\lambda_2^g \qquad (4)$$

a and b are parameters defined by the initial conditions, i.e. by the values of α for generations 0 and 1, and λ_1 and λ_2 are the roots of the equation:

$$\lambda^2 - \left(1 - \frac{1}{N_e}\right)\lambda - \frac{1}{2N_e} = 0.$$

Thus:

$$\lambda_1 = \frac{1}{2N_e}(N_e - 1 - \sqrt{1 + N_e^2})$$

$$\lambda_2 = \frac{1}{2N_e}(N_e - 1 + \sqrt{1 + N_e^2}).$$

One of these roots is positive and one is negative. Both are increasing functions of N_e, and are less than 1 in absolute value (Fig. 8.2).

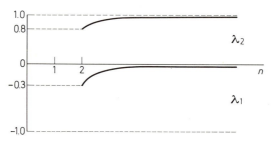

Fig. 8.2. Values of λ_1 and λ_2 for different values of N

As N_e increases from its minimum possible value $N_e = 2$ (one male and one female):

λ_1 increases from $\dfrac{1-\sqrt{5}}{4} \approx -0.3$ to 0

λ_2 increases from $\dfrac{1+\sqrt{5}}{4} \approx 0.8$ to 1.

As the generation number g rises, λ_1^g and λ_2^g tend towards zero, $\alpha_g - 1$ tends towards zero, and α_g towards 1.

In the final equilibrium state, the population therefore consists wholly of individuals whose two genes are identical, and who are therefore all homozygous.

The relation $\beta_{g-1} = \alpha_g$ shows that β also tends towards 1, so that, in the final state, two genes from two individuals are always identical.

In the limit, therefore, there is only one gene at the locus in the population. We can summarise these conclusions as follows:

A population of constant finite size, in which the other conditions for Hardy-Weinberg equilibrium for all loci are satisfied, tends towards a state of fixation at every locus.

Whatever the initial genic structure, the final equilibrium structure is of the form $\hat{s} = (0, \ldots, 0, 1, 0, \ldots, 0)$, with only one allele present.

This process of change in gene frequencies purely as a result of chance events, rather than of any definite force such as selection or mutation, is called *"genetic drift"* (Wright, 1931).

1.3.1. The case when the initial population is not inbred. If

$$\alpha_0 = \alpha_1 = 0$$

the constants a and b of Eq. (4) are:

$$a = \frac{\lambda_2 - 1}{\lambda_2 - \lambda_1}, \qquad b = \frac{1 - \lambda_1}{\lambda_2 - \lambda_1}$$

and α_g is given by:

$$1 - \alpha_g = \frac{1}{\lambda_2 - \lambda_1} \{ (\lambda_2 - 1) \lambda_1^g + (1 - \lambda_1) \lambda_2^g \}. \tag{5}$$

This equation shows that the approach to the final, homozygous equilibrium state is very slow if N_e is at all large. If $1/N_e^2$ can be neglected compared with $1/N_e$, we can write:

$$\lambda_1 = \tfrac{1}{2} \left(1 - \frac{1}{N_e} - \sqrt{1 + \frac{1}{N_e^2}} \right) \approx -\frac{1}{2 N_e},$$

$$\lambda_2 = \tfrac{1}{2} \left(1 - \frac{1}{N_e} + \sqrt{1 + \frac{1}{N_e^2}} \right) \approx 1 - \frac{1}{2 N_e}.$$

Hence:

$$1 - \alpha_g \approx \left(-\frac{1}{2 N_e} \right)^{g+1} + \left(1 + \frac{1}{2 N_e} \right) \left(1 - \frac{1}{2 N_e} \right)^g.$$

If g is large, this becomes:

$$1 - \alpha_g \approx \left(1 + \frac{1}{2 N_e} \right) \left(1 - \frac{1}{2 N_e} \right)^g$$

or:

$$\alpha_g \approx 1 - \left(1 + \frac{1}{2 N_e} \right) e^{-g/2 N_e}. \tag{6}$$

The number of generations required for the population to approach fixation to a significant extent is therefore of the same order of magnitude as the effective population size. $1 - \alpha_g$ is reduced to half its former level (in other words, half of all loci remaining polymorphic become fixed) in a number of generations G, which is such that:

$$e^{-G/2 N_e} = \tfrac{1}{2}.$$

Hence:

$$G = 2 N_e \log_e 2 \approx 1.4 N_e.$$

For example, for a human population consisting of 50 men and 50 women of reproductive age in every generation ($N_e = 100$), it would take 140 generations (i. e. about three or four thousand years) to eliminate half of the polymorphisms that existed in the initial population, assuming no mutation or selection.

1.4. Changing Effective Population Size

The difference equation (2) enables us to calculate α_g, given the set of values of $N_e^{(g)}$ of the effective population size in every generation. We shall consider the case when N_e remains finite, and also the case when N_e increases without limit.

a) First, suppose that N_e remains finite. Malécot (1966) has shown that, in this case, α_g tends towards 1. The proof is as follows. We make the following change of variable:

$$1 - \alpha_g = k_1 \times k_2 \times \cdots \times k_g.$$

The new variables k_i are all positive, because α_g is a probability, and therefore has a value between 0 and 1 in every generation. Furthermore, from Eq. (2), the values of k_i satisfy the relation:

$$k_g = 1 - \frac{1}{N_e^{(g-2)}} + \frac{1}{2 k_{g-1} N_e^{(g-2)}}. \tag{7}$$

If we know k_1, we can therefore calculate k_g.

But k_1 is assumed to be non-zero (because $k_1 = 0$ would imply $\alpha_1 = 1$, which would mean that the population was already at fixation in the initial generation). Hence Eq. (7) shows that:

$$k_{g-1} > 1 - \frac{1}{N_e^{(g-3)}}$$

whence

$$k_g = 1 - \frac{1}{N_e^{(g-2)}} + \frac{1}{2 N_e^{(g-2)} k_{g-1}}$$

$$< 1 - \frac{1}{N_e^{(g-2)}} + \frac{1}{2 \left(1 - \dfrac{1}{N_e^{(g-3)}}\right) N_e^{(g-2)}}$$

$$= 1 - \frac{(N_e^{(g-3)} - 2)}{2 N_e^{(g-2)} (N_e^{(g-3)} - 1)}$$

so that $k_g < 1 - a$ where a is strictly positive when $N_e^{(g-3)} > 2$.

Thus $1 - \alpha_g$ is equal to the product of g terms k_i which are all less than 1; this product therefore decreases as g increases, and tends towards a limit α_1, which is given by

$$\log(1 - \alpha_1) = \sum_0^\infty \log k_i.$$

Since the k_i are less than 1, $\log k_i < 0$ and the series $\log k_i$ is divergent: $\log(1 - \alpha_1)$ is therefore $-\infty$. Hence:

$$\alpha_1 = 1.$$

A population with finite, but changing, effective population size therefore tends towards fixation.

b) If, however, the effective population size $N_e^{(g)}$ increases indefinitely as g increases, Eq. (7) shows that k_g tends towards 1. To prove this, we make another change of variable:

$$k_g = 1 - u_g.$$

Eq. (7) then becomes:

$$u_g = \frac{1 - 2u_{g-1}}{2(1 - u_{g-1}) N_e^{(g-2)}} \tag{8}$$

which shows that u_g becomes infinitely small $\left(\text{of the order of } \dfrac{1}{2N_e^{(g-2)}} \right)$ as g increases.

The limit of α is then given by:

$$\log(1 - \alpha_l) = \sum_0^\infty \log k_g = \sum_0^\infty \log(1 - u_g).$$

If the series u_g is convergent, the series $\log(1 - u_g)$ is also convergent. This is the case when the series $\dfrac{1}{2N_e^{(g-2)}}$ is convergent. Therefore:

1. If $N_e^{(g)}$ increases at most as a linear function of g, the series $\log k_g$ diverges, and $\alpha \rightarrow 1$;

2. If $N_e^{(g)}$ increases at least in proportion to $g^{1+\varepsilon}$ ($\varepsilon > 0$), the series $\log k_g$ converges, and α tends to a limit other than 1.

We can summarise the results of this section as follows:

The mean inbreeding coefficient of a random mating population, assuming no selection or mutation, tends towards 1 if either the effective population size remains finite, or if it increases at most as a linear function of generation number.

In other words, every locus will become fixed for one or other allele in the long term, unless the population is growing at a rate which itself increases from generation to generation [1].

1.5. Relations between Relatives in a Finite Population

1.5.1. Coefficients of kinship. The genetic relation between two individuals A and B can be measured by their coefficient of kinship, Φ_{AB}, which is defined as the probability that two genes, one taken at random from A, and one from B, are identical by descent.

---

[1] This result appears to contradict the theorem of Lotka that in a "stable" population (i.e. with a constant geometric rate of increase) the probability of extinction of a name is 1 only if the population is "stationary", i.e. of constant size. A constant rate of linear increase (as in our result) corresponds, in the limit, to a zero rate of geometric increase (as in Lotka's result) and the two results coincide.

As we saw in Section 1.3.1 of Chapter 6, Φ_{AB} can be calculated from the formula:

$$\Phi_{AB} = \sum_i (\tfrac{1}{2})^{n_i + p_i + 1}(1 + f_{c_i})$$

where the summation is over all the lines of descent which go back from A and B to a common ancestor; the number of links between A and the common ancestor C_i is designated by n_i, and the number of links between B and C_i is p_i; the inbreeding coefficient of C_i is f_{c_i}.

This formula is valid only when the various common ancestors C_i are unrelated to one another. It cannot therefore be used when the effective population size is finite.

For example, consider a brother B and sister S, whose parents, M and P, have no *known* common ancestors (Fig. 8.3).

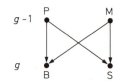

Fig. 8.3. The genetic relationship between full sibs

Despite the apparent lack of relationship between M and P, there is a certain probability that their genes are identical by descent, simply because they are members of the same, finite population. We have designated this probability as β_{g-1}, where g is the generation to which B and S belong.

It is easy to calculate the coefficient of kinship Φ_{BS} of B and S; we find:

$$\Phi_{BS}^{(g)} = \tfrac{1}{4}(\tfrac{1}{2} + \tfrac{1}{2}f_P) + \tfrac{1}{4}(\tfrac{1}{2} + \tfrac{1}{2}f_M) + \tfrac{1}{2}\beta_{g-1}$$

$$= \tfrac{1}{4}(1 + 2\alpha_g + \alpha_{g-1})$$

since:

$$f_P = f_M = \alpha_{g-1} \quad \text{and} \quad \beta_{g-1} = \alpha_g.$$

By a similar method, if we consider a pair of first-cousins F and C, belonging to generation g of a finite population, we find:

$$\Phi_{FC}^{(g)} = \tfrac{1}{4}\Phi_{BS}^{(g-1)} + \tfrac{3}{4}\beta_{g-1}$$

$$= \tfrac{1}{16}(1 + 12\alpha_g + 2\alpha_{g-1} + \alpha_{g-2}). \tag{9}$$

For double first cousins, we find:

$$\Phi_{DC}^{(g)} = \tfrac{1}{2}\Phi_{BS}^{(g-1)} + \tfrac{1}{2}\beta_{g-1}$$

$$= \tfrac{1}{8}(1 + 4\alpha_g + 2\alpha_{g-1} + \alpha_{g-2}). \tag{10}$$

Finally, for a father and son, P and O:

$$\Phi_{PO}^{(g)} = \frac{1}{2}(\frac{1}{2} + \frac{1}{2}\alpha_{g-1}) + \frac{1}{2}\beta_{g-1}$$
$$= \frac{1}{4}(1 + 2\alpha_g + \alpha_{g-1}) \tag{11}$$

which is the same as the formula for full sibs.

When the population size has been fairly large for many generations, we can consider the coefficients α_g, α_{g-1} and α_{g-2} ... to be approximately equal. Eqs. (9) to (11) become:

$$\Phi_{FC} = \frac{1}{16} + (1 - \frac{1}{16})\,\alpha,$$

$$\Phi_{DC} = \frac{1}{8} + (1 - \frac{1}{8})\,\alpha,$$

$$\Phi_{PO} = \frac{1}{4} + (1 - \frac{1}{4})\,\alpha.$$

In general, we have:

$$\Phi_T = \Phi_O + (1 - \Phi_O)\,\alpha$$

where Φ_O is the coefficient of kinship given by the pedigree alone and Φ_T is the "total" coefficient of kinship, taking account of both the known genealogy and also of the consanguinity due to finite population size. This formula thus takes account of the fact that two genes can be identical by descent either because they are both copies of a gene in a known common ancestor, which has probability Φ_O, or because, though not descended from the same gene in a known common ancestor (probability $1 - \Phi_O$), they are identical because of the population's finite size (probability α). This relation is often written in the form (Wright, 1951):

$$1 - \Phi_T = (1 - \Phi_O)(1 - \alpha).$$

1.5.2. Some examples. In order to see the effect of the "background" level of inbreeding due to finite population size, we shall examine two examples.

a) Suppose we know that a population has been totally isolated for ten generations, with an effective population size of about 50 throughout this period. By Eq. (6), the mean inbreeding coefficient is:

$$\alpha_g \approx 1 - (1 + \frac{1}{100})\,e^{-g/100}$$

which gives: $\alpha_8 = 0.077$, $\alpha_9 = 0.086$, $\alpha_{10} = 0.095$. Eqs. (9) and (10) give:

$$\Phi_{FC}^{(10)} = \frac{2.39}{16}, \qquad \Phi_{DC}^{(10)} = \frac{1.63}{8}.$$

Thus in this population, inbreeding due to remote ancestors is equal to 139% of the value of the coefficient of kinship due to the known relationship, for first cousins; for double first cousins, the figure is approximately 63%.

b) Now suppose that a population has remained isolated, with an effective size of about 100, for a thousand years, i.e. about 40 generations for a human population. The mean inbreeding coefficient is given by:

$$\alpha \approx 1 - (1 + \tfrac{1}{200}) \, e^{-40/200} \approx 1 - \left(\frac{1}{e}\right)^{1/5} = 0.18.$$

In this population, the coefficients of kinship would be:

First cousins: $\Phi_{FC} = \dfrac{3.7}{16} = \dfrac{0.9}{4}$

Double first cousins: $\Phi_{DC} = \dfrac{2.3}{8} = \dfrac{1.15}{4}.$

In such a population, the coefficient of kinship between first cousins is almost as large as that between sibs in an infinite population (for whom $\Phi_{FS} = \tfrac{1}{4}$); for double first cousins, the value is higher than $\tfrac{1}{4}$.

1.5.3. "Remote" and "absolute" consanguinity. These examples show that the concept of inbreeding can be looked at in two rather different ways, depending on the type of information with which we are dealing. First we may know the pedigree of two individuals and be able to calculate their coefficients of kinship. Second, we may know the effective population size of the population to which two individuals belong; this determines the relationship between the individuals which is due to "remote consanguinity".

Many authors have remarked on the fact that the coefficient of kinship between individuals, based on their pedigree, is only a partial measure of their relationship. The question then arises: given that a pedigree is always limited in its scope, and sometimes is very limited indeed, what fraction of the total inbreeding does it measure? What should be added to the measure of inbreeding calculated from a pedigree, in order to take into account the existence of remote consanguinity?

Since we have been treating inbreeding coefficients as probabilities, this question has, in formal terms, no meaning: a probability is always implicitly dependent on the information which we possess, and the assumptions we have made about the situation with which we are dealing.

There is, however, a real practical problem. We always know that, in addition to the relationships which we know of from a pedigree, other relationships exist, which are due to the effect of finite population size during past generations. We would like to be able to take this into account in the calculation of coefficients of kinship.

Eqs. (9), (10) and (11) above show how we can "improve" our estimates of coefficients of kinship.

However, the "improved" coefficients should not be considered to be more "exact", or closer to "reality" than those based on pedigrees alone. They simply correspond to a different set of information.

A coefficient of kinship is not an intrinsic property of a given pair of individuals: it is a measure associated with the information we have concerning them. If we obtain extra information, the value of the coefficient will obviously change, but the reality of which it is a measure is clearly not modified at all.

Furthermore, a too absolute interpretation of coefficients of kinship leads to insoluble paradoxes. For example, we can prove that the coefficient of kinship of any two individuals is 1, without invoking the myth of Adam and Eve, as follows. Every person has two parents, four grandparents, ... 2^n ancestors n generations ago; thus, ten centuries ago, i.e. about 35 generations ago, everyone has $2^{35} = 30\,000\,000\,000$ ancestors. Since the total human population size of the world in the tenth century was probably only several hundred millions, the probability that the two genes at a locus in any individual are identical is close to 1. Therefore every individual has an inbreeding coefficient close to 1, and this measure becomes useless.

It is obvious that, from this point of view, an inbreeding coefficient cannot be regarded as an estimate of any real quantity, but is simply a measure of information.

1.6. The Effect of Variance in Number of Offspring on the Effective Population Size

In Section 1.2, we assumed that every parent had the same chance of producing offspring, in other words that individuals in a given generation have an equal likelihood of having as parents any of the individuals of the preceding generation. If, however, there are fertility differences a member of a certain generation will have a greater chance of having as his parent an individual who has a large expected number of offspring. We can carry through arguments analogous to those of Section 1.2 with these new assumptions. But first we must define more clearly what is meant by the number of offspring of an individual.

1.6.1. Number of successful offspring. From the point of view of the transmission of genes, and of the genetic endowment of a population, a child who dies before adulthood is equivalent to a child who is born dead, or to an aborted offspring; thus, the only offspring we need to take into account are those who take part in reproduction during the next generation, and transmit the genes which they have received to future generations. We shall call these "successful offspring". It is obvious that the number of successful offspring of a mating couple will not in general be the same as the number of offspring who are born to them.

In order to characterise the genetically relevant breeding behaviour of a population, we need to know the distributions of numbers of successful offspring of both males and females.

Let us write $_m\lambda_x^{(g)}$ and $_f\lambda_x^{(g)}$ for the proportions of men and women who have x successful offspring, in generation g.

The numbers of offspring of a man or woman chosen at random are thus random variables, with means:

$$\bar{x}_m^{(g)} = \sum_x x \;_m\lambda_x^{(g)}$$

$$\bar{x}_f^{(g)} = \sum_x x \;_f\lambda_x^{(g)}$$

(12)

and variances:

$$V_m^{(g)} = \sum_x x^2 \;_m\lambda_x^{(g)} - [\bar{x}_m^{(g)}]^2$$

$$V_f^{(g)} = \sum_x x^2 \;_f\lambda_x^{(g)} - [\bar{x}_f^{(g)}]^2.$$

(13)

Notice that the treatment of effective population size in Section 1.2 is equivalent to assuming that the number of offspring is a binomial variate with the same parameter in both sexes.

Also, we have the following obvious relation between the population sizes in successive generations:

$$N^{(g)} = N_m^{(g)} + N_f^{(g)}$$

$$= \bar{x}_m^{(g-1)} N_m^{(g-1)}$$

(14)

$$= \bar{x}_f^{(g-1)} N_f^{(g-1)}.$$

1.6.2. The general case. Let F be the father of an individual A chosen at random from generation g. The probability that F's father FF had x offspring is proportional to the number of individuals who come from sibships of size x; let us denote this probability by $Kx \;_m\lambda_x^{(g-2)}$, where K is a constant such that $K \sum_x x \;_m\lambda_x^{(g-2)} = 1$, so that, from Eq. (12):

$$K = \frac{1}{\bar{x}_m^{(g-2)}}.$$

If F is a member of a family of size x, which will happen with probability

$$\frac{x \;_m\lambda_x^{(g-2)}}{\bar{x}_m^{(g-2)}}$$

the number of other members of the population in generation $g-1$ who are also children of FF is equal to $x-1$; of these, a proportion

$$\frac{N_f^{(g-1)}}{N_m^{(g-1)} + N_f^{(g-1)} - 1}$$

are expected to be females. The probability that A's mother M is also a child of FF is thus, taking account of Eqs. (14):

$$\frac{1}{N_f^{(g-1)}} (x-1) \frac{N_f^{(g-1)}}{N_m^{(g-1)} + N_f^{(g-1)} - 1} = \frac{x-1}{\bar{x}_m^{(g-2)} N_m^{(g-2)} - 1} .$$

The probability that M and F have the same father is therefore:

$$\sum_x \frac{x \,_m\lambda_x^{(g-2)}}{\bar{x}_m^{(g-2)}} \left[\frac{x-1}{\bar{x}_m^{(g-2)} N_m^{(g-2)} - 1} \right] = \sum_x \frac{x^2 \,_m\lambda_x^{(g-2)} - x \,_m\lambda_x^{(g-2)}}{\bar{x}_m^{(g-2)} [N_m^{(g-2)} \bar{x}_m^{(g-2)} - 1]}$$

$$= \frac{1}{N_m^{(g-2)} - \frac{1}{\bar{x}_m^{(g-2)}}} \left[\frac{V_m^{(g-2)}}{\{\bar{x}_m^{(g-2)}\}^2} - \frac{1}{\bar{x}_m^{(g-2)}} + 1 \right]$$

from Eqs. (12) and (13).

Similarly, replacing the subscript m by f, we obtain the probability that M and F have the same mother.

Finally, the probability $\dfrac{1}{N_e^{\prime (g-2)}}$ that A received both of his genes from the same individual in generation $g-2$ is:

$$\frac{1}{N_e^{\prime (g-2)}} = \frac{1}{4\left[N_m^{(g-2)} - \frac{1}{\bar{x}_m^{(g-2)}}\right]} \left[\frac{V_m^{(g-2)}}{\{\bar{x}_m^{(g-2)}\}^2} - \frac{1}{\bar{x}_m^{(g-2)}} + 1 \right]$$

$$+ \frac{1}{4\left[N_f^{(g-2)} - \frac{1}{\bar{x}_f^{(g-2)}}\right]} \left[\frac{V_f^{(g-2)}}{\{\bar{x}_f^{(g-2)}\}^2} - \frac{1}{\bar{x}_f^{(g-2)}} + 1 \right]. \qquad (15)$$

This result can be used in the arguments of Section 1.2, replacing N_e by the new effective population size N_e' which we have just obtained; this determines the rate of increase of the inbreeding coefficient of the population.

Eq. (15) is complex, because it takes into account all the parameters which determine the effect of genetic drift. In human populations, polygamy is not uncommon: the mean numbers of offspring of men can therefore differ quite considerably from those of women, and the variances will also differ. Eq. (15) enables us to take into account all these parameters.

1.6.3. The case of constant population size. The simplest case is when the total population size $N = N_m + N_f$ is constant, and the means and variances $\bar{x}_m$, $\bar{x}_f$, V_m and V_f are also constant. From Eqs. (14) we obtain:

$$N = N_m + N_f = \bar{x}_m N_m = \bar{x}_f N_f$$

and Eq. (15) can then be written in the form:

$$\frac{1}{N_e'} = \left[\frac{1}{4 N_m} \left(\frac{V_m}{\bar{x}_m^2} - \frac{1}{\bar{x}_m} + 1 \right) + \frac{1}{4 N_f} \left(\frac{V_f}{\bar{x}_f^2} - \frac{1}{\bar{x}_f} + 1 \right) \right] \frac{N}{N-1}. \quad (16)$$

Some special cases. When the numbers of successful offspring are such that, for both sexes, the variances are equal to the means (which is notably the case when the distributions of numbers of successful offspring follow Poisson distributions) the expressions in brackets in Eq. (16) are equal to 1, and Eq. (16) becomes:

$$\frac{1}{N_e'} = \frac{N}{N-1} \left[\frac{1}{4 N_m} + \frac{1}{4 N_f} \right] = \frac{N}{N-1} \cdot \frac{1}{N_e}$$

or, in practice $N_e' \approx N_e$; a Poisson distribution of offspring number thus gives almost the same effective population size as the treatment of this problem in Section 1.2.

In the case when the distributions of numbers of offspring have zero variance, i.e. when all individuals have the same number of offspring, we obtain the expression:

$$\frac{1}{N_e'} = \frac{N}{N-1} \left[\frac{1}{4 N_m} \left(1 - \frac{N_m}{N} \right) + \frac{1}{4 N_f} \left(1 - \frac{N_f}{N} \right) \right]$$

$$= \frac{1}{4(N-1)} \left[\frac{N_f}{N_m} + \frac{N_m}{N_f} \right]$$

which, when the numbers of males and females are equal, reduces to:

$$N_e' = 2(N-1)$$

or, in practice $N_e' \approx 2 N_e$. When population size is constant, complete uniformity in family size thus leads to an effective population size which is twice the value obtained in Section 1.2.

Finally, when the numbers of males and females are the same, and the distributions of offspring number are the same for both sexes, we have:

$$\bar{x}_m = \bar{x}_f = 2$$

$$N_m = N_f = \tfrac{1}{2} N$$

$$V_m = V_f = V$$

and Eq. (16) becomes:

$$\frac{1}{N'_e} = \frac{1}{4(N-1)} \cdot (V+2)$$

which gives:

$$N'_e = \frac{4(N-1)}{V+2}.$$

1.6.4. The importance of variance in the number of successful offspring. To summarise, the variance in the number of successful offspring has a strong effect on effective population size, and is thus important in determining the effect of genetic drift. The greater the variance in offspring number, the smaller the effective population size will be, and consequently the greater will be the effect of random factors in changing the genetic make-up of the population.

In order to understand this effect intuitively, we can consider an extreme example. Consider two populations A and B of constant size, in which all mating couples have on average two successful offspring. In population A, all families are assumed to consist of two offspring, so that the variance of number of offspring is zero, whereas in population B half the families are of size 4, while the other half have no offspring, so that the variance is equal to 4. Consider a gene carried by a particular individual, at a given locus. When the individual in question reproduces, the probability that he does not transmit this gene is $\frac{1}{2}$; if he has x successful offspring, the probability that none of them receives this gene is $(\frac{1}{2})^x$. In population A, the probability that a gene taken at random is not transmitted is therefore $(\frac{1}{2})^2 = 0.25$. In population B, the chance that the gene is carried by an individual who has no offspring is $\frac{1}{2}$, and the chance that it is carried by an individual who has 4 offspring is also $\frac{1}{2}$. The probability of non-transmission is therefore $\frac{1}{2} \times (\frac{1}{2})^0 + \frac{1}{2} \times (\frac{1}{2})^4 = 0.53$. This example shows that the role of chance in the process of transmission of genes from one generation to the next is twice as great in population B as in population A.

1.7. The Effect of the Prohibition of Incest on Effective Population Size

Up to now, we have assumed that the mother and father of an individual A could, by reason of random mating, have the same parents. In most human societies, however, marriages between brothers and sisters are forbidden, and this has probably been the case over a long period of time. In order to have a fully realistic model of the action of genetic drift in human populations, we ought to take this fact into account.

Let us therefore consider a population whose members mate at random, except that brothers and sisters do not mate with one another.

In order to simplify the reasoning, without great loss of generality, we shall assume that the population is of constant size N, with equal numbers of both sexes. The mean number of offspring per couple is thus $\bar{x}=2$. We shall also assume that the variance V of offspring number is the same in every generation. We can therefore drop the superscript denoting generation. Writting λ_x for the proportion of couples which have x offspring, we have:

$$\bar{x}=\sum_x x\,\lambda_x=2$$

$$V=\sum_x x^2\,\lambda_x-4.$$

As we saw in Section 1.6.2, the probability that an individual taken at random belongs to a family of size x is:

$$\frac{x\,\lambda_x}{\bar{x}}=\tfrac{1}{2}\,x\,\lambda_x$$

his number of sibs is therefore equal to $x-1$. The expectation S of the number of sibs of an individual is therefore:

$$S=\sum_x \frac{1}{\bar{x}}\,(x-1)\,x\,\lambda_x.$$

In the case we are considering, with $\bar{x}=2$, we therefore have:

$$S=\tfrac{1}{2}\sum_x x^2\,\lambda_x-\tfrac{1}{2}\sum_x x\,\lambda_x$$

$$=1+\frac{V}{2}.$$

Let A (Fig. 8.1) be an individual taken at random in generation g. His father F and his mother M belong to generation $g-1$; we know that M and F are not sibs, since they are married, and we have assumed that sib matings are prohibited. The grandparents FF, FM, MF and MM of A are thus four distinct members of generation $g-2$.

A's two genes may have been transmitted to him either by FF and MF, or by FF and MM, or by FM and MF or by FM and MM, the probability of each of these events being $\tfrac{1}{4}$.

In each of these possible cases, the two grandparents of A which are involved may either be:

a) Sibs, with probability $\dfrac{S}{2}\Big/\Big(\dfrac{N}{2}-1\Big)=\dfrac{S}{N-2}$. According to the results obtained in Section 1.5.1, the probability that the genes transmitted by these grandparents are identical is:

$$\Phi_{\text{FS}}^{(g-2)}=\tfrac{1}{4}(1+2\alpha_{g-2}+\alpha_{g-2}).$$

b) Not sibs, with probability $1 - \dfrac{S}{N-2}$; in this case, the probability that the genes transmitted by these grandparents are identical is α_{g-1}, since individuals in generation $g-1$ receive their two genes from two non-sib individuals of generation $g-2$.

Finally, therefore, we have:

$$\alpha_g = \frac{1}{4} \frac{S}{N-2} (1 + 2\alpha_{g-2} + \alpha_{g-3}) + \left(1 - \frac{S}{N-2}\right) \alpha_{g-1}.$$

This linear difference equation can be put in the homogeneous form:

$$1 - \alpha_g = \left(1 - \frac{S}{N-2}\right)(1 - \alpha_{g-1}) + \frac{1}{2} \frac{S}{N-2}(1 - \alpha_{g-2})$$
$$+ \frac{1}{4} \frac{S}{N-2}(1 - \alpha_{g-3})$$

which has the solution:

$$1 - \alpha_g = \alpha \, \lambda_1^g + b \, \lambda_2^g + c \, \lambda_3^g$$

where a, b and c are parameters whose values depend on the values α_0, α_1 and α_2 of the inbreeding coefficients in the first three generations; λ_1, λ_2 and λ_3 are the roots of the equation

$$f(\lambda) = \lambda^3 - \left(1 - \frac{S}{N-2}\right)\lambda^2 - \frac{1}{2}\frac{S}{N-2}\lambda - \frac{1}{4}\frac{S}{N-2} = 0. \quad (17)$$

It is easy to see that the largest root of Eq. (17) is the single positive root λ_m. If N is sufficiently large for $1/N^2$ to be neglected, compared with $1/N$, this positive root can be approximated by assimilating the curve representing $f(\lambda)$ to its tangent at the point $\lambda = 1$, $f(1) = \dfrac{S}{4(N-2)}$. We find:

$$\lambda_m = 1 - \frac{S}{2(2N - 4 + 3S)} = 1 - \frac{1 + V/2}{2(2N - 1 + 3\,V/2)}$$

for the particular case which we have been studying.

With the passage of the generations, α_g therefore tends towards 1 at a rate which is determined by the value of the largest root λ_m:

$$1 - \alpha_g \approx a \, \lambda_m^g.$$

This result is analogous to the result we obtained for the case when mating is at random, except that λ_m replaces $1 - 1/2\,N_e$.

It follows from the above results that, in a population which prohibits sib-mating, the effect of genetic drift is the same as in a random-mating population of N'_e individuals, with:

$$N'_e = \frac{2N - 1 + 3V/2}{1 + V/2}.$$

When the population size is sufficiently large for the variance V to be negligible compared with N, this result shows that the effective population size, with prohibition of incest, is of the order of $\frac{4N}{2+V}$, exactly the same as in the case when incest was not forbidden. The prohibition of incest therefore has a significant effect only when the population is very small (of the order of 10 individuals).

It should be mentioned that this result contradicts that of MacCluer and Schull (1970). These authors concluded from the results of numerous simulations that: "the avoidance of close consanguinity (...) can increase the effective population size considerably". It would be useful to find the reason for this disagreement, which is probably due to differences in the assumptions made.

Some special cases. If all couples have the same number of offspring, i.e. if $V = 0$, it follows that $N'_e = 2N - 1$, which is essentially identical with the result we obtained assuming marriage to take place at random.

If the variance is equal to the mean (the case of a Poisson distribution of family size), we have $V = 2$ and $N'_e = N + 1$.

Finally, we may note that when $N = 4$ and $V = 0$, all marriages are between double first-cousins. In this case Eq. (17) becomes:

$$\lambda^3 - \tfrac{1}{2}\lambda^2 - \tfrac{1}{4}\lambda - \tfrac{1}{8} = 0.$$

This result was first obtained by Wright (1921) in his study of systems of mating.

1.8. Effective Population Size in Populations with Overlapping Generations

The models which we have considered up to now have all assumed that members of different generations do not interbreed. The case of overlapping generations has been considered by Felsenstein (1971) and by Crow and Kimura (1972), who have been able to derive approximate expressions for the effective population size in such populations.

2. Changes in the Genotypic Probability Structure

We have seen that, in order to characterise a population with respect to the various possible combinations of alleles at a given locus, we need to know the frequencies P_{ij} of each of the genotypes $(A_i A_j)$.

If there are n alleles at the locus, the set of $\dfrac{n(n+1)}{2}$ frequencies P_{ij} constitute a trimat which is called the genotypic structure of the population, S.

In a finite population, the coefficient of kinship β and the inbreeding coefficient α tend towards 1 as the generation number increases; regardless of the initial genotypic structure S_0, the population moves towards an equilibrium structure S_e, which consists only of homozygotes for one of the alleles. S_e is therefore a trimat whose elements are all zero, except for one element on the hypotenuse, whose value is 1. For example, in the case when there are two alleles A_1 and A_2 in the initial population, and when only A_1 remains present in the equilibrium population,

$$S_e = \begin{array}{|cc} 1 & \\ 0 & 0 \end{array}$$

In this section, we shall study the changes in the genotypic structure of a population, as it moves from S_0 to S_e.

We assume that we know only the genotypic structure S_0, and the numbers N_m and N_f of males and females in the population. N_m and N_f are assumed to remain constant from generation to generation. If we consider a future generation g, we cannot know its actual genotypic structure S_g, but we can work in terms of the genotypic probability structure $\mathscr{S}_g$; this was defined in Section 3.2 of Chapter 2 as "the set of probabilities attached to the various possible genotypes that an individual taken at random from the population might be found to have".

We shall now try to find an expression for $\mathscr{S}_g$, knowing that as g increases $\mathscr{S}_g$ tends towards a trimat with zero probabilities for all the heterozygous genotypes.

2.1. The Difference Equation for Genotypic Probability Structure

Consider a population of N_m males and N_f females, constant from generation to generation.

As we saw in Section 1.2 of this chapter (see Fig. 8.1), an individual A in generation g has two genes whose origins may be:

1. Both from the same grandparent in generation $g - 2$. This has the probability:

$$\frac{1}{4N_m} + \frac{1}{4N_f} = \frac{1}{N_e}$$

where N_e is the effective population size.

2. From two different grandparents. This has the probability $1 - \dfrac{1}{N_e}$.

In the first case there are two possibilities:

1. Both of A's genes are copies of the same gene of this grandparent (probability $\frac{1}{2}$). In this case they are identical and A's genotype is $(A_i A_i)$ with probability $\pi_i^{(g-2)}$, the probability that a gene taken at random from an individual in generation $g-2$ is A_i.

2. A's genes are copies of the two genes of this grandparent (probability $\frac{1}{2}$). In this case, A's genotype is the same as that of his grandparent; the probability that it is $(A_i A_j)$ is therefore equal to $\Pi_{ij}^{(g-2)}$, the probability that an individual in generation $g-2$ has this genotype.

In the second case, assuming random mating, A's genes are two genes taken at random from two individuals in generation $g-2$. The probability that they are an A_i and an A_j gene is equal to the probability that an individual of generation $g-1$ has the genotype $(A_i A_j)$, i.e. $\Pi_{ij}^{(g-1)}$.

Gathering up the appropriate probabilities, we therefore have:
Probability that A is homozygous $(A_i A_i)$ is:

$$\Pi_{ii}^{(g)} = \frac{1}{2 N_e} (\pi_i^{(g-2)} + \Pi_{ii}^{(g-2)}) + \left(1 - \frac{1}{N_e}\right) \Pi_{ii}^{(g-1)}. \qquad (18)$$

Probability that A is heterozygous $(A_i A_j)$ is:

$$\Pi_{ij}^{(g)} = \frac{1}{2 N_e} \Pi_{ij}^{(g-2)} + \left(1 - \frac{1}{N_e}\right) \Pi_{ij}^{(g-1)}. \qquad (19)$$

In the condensed notation of trimats, these equations become:

$$\mathscr{S}_g = \frac{1}{2 N_e} \mathscr{S}_H^{(g-2)} + \frac{1}{2 N_e} \mathscr{S}_{g-2} + \left(1 - \frac{1}{N_e}\right) \mathscr{S}_{g-1} \qquad (20)$$

where $\mathscr{S}_H^{(g-2)}$ is the structure of the fully homozygous population corresponding to the genic structure s_{g-2}, in generation $g-2$, i.e.

$$\mathscr{S}_H^{(g-2)} = \begin{vmatrix} \pi_1^{(g-2} \\ \vdots \\ 0 \dots \pi_n^{(g-2)} \end{vmatrix}$$

If we know the genic and genotypic probability structures in generations 0 and 1, Eq. (20) enables us to determine $\mathscr{S}_g$ for subsequent generations, as follows:

a) Assuming the initial genotypic structure is known, we have: $\mathscr{S}_0 = S_0$.

b) The genotypic probability structure of an individual taken at random in generation 1 is given by the genetic product of the genic structures of his two parents. Since we are assuming random mating, the parents both have the same genic probability structure, s_0. Hence:

$$\mathscr{S}_1 = \vec{s}_0^2 = S_P \tag{21}$$

where S_P is the structure of the population in Hardy-Weinberg equilibrium which has the same genic structure as the initial generation.

c) We next investigate the changes in the genic probability structure s_g; taking the contracted vector of each term of Eq. (20) we have:

$$s_g = \frac{1}{2N_e} s_{g-2} + \frac{1}{2N_e} s_{g-2} + \left(1 - \frac{1}{N_e}\right) s_{g-1}.$$

Hence:

$$s_g - s_{g-1} = \left(-\frac{1}{N_e}\right) (s_{g-1} - s_{g-2}) = \left(-\frac{1}{N_e}\right)^{g-1} (s_1 - s_0).$$

But, taking the contracted vectors in Eq. (21), $s_1 = s_0$. Hence:

$$s_g = s_{g-1} = s_0.$$

Thus the genic probability structure remains equal to that in the initial generation. Therefore:

$$\mathscr{S}_H^{(g-2)} = S_H^{(0)}.$$

We can therefore drop the superscript for generation number. Then Eq. (20) can be written in the homogeneous form:

$$\mathscr{S}_g - S_H = \left(1 - \frac{1}{N_e}\right) (\mathscr{S}_{g-1} - S_H) + \frac{1}{2N_e} (\mathscr{S}_{g-2} - S_H).$$

The solution of this equation is:

$$\mathscr{S}_g = S_H + A \lambda_1^g + B \lambda_2^g \tag{22}$$

where A and B are trimats determined by the initial conditions, and λ_1 and λ_2 are the roots of:

$$\lambda^2 - \left(1 - \frac{1}{N_e}\right) \lambda - \frac{1}{2N_e} = 0.$$

The values of λ_1 and λ_2 are the same as those we found for the difference equation for the inbreeding coefficient, in Section 1.3 of this chapter.

Since the roots of this equation are both less than 1 in absolute magnitude, λ_1^g and λ_2^g both tend towards zero as g increases, and $\mathscr{S}_g$ therefore tends towards S_H. In other words: *the genotypic probability structure tends towards the homozygous structure corresponding to the initial population*. This confirms, in a more explicit way, the result we obtained in Section 1.

2.2. The Genotypic Probability Structure at Intermediate Stages

If we replace A and B in Eq. (22) by their values as determined from the initial conditions, $\mathscr{S}_0 = S_0$ and $\mathscr{S}_1 = S_P$, we have:

$$\mathscr{S}_g = \left\{ 1 + \frac{\lambda_1^g(1-\lambda_2) - \lambda_2^g(1-\lambda_1)}{\lambda_2 - \lambda_1} \right\} S_H + \frac{(\lambda_2 \lambda_1^g - \lambda_1 \lambda_2^g)}{\lambda_2 - \lambda_1} S_0$$

$$+ \frac{\lambda_2^g - \lambda_1^g}{\lambda_2 - \lambda_1} S_P. \tag{23}$$

Comparing this with Eq. (5) of Section 1, it is obvious that the coefficient of S_H in Eq. (23) is equivalent to α_g, under the assumption that the initial population is not inbred.

If we also put:

$$S_0 = S_P + \Delta_0$$

where Δ_0 represents the difference between the initial structure and the corresponding Hardy-Weinberg equilibrium structure, Eq. (23) becomes:

$$\mathscr{S}_g = \alpha_g S_H + (1 - \alpha_g) S_P + \gamma_g \Delta_0 \tag{24}$$

where:

$$\gamma_g = \frac{\lambda_2}{\lambda_2 - \lambda_1} \lambda_1^g - \frac{\lambda_1}{\lambda_2 - \lambda_1} \lambda_2^g \tag{25}$$

is a coefficient which measures the influence of the initial structure S_0 on the structure $\mathscr{S}_g$.

Note that γ_g satisfies the same difference equation as $1 - \alpha_g$, i.e.:

$$\gamma_g = \left(1 - \frac{1}{N_e}\right) \gamma_{g-1} + \frac{1}{2N_e} \gamma_{g-2} \tag{26}$$

but with the initial conditions:

$$\gamma_0 = 1, \quad \gamma_1 = 0.$$

Eq. (24) shows that the genotypic probability structure can be analysed into three components:

1. The corresponding homozygous structure S_H.
2. The corresponding Hardy-Weinberg equilibrium structure S_P.
3. The difference Δ_0 between the initial structure S_0 and the corresponding Hardy-Weinberg proportions.

The probability structure in any generation g is thus a function of the "weights" attached to these three components, i.e.:

$$\alpha_g, \quad 1 - \alpha_g \quad \text{and} \quad \gamma_g.$$

We have seen that, if N_e is sufficiently large for $1/N_e^2$ to be negligible compared with $1/N_e$, $1 - \alpha_g$ is given by:

$$\left(1 + \frac{1}{2N_e}\right) e^{-g/2N_e}$$

which tends to zero as g increases. A similar argument shows that:

$$\gamma_g \approx \frac{1}{2N_e} e^{-g/2N_e}$$

and consequently, for sufficiently large g:

$$\frac{1 - \alpha_g}{\gamma_g} \approx 2N_e + 1.$$

This shows that the contribution of S_P to $\mathscr{S}_g$ soon becomes much greater than the contribution of Δ_0.

2.3. The Stages of Change in Genotypic Structure

In summary, the genotypic probability structure of a finite population can be thought of as passing through several stages:

1. If the initial structure is S_0, the population's genotypic probability structure becomes equal to the corresponding Hardy-Weinberg equilibrium structure S_P, in the first generation.

2. In the second generation, the "memory" of the initial structure brings the population's genotypic probability structure $\mathscr{S}_2$ back towards S_0.

3. The weight attached to this "memory" gradually becomes smaller, and $\mathscr{S}_g$ approaches S_P, as the generations pass.

4. But, simultaneously, the mean inbreeding coefficient increases towards 1. When the effect of increasing α_g begins to outweigh the effect of S_P, $\mathscr{S}_g$ moves away from S_P and towards S_H. Fig. 8.4 shows these changes in schematic form.

These considerations show that finite population size has two consequences:

1. The tendency towards a homozygous structure.

2. The gradual loss of the "memory" of the initial structure. In a random-mating population of infinite size, this "memory" is lost in the first generation of random mating.

If the initial genotypic proportions are those given by the Hardy-Weinberg formula, $\Delta_0 = 0$. In this case, but only in this case, Eq. (24) can be written in the form:

$$\mathscr{S}_g = \alpha_g S_H + (1 - \alpha_g) S_P. \tag{24b}$$

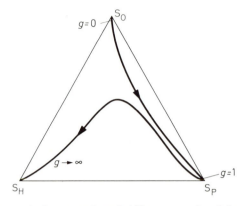

Fig. 8.4. The changes in the genotypic probability structure in a finite population

A formula equivalent to this was derived by Wright (1922), using a different notation.

Note. One might be tempted to try to prove Eq. (24b) by the following reasoning. An individual is taken at random from generation g. The probability that his two genes are identical is α_g, and the probability that they are non-identical is $1 - \alpha_g$. In the first case, he is a homozygote, and the probability that his genotype is $(A_i A_i)$ is p_i, the frequency of the allele A_i. In the second case, his genes are independent; the probability that his genotype is $(A_i A_i)$ is p_i^2; the probability that he is $(A_i A_j)$ is $2 p_i p_j$. This is equivalent to (24b).

The flaw in this argument is that it equates non-identical genes with independent genes. As we saw in Section 2.2 of Chapter 3, this is valid only in infinite populations; when the population size is finite, the genes are not independent, even if they are not identical by descent, because the initial structure is a constraint on the structures in all subsequent generations. This dependence disappears only if the initial structure is the same as that which would result from choosing pairs of independent genes, in other words, if $S_0 = S_P$.

2.4. Genetic Drift

The existence of dependence between the genotypes of the members of a finite population, which increases with generation number, has sometimes been used to draw wrong conclusions.

The fundamental Eq. (24):

$$\mathcal{S}_g = \alpha_g S_H + (1 - \alpha_g) S_P + \gamma_g \Delta_0$$

where α_g tends to 1 as g increases, gives a way of finding the genotypic probability structure of successive generations, given certain information about the initial generation.

Another way of expressing this is to say that $\mathscr{S}_g$ characterises the ensemble of outcomes which are possible for the population after the passage of g generations. But if we actually determine the genotypic proportions in a real population in a given generation, i.e. S_g, we cannot equate these with $\mathscr{S}_g$, and use them in the fundamental equation. In the language of probability theory, S_g is the "result of a trial" which consists of allowing g generations to pass; this result cannot tell us anything about the probability distribution for this type of trial [2].

In particular, if we find the gene frequencies in generation g, i.e. if we know s_g, the results derived in Chapter 3 for populations obeying the first five conditions for the establishment of equilibrium show that:

$$\mathscr{S}_{g+1} = \bar{s}_g^2 = S_P^{(g)}.$$

The genotype proportions will be given by the Hardy-Weinberg formula. Some authors have stated that there will be a deviation from these proportions, in a finite population. However, this is not so, and any deviation that may be observed must result from the failure of one of the other conditions for the equilibrium to be satisfied, e.g. the conditions of random mating and no selection; they are not caused by the inbreeding due to the finite population size.

Since, for any generation, $\mathscr{S}_g = \bar{s}_{g-1}^2$, the fact that $\mathscr{S}_g$ can change proves that s_g must also change. This is the phenomenon of "genetic drift". In particular, when the final, equilibrium state is reached, the genotypic probability structure is:

$$\hat{\mathscr{S}} \equiv \begin{vmatrix} p_1 \\ \vdots \\ \cdots \cdots \\ \vdots \\ 0 \ldots p_i \\ \vdots \quad \vdots \\ \cdots \cdots \cdots \\ \vdots \quad \vdots \\ 0 \ldots 0 \ldots p_j \\ \vdots \quad \vdots \quad \vdots \\ \cdots \cdots \cdots \cdots \\ \vdots \quad \vdots \quad \vdots \\ 0 \ldots 0 \ldots 0 \ldots p_n \end{vmatrix}$$

[2] Cf. Malécot (1969), pp. 34–35.

Therefore the only possible genotypic structures are of the form:

$$S_{ii} \equiv \begin{vmatrix} 0 & \ddots & \\ \vdots & \ddots 1 & \ddots \\ 0 & \cdots\cdots & 0 \end{vmatrix}$$

and the only genic structures that are possible are:

$$s_i \equiv (0, \ldots, 1, \ldots, 0).$$

Thus a real population becomes fixed for one allele, though the genic probability structure s_g remains equal to s_0.

For example, in the simple case of two alleles A_1 and A_2, with initial frequencies of p and q, the probability of the equilibrium state with all the members of the population $(A_1 A_1)$ homozygotes is p, and the probability of the $(A_2 A_2)$ equilibrium state is q:

$$\mathscr{S}_g \to S_H^{(0)} = \begin{vmatrix} p & \\ 0 & q \end{vmatrix} = p \begin{vmatrix} 1 & \\ 0 & 0 \end{vmatrix} + q \begin{vmatrix} 0 & \\ 0 & 1 \end{vmatrix}$$

In other words, the parameters p and q, which represent frequencies in the initial population, represent probabilities when we are considering the final, equilibrium state (in which the only frequencies that are possible are 0 and 1).

2.5. The Disappearance of Heterozygotes

Let p_i and p_j be the frequencies of alleles A_i and A_j in the initial population, and let $P_{ij}^{(0)}$ be the frequency of heterozygotes $(A_i A_j)$ in this population. Applying Eq. (24) to this genotype, we have:

$$\Pi_{ij}^{(g)} = (1 - \alpha_g - \gamma_g) \times 2 p_i p_j + \gamma_g P_{ij}^{(0)}. \tag{27}$$

We will write H_g for the probability that an individual of generation g is heterozygous, H_0 for the proportion of heterozygotes in the initial population, and $H_P^{(0)}$ for this proportion in a population in Hardy-Weinberg equilibrium with the same genic structure. Summing Eq. (27) term by term for all heterozygous genotypes, we obtain:

$$H_g = (1 - \alpha_g - \gamma_g) H_P^{(0)} + \gamma_g H_0. \tag{28}$$

But we have already seen (Eqs. (5) and (25)) that the values of α_g and γ_g are given by:

$$1 - \alpha_g = \frac{1}{\lambda_2 - \lambda_1} \{ (\lambda_2 - \lambda_1) \lambda_1^g + (1 - \lambda_1) \lambda_2^g \}$$

$$\gamma_g = \frac{1}{\lambda_2 - \lambda_1} (\lambda_2 \lambda_1^g - \lambda_1 \lambda_2^g)$$

where λ_1 and λ_2 are the roots of the equation:

$$\lambda^2 - \left(1 - \frac{1}{N_e}\right)\lambda - \frac{1}{2N_e} = 0.$$

The value of the positive root λ_2 is always considerably higher than that of the negative root λ_1. Therefore, as g increases, λ_1^g soon becomes negligible compared with λ_2^g, and Eq. (28) can be written:

$$H_g \approx \frac{\lambda_2^g}{\lambda_2 - \lambda_1}(H_P^{(0)} - \lambda_1 H_0).$$

Thus, between two successive generations, the probability that an individual is a heterozygote is reduced in the ratio:

$$\frac{H_{g+1}}{H_g} = \lambda_2 = \frac{1}{2N_e}(N_e - 1 + \sqrt{1 + N_e^2}). \tag{29}$$

This equation is valid for sufficiently high values of g, whatever the numbers N_m and N_f of males and females in the population.

When the effective population size is large enough for $1/N_e^2$ to be negligible compared with $1/N_e$, Eq. (29) becomes:

$$\frac{H_{g+1}}{H_g} \approx 1 - \frac{1}{2N_e}. \tag{30}$$

Note. This equation can also be derived by replacing the expressions for α_g and γ_g (Eqs. (5) and (25)) in Eq. (28) by the approximate expressions, valid for the case when $1/N_e^2$ is negligible compared with $1/N_e$. These are:

$$1 - \alpha_g \approx \left(1 + \frac{1}{2N_e}\right)e^{-g/2N_e}$$

$$\gamma_g \approx \frac{1}{2N_e}e^{-g/2N_e}.$$

Substituting in Eq. (28) we have:

$$H_g \approx e^{-g/2N_e}\left(H_P^{(0)} + \frac{1}{2N_e}H_0\right). \tag{31}$$

Hence:

$$\frac{H_{g+1}}{H_g} \approx e^{-\frac{1}{2}N_e} \approx 1 - \frac{1}{2N_e}.$$

Thus the probability of encountering a heterozygous individual, in a finite population, is reduced by a factor of $1/2N_e$ each generation. Notice that this result is independent of the number of alleles at the locus.

Examples. In the case of a population consisting, in each generation, of just one male and one female (sib-mating), we have $N_e = 2$, and

$\lambda_2 = \dfrac{1+\sqrt{5}}{4} = 0.809$. Thus the proportion of heterozygotes is reduced by about 19 % each generation.

If $N_m = N_f = 10$, we have $N_e = 20$, and $\lambda_2 = 0.975$. In this case, the proportion of heterozygotes is reduced by 2.5 % each generation. This example shows that the approximation, $\lambda_2 \approx 1 - \dfrac{1}{2N_e}$ is adequate, even with very small population sizes.

If the numbers of males and females are unequal, the smaller of the two numbers determines the rate of loss of heterozygosity. For example, if $N_m = 1$ and $N_f = 100$, $N_e = \dfrac{4 N_m N_f}{N_m + N_f} = 3.96$ and $\lambda_2 = 0.889$. The proportion of heterozygotes is reduced by 11 % each generation, despite the relatively large numbers of one sex.

2.6. Sib-Mating

Populations consisting of two individuals, one of each sex, have the smallest population size possible for sexually-reproducing species. Many studies of such sib-mating populations have been made.

The problem has usually been treated in terms of the case when the locus in question is initially segregating for two alleles. The starting population in this case consists either of two homozygotes, or two heterozygotes, and this simplifies the calculations.

The equations which we have derived above allow us to treat this case for any number of alleles, whatever the initial genotypic structure. Sib-mating is simply the special case when $N_e = 2$.

When $N_e = 2$, λ_1 and λ_2 are:

$$\lambda_1 = \frac{1-\sqrt{5}}{4}, \qquad \lambda_2 = \frac{1+\sqrt{5}}{4}.$$

Under the assumption that $\alpha_0 = \alpha_1 = 0$, the parameters a and b of Eq. (4) are:

$$a = \frac{5-3\sqrt{5}}{10}, \qquad b = \frac{5+3\sqrt{5}}{10}.$$

Hence:

$$1 - \alpha_g = \frac{5-3\sqrt{5}}{10} \left(\frac{1-\sqrt{5}}{4} \right)^g + \frac{5+3\sqrt{5}}{10} \left(\frac{1+\sqrt{5}}{4} \right)^g \qquad (32)$$

and, similarly:

$$\gamma_g = \frac{5+\sqrt{5}}{10} \left(\frac{1-\sqrt{5}}{4} \right)^g + \frac{5-\sqrt{5}}{10} \left(\frac{1+\sqrt{5}}{10} \right)^g. \qquad (33)$$

By substituting these values in Eq. (24), we can calculate $\mathscr{S}_g$.

For the values of α_g and γ_g for the initial generations, we note that the equation giving α_g can be written:

$$1-\alpha_g=\tfrac{1}{2}(1-\alpha_{g-1})+\tfrac{1}{4}(1-\alpha_{g-2})$$

and similarly:

$$\gamma_g=\tfrac{1}{2}\gamma_{g-1}+\tfrac{1}{4}\gamma_{g-2}.$$

The solution of this form of difference equation is known. It is called the Fibonacci series, and the successive terms are calculated by the rule that the g-th term is of the form:

$$\frac{A_g}{2^g} \quad \text{where} \quad A_g=A_{g-2}+A_{g-1}.$$

Table 8.1 gives some values for the coefficients.

Table 8.1. Sib-mating

Generation	$1-\alpha_g$	α_g	γ_g
0	1	0	1
1	1 (=2/2)	0	0
2	3/4	1/4	1/4
3	5/8	3/8	1/8
4	8/16	8/16	2/16
5	13/32	19/32	3/32
6	21/64	43/64	5/64
7	34/128	94/128	8/128
8	55/256	201/256	13/256
9	89/512	423/512	21/512
10	144/1024	880/1024	34/1024
∞	0	1	0

This table shows that, by the fifth generation of sib-mating, the weight attached to the corresponding homozygous structure $S_H^{(0)}$ is almost $\tfrac{2}{3}$, while the weight attached to the initial deviation from Hardy-Weinberg proportions is less than one tenth. By the tenth generation, the weighting factors have changed to about 86 % and 3 %.

Despite this rapid decrease in the weighting factor, it can sometimes be important to take the initial deviation from Hardy-Weinberg proportions into account, as we shall see if we study the rate of loss of heterozygosity in some particular cases.

a) Initial pair both heterozygotes ($A_1 A_2$). In this case,

$$H_0=1, \quad p_1=p_2=\tfrac{1}{2}, \quad H_P^{(0)}=2p_1 p_2=\tfrac{1}{2}.$$

Eq. (28) then becomes:

$$H_g = \tfrac{1}{2}(1 - \alpha_g + \gamma_g).$$

But it is easy to see from Eqs. (32) and (33) that:

$$1 - \alpha_g + \gamma_g = 1 - \alpha_{g-1}.$$

Hence:

$$H_g = \tfrac{1}{2}(1 - \alpha_{g-1}).$$

b) Initial pair both homozygotes $(A_1 A_1)$ and $(A_2 A_2)$. In this case, Eq. (24) is not valid, since S_1 is not equal to S_P, because of the asymmetry between the genetic structures in the two sexes; in fact, $S_1 = \begin{vmatrix} 0 \\ 1 & 0 \end{vmatrix}$.

Thus generation 1, for this case, is the same situation as generation 0 of the preceding case, so we can write for $g > 1$:

$$H_g = \tfrac{1}{2}(1 - \alpha_{g-2}).$$

In these two cases, H_g can be expressed solely as a function of α_g, and γ_g apparently disappears. This appearance is, however, deceptive. The "memory" of the initial structure plays a very important role in these cases; in the first case, it delays the decrease of H_g by one generation, compared with that of $1 - \alpha_g$; in the second case, the decrease is delayed by two generations.

2.7. Summary

We have found that the genotypic probability structure of a future generation is a function of three fundamental genotypic structures:

1. The initial structure, S_0.
2. The corresponding Hardy-Weinberg equilibrium structure, S_P.
3. The corresponding homozygous structure $S_H^{(0)}$.

The "weights" attached to these three structures are functions of the mean inbreeding coefficient, which tends towards 1 as generation number increases, and also of a coefficient which measures the effect of the initial departure from Hardy-Weinberg proportions; this coefficient tends towards zero. This coefficient enables us to describe the change of the structure of a population, in probabilistic terms, whatever its initial structure.

Finally, we have discussed the significance of the results we obtained. In discussing the concept of genetic drift, we have laid stress on the fact that finite population size leads to increased consanguinity between members of the population, and thus to an increased mean inbreeding coefficient; this coefficient enters into the expressions for the probability structure of the population, but does not affect the observed structure of a population, so long as mating is at random. The Hardy-Weinberg proportions, which link the gene and genotype frequencies in a given generation, are affected by limitation of the population size only in a probabilistic sense; in any real population, they continue to hold, whatever the level of inbreeding due to finite population size. This is not always clearly understood, and is sometimes considered to be a paradox. However, if the definitions of the concepts are clearly remembered, there is no paradox, and what we have just said seems obvious.

3. The Transmission of Genes from One Generation to the Next

We can consider the phenomenon of genetic drift from another quite different viewpoint: by analysing the effects of change during the transmission of genes from one generation to the next.

3.1. The Probability Distribution of the Number of Genes Transmitted

Consider a population with N individuals in each generation. Assume that the numbers of the two sexes are the same; then there are $N/2$ mating couples. Let $p_i^{(g)}$ be the frequency of allele A_i in generation g. Each time that a gene is transmitted to the next generation, $g+1$, the probability that it is a copy of A_i is therefore $p_i^{(g)}$; the probability that x of the $2N$ genes which are transmitted to the next generation are A_i genes is:

$$P(x) = \frac{2N!}{x!(2N-x)!} (p_i^{(g)})^x (1-p_i^{(g)})^{2N-x}. \tag{34}$$

Thus the number of A_i genes present in the population in generation $g+1$ is a random variable, and follows the binomial distribution given by Eq. (34).

This equation shows that x can take any value from 0 to $2N$, and that the probabilities of the extreme values are very small if the population size N is large.

The probability that no A_i gene is transmitted is $P(0) = (1-p_i^{(g)})^{2N}$. With $N=5$, for example, and $p_i^{(g)} = \frac{1}{2}$, $P(0)$ is equal to 10^{-3}.

Similarly, the probability that all the genes which are transmitted are A_i is $P(2N) = (p_i^{(g)})^{2N} = 10^{-3}$, in the example just given.

The value of x which has the highest probability is $x_m = 2N p_i^{(g)}$. In other words the "most probable" event is that the frequency of A_i in generation $g+1$ is the same as its frequency in generation g. However, this most probable event is far from a certainty. For the example given above, its probability is:

$$P(2N\,p_i^{(g)}) = \frac{10!}{5!\,5!}\left(\frac{1}{2}\right)^{10} = \frac{252}{1\,024}.$$

Thus the probability that the gene frequency remains exactly the same is only $\frac{1}{4}$.

If we consider a large number of populations, each consisting of five individuals, and all having a frequency of 0.5 for the same allele A_i at a particular locus, we would expect that one in a thousand of the populations would have lost the gene in the next generation; in one in a thousand populations this gene would have replaced the other alleles, and in one in four of the populations it would still be at the same frequency; in the other populations, its frequency would have changed somewhat.

Exactly the same arguments apply if we consider the changes in gene frequency at a large number of independent loci in one population, instead of considering one locus in many populations.

3.2. Changes in Gene Frequencies

The frequency distribution given in Eq. (28) can, like any frequency distribution, be characterised by its mean and variance.

The mean is:

$$E(x) = \sum_x x\,P(x) = 2N\,p_i^{(g)}$$

and the variance is:

$$V(x) = \sum_x \{x - E(x)\}^2\,P(x) = 2N\,p_i^{(g)}(1 - p_i^{(g)})$$

according to the well known formulae for a binomial distribution.

However, we are usually interested in the gene frequency $p = x/2N$, rather than the number of genes, x. It is easy to show that its mean and variance are:

$$E\{p_i^{(g+1)}\} = p_i^{(g)}$$

$$V\{p_i^{(g+1)}\} = \frac{p_i^{(g)}(1 - p_i^{(g)})}{2N}. \tag{35}$$

The expectation of the gene frequency in any generation is therefore p_i, the initial frequency. The spread of the frequencies around the mean value p_i decreases with decreasing variance, i.e. with increasing population size; in the limit when N is infinite, the variance is zero and there are no populations with changed frequency. This is the result we obtained in Chapter 3 for populations satisfying the conditions for Hardy-Weinberg equilibrium.

Table 8.2. The probability distributions of gene frequency in the offspring of a group of individuals of size N, in which the gene frequency is $\frac{1}{2}$

Population size (N)	Frequency							
	0.35	0.35 to 0.40	0.40 to 0.45	0.45 to 0.50	0.50 to 0.55	0.55 to 0.60	0.60 to 0.65	0.65
2	0.31	—	—	0.38		—	—	0.31
4	0.14	0.22	—	0.28		—	0.22	0.14
10	0.09	0.10	0.14	0.17	0.17	0.14	0.10	0.09
20	0.03	0.08	0.16	0.23	0.23	0.16	0.08	0.03
50	0.002	0.02	0.14	0.34	0.34	0.14	0.02	0.002
100	0.00003	0.002	0.07	0.43	0.43	0.07	0.002	0.00003

As an example of the effect of population size on the dispersion of gene frequencies, consider the case when a gene has the initial frequency $p_i^{(0)} = \frac{1}{2}$. Table 8.2 shows the proportions of populations with various frequencies in the next generation, for a range of values of N.

If the population size is at all large, the change from one generation to the next is very small: in a population of 100 individuals, the chance that a frequency of 0.5 is changed by more than 0.1 is less than 4 in 1000. A population of 10 individuals, however, has a chance of 38 % of undergoing as large a change as this in one generation. For small populations, therefore, there is a high probability of undergoing a large change in frequency.

3.3. Genetic Drift

We have seen that the transition from one generation to the next causes a random, unpredictable change in the frequencies of genes, and that these changes are biggest in small populations.

These changes continue so long as no other force intervenes to bring the frequencies back towards their initial values. If the frequency of a gene has changed from $p_i^{(g)}$ to some lower value $p_i^{(g+1)}$, there is no reason why the frequency in the next generation should be higher than $p_i^{(g+1)}$ and thus approach $p_i^{(g)}$ again. In each generation, the change in frequency which will take place is simply a function of the genic structure in that generation, and is unaffected by the frequencies in any earlier generation; in other words, this process is "strictly Markovian".

Wright (1931) introduced the term "genetic drift" for this process. Genetic drift causes the frequency of a gene to change slightly every generation; sometimes it will increase, and sometimes decrease, but there is no relation between the changes in different generations, since they are purely random changes. It might therefore appear that there would be no long-term trends in gene frequency, under these conditions; one might expect that the genes would take all possible frequencies without there being any tendency towards a limiting distribution of gene frequencies.

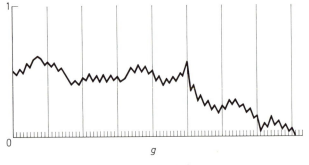

Fig. 8.5. Genetic drift

However, we have to take another factor into account: if, by the process of genetic drift, the frequency of an allele A_i becomes very low, of the order of $1/N$, the number of these genes in the population becomes small. The probability that no A_i gene is transmitted to the next generation is then quite large. For example, an allele which is represented only once among the males, and not at all among the females, has a probability of $p_e = (1 - 1/N)^N$ of being absent altogether in the next generation[3]; if N is sufficiently large, $p_e \approx e^{-1}$, so that the probability that a gene which is represented only once in one generation will be lost by the next generation is approximately 0.4.

Thus the action of genetic drift can result in the loss of an allele. Once this has happened, the lost allele cannot be restored to the population (since we are assuming that mutation does not happen); from this time onwards, its frequency remains constant at 0. In the symmetrical situation, when an allele A_i reaches a frequency close to 1, as a result of drift, the number of examples of alleles other than A_i becomes small, and these have a high chance of being lost in the passage to the next generation. If this happens, A_i is the only allele left at the locus.

As the number of generations increases, fixation of alleles at frequencies of 0 or 1 becomes more and more probable. Changes in gene frequency come to an end only when one allele is fixed, at each locus. The population then consists only of homozygotes.

---

[3] In order for the gene to be lost, it must not be transmitted to any of the N members of the next generation. Since the $N/2$ males carry N genes, the probability that the gene is not transmitted to a given individual of the next generation is $1 - 1/N$.

3.4. The Rate of Attainment of Homozygosity

Consider a population with a constant size of N individuals, in which two alleles A_1 and A_2 at a locus are segregating, each with a frequency of 0.5 in the initial generation. The frequency p of A_1 will fluctuate from generation to generation, until it becomes equal to 0 or 1. The frequency p many generations after the initial one is a random variable with a probability distribution of the form shown in Fig. 8.6; values of 0 and 1 have a high probability, while intermediate frequencies have low probabilities. We shall assume, as Wright did, that the intermediate frequencies are equiprobable; in other words, the probability distribution of the random variable "frequency of A_1 in a distant generation" is defined by:

$$P_g(0) = P_g(1) = k_g, \qquad P_g(p \neq 0 \text{ or } 1) = k'_g.$$

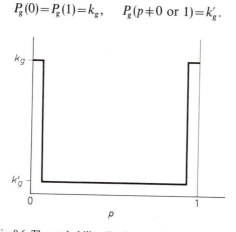

Fig. 8.6. The probability distribution of gene frequency

This assumption enables us to study this situation simply, in an approximate fashion. The expectation and variance of this frequency distribution are:

$$E_g = \tfrac{1}{2}, \qquad V_g = \sum P_g \, p^2 - \tfrac{1}{4}.$$

If we now consider the next generation, the distribution will be of the same form, but the probability k_{g+1} of the frequencies 0 and 1 will be greater, while the probability of the intermediate frequencies, k'_{g+1}, will be less than k'_g.

The probability that the locus has become fixed changes from $2k_g$ to $2k_{g+1}$. The relative decrease in the probability that the locus remains polymorphic is:

$$\lambda = \frac{2k_{g+1} - 2k_g}{1 - 2k_g}.$$

We shall now try to derive an expression for the parameter λ.

The variance of the p values in generation $g+1$ is related to that in the previous generation by:

$$V_{g+1} = (1-\lambda)\, V_g + \frac{\lambda}{2}\, (0-\tfrac{1}{2})^2 + \frac{\lambda}{2}\, (1-\tfrac{1}{2})^2$$

$$= (1-\lambda)\, V_g + \frac{\lambda}{4}.$$

Hence:

$$V_{g+1} - V_g = \lambda\, (\tfrac{1}{4} - V_g). \tag{36}$$

This increase in variance can be regarded as the result of the chance gene frequency changes due to the random nature of the transmission of genes. We have already seen that, if the frequency of A_1 in generation g is $p_1^{(g)}$, the variance due to these chance changes is $V(p_1^{(g)}) = \dfrac{p_1^{(g)}(1-p_1^{(g)})}{2N}$. The total increase in variance is therefore:

$$V_{g+1} - V_g = \sum P_g\, V(p_1^{(g)})$$

$$= \sum P_g\, \frac{p_1^{(g)}(1-p_1^{(g)})}{2N}$$

$$= \frac{1}{2N}\, (E_g - V_g - \tfrac{1}{4}) \tag{37}$$

$$= \frac{1}{2N}\, (\tfrac{1}{4} - V_g).$$

Comparison of Eqs. (36) and (37) shows that $\lambda = 1/2N$.

In other words, the probability that a locus remains polymorphic is decreased by a factor of $1/2N$ in each generation.

If we write l_g for the number of loci with more than one allele in generation g, in a population of size N (or, equivalently, the number of populations of size N which have more than one allele at a given locus), we can write:

$$\Delta l_g = -\frac{1}{2N}\, l_g.$$

Therefore:

$$l_g = l_0\, e^{-g/2N}. \tag{38}$$

To summarise, we have seen by two independent arguments that a finite population becomes progressively more homozygous, at more and more loci. The action of chance events alone results in the elimination of all but one of the alleles present at a locus; finally, all members of the population will be homozygous, and their two genes will be identical by descent. The speed with which this state is reached is inversely proportional to the population size.

These conclusions are valid only insofar as our initial assumptions are valid. In the whole of this chapter we have assumed that there is no selection and no migration. In later chapters, we shall see how these factors act to maintain genetic variation.

4. Matings between Relatives in a Finite Population

As we have seen, the assumption that mating takes place "at random" means that mating will sometimes take place between a brother and sister, or between cousins. The absence of an element of choice of mates does not mean that no consanguineous matings take place. If, however, the population in question is in a state of Hardy-Weinberg equilibrium, condition ⑥ of Chapter 3 (infinite population size) has the consequence that the probability of matings between relatives is "infinitely small", and can be considered to be zero. This is obviously not valid when the population is finite.

We shall next study the frequencies of matings between individuals who are related in a particular way, in a population of size N, consisting of $N/2$ males and $N/2$ females who mate at random.

4.1. Matings between Sibs

Brother-sister marriages are prohibited in almost all human societies. In a panmictic population, however, they could, theoretically, occur; what will be the frequency of such sib-matings?

Consider a population in which the proportion of couples in generation g that have x successful offspring (see Section 1.6) is $\lambda_x^{(g)}$; the mean and variance of this number will be:

$$\bar{x}_g = \sum_x x\, \lambda_x^{(g)},$$
$$V_g = \sum_x x^2\, \lambda_x^{(g)} - \bar{x}_g^2. \tag{39}$$

By a similar argument to that in Section 1.7, we find that S_g, the expected number of sibs of a randomly-chosen individual A, in generation g is:

$$S_g = \frac{1}{\bar{x}_{g-1}}\left(V_{g-1} + \bar{x}_{g-1}^2 - \bar{x}_{g-1}\right)$$

of whom half are of the opposite sex to A.

If the numbers of males and females in the population are the same, the total number of individuals of opposite sex to A is $\frac{1}{2} N_g$, so that the proportion of brother-sister matings will be:

$$\mu_s^{(g)} = \frac{V_{g-1} + \bar{x}_{g-1}^2 - \bar{x}_{g-1}}{\bar{x}_{g-1} N_g}$$

$$= 2 \frac{(V_{g-1} + \bar{x}_{g-1}^2 - \bar{x}_{g-1})}{\bar{x}_{g-1}^2 N_{g-1}}.$$

If we define an "index of family structure" in generation g by:

$$F_g = \frac{1}{N_g} \left(1 - \frac{1}{\bar{x}_g} + \frac{V_g}{\bar{x}_g^2} \right) \tag{40}$$

it is easy to see that:

$$\mu_s^{(g)} = 2 F_{g-1}. \tag{41}$$

This can be stated as follows: *the expectation of the frequency of brother-sister matings in a panmictic population is equal to twice the index of family structure of the preceding generation.*

4.2. Matings between First Cousins

In some human societies, first-cousin marriages are discouraged (e.g. in Western Europe under the Catholic church), but this is by no means generally true, and this type of marriage is not at all uncommon in many societies.

We can calculate the frequency of first-cousin matings under random mating, by an argument similar to the one we used for sib-mating.

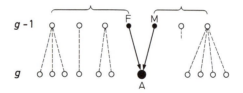

Fig. 8.7. The expectation of the number of first-cousins

Let A be a member of the population in generation g. We want to find the expectation of the number of his cousins.

First consider his paternal cousins, i.e. the offspring of the brothers and sisters of A's father F (see Fig. 8.7).

If we assume that the number of offspring of these brothers and sisters is independent of the size of family they came from (this is equivalent to assuming that family size is not a hereditary character) we can write:

$$E(N_{PC}) = E(N_{PU} \times n) = E(N_{PU}) \times E(n)$$

where N_{PC} is the number of paternal cousins, N_{PU} is the number of paternal uncles and aunts, and n is the mean number of offspring of the uncles and aunts.

But $E(n) = \bar{x}_{g-1}$, the mean number of offspring per family in generation $g-1$. N_{PU} is equal to the number of offspring of F's parents, minus 1. But the probability that F came from a sibship of size x is $K x \lambda_x^{(g-2)}$, where $\lambda_x^{(g-2)}$ is the proportion of couples in generation $g-2$ that have x offspring, and K is a constant such that $\sum K x \lambda_x^{(g-2)} = 1$. Hence, from Eq. (39):

$$K = \frac{1}{\bar{x}_{g-2}}.$$

The expectation of the number of paternal uncles and aunts is therefore

$$E(N_{PU}) = \sum_x \left[(x-1) \frac{1}{\bar{x}_{g-2}} \right] \times \lambda_x^{(g-2)}$$

$$= \frac{1}{\bar{x}_{g-2}} (V_{g-2} + \bar{x}_{g-2}^2 - \bar{x}_{g-2}).$$

Finally, the expectation of the number of paternal cousins is:

$$E(N_{PC}) = \bar{x}_{g-1} (\bar{x}_{g-2} - 1 + V_{g-2}/\bar{x}_{g-2}).$$

An identical argument gives the expected number of maternal cousins.

If we count double cousins twice over (these are cousins of A on both the mother's and the father's side) we can write the expected number of first cousins C of an individual A in generation g as:

$$C_g = 2 \bar{x}_{g-1} \left(\bar{x}_{g-2} - 1 + \frac{V_{g-2}}{\bar{x}_{g-2}} \right). \tag{42}$$

Half of these will be of opposite sex to A.

But the total number of individuals of opposite sex to A is:

$$\tfrac{1}{2} N_g = \tfrac{1}{4} \bar{x}_{g-1} N_{g-1} = \tfrac{1}{8} \bar{x}_{g-1} \bar{x}_{g-2} N_{g-2}.$$

The probability of a mating between cousins is therefore:

$$\mu_{FC}^{(g)} = \frac{8}{N_{g-2}} \left(1 - \frac{1}{\bar{x}_{g-2}} + \frac{V_{g-2}}{\bar{x}_{g-2}^2} \right)$$

or, in terms of the index of family structure, F:

$$\mu_{FC}^{(g)} = 8 F_{g-2}. \tag{43}$$

This can be stated as follows: *the expectation of the frequency of matings between first cousins in a panmictic population in generation g is equal to eight times the index of family structure of generation $g-2$.*

A similar argument gives the following result for second cousins:

$$\mu_{\text{SC}}^{(g)} = 32 F_{g-3}. \tag{44}$$

4.3. The Role of the Variance in Number of Offspring

The results expressed in Eqs. (40) to (44) represent a more complete treatment of a problem that was partially solved by Dahlberg (1948). Dahlberg, however, made the assumption that all the sibships in a given generation were the same size. Under this assumption V_g would be equal to zero, and the formula for cousin marriages, for example, becomes:

$$\mu_{\text{FC}}^{(g)} = \frac{8}{N_{g-2}} \left(1 - \frac{1}{\bar{x}_{g-2}} \right)$$

$$= \frac{2\bar{x}_{g-2}(\bar{x}_{g-2}-1)}{N_g} \tag{45}$$

and, for second cousins:

$$\mu_{\text{SC}}^{(g)} = \frac{32}{N_{g-3}} \left(1 - \frac{1}{\bar{x}_{g-3}} \right)$$

$$= \frac{4\bar{x}_{g-1}\bar{x}_{g-2}(\bar{x}_{g-3}-1)}{N_g}.$$

However, the term for the variance in family size can be very important in practice. In an intuitive way, it is easy to understand this by considering the extreme case of a population which, in each generation, consists of families half of whom have 4 successful offspring, while the other half have none; the mean sibship size in such a population is therefore 2. The father and the mother of each individual must each have been one of a set of four sibs, since it is impossible to belong to a sibship of size zero. Therefore each individual would have $2(4-1)=6$ uncles and aunts, and the expected number of first cousins would be $6 \times 2 = 12$, whereas Dahlberg's formula gives an expected number of 4.

This result agrees with Eq. (42), since we have $\bar{x}=2$, and $V = \frac{1}{2}(4-2)^2 + \frac{1}{2}(0-2)^2 = 4$, so that:

$$C_g = 2 \times 2 (2-1+\tfrac{4}{2}) = 12.$$

No real situation is as extreme as this example, so that the correct expectation of the number of cousins will not differ from the one given by Dahlberg's formula by as much as in this example. But it will always be higher.

A case which is often studied is to assume that sibship size follows a Poisson distribution:

$$\lambda_x = e^{-m} \frac{m^x}{x!}$$

where m is the mean size of sibships ($\bar{x}$). In this case, $V_g = \bar{x}_g$ and the index of family structure, F, becomes:

$$F_g = \frac{1}{N_g} \left(1 - \frac{1}{\bar{x}_g} + \frac{\bar{x}_g}{\bar{x}_g^2} \right) = \frac{1}{N_g}.$$

The frequency of cousin marriages will then be:

$$\mu_{FC}^{(g)} = \frac{8}{N_{g-2}}$$

whereas Dahlberg's formula (45) gives:

$$\mu_{FC}^{(g)} = \frac{8}{N_{g-2}} \left(1 - \frac{1}{\bar{x}_{g-2}} \right)$$

which, if $\bar{x}_{g-2} = 2$, gives:

$$\mu_{FC}^{(g)} = \frac{4}{N_{g-2}}$$

which is half the true value.

Notes. a) Formula (45) has been used mainly to estimate the population size N from the number of marriages between first cousins, on the assumption that mating is at random. The discussion of the effect of variance of family size above shows that omission of this variance term can lead to estimates of population size which are lower than they should be.

b) The variance term may also help to explain how a sudden, striking change in the frequencies of consanguineous matings may occur in a small population, without any detectable change in population size. This could sometimes be due to a past change in the mating structure of the population, since the variance term for past generations enters into the expression for the expected number of relatives of a particular kind; for example, the variance term for generation $g-2$ enters into the equation for first cousins, and, for second cousins, V_{g-3} enters.

The variance term for a past generation could be changed as a result of a change in the social and biological factors which help to maintain an approximately constant population size. For example, constant size might be maintained, at one period, purely by the equilibrium between births and infant mortality; at another period, emigration might serve to stabilise population size; at another period again, this might be achieved by limitation of family size. These mechanisms of maintaining stable population size lead to quite different variances in sibship size. The change-over from one mechanism to another would change the frequencies of the various types of consanguineous matings, after a delay of one or more generations.

5. Observations on Human Populations

The results we have just derived give us a method of estimating N from the numbers of consanguineous marriages in a population. If the population in question is a closed isolate, as we have been assuming in this chapter, N estimates the population size, and we could use this in the equations of the earlier sections of this chapter, to predict the effects which the small size of the population will have on its genetic properties.

However, most human populations are not isolated from others, but exchange migrants with one another, or even form a continuous population covering a certain area, e.g. in a city. Individuals in populations of this sort will tend to marry people who live nearby, but the area from which an individual's mate will probably come cannot be considered to be an isolate. Wright (1943) introduced the term "neighbourhood" for this concept. The genetic consequences of neighbourhood size have been studied extensively (for a review, see Wright, 1969). In particular, Wright has shown that significant genetic differentiation occurs if the neighbourhood size is less than 200. It can be shown that, in a continuous population, the value of N which we calculate from the frequencies of consanguineous marriages is an estimate of the neighbourhood size.

The formulae for the expected numbers of the different types of consanguineous marriages under random mating can be used to detect departures from random mating. In an isolate (or a continuous population) for which we have estimated the isolate size (or neighbourhood size) from the frequency of a particular type of consanguineous marriage, we can predict the expected frequencies of other types of consanguineous marriage. The expected ratios of the different types of consanguineous marriages, on the assumption of random mating, can be compared with the observed ratios in the population in question.

We shall now describe some observations of this sort on human populations. We shall see that non-random mating occurs in human populations.

5.1. The Frequency of Consanguineous Marriages

It is important to understand clearly that our ideas about the consanguinity of matings are a function of the information we possess; if pedigrees are available which go back to very distant generations, practically all marriages will be consanguineous, often through several lines of descent. In the limit, the coefficient of kinship for any two individuals is equal to 1.

If, therefore, we want to compare two populations with respect to their inbreeding, we must use comparable measures of the inbreeding.

This is usually done by studying the lines of descent for only three generations back from the present populations (i.e. taking into account the population's parents, grandparents and great-grandparents, but not its more distant ancestors). The most distant relatives to be considered are therefore second-cousins.

This method of measuring inbreeding is relatively easy to carry out in Catholic countries, because the Catholic church requires a special dispensation for marriages between second cousins or closer relatives, and a register of these dispensations is maintained by each diocese. When Catholics form a reasonably large proportion of the population (as in Belgium, Brazil, France and Italy), it is probably valid to assume that the proportions of the various types of consanguineous marriage are the same in the rest of the population. In other countries, direct enquiries have to be made, and this is necessarily limited to a sample of supposedly typical couples.

The types of consanguineous marriages which are possible if we trace back the lines of descent for three generations are as follows:

Uncle-niece or aunt-nephew marriages (called 3rd degree).
First cousin marriages (4th degree).
Marriages between first cousins once removed (5th degree).
Second cousin marriages (6th degree).
Marriages between double first cousins.

We are sometimes interested in the frequencies of these five types of marriage themselves, which we will designate by μ_3, μ_4, μ_5, μ_6 and μ_d. Often, however, we will find it useful to calculate an index which represents the mean consanguinity due to marriages of all these types. We can do this by using the coefficients of kinship corresponding to the types of relationship involved in the marriages; the resulting coefficient c will be referred to as the "*apparent consanguinity*", and is given by the formula:

$$c = \tfrac{1}{8}\mu_3 + \tfrac{1}{16}\mu_4 + \tfrac{1}{32}\mu_5 + \tfrac{1}{64}\mu_6 + \tfrac{1}{8}\mu_d.$$

It is obvious that c is not the same as the mean coefficient of kinship α defined in Chapter 6. The two are equivalent only if the information we take into consideration in calculating α is limited to the three generations preceding the one in question. If, however, α is based on the fact that the population size has been small for many generations, there is no clear relation between α and c; at most, we can regard variation in c as being indicative of variation in α. In reasonably stable populations it is probably also valid to assume that differences in c indicate differences in the amount of inbreeding due to non-random mating (i.e. due to the parameter δ as defined in Section 1.1 of the next chapter), while the contribution due to finite population size remains approximately constant.

This discussion shows how careful we must be when we try to interpret different values of c in the history of a population, or in different populations. It is always important to distinguish between α and c, so we shall continue to use different symbols and different names for them.

5.2. Consanguineous Marriages in France

The data collected by Sutter and Tabah (1948) and Sutter and Goux (1962) for the population of France are some of the most complete to have been published.

a) Data are available for all departments during the period 1926–1958. During these thirty-two years, the frequency of all types of consanguineous marriages has steadily decreased. The value of the coefficient of apparent consanguinity was 86.1×10^{-5} in 1926, and had fallen to 23.2×10^{-5} by the end of this period (see Table 8.3).

A similar decrease was found to have occurred in nearly all the individual departments, and the decrease was usually most rapid in areas where the value was high at the start of the period. The result was that the inbreeding of the various departments became more uniform, and this is reflected in a decreased value of the variance of c (Table 8.4).

Table 8.3. Frequencies of consanguineous marriages in France (per thousand marriages)

	1926 to 1930	1931 to 1935	1936 to 1940	1941 to 1945	1946 to 1950	1951 to 1955	1956 to 1958
Uncle-niece or aunt-nephew	0.20	0.16	0.14	0.06	0.12	0.13	0.06
First-cousins	9.1	7.2	5.8	5.0	3.4	2.4	2.1
First-cousins once removed	2.9	2.2	1.8	1.8	1.3	1.0	0.8
Second-cousins	11.0	9.6	6.9	6.7	5.9	4.6	4.1
$c \times 10^5$	86	68	57	52	36	27	23

Table 8.4. Variation in apparent consanguinity in France

	1926 to 1930	1931 to 1935	1936 to 1940	1941 to 1945	1946 to 1950	1951 to 1955	1956 to 1958
Mean $c \times 10^5$	86	68	57	52	36	27	23
Standard deviation of c between departments	44	36	32	26	21	13	10
Coefficient of variation (%)	51	53	56	50	58	48	43

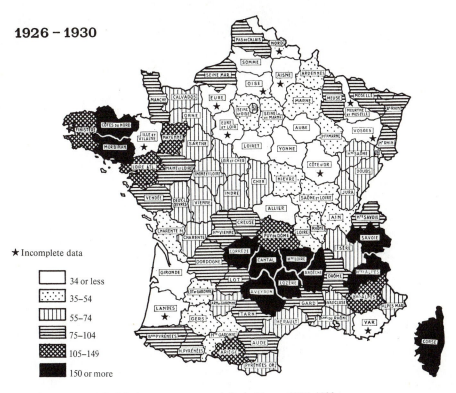

1926 – 1930

★ Incomplete data

☐	34 or less
⬚	35–54
⊞	55–74
▤	75–104
▨	105–149
■	150 or more

Fig. 8.8. Apparent consanguinity in France 1926–1930

This increase in homogeneity is very clearly shown in the two maps (Figs. 8.8 and 8.9). These show the values of the apparent consanguinity c in the different departments, in 1926–30 and in 1956–58.

In the earlier period, Fig. 8.8 shows that there were only 6 departments in the lowest category ($c < 34 \times 10^{-5}$), while Fig. 8.9 shows that all but seven of them were in this category in the later period.

It is also interesting to compare the frequencies of first- and second-cousin marriages. From the results of Section 4.2, we expect that in a random-mating population of constant size, the frequency of second-cousin marriages should be four times the frequency of second-cousin marriages. In practice, this is far from being the case; Table 8.3 shows that marriages between second cousins are barely twice as frequent as marriages between first cousins. This shows that these populations cannot be assumed to be random-mating, and that there is a large measure of choice of mates according to relatedness.

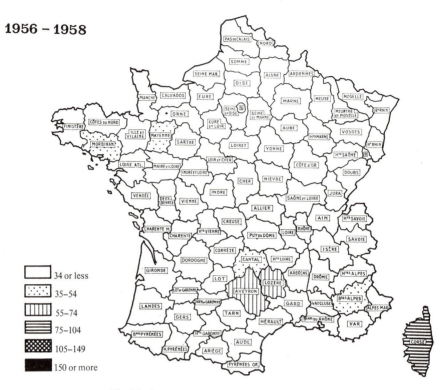

1956 - 1958

	34 or less
	35–54
	55–74
	75–104
	105–149
	150 or more

Fig. 8.9. Apparent consanguinity in France 1956–1958

It is probable that the assumption of randomness is reasonably accurate for distant relatives, for example second cousins, who might not even be aware of the relationship. In many areas, brothers and sisters prefer to marry their children to one another, so as to avoid having to divide up the family land or sell off possessions[4]; it is therefore not surprising that there is an excess of marriages between first cousins.

If, therefore, we want to make use of Eqs. (43) and (44) to estimate the size of an isolate or neighbourhood, it would seem preferable to use the frequency of second-cousin marriages rather than first-cousin marriages, otherwise the assumption that the matings take place at random may be invalid. If we assume that family size follows a Poisson distribution, V is equal to $\bar{x}$, and the index of family structure is $F = 1/N$. Eq. (44) for the size of the isolate or neighbourhood then becomes $N = 32/\mu_6$.

---

[4] The size of holdings in a wine-growing area of Loiret fell from 14000 hectares in 1830 to 8000 in 1930, simply as a result of their being divided up in legacies.

Using the data given in Table 8.3, we obtain the results shown in Table 8.5 for neighbourhood sizes in France.

Table 8.5. Neighbourhood sizes in France

	N
1926–1930	2900
1931–1935	3300
1936–1940	4600
1941–1945	4800
1946–1950	5400
1951–1955	6900
1956–1958	7800

The frequency of marriages between first cousins, on the other hand, can be used to give an indication of the extent to which the population's mating behaviour deviates from random-mating. Using the data of Table 8.3, we can calculate the excess of first-cousin marriages over the expected proportions. Table 8.6 shows that this excess has decreased sharply since the period 1941–45.

Table 8.6. Excess proportions of first-cousin marriages in France

	1926 to 1930	1931 to 1935	1936 to 1940	1941 to 1945	1946 to 1950	1951 to 1955	1956 to 1958
Actual proportion / Random-mating proportion	3.3	3.0	3.3	3.0	2.3	2.1	2.0

b) The data we have just given cover the whole of France, but this full data is unfortunately available for only a relatively short period of time. For the department of Loir-et-Cher, the excellent records which have been kept by the diocese of Blois enable one to calculate the frequencies of the various types of consanguineous marriages since 1824.

During this period of a century and a half, inbreeding has not declined steadily, as one might have expected. Fig. 8.10 shows that the frequency of consanguineous marriages was quite low in the early 19th century, and then increased steadily to a maximum during the decade 1890–1900, after which it fell rapidly; by 1946–50 the frequency of consanguineous marriages was very low.

This shows that populations in the past were not always more inbred than present-day populations, as has often been thought. Similar results have also been obtained in Italy (Moroni, 1964).

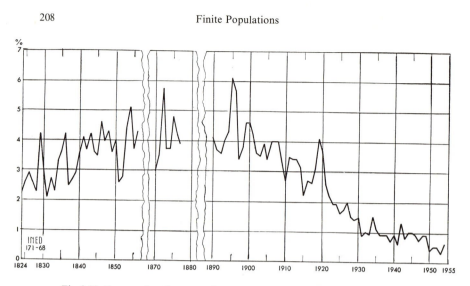

Fig. 8.10. Frequencies of consanguineous marriages in Loir-et-Cher 1824–1955

5.3. Consanguineous Marriages in Several Catholic Countries

Studies of consanguineous marriages have been carried out in several other Catholic countries, using the registers of dispensations of marriage as the source of data. We shall study some results from Italy, Brazil and Belgium, and from the province of Quebec, in Canada.

a) *Italy*. The study of Serra and Soini (1959) of inbreeding in the provinces of Como, Milan and Varese, which have a combined population of over three million people, demonstrated a large difference between Milan and the other areas (Table 8.7).

Table 8.7. Apparent consanguinity in the archdiocese of Milan

$c \times 10^5$	1903–1923	1926–1931	1933–1953
City of Milan	143	77	48
Other areas	355	213	94
All areas	285	163	75

This heterogeneity is also shown when the communities other than Milan are classified according to the sizes of their populations. Over the whole of the period studied, the apparent consanguinity in communities of less than 300 inhabitants was 430×10^{-5}; the value decreased as population size increased, down to 139×10^{-5} for communities of over 2000.

In this data, as in France, inbreeding has decreased greatly during the past 50 years. Moroni (1964) has shown that this decrease was preceded, as in France, by an increase during the 19th century. For example, in the diocese of Parma, the percentage of marriages requiring a dispensation was approximately 1 % in 1850, then rose to above 3 % in 1900, but in 1960 it was down to about 1 % again.

b) *Quebec.* Data from Quebec (Laberge, 1967) also show an effect of population size on the rate of inbreeding. In this study, the frequency of consanguineous marriages was found to be distinctly higher in rural regions than in the towns, due to the small size and the isolation of the country communities. This difference, however, decreased along with the decrease in the mean rate of inbreeding which occurred during the period studied (Table 8.8).

Despite the decrease, the apparent consanguinity in Quebec in 1955–65 was about the same as the value for France in 1925.

Table 8.8. Apparent consanguinity in Quebec

$c \times 10^5$	1885–1895	1915–1925	1945–1955	1955–1965
Urban areas	190	190	72	41
Rural areas	303	338	146	94
All areas	286	316	134	86

c) *Belgium.* Twiesselmann, Moureau and François (1962) give data on the various different kinds of consanguineous marriages in Belgium. Table 8.9 gives a summary of their findings, and shows that the values are very similar to those for France, both in terms of overall values, and for the different particular types of consanguineous marriages.

Table 8.9. Consanguineous marriages in Belgium (per thousand marriages)

	1918 to 1919	1925 to 1929	1935 to 1939	1945 to 1949	1955 to 1959
Uncle-niece or aunt-nephew	0.4	0.2	0.1	0.1	0.1
First-cousins	9.8	6.0	4.9	3.8	2.2
First-cousins once removed	2.4	1.9	1.5	1.5	1.0
Second-cousins	9.9	8.8	7.9	6.6	6.1
$c \times 10^5$	92	62	51	42	29

d) *Brazil.* The data of Freire-Maia (1957) show extreme differences in the rates of inbreeding in different provinces. In 1950–1960, the apparent consanguinity for the whole population was 200×10^{-5}; in the state of Piaui (North-East Brazil) it was 442×10^{-5}, while in the federal district (which includes Rio-de-Janeiro) it was only 40×10^{-5}.

5.4. Consanguineous Marriages in some Non-Catholic Countries

In the absence of a register of consanguineous marriages such as those kept by the Catholic church, we can obtain an estimate of the frequency of these marriages from studies on representative samples of the population.

a) *Japan.* In Japan, these studies are helped by the existence of "Koseki" books. These keep a record of all the important events concerning a family. Schull (1958) has published very full data based on 36 286 couples from the prefecture of Hiroshima, and 33 319 couples from the prefecture of Nagasaki. Table 8.10 shows the very high rates of

Table 8.10. Consanguineous marriages in Japan (per thousand marriages)

	Hiroshima	Nagasaki
First-cousins	35.2	48.1
First-cousins once removed	12.1	12.4
Second-cousins	14.7	19.5
$c \times 10^5$	281	370

inbreeding that were found. These high rates are mainly due to the high frequency of marriages between first cousins, which is approximately 2.5 times higher than the frequency of marriages between second cousins; this should be compared with the expectation on the assumption of random-mating, that first-cousin marriages should be four times *less* frequent than second-cousin marriages. This deviation shows the strength of social pressure in favour of first cousin marriages in the Japanese population.

b) *India.* Data obtained by direct questioning of a population sample are given by Dronamraju (1964) for the population of Andhra Pradesh (East Coast). The frequency of consanguineous marriages was found to be very high, because the caste system imposed considerable restrictions on the choice of marriage partner (Table 8.11).

Table 8.11. Consanguineous marriages in Andhra Pradesh (India)
(per thousand marriages)

Uncle-niece	89
First-cousins	165
First-cousins once removed	30
Second-cousins	27
$c \times 10^5$	2 280

The value of the apparent consanguinity found in this study is by far the highest recorded for a large population. Unfortunately the sample was small (327 couples) so that the values are not very precise.

c) *Senegal.* According to Cantrelle (1960), the high rate of inbreeding in the Peul tribesmen of the Fouta-Djalon, Senegal is also due to a caste system, whose barriers are more rigidly-fixed even than those between races. In addition, the high frequency of first-cousin marriages, which is as high as that of marriages between second-cousins, shows that this type of marriage is favoured in this population (Table 8.12).

Table 8.12. Consanguineous marriages in the Fouta tribesmen of Senegal
(per thousand marriages)

First-cousins	113
First-cousins once removed	50
Second cousins	113
$c \times 10^5$	1 040

These estimates are based on a sample of 6 208 couples, and are therefore quite precise. The rate of inbreeding in this population is ten times higher than that in France or Belgium 50 years ago.

5.5. Mating between Relatives in Populations with Overlapping Generations

The models which we have considered so far have assumed that the generations are distinct. This is clearly unrealistic for human populations. Hajnal (1963) has studied the case of overlapping generations, taking into account age-preferences in the choice of mates. This model leads to predictions of the frequencies of the different types of consanguineous matings, which are more like those observed in human populations than those predicted by models which assume non-overlapping generations. For instance, first-cousins are often of suitable ages to marry one another, so that a higher frequency of first-cousin marriages than predicted by a model with distinct generations is to be expected in human populations. Cavalli-Sforza *et al.* (1966) have also studied this question, and have compared the observed frequencies of consanguineous marriages with those to be expected in age-structured populations.

6. Subdivision of a Population

It is a common event in the evolution of a species, including the human species, for populations to become subdivided into a number of subgroups; after such splitting, each sub-population experiences the effects of genetic drift, which are especially marked when the group is small. Genetic drift has the following two effects: first, the mean inbreeding coefficient of the population (measured in terms of the α coefficients of the groups) are increased; and second, the genic structures of the different groups diverge.

These two effects are simply different aspects of the same phenomenon. We shall next see how the parameters which measure them can be related to one another.

6.1. Changes in Gene Frequencies and Coefficients of Kinship

Consider a homogeneous population of size N in Hardy-Weinberg equilibrium. Let its genic structure at a given locus be

$$s_0 = \{p_i^{(0)}, \ldots, p_i^{(0)}, \ldots, p_n^{(0)}\}.$$

Now suppose that the population becomes divided into t groups $G_1, \ldots, G_r, \ldots, G_t$, where the size of the r-th group is N_r, and with $\sum_r N_r = N$. To simplify matters, we shall assume that these groups remain totally separate from one another.

Since the groups are finite in size, genetic drift will occur in each one, and the gene frequencies, which start the same in all groups, gradually change within the groups, which thus diverge from one another. We have seen that the frequency $p_{ir}^{(g)}$ of allele A_i in group G_r, in generation g, is a random variable whose expectation, in the absence of selection and mutation, is equal to the initial frequency:

$$E[p_{ir}^{(g)}] = p_i^{(0)}.$$

What is the second moment of this variable? Since each group remains panmictic, we have

$$E[p_{ir}^{(g)}]^2 = E[P_{ii,r}^{(g)}], \tag{46}$$

$P_{ii,r}^{(g)}$ is the frequency of the genotype $(A_i A_i)$ in the r-th group, in generation g. Now we know from Eq. (24b) that this frequency is such that:

$$E[P_{ii,r}^{(g)}] = \alpha_r^{(g)} p_i^{(0)} + [1 - \alpha_r^{(g)}][p_i^{(0)}]^2,$$ (47)

where $\alpha_r^{(g)}$ is the mean inbreeding coefficient of the r-th group, in generation g.

The mean frequency $\bar{p}_i^{(g)}$ of the allele A_i in generation g, taken over all the groups, is similarly a random variable with expectation:

$$E[\bar{p}_i^{(g)}] = \sum_r \frac{N_r}{N} E[p_{ir}^{(g)}] = p_i^{(0)}.$$

Taking account of the property that the gene frequencies in the different groups are independent, because the groups are completely isolated, the second moment is:

$$E[\bar{p}_i^{(g)}]^2 = \sum_r \frac{N_r^2}{N^2} [E\{p_{ir}^{(g)}\}^2 - \{p_i^{(0)}\}^2] + [p_i^{(0)}]^2,$$

which, by Eqs. (46) and (47), becomes:

$$E[\bar{p}_i^{(g)}]^2 = \sum_r \frac{N_r^2}{N^2} \alpha_r^{(g)} \{p_i^{(0)}[1 - p_i^{(0)}]\} + [p_i^{(0)}]^2.$$

Hence, putting:

$$\tilde{\alpha}_g = \sum_r \frac{N_r^2}{N^2} \alpha_r^{(g)},$$

we obtain:

$$E[\bar{p}_i^{(g)}]^2 = \tilde{\alpha}_g p_i^{(0)} + (1 - \tilde{\alpha}_g)[p_i^{(0)}]^2.$$ (48)

At any given time after the splitting of the population, the frequency of allele A_i is different from that in other groups. The dispersion of allele frequency is measured by the variance:

$$V_i^{(g)} = \sum_r \frac{N_r}{N} [p_{ir}^{(g)}]^2 - [\bar{p}_i^{(g)}]^2.$$

This variance is itself a random variable whose expectation is:

$$E[V_i^{(g)}] = \sum_r \frac{N_r}{N} E[p_{ir}^{(g)}]^2 - E[\bar{p}_i^{(g)}]^2,$$

or, by Eq. (47):

$$E[V_i^{(g)}] = \sum_r \frac{N_r}{N} \{\alpha_r^{(g)} p_i^{(0)} + [1 - \alpha_r^{(g)}][p_i^{(0)}]^2\} - E[\bar{p}_i^{(g)}]^2.$$

Putting:

$$\bar{\alpha}_g = \sum_r \frac{N_r}{N} \alpha_r^{(g)},$$

and taking account of Eq. (48), we obtain:

$$E[V_i^{(g)}] = (\bar{\alpha}_g - \tilde{\alpha}_g) \, p_i^{(0)} [1 - p_i^{(0)}] . \tag{49}$$

A similar train of reasoning leads to an expression for the expectation of the covariance of the frequencies of A_i and A_j, in generation g. We remember that, since the groups are panmictic, the frequency of heterozygotes $(A_i A_j)$ in the r-th group in generation g is:

$$ij(P_{ij,r}^{(g)}) = 2 p_{ir}^{(g)} p_{jr}^{(g)} .$$

Furthermore, by Eq. (24)

$$E[P_{ij,r}^{(g)}] = 2 [1 - \alpha_r^{(g)}] \, p_i^{(0)} p_j^{(0)} .$$

We can thus obtain the result:

$$E[\text{Cov}_{ij}^{(g)}] = -(\bar{\alpha}_g - \tilde{\alpha}_g) \, p_i^{(0)} p_j^{(0)} . \tag{50}$$

Finally, the "standardised variances and covariances":

$$\frac{V_i^{(g)}}{p_i^{(0)} [1 - p_i^{(0)}]} \quad \text{and} \quad \frac{-\text{Cov}_{ij}^{(g)}}{p_i^{(0)} p_j^{(0)}} , \tag{51}$$

are random variables with the same expectation, provided that the assumptions of our model are met (i.e. no selection or mutation, total isolation of the groups, and random mating within the groups). This expectation is given by:

$$\bar{\alpha}_g - \tilde{\alpha}_g ,$$

where $\bar{\alpha}_g$ is the mean inbreeding coefficient of the overall population, and $\tilde{\alpha}_g$ is the quantity defined above.

When each group is a small fraction of the total population, $\tilde{\alpha}_g$ can be neglected, compared with $\bar{\alpha}_g$, and the standardised variances and covariances defined above provide estimates of the mean inbreeding coefficient of the population, $\bar{\alpha}_g$.

The formula:

$$\bar{\alpha} \approx \frac{E(V_i)}{p_i(1 - p_i)}$$

expresses the desired relation between the two effects of genetic drift which we mentioned at the beginning of this section: the increased consanguinity between individuals, and the differentiation between the genic structures of the groups.

Note, however, that it is not the "standardised variance" itself, but its expectation, which is equal to $\bar{\alpha}$. If we want to estimate $\bar{\alpha}$ by means of this expectation, we therefore need to know V_i and p_i for many genes, and to find the mean value. In practice, the standardised variances calculated for different genes are often very different. This has sometimes been considered to indicate that one or more of the assumptions of the model is invalid, for example that selection has taken place. Such a conclusion, however, is valid only if we know that the differences are statistically significant; we must realise, in this context, that our data is limited, being based on a sample from each group, and on only a few of the groups.

6.2. Effect of Limited Sample Sizes

The calculation of $\bar{\alpha}_g$ using Eqs. (49) or (50) assumes that we have exact knowledge of the frequencies of the allele in question, in the different groups. In practice, these frequencies are usually estimated from samples of individuals, so that, even if the frequencies were actually all the same, the estimates would show a certain scattering due to sampling; this would be bigger, the smaller the samples.

Let N_r^* be the number of individuals studied in group G_r. Assuming that the frequency of an allele, A_i, is the same $(\bar{p}_i)$ in every group, the observed frequency p_{ir}^* is a binomial variate with mean $\bar{p}_i$ and variance:

$$\frac{\bar{p}_i(1-\bar{p}_i)}{2N_r^*}.$$

If N_r^* is sufficiently large, the variable:

$$\frac{p_{ir}^* - \bar{p}_i}{\sqrt{\dfrac{\bar{p}_i(1-\bar{p}_i)}{2N_r^*}}}$$

can be considered as a standardised normal variate. The expression:

$$X = \sum_{r=1}^{t} \frac{2N_r^*(p_{ir}^* - \bar{p}_i)^2}{\bar{p}_i(1-\bar{p}_i)}$$

is thus the sum of t standardised normal variates; it is therefore a χ^2 with t degrees of freedom. Its mean is t and its variance is equal to $2t$. We therefore have:

$$E(X) = \frac{2}{\bar{p}_i(1-\bar{p}_i)} \sum_r E[N_r^*(p_{ir}^* - \bar{p}_i)^2] = t.$$

If approximately the same fraction of each sub-population is sampled, the variance of the observed frequencies is, by definition:

$$V_i = \sum_r \frac{N_r^*}{N^*}(p_{ir}^* - \bar{p}_i)^2.$$

The expectation of V_i, on the assumption that the populations are homogeneous, is therefore:

$$E(V_i|\text{homogeneous}) = \frac{1}{N^*} \sum_r E[N_r^*(p_{ir} - \bar{p}_i)^2]$$

$$= \frac{t\,\bar{p}_i(1 - \bar{p}_i)}{2N^*}.$$

(52)

To allow for the fact that we do not know the true value of $\bar{p}_i$, but have to use our estimated value $\bar{p}_i^*$ in Eq. (52), we reduce the number of degrees of freedom of the variable X from t to $t-1$. Thus, the following estimate of the component of variance due to sampling is obtained:

$$\frac{t-1}{2N^*} p_i^* (1 - \bar{p}_i^*).$$

(53)

Finally, we find the value of $\hat{\alpha}$, the estimated mean inbreeding coefficient of the population based on the dispersion of gene frequencies, by the formula:

$$\hat{\alpha} = \frac{V_i^*}{\bar{p}_i^*(1 - \bar{p}_i^*)} - \frac{t-1}{2N^*}.$$

(54)

6.3. Sampling Variance of α

When we calculate the standardised variance in gene frequency based on different alleles at the same locus, or on alleles at different loci, the different calculations can be considered to be based on sets of distinct groups, one set for each allele considered. This can by itself result in different estimates of α, when different genes are used. We shall now consider how to estimate this variance.

In the set of t groups, the gene frequency has a distribution which can be characterised by its moments:

$$M_1 = \bar{p},$$

$$M_2 = \sigma^2,$$

$$M_3 = E(p_r - \bar{p})^3,$$

$$M_4 = E(p_r - \bar{p})^4.$$

Suppose that we take observations on a sub-set of n groups; we shall assume for simplicity that the sizes of the groups in this set are equal. The mean frequency of the allele in question in the j-th sub-set is:

$$\bar{p}_j = \frac{1}{n} \sum p_r,$$

and the estimate of α for this sub-set is:

$$\hat{\alpha}_j = \frac{1}{n\,\bar{p}_j(1-\bar{p}_j)}\sum(p_r-\bar{p}_j)^2.$$

If the j-th sub-set is a large fraction of the total of all groups, we may assume that the mean gene frequency, $\bar{p}_j$, which we find is close to $\bar{p}$. Then the estimate:

$$\hat{\alpha}_j \approx \frac{1}{n\,\bar{p}(1-\bar{p})}\sum(p_r-\bar{p})^2$$

is a random variable whose mean and variance we can calculate. The mean is given by:

$$E(\hat{\alpha}_j) = \frac{1}{n\,\bar{p}(1-\bar{p})}\sum E(p_r-\bar{p})^2$$

$$= \frac{\sigma^2}{\bar{p}(1-\bar{p})}.$$

(55)

Assuming that the different samples are independent, the variance is given by:

$$V(\hat{\alpha}_j) = \frac{1}{n^2\,\bar{p}^2(1-\bar{p})^2}\sum V(p_r-\bar{p})^2.$$

But

$$V[(p_r-\bar{p})^2] = E(p_r-\bar{p})^4 - [E(p_r-\bar{p})^2]^2$$

$$= M_4 - \sigma^4.$$

Hence:

$$V(\hat{\alpha}_j) = \frac{M_4-\sigma^4}{n\,\bar{p}^2(1-\bar{p})^2}$$

$$= \frac{\dfrac{M_4}{\sigma^4}-1}{n}[E(\hat{\alpha}_j)]^2.$$

If the number of sub-sets is sufficiently large, we can equate $E(\hat{\alpha}_j)$ with the observed mean $\bar{\hat{\alpha}}$. Making this substitution, and also writing

$$k = \frac{M_4}{\sigma^4}-1,$$

we obtain:

$$V(\hat{\alpha}_j) = \frac{k}{n}(\bar{\hat{\alpha}})^2.$$

(56)

If we know k, we can use this equation to test whether the observations deviate from the predictions of the model. The value of k depends on the distribution we assume for p_r. For example, a binomial distribution with parameters N and Θ gives:

$$k = 2 + \frac{1 - 6\,\Theta(1-\Theta)}{N\Theta(1-\Theta)};$$

a normal distribution gives $k=2$; a uniform distribution between 0 and 1 gives $k=0.8$; a beta distribution with parameters 0.5 and 0.5 (U-shaped) gives $k=0.51$; with parameters 100 and 100, the beta distribution gives $k=2.2$; finally a Poisson distribution with parameter μ gives

$$k = 2 + \frac{1}{\mu}.$$

Lewontin and Krakauer (1973) have simulated the process of sampling a set of populations, with different distributions of p_r, and found the values of $V(\hat{\alpha}_j)$ which resulted; these were used to find values of k, by Eq. (52), and agreement with the values given here was close. Overall, k was rarely greater than 2. When distinctly higher values are found, it is tempting to think that the mathematical model which we have used here is unrealistic in some way; in particular that selection may have taken place in the history of the populations. We shall return to this topic in Chapter 10.

6.4. The Effect of Relationship between Groups

As before, we consider a population which becomes subdivided into a set of groups $G_1, \ldots, G_r, \ldots, G_t$. But now we shall assume that these groups are not totally isolated, but that migration between them can occur.

The best measure of the relatedness of two groups, G_r and G_s, is the mean coefficient of kinship Φ_{rs}, which is defined as the probability that a gene taken at random from the group G_r and a gene at the same locus taken at random from the group G_s are identical by descent. To calculate the value of Φ_{rs}, we need to know the genealogies of the individual members of the two groups, going back to the time when the populations split. This type of information is usually not available. An indirect method, based on the relation between Φ_{rs} and the gene frequencies has been proposed by Morton et al. (1971). The reasoning involved is as follows.

Let $p_{ir}^{(g)}$ be the frequency of allele A_i in the r-th group, in generation, and let $p_i^{(0)}$ be its initial frequency in the original population. If we assume that there has been no selection or mutation, we have:

$$E[p_{ir}^{(g)}] = p_i^{(0)}.$$

We now want to find the expectation of the product $p_{ir}^{(g)} \times p_{is}^{(g)}$. This represents the expectation of the frequency of $(A_i A_i)$ homozygotes in an imaginary group, group "$r+s$", formed by mating each individual from G_r with one from G_s. We have:

$$E[p_{ir}^{(g)} \times p_{is}^{(g)}] = E[P_{ii,\,r+s}^{(g)}].$$

But:

$$E[P_{ii,\,r+s}^{(g)}] = \Phi_{rs}\, p_i^{(0)} + (1 - \Phi_{rs})\, E[p_{ir}^{(g)}] \times [p_{is}^{(g)}].$$

Hence:

$$\Phi_{rs} = \frac{E[p_{ir}^{(g)} \times p_{is}^{(g)}] - [p_i^{(0)}]^2}{p_i^{(0)}[1 - p_i^{(0)}]}. \tag{57}$$

Of course, the initial frequencies $p_i^{(0)}$ are unknown. If, however, the population is large, we can use the mean frequency $\bar{p}_i^{(g)}$ found from the set of sub-groups, as an approximate estimate of $p_i^{(0)}$.

Eq. (57) therefore expresses the desired relation between Φ_{rs} and the observed frequencies. As in the previous section, however, we must remember that the operation of taking expectations means that the estimates obtained are very imprecise. It would be preferable to attach to this estimate a confidence interval; to calculate this, we would need to know the value of the variance:

$$V[p_{ir}^{(g)} \times p_{is}^{(g)}].$$

Chapter 9

Deviations from Random Mating

Human beings (and probably most animals) do not mate completely at random. There is always an element of conscious choice, based on various considerations, though the importance of this element varies from one individual to another.

It is obvious that the fact that the members of a population choose their mates can affect the genetic structure of the next generation. For example, the genetic make-up of the next generation would be affected by the prohibition of marriages between sibs, or by any tendency for individuals to marry, or avoid marrying, individuals similar to themselves with respect to some character.

If all the other conditions for Hardy-Weinberg equilibrium hold, there can be two basic types of non-random mating due to choice of mates:

1. Choice of mates according to the individuals' phenotypes: marriages between people who are felt to be "suitable" for one another, according to the criteria of the culture in which they live, are more likely to occur than other marriages. For example, it is well known that, in Western societies, tall people tend to marry tall people, and that people with high intelligence tend to marry one another. This is called "*assortative mating*".

2. Choice of mates according to the relationship between the individuals: two people may consider one another as possible marriage partners, or not, according to how they are related. This can be due to social conventions (e.g. the prohibition of certain marriages as "incestuous", for example brother-sister marriages or cousin marriages) or for economic reasons (because marriages between relatives mean that the family can keep its inheritance intact).

Non-random mating on the basis of relatedness and assortative mating have different effects on the genetic make-up of populations. Before we study these two cases we will investigate the effect on the Hardy-Weinberg equilibrium of dropping the assumption that mates are genetically independent of one another.

1. Genotype Frequencies Among the Offspring of Consanguineous and Assortative Matings

In Chapter 3, we obtained the probabilities of the various possible genotypes among the offspring of a mating. These probabilities were given by the formulae:

$$\Pi_{ii} = {}_F\pi_i \times {}_M\pi_{i|i}$$
$$\Pi_{ij} = {}_F\pi_i \times {}_M\pi_{j|i} + {}_F\pi_j \times {}_M\pi_{i|j} \tag{1}$$

where ${}_F\pi_i$ is the probability that the father transmits an A_i gene, and ${}_M\pi_{i|j}$ is the probability that the mother transmits an A_i gene, *given that the father has transmitted an A_j gene.*

If we assume that the parents are genetically independent for the locus in question, which is true in the important case when mating is "at random", we can leave out the phrase "*given that ...*", and simply put ${}_M\pi_{i|j} = {}_M\pi_i$. We then obtain the Hardy-Weinberg genotype frequencies:

$$\Pi_{ii} = \pi_i^2, \qquad \Pi_{ij} = 2\pi_i \pi_j.$$

If, however, we cannot make this assumption, Eqs. (1) can be used to find the genotype frequencies only if we know the set of conditional probabilities $\pi_{i|j}$.

1.1. An Example of Non-Independence between Mates

We could imagine an infinite number of cases, each corresponding to a set of $\pi_{i|j}$ values. The simplest case is when the probability that the mother transmits a different gene from the father is determined by a factor δ, which is the same for all alleles; in this case, we can write, for all values of i not equal to j:

$$\pi_{i|j} = (1 - \delta)\pi_i. \tag{2}$$

But we must have $\sum_i \pi_{i|j} = 1$, hence:

$$\pi_{j|j} = 1 - \sum_{i \neq j} \pi_{i|j} = 1 - (1 - \delta)\sum_{i \neq j} \pi_i$$
$$= 1 - (1 - \delta)(1 - \pi_j) = \delta + (1 - \delta)\pi_j. \tag{3}$$

The $\pi_{i|j}$ values are probabilities, and therefore have values between 0 and 1; this imposes restraints on the possible values of δ. The maximum value of δ is clearly $\delta = +1$ (which implies $\pi_{i|j} = 0$ and $\pi_{j|j} = 1$). The minimum value follows from the condition $\pi_{j|j} \geq 0$, which, by Eq. (3)

shows that:

$$\delta(1-\pi_j)+\pi_j \geqq 0, \quad \text{hence} \quad \delta \geqq \frac{-\pi_j}{1-\pi_j}.$$

Now, the right-hand side of this inequality is a decreasing function of π_j, so that the lower limit of δ is determined by the smallest frequency π_j. This limit is attained when there are only two alleles, with equal frequencies; the minimum possible value of δ is therefore $\delta_m = -1$.

The property of taking all values between $+1$ and -1 is also a property of correlation coefficients. The classical approach to the problem of non-random mating (Wright, 1921) consisted of characterising the population by the correlation between the uniting gametes which would form the next generation. It is, however, obvious that we cannot consider the δ values used in the present treatment as correlation coefficients; in most cases, δ does not take any value between -1 and $+1$; for example, when there are several alleles, and the rarest of them has a frequency of 0.2, δ varies between $-\dfrac{0.2}{0.8} = -0.25$ and $+1$, and this is quite unlike a correlation coefficient. We shall use the term "*coefficient of deviation from random-mating*" for δ; it measures the effect of the choice of mates on the probabilities of the various alleles in uniting gametes.

Under the assumptions about the choice of mates which are summed up in Eqs. (2) and (3), the probabilities of the various genotypes in the next generation which are given by Eq. (1) become:

$$\Pi_{ii}=(1-\delta)\,\pi_i^2 + \delta\,\pi_i \tag{4}$$

$$\Pi_{ij}=2(1-\delta)\,\pi_i\,\pi_j.$$

(Equations of the same form, with δ replaced by F, the "fixation index", are given by Wright, 1969.) This can be written in the condensed notation of trimats, as follows:

$$\mathscr{S}_\delta=(1-\delta)\,S_P+\delta S_H \tag{5}$$

where $\mathscr{S}_\delta$ is the set of the probabilities of the various genotypes among the offspring of a couple formed as a result of the type of choice expressed in Eqs. (2) and (3), and S_H and S_P are the structures of the corresponding homozygous and Hardy-Weinberg populations (as defined in Chapters 6 and 8) having the same gene frequencies as the population in question:

$$S_P \equiv \begin{vmatrix} p_1^2 \\ \cdots\cdots \\ 2\,p_i\,p_1 \cdots p_i^2 \\ \cdots\cdots\cdots \\ 2\,p_n\,p_1 \cdots p_n^2 \end{vmatrix} \qquad S_H \equiv \begin{vmatrix} p_1 \\ \cdots\cdots \\ 0 \cdots p_i \\ \cdots\cdots\cdots \\ 0 \cdots 0 \cdots p_n \end{vmatrix}$$

The model of mating-choice expressed in Eqs. (2) and (3) may seem to be a very special case. In fact, however, it covers several very different cases, in particular the case when the choice of mates is based on their relationship.

1.2. The Offspring of a Consanguineous Mating

1.2.1. Increase in the proportion of homozygotes. In Chapter 6, we saw that the genic probability structure $s_{B|A}$ of an individual B, given that of a non-inbred individual A, is given by:

$$s_{B|A} = 2\Phi_{AB} s_A + (1 - 2\Phi_{AB}) s \tag{6}$$

where s is the genic structure of the population $s = (p_1, \ldots, p_i, \ldots, p_n)$ and Φ_{AB} is the coefficient of kinship between A and B.

If A and B mate and produce an offspring, and we know that A has transmitted allele A_i to this offspring, then A's genic probability structure is:

$$s_A = \tfrac{1}{2} i + \tfrac{1}{2} s$$

where i is the vector whose elements are all zero except for the i-th element, which is equal to 1. Eq. (6) then gives:

$$s_{B|A_i} = \Phi_{AB} i + (1 - \Phi_{AB}) s.$$

Therefore the probability that B transmits:

1. Allele A_i, the same allele as A, is:

$$\pi_{i|i} = \Phi_{AB} + (1 - \Phi_{AB}) p_i.$$

2. Allele A_j, different from the gene transmitted by A, is:

$$\pi_{j|i} = (1 - \Phi_{AB}) p_j.$$

These equations are the same as Eqs. (3) and (2), replacing the coefficient of deviation from random mating by the coefficient of kinship of A and B.

The genotypic probability structure of the offspring of such a mating is therefore:

$$\mathscr{S}_\Phi = (1 - \Phi) S_P + \Phi S_H. \tag{7}$$

This shows that the proportion of homozygotes will increase as the coefficient of kinship increases, since, unlike the coefficient of deviation from random mating δ in Eq. (5), the coefficient of kinship Φ is always positive because it is a probability; thus the existence of the term ΦS_H in Eq. (7) means that, unless $\Phi = 0$, there will be an excess of homozygotes in $\mathscr{S}_\Phi$ compared with the corresponding Hardy-Weinberg structure.

1.2.2. The two-allele case. When there are only two alleles, A_1 and A_2, with frequencies p and q, the frequencies of the three genotypes can be designated by D (for $A_1 A_1$), $H (A_1 A_2)$ and $R (A_2 A_2)$.

In this case, Eq. (7) becomes:

$$D = (1 - \Phi) p^2 + \Phi p = p^2 + \Phi p q$$
$$H = 2(1 - \Phi) p q = 2 p q - 2 \Phi p q \qquad \qquad (8)$$
$$R = (1 - \Phi) q^2 + \Phi q = q^2 + \Phi p q .$$

These equations show very clearly the increase in both of the homozygous genotypes.

In Chapter 3 we saw that the genotypic structure of a population, when there are two alleles, can be represented on a triangular graph. The genotypic structure of a population in Hardy-Weinberg equilibrium lies on a parabola with the equation $H^2 = 4 D R$, which is called the "parabola of de Finetti". The offspring of a consanguineous mating do not satisfy this equation, but it is easy to prove, by eliminating p and q from the set of Eqs. (8), that

$$4 D R = H^2 + 2 \frac{\Phi}{1 - \Phi} H .$$

Thus the points corresponding to the offspring of this type of mating lie on a parabola below the parabola of de Finetti (Fig. 9.1).

Direct proof of Eq. (7). Eq. (7) can be proved directly by using the definition of the coefficient of kinship Φ: the coefficient of kinship is the probability that a gene from A is *identical by descent* with a gene at the same locus in B.

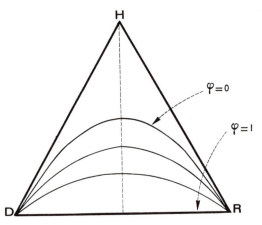

Fig. 9.1. The genotypic structure of the offspring of a consanguineous marriage

Consider the two genes received by an offspring of A and B. Either:

1. The two genes are identical (with probability Φ). Then the gene from A will be allele A_i with probability p_i, the frequency of this allele in the population. The gene from B will be the same (since we are assuming that mutations do not occur, and because the gene from B is identical with the gene from A).

2. The two genes are not identical (with probability $1 - \Phi$). They are therefore independent (since we are assuming that the population is in Hardy-Weinberg equilibrium), and the genotype frequencies are therefore:

$$\Pi_{ii} = \Phi p_i + (1 - \Phi) p_i^2 \qquad \text{for all } i$$

$$\Pi_{ij} = 2(1 - \Phi) p_i p_j \qquad \text{for } i \neq j$$

which is equivalent to Eq. (7).

1.3. The Biological Consequences of Consanguineous Mating

The results of the last section can be summarised by the statement that the proportion of homozygotes among the offspring of consanguineous matings is higher than for non-consanguineous matings. The amount of this excess increases as the coefficient of kinship increases; this effect is also more pronounced when the allele in question has a low frequency in the population than when it is common. We can see this as follows. For an allele A_2 whose frequency is q:

$$\Pi_{11}(\Phi) = (1 - \Phi) q^2 + \Phi q, \qquad \Pi_{11}(0) = q^2.$$

Hence:

$$\Pi_{11}(\Phi) = \left(1 - \Phi + \frac{\Phi}{q}\right) \Pi_{11}(0).$$

The term in brackets, by which $\Pi_{11}(0)$ is multiplied, increases as the frequency q decreases. This completes the proof.

Many diseases are due to recessive genes, and are manifest only by homozygotes for the genes. Cystic fibrosis has been mentioned earlier in this book, as a disease caused by homozygosity for a recessive gene whose frequency is about 2 % in Europe. The chance that an offspring of a non-consanguineous marriage will have this disease is therefore $(\frac{2}{100})^2 = 4 \times 10^{-4}$; for a child of a first-cousin marriage (for which the coefficient of kinship $\Phi = \frac{1}{16}$), the probability becomes:

$$\tfrac{15}{16} = (\tfrac{2}{100})^2 + \tfrac{1}{16} \times \tfrac{2}{100} = 16.2 \times 10^{-4}.$$

This is four times the value for an offspring of a non-consanguineous marriage.

In the case of a rarer gene, the increase in frequency among the offspring of consanguineous marriages is even greater. For a gene whose frequency $q = \frac{2}{1000}$, the proportion of affected individuals is $(\frac{2}{1000})^2 = 4 \times 10^{-6}$ among the offspring of non-consanguineous marriages, and for first-cousin marriages:

$$\frac{15}{16}\left(\frac{2}{1000}\right)^2 + \frac{1}{16} \times \frac{2}{1000} = 129 \times 10^{-6}.$$

This is an increase of 32 times.

The effect is even more marked when the disease is due to simultaneous homozygosity at two loci. Let us assume that the loci are not linked, and let q_1 and q_2 be the frequencies of the two recessive genes in question. The probability of having the disease is the probability of being a double homozygote, which is: $\Pi(0) = q_1^2 \times q_2^2$ for the offspring of a non-consanguineous marriage, and

$$\Pi(\Phi) = \{(1 - \Phi)\,q_1^2 + \Phi q_1\}\,\{(1 - \Phi)\,q_2^2 + \Phi q_2\}$$
$$= q_1^2\,q_2^2 + \Phi q_1\,q_2\,\{(1 - q_1)\,q_2 + (1 - q_2)\,q_1\} + \Phi^2\,q_1\,q_2\,(1 - q_1)(1 - q_2)$$

for the offspring of a marriage between individuals whose coefficient of kinship is Φ.

When $q_1 = q_2 = \frac{1}{100}$, $\Pi(0)$ is therefore equal to 10^{-8} and

$$\Pi(\Phi) = 10^{-8}(1 + 198\,\Phi + 9801\,\Phi^2)$$

which, for the case of first cousins, gives $\Pi(\frac{1}{16}) = 51 \times 10^{-8}$. The risk of having the trait is therefore 51 times higher for a child of a cousin marriage than for a child of a non-consanguineous marriage.

Independently of the fact that the offspring of consanguineous matings have an increased risk of being affected by one of the serious genetic diseases such as we have considered above (whose frequencies, however, are all low), we might also expect that they would suffer a disadvantage compared with the offspring of non-consanguineous matings, because of their higher general level of homozygosity. There is much evidence that, for many loci, homozygotes are at a disadvantage compared with heterozygotes; this may be either because of homozygosity for new, deleterious mutations, or to homozygosity for genes which are maintained in the population by heterozygous advantage. We have seen that the proportion of loci that are heterozygous in a child of a consanguineous marriage is reduced (by a factor of Φ, which is $\frac{1}{16}$ if the parents are first-cousins, for example). Such families ought therefore to have higher mortality and morbidity rates, as well as decreased longevity, compared with non-consanguineous families.

Note. Sex-linked genes. The effect of consanguinity of marriages on the frequency among their offspring of disorders due to sex-linked genes, depends on the sexes of the individual members of the mates' pedigrees. Referring to Fig. 6.8, we can consider the four possible types of cousin. Consider a sex-linked gene A_1, whose frequency is p. Then the probabilities that a child of a marriage between the four types of cousins will be a homozygote $(A_1 A_1)$ are:

$$P(A_1 A_1) = p^2 \qquad\qquad \text{for cases I and III}$$

$$P(A_1 A_1) = p^2 + \tfrac{3}{16}(1-p)\, p \quad \text{for case II}$$

$$P(A_1 A_1) = p^2 + \tfrac{1}{8} p(1-p) \quad \text{for case IV}$$

1.4. The Frequency of Consanguineous Marriages among the Parents of Children Affected with Genetic Disorders

We have just seen that the frequency of homozygous children is higher among the offspring of consanguineous marriages than the offspring of unrelated parents. Conversely, the finding that a child is homozygous at a particular locus increases the probability that his parents are related.

In order to determine what this increase in probability is, we make use of Bayes' theorem (see Chapter 2). For example, let us consider the case of first-cousin marriages, and write:

C for the event: the parents are cousins; and

R for the event: the child manifests a character which is due to a recessive gene (he is therefore homozygous at the locus in question).

We can write:

$$\begin{aligned}
\text{Prob}(C \cap R) &= \text{Prob}(C) \times \text{Prob}(R \mid C) \\
&= \text{Prob}(R) \times \text{Prob}(C \mid R).
\end{aligned} \tag{9}$$

Let q be the frequency of the recessive gene, c be the frequency of marriages between first-cousins in the population, and c' be this frequency, knowing that a child of the marriage is a homozygote. Since $\Phi = \tfrac{1}{16}$ is the coefficient of kinship between first-cousins, we have:

$$\text{Prob}(C) \ = c$$

$$\text{Prob}(R \mid C) = q^2 + \tfrac{1}{16} q(1-q) = \frac{q}{16}(1 + 15q)$$

$$\text{Prob}(R) \ = q^2$$

$$\text{Prob}(C \mid R) = c'.$$

Eq. (9) can therefore be written:

$$c' = c \frac{1 + 15q}{16q}. \tag{10}$$

When the gene is rare, the parents of homozygous children are thus much more often related to one another than is the case in the general population. With $q = 1\%$, for example, c' is found to be equal to $7c$.

Note. Eq. (10) is slightly different from the formula which is generally given, and which is due to Dahlberg (1948). The usual formula uses the value of $\text{Prob}(R)$ derived by the following reasoning. Either the parents are first cousins (probability c), and their child is a homozygote $\left(\text{probability } \dfrac{q}{16}(1 + 15q)\right)$, or else they are not cousins (probability $1 - c$), and their child is a homozygote (probability q^2). Hence:

$$\text{Prob}(R) = c \frac{q}{16}(1 + 15q) + (1 - c)q^2$$

$$= q^2 + \frac{cq}{16}(1 - q)$$

which leads to:

$$c' = c \frac{1 + 15q}{16q + c(1 - q)}.$$

In a finite population, however, we must remember that random mating can lead to consanguineous marriages. If we assume that the population is panmictic, it is therefore more correct to write $\text{Prob}(R) = q^2$ than $\text{Prob}(R) = q^2 + \dfrac{cq}{16}(1 - q)$. Eq. (10) is thus both simpler than the usual formula, and also more rigorous. Dahlberg's formula is, however, justified if the proportion of marriages between first-cousins is far from the value expected under random mating. Notice also that the two formulae are essentially equivalent when c is small.

2. Choice of Mates Based on Relatedness

In Section 1, we have studied the effect of the consanguinity of parents on the genetic make-up of their offspring. However, our basic aim is to study the evolution of populations, and not merely the results in single families. We therefore have to know the conditions which govern the frequencies of the various possible kinds of consanguineous matings. In other words, we have to make some assumptions about the mating behaviour of the population. As before, we shall start by investigating some extreme cases, which are highly unrealistic, but which will help us to set some "bounds" to the possible outcomes; this is the most we can expect to be able to do when the situation is too complex to be described in terms of a few parameters. The extreme cases are panmixia (i.e. no choice of mate) and, on the other hand, some of the regular systems of non-random mating (all matings taking place between individuals who have a particular relationship). Panmixia has been considered earlier, in Chapter 8.

In any real human population, marriages take place between individuals who are related in nearly every conceivable way, whatever the social conventions which may limit the frequency of some of them. It might therefore seem highly unrealistic to study cases when all matings are between relatives of a given type. However, these models can be useful guides to what may happen in the real situation.

2.1. Sib-Mating

If we assume that all matings in a population are between sibs, then it is obvious that the population will be effectively split up into a set of isolates of two individuals each, without genetic exchanges between the isolates. The population will behave exactly like a set of independent lines.

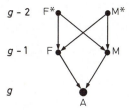

Fig. 9.2. Sib-mating

We have already studied this type of mating system in Chapter 8, as the special case of a finite population when $N = 2$. We therefore already know the principal results concerning the genetic structure of this type of population. However, we shall derive the equations for the changes in genetic structure again in this chapter, by reasoning based directly on the pedigree for this type of mating (Fig. 9.2).

Let A be an individual of generation g. His parents, F and M are brother and sister; they are both offspring of F* and M* in generation $g - 2$. A's two genes may have come either:

1. both from his grandfather F* (probability $\frac{1}{4}$);
2. both from his grandmother M* (probability $\frac{1}{4}$); or
3. one from F* and one from M* (probability $\frac{1}{2}$).

In the first case, A's genes may be either:

a) both copies of one of F*'s genes; in this case they are identical by descent and both are copies of allele A_i with probability equal to $\pi_i^{(g-2)}$, the probability of encountering an A_i gene among the genes in generation $g-2$; or

b) they may be copies of the two genes of F*; in this case A has the same genotype as F*. These two possibilities have the same probability, therefore:

$$\mathscr{S}_{A\,|\,\text{case}\,(1)} = \tfrac{1}{2}\,\mathscr{S}_H^{(g-2)} + \tfrac{1}{2}\,\mathscr{S}_{g-2}$$

where $\mathscr{S}_H^{(g-2)}$ is the corresponding homozygous structure for generation $g-2$ (see Section 2.1 of Chapter 8).

For the second case, a similar argument leads to:

$$\mathscr{S}_{A\,|\,\text{case}\,(2)} = \tfrac{1}{2}\,\mathscr{S}_H^{(g-2)} + \tfrac{1}{2}\,\mathscr{S}_{g-2}.$$

Finally, in the third case, A's two genes come from two individuals in generation $g-2$, just as if he was a member of generation $g-1$. Hence, in this case:

$$\mathscr{S}_{A\,|\,\text{case}\,(3)} = \mathscr{S}_{g-1}.$$

Combining the three cases, we therefore have:

$$\mathscr{S}_g = \tfrac{1}{2}\,\mathscr{S}_{g-1} + \tfrac{1}{4}\,\mathscr{S}_{g-2} + \tfrac{1}{4}\,\mathscr{S}_H^{(g-2)}. \tag{11}$$

Taking the contracted vector of each term of this equality, we have:

$$\mathbf{s}_g = \tfrac{1}{2}\,\mathbf{s}_{g-1} + \tfrac{1}{2}\,\mathbf{s}_{g-2}$$

which leads to:

$$\mathbf{s}_g - \mathbf{s}_{g-1} = (-\tfrac{1}{2})(\mathbf{s}_{g-1} - \mathbf{s}_{g-2})$$
$$= (-\tfrac{1}{2})^{g-1}(\mathbf{s}_1 - \mathbf{s}_0)$$
$$= 0, \quad \text{since } \mathbf{s}_1 = \mathbf{s}_0.$$

Therefore the genic probability structure remains constant from one generation to the next. The corresponding homozygous structure must also, therefore, be constant:

$$\mathscr{S}_H^{(g)} = S_H.$$

Eq. (11) can thus be written:

$$\mathscr{S}_g - S_H = \tfrac{1}{2}(\mathscr{S}_{g-1} - S_H) + \tfrac{1}{4}(\mathscr{S}_{g-2} - S_H) \tag{12}$$

which shows that the deviation between the probability structure and the corresponding homozygous structure decreases progressively, as a Fibonacci series.

In particular, if we write $\mathcal{H}_g$ for the sum of the probabilities of the various heterozygous genotypes, Eq. (12) shows that:

$$\mathcal{H}_g = \tfrac{1}{2}\mathcal{H}_{g-1} + \tfrac{1}{4}\mathcal{H}_{g-2}. \tag{13}$$

For sufficiently high values of g, the rate of decrease of $\mathcal{H}_g$ will tend towards a value λ_1 such that:

$$\frac{\mathcal{H}_g}{\mathcal{H}_{g-1}} = \frac{\mathcal{H}_{g-1}}{\mathcal{H}_{g-2}} = \lambda_1.$$

From Eq. (13), λ_1 is the largest root of the equation:

$$\lambda^2 - \tfrac{1}{2}\lambda - \tfrac{1}{4} = 0.$$

Hence:

$$\lambda_1 = \frac{1+\sqrt{5}}{4} = 0.809.$$

Thus the proportion of heterozygous individuals decreases by 19 % in each generation. Convergence towards homozygosity at all loci is therefore rapid, since the values of successive powers of λ_1 are as follows:

g	1	2	3	4	5	... 10	... 20	... ∞
λ_1^g	0.809	0.654	0.529	0.428	0.346	... 0.120	... 0.014	... 0

These figures show that nine out of ten individuals will be homozygous after 10 or 11 generations, and nearly 99 % of individuals will be homozygous after 20 generations.

Note. Transforming Eq. (13), we can write:

$$\begin{aligned}
\mathcal{H}_g &= \mathcal{H}_{g-1} - \tfrac{1}{2}\mathcal{H}_{g-1} + \tfrac{1}{4}\mathcal{H}_{g-2} \\
&= \mathcal{H}_{g-1} - \tfrac{1}{2}(\tfrac{1}{2}\mathcal{H}_{g-2} + \tfrac{1}{4}\mathcal{H}_{g-3}) + \tfrac{1}{4}\mathcal{H}_{g-2} \\
&= \mathcal{H}_{g-1} - \tfrac{1}{8}\mathcal{H}_{g-3}
\end{aligned}$$

which shows very clearly the continually decreasing nature of $\mathcal{H}_g$.

2.2. Parent-Offspring Mating

Now let us suppose that all marriages take place between a father and his daughter or a mother and her son. In this model, the idea of generation number loses its meaning, since each individual is both the child and the grandchild of the same individual. We shall adopt the convention that, of the two parents of an individual A in generation g, his mother will belong to generation $g-1$, and his father to generation

$g-2$; one generation separates a mother and her son, or a father and his daughter, but two generations come between a father and his son, or a mother and her daughter (Fig. 9.3).

A's two genes may have come either:

1. both from F (probability $\frac{1}{2}$); or
2. one from F and one from M* (probability $\frac{1}{2}$).

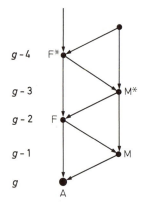

Fig. 9.3. Regular parent-offspring mating

Reasoning exactly as in the previous section, we find:

$$\mathscr{S}_{A\,|\,\text{case (1)}} = \tfrac{1}{2}\mathscr{S}_{g-2} + \tfrac{1}{2}\mathscr{S}_H^{(g-2)}$$

$$\mathscr{S}_{A\,|\,\text{case (2)}} = \mathscr{S}_{g-1}.$$

Hence:

$$\mathscr{S}_g = \tfrac{1}{2}\mathscr{S}_{g-1} + \tfrac{1}{4}\mathscr{S}_{g-2} + \tfrac{1}{4}\mathscr{S}_H^{(g-2)} \tag{14}$$

which is the same as Eq. (11) which we derived for the model of sib-mating. This type of population also tends towards homozygosity, with the proportion of heterozygotes decreasing by a factor of 0.809 each generation, but it is important to realise that the word "generation" has a quite different meaning for the two models.

2.3. Half-Sib Mating

Fig. 9.4 shows an example of a population in which all matings are between half-sibs (i.e. between the males and their half-sisters, and between the females and their half-brothers). Each male is assumed to mate with a group of n females who are all half-sisters of one another, and of the male in question. He is assumed to produce a family of one male and $n-1$ females.

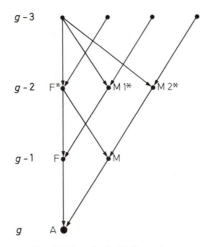

Fig. 9.4. Regular half-sib mating

The genes of A, a member of generation g, can be either:

1. both from F* (probability $\frac{1}{4}$);
2. one from F* and the other from M_1^* or M_2^* (probability $\frac{1}{2}$); or
3. one from M_1^* and the other from M_2^* (probability $\frac{1}{4}$).

It is not difficult to show that:

$$\mathscr{S}_{A\mid case\,(1)} = \tfrac{1}{2}\mathscr{S}_{g-2} + \tfrac{1}{2}\mathscr{S}_H^{(g-2)}$$

$$\mathscr{S}_{A\mid case\,(2)} = \mathscr{S}_{A\mid case\,(3)} = \mathscr{S}_{g-1}.$$

Therefore:

$$\mathscr{S}_g = \tfrac{3}{4}\mathscr{S}_{g-1} + \tfrac{1}{8}\mathscr{S}_{g-2} + \tfrac{1}{8}\mathscr{S}_H^{(g-2)}. \tag{15}$$

In this case too, we can take the contracted vector of both sides of this equality, and prove that $s_g = s_0$. We can thus write Eq. (15) in the form:

$$\mathscr{S}_g - S_H = \tfrac{3}{4}(\mathscr{S}_{g-1} - S_H) + \tfrac{1}{8}(\mathscr{S}_{g-2} - S_H) \tag{16}$$

which shows that in this case too the deviation between the probability structure and the corresponding homozygous structure decreases steadily.

The proportion of heterozygotes satisfies the difference equation:

$$\mathscr{H}_g = \tfrac{3}{4}\mathscr{H}_{g-1} + \tfrac{1}{8}\mathscr{H}_{g-2}.$$

It follows (see Appendix A) that $\mathscr{H}_g$ tends towards zero as λ_1^g, where λ_1 is the largest root of the equation:

$$\lambda^2 - \tfrac{3}{4}\lambda - \tfrac{1}{8} = 0$$

which gives:

$$\lambda_1 = \tfrac{1}{8} + (3 + \sqrt{17}) = 0.890.$$

In this case, therefore, the rate of approach towards homozygosity at all loci is somewhat less than for sib-mating, as one might expect; the proportion of heterozygotes decreases by 11 % per generation, compared with 19 % for sib-mating, and this proportion reaches $\tfrac{1}{10}$ of its initial value after about 20 rather than 10 generations.

g	1	2	3	4	5	... 10	... 20	... ∞
λ_1^g	0.890	0.792	0.705	0.628	0.558	... 0.312	... 0.097	... 0

2.4. Double First-Cousin Mating

Just as sib-mating splits up the population into a set of independent lines each consisting of two individuals in each generation, so mating exclusively between individuals who are double first-cousins splits up the population into lines consisting of four individuals in each generation. (Double first cousins are related as follows: each of the parents of one is a sib of one of the parents of the other. They therefore have only four grandparents. See Fig. 9.5.)

Consider an individual A in generation g; his parents F and M are each the offspring of a couple whose members are both sibs of the members of the other couple (in Fig. 9.5, F_1^* is a brother of F_2^*, and M_1^* is a sister of M_2^*).

A can have received his two genes either:

1. both from the same great-grandparent F_1^{**}, M_1^{**}, F_2^{**} or M_2^{**} (probability $\tfrac{1}{4}$);

2. from two great-grandparents, who are a mating pair, e.g. F_1^{**} and M_1^{**} or F_2^{**} and M_2^{**} (probability $\tfrac{1}{4}$); or

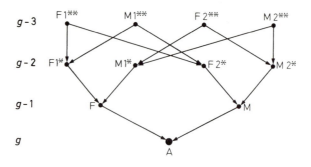

Fig. 9.5. Double first-cousin mating

3. from two great-grandparents who are not members of a mating pair — e.g. F_1^{**} and F_2^{*}, F_1^{**} and M_2^{*}, M_1^{**} and F_2^{**} or M_1^{**} and M_2^{**} (probability $\frac{1}{2}$).

The probabilities for the various genotypes that A could have, in these three cases, are as follows:

$$1. \quad \mathscr{S}_{A\,|\,\text{case (1)}} = \tfrac{1}{2}\mathscr{S}_H^{(g-3)} + \tfrac{1}{2}\mathscr{S}_{g-3}$$

since in this case A's two genes may both be copies of a single gene in one of his great-grandparents, or copies of the two different genes of a great-grandparent.

$$2. \quad \mathscr{S}_{A\,|\,\text{case (2)}} = \mathscr{S}_{g-2}$$

since in this case A's two genes are copies of two genes, one from each member of a mating pair in generation $g-3$, just as for an individual of generation $g-2$.

$$3. \quad \mathscr{S}_{A\,|\,\text{case (3)}} = \mathscr{S}_{g-1}$$

since A's two genes come from two individuals in generation $g-3$ who are not mates, just as for an individual of generation $g-1$.

Finally, therefore

$$\mathscr{S}_g = \tfrac{1}{8}\mathscr{S}_H^{(g-3)} + \tfrac{1}{8}\mathscr{S}_{g-3} + \tfrac{1}{4}\mathscr{S}_{g-2} + \tfrac{1}{2}\mathscr{S}_{g-1}.$$

As in the case of the systems of mating we have already studied, s_g can be shown to be constant, and therefore the above equation can be written in the form:

$$\mathscr{S}_g - S_H = \tfrac{1}{2}(\mathscr{S}_{g-1} - S_H) + \tfrac{1}{4}(\mathscr{S}_{g-2} - S_H) + \tfrac{1}{8}(\mathscr{S}_{g-1} - S_H). \tag{17}$$

In this case again, therefore, the deviation between the genotypic probability structure and the corresponding homozygous structure decreases steadily. When g is sufficiently high the coefficient of reduction of heterozygosity from one generation to the next is given by λ_1, the largest root of the equation:

$$\lambda^3 - \tfrac{1}{2}\lambda^2 - \tfrac{1}{4}\lambda - \tfrac{1}{8} = 0$$

which gives $\lambda_1 = 0.920$. The powers of this number are as follows:

g	1	2	3	4	5	... 10	... 20	... ∞
λ_1^g	0.920	0.846	0.778	0.715	0.658	... 0.432	... 0.187	... 0

The figures show that the proportion of heterozygotes decreases, but that this process is slow compared with the cases we have studied before; after twenty generations, the proportion of heterozygotes is still 20 % of its original value.

2.5. First-Cousin Mating

If all the matings in a population are between first cousins (i.e. between individuals with two of their grandparents in common) it is easy to show that the pedigree of an individual in generation g will contain $2i$ ancestors in generation $g-i$, compared with 2^i when all matings are between unrelated individuals. The demonstration is as follows. An individual A (Fig. 9.6), for example, has four grandparents in generation $g-2$, but two of them are brother and sister, so A has only six great-grandparents. A's father F and mother M are also offspring of cousin marriages, so that A cannot have more than eight great-great-grandparents.

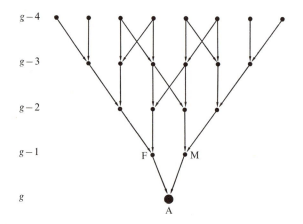

Fig. 9.6. Regular first-cousin mating

The somewhat complex form of Fig. 9.6 can be made simpler by representing the pedigree as the "dual" of its usual form; instead of representing individuals as points, and the connection between parent and offspring as lines, we represent individuals by lines, and matings by points. Each individual is represented as a line starting from the point representing the mating of his parents. Two individuals which are represented by two lines starting from the same point are therefore sibs.

In a system of regular cousin marriages, two of A's grandparents are sibs, and are therefore represented as two lines starting from the same point; the same is true for his father F and mother M. Considering successively more distant generations, we thus obtain the simple network of Fig. 9.7.

Provided that enough generations have passed since the establishment of this system of mating, we can see from Fig. 9.7 that any pair of ancestors of A are related to one another. This fact, and the fact that the number of ancestors increases linearly the further back one goes, mean that we cannot find a recurrence relation for the genotypic structure $\mathscr{S}_g$.

For example, considering generation $g-2$, we can see that the two genes of A may have come either:

1. from the sibs M_1^* and F_2^* (probability $\frac{1}{4}$);
2. from the two first cousins F_1^* and F_2^* or M_1^* and M_2^* (probability $\frac{1}{2}$); or
3. from the two second cousins F_1^* and M_2^* (probability $\frac{1}{4}$).

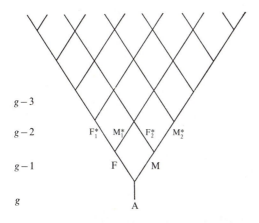

Fig. 9.7. The dual method of representing regular first-cousin mating

It is easy to show that:

$$\mathscr{S}_{A|\text{case}(1)} = \tfrac{1}{4}\,S_H + \tfrac{1}{4}\,\mathscr{S}_{g-3} + \tfrac{1}{2}\,\mathscr{S}_{g-2}$$

$$\mathscr{S}_{A|\text{case}(2)} = \mathscr{S}_{g-1}$$

but

$$\mathscr{S}_{A|\text{case}(3)},$$

which involves second cousins, cannot be found without going back to generation $g-3$. This, however, would involve third cousins, so we would have to go back to generation $g-4$, and so on for ever.

The only way to find the genetic structure in a generation is to start at the initial generation and work out the structure in each successive generation.

Before we proceed, it will be useful to make a generalisation of the notation $\mathscr{S}$ for a genotypic probability structure. We imagine that a gene is taken from an individual in generation $g-1$, and another from a second cousin of the first individual; we will write $\mathscr{S}_2^{(g)}$ for the set of the probabilities of each of the possible combinations of genes that might result. Similarly we can write $\mathscr{S}_i^{(g)}$ for the set of probabilities of the combinations of genes obtained by taking a gene from each of two i-th cousins in generation $g-1$. We shall also adopt the convention that sibs are cousins of 0th degree (0th cousins). With this notation, we can define the set of genotypic structures: $\mathscr{S}_0^{(g)}, \mathscr{S}_1^{(g)}, \ldots, \mathscr{S}_i^{(g)}, \ldots$ in each generation; since all matings are assumed to be between first cousins, only one of these genotypic structures, $\mathscr{S}_1^{(g)}$, corresponds to the genotypic probability structure of the population we are considering here.

The reasoning we employed above when we considered the four grandparents of A leads, in this notation, to:

$$\mathscr{S}_1^{(g)} = \tfrac{1}{4} \mathscr{S}_0^{(g-1)} + \tfrac{1}{2} \mathscr{S}_1^{(g-1)} + \tfrac{1}{4} \mathscr{S}_2^{(g-1)}.$$

If we now consider the second cousins F_1^* and M_2^*, we can write:

$$\mathscr{S}_2^{(g-1)} = \tfrac{1}{4} \mathscr{S}_1^{(g-2)} + \tfrac{1}{2} \mathscr{S}_2^{(g-2)} + \tfrac{1}{4} \mathscr{S}_3^{(g-2)}$$

which now involves generation $g-3$. We can carry on from each generation to the one before in this way, each of the equations having the form:

$$\mathscr{S}_i^{(g)} = \tfrac{1}{4} \mathscr{S}_{i-1}^{(g-1)} + \tfrac{1}{2} \mathscr{S}_i^{(g-1)} + \tfrac{1}{4} \mathscr{S}_{i+1}^{(g-1)} \tag{18}$$

which is valid for cousins of all degrees such that $1 \leq i \leq \dfrac{G}{2}$, where G is the number of generations since systematic cousin-marriage began.

For the case of $i=0$, which gives the probabilities of the various possible combinations of two genes that were taken from sibs, we need another formula. By considering a pair of sibs in generation $g-1$, it is easy to establish that:

$$\mathscr{S}_0^{(g)} = \tfrac{1}{4} S_H + \tfrac{1}{2} \mathscr{S}_1^{(g-1)} + \tfrac{1}{4} \mathscr{S}_1^{(g-2)}. \tag{19}$$

This can be shown as follows: the two genes have a probability of $\tfrac{1}{4}$ of being copies of the same gene, a probability of $\tfrac{1}{2}$ of being copies of two genes from two first cousins in generation $g-2$, and a probability of $\tfrac{1}{4}$ of being copies of two genes of one individual in generation $g-2$; in this last case, they must have come from two first cousins in generation $g-3$.

Eqs. (18) and (19) enable us to calculate the genotypic structures $\mathscr{S}_i^{(g)}$, and hence the structure $\mathscr{S}_1^{(g)}$ corresponding to systematic cousin marriage.

Loss of heterozygosity. If we write $\mathscr{H}_i^{(g)}$ for the sum of the probabilities of all the heterozygous combinations in the probability structure $\mathscr{S}_i^{(g)}$,

Eqs. (18) and (19) give:

$$\text{for } 1 \leq i \leq \frac{G}{2}: \quad \mathcal{H}_i^{(g)} = \tfrac{1}{4} \mathcal{H}_{i-1}^{(g-1)} + \tfrac{1}{2} \mathcal{H}_i^{(g-1)} + \tfrac{1}{4} \mathcal{H}_{i+1}^{(g-1)},$$

$$\text{for } i=0: \quad \mathcal{H}_0^{(g)} = \tfrac{1}{2} \mathcal{H}_1^{(g-1)} + \tfrac{1}{4} \mathcal{H}_1^{(g-2)}.$$

(20)

If $\mathcal{H}_0$ is the initial proportion of heterozygotes, and if we write $\mathcal{H}_i^{(g)} = (1 - \alpha_i^{(g)}) \mathcal{H}_0$, the pair of Eqs. (20) become:

$$\text{for } 1 \leq i \leq \frac{G}{2}: \quad \alpha_i^{(g)} = \tfrac{1}{4} \alpha_{i-1}^{(g-1)} + \tfrac{1}{2} \alpha_i^{(g-1)} + \tfrac{1}{4} \alpha_{i+1}^{(g-1)},$$

$$\text{for } i=0: \quad \alpha_0^{(g)} = \tfrac{1}{4} + \tfrac{1}{2} \alpha_1^{(g-1)} + \tfrac{1}{4} \alpha_1^{(g-2)}.$$

(21)

These equations enable us to calculate the values of $\alpha_i^{(g)}$, the coefficients of reduction of the proportion of heterozygous loci. Some of these values are given in Table 9.1.

Table 9.1. Values of $\alpha_i^{(g)}$

g	$i=0$	1	2	3
0	0.2500	0	0	0
1	0.2500	0.0625	0	0
2	0.2812	0.0937	0.0156	0
3	0.3125	0.1211	0.0312	0.0039
4	0.3340	0.1465	0.0469	0.0098
5	0.3535	0.1685	0.0625	0.0169
6	0.3709	0.1883	0.0776	0.0248
7	0.3863	0.2062	0.0920	0.0332
8	0.4002	0.2227	0.1059	0.0419
9	0.4129	0.2379	0.1191	0.0508
10	0.4246	0.2519	0.1317	0.0597
20	0.5076	0.3543	0.2319	0.1416
50	0.6204	0.5031	0.3951	0.3017
∞	1.0	1.0	1.0	1.0

Note. The value $\alpha_i^{(g)} = 1$ for $g \to \infty$ can be proved as follows: if there is an equilibrium state, $\alpha_i^{(g)} = \alpha_i^{(g-1)} = \hat{\alpha}_i$; from Eq. (21), we then have:

$$\tfrac{1}{2} \hat{\alpha}_i = \tfrac{1}{4} \hat{\alpha}_{i-1} + \tfrac{1}{4} \hat{\alpha}_{i+1}.$$

Hence:

$$\hat{\alpha}_{i+1} - \hat{\alpha}_i = \hat{\alpha}_i - \hat{\alpha}_{i-1}.$$

Since all values of α_i are less than or equal to 1, this difference equation shows that all the α_i tend towards the same limit whatever the value of $i \geq 1$. Let μ be this common limit. Then we have:

$$\mu = \tfrac{1}{4}\hat{\alpha}_0 + \tfrac{3}{4}\mu = \tfrac{1}{4}(\tfrac{1}{4} + \tfrac{3}{4}\mu) = \tfrac{1}{16} + \tfrac{15}{16}\mu.$$

Hence $\mu = 1$.

This shows that in a population in which all marriages are between first cousins, heterozygotes will disappear and the population will become composed of nothing but homozygotes. However, this process is extremely slow: after ten generations, the proportion of heterozygotes will have decreased by only one quarter, and after twenty generations it will have decreased by a third; it will not be decreased by a half until this system of mating has been practised for 50 generations.

2.6. Second-Cousin Mating

Now let us study what happens in a population in which all matings are between second cousins. Consider an individual A in generation g. His parents F and M (see Fig. 9.8) have four parents, and two of these,

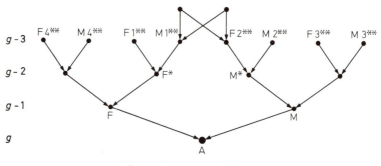

Fig. 9.8. Second-cousin mating

F* and M*, are again first-cousins; in other words the parents, F_1^{**} and M_1^{**}, of F* and M* are sibs. Let the other paternal great-grandparents of A be F_1^{**}, M_4^{**} and F_4^{**}, and let his other maternal grandparents be M_2^{**}, F_3^{**} and M_3^{**}. Then A's two genes may have come:

1. from F_2^{**} and M_1^{**} (probability $\tfrac{1}{16}$);
2. from M_1^{**} and from the mate M_2^{**} of F_2^{**}, or from F_2^{**} and from the mate F_1^{**} of M_1^{**} (probability $\tfrac{2}{16}$); or
3. from other ancestors (probability $\tfrac{13}{16}$).

Then we have:

$$1. \quad \mathscr{S}_{A\,|\,\text{case}\,(1)} = \tfrac{1}{4}\mathscr{S}_H^{(g-4)} + \tfrac{1}{2}\mathscr{S}_{g-3} + \tfrac{1}{4}\mathscr{S}_{g-4}$$

as in the section on matings between first cousins.

2. $\mathscr{S}_{A|\text{case}(2)} = \mathscr{S}_{g-2}$,

since F_2^{**} is the brother of M_1^{**}, and these are second cousins of F_1^{**}, and similarly M_2^{**} and M_1^{**} are second cousins.

3. $\mathscr{S}_{A|\text{case}(3)} = \mathscr{S}_P^{(g-3)} = s_{g-3}^2$,

since the two ancestors in generation $g-2$ from whom A received his genes are not related, so that their genic structures are independent.

Finally, therefore, we obtain:

$$\mathscr{S}_g = \tfrac{1}{64}(\mathscr{S}_H^{(g-4)} + \mathscr{S}_{g-4} + 2\mathscr{S}_{g-3} + 8\mathscr{S}_{g-2} + 52\mathscr{S}_P^{(g-3)}).$$

Taking the contraction of each term of this equality, we can show that the genic probability structure s_g is constant. Therefore the two corresponding structures $\mathscr{S}_P^{(g)}$ and $\mathscr{S}_H^{(g)}$ are also constant, and we can write:

$$\mathscr{S}_g = \tfrac{1}{64}(S_H + \mathscr{S}_{g-4} + 2\mathscr{S}_{g-3} + 8\mathscr{S}_{g-2} + 52 S_P). \tag{22}$$

If we write $\hat{S} = \tfrac{1}{53}(S_H + 52 S_P)$, Eq. (22) becomes:

$$\mathscr{S}_g - \hat{S} = \tfrac{1}{64}(\mathscr{S}_{g-4} - \hat{S}) + \tfrac{1}{32}(\mathscr{S}_{g-3} - \hat{S}) + \tfrac{1}{8}(\mathscr{S}_{g-2} - \hat{S}). \tag{23}$$

This equation shows that the deviation of the probability structure $\mathscr{S}_g$ from the structure $\hat{S}$ tends towards zero as g increases. If g is sufficiently large, the rate of decrease of this deviation is approximately equal to λ_1, the largest root of the equation:

$$\lambda^4 - \tfrac{1}{8}\lambda^2 - \tfrac{1}{32}\lambda - \tfrac{1}{64} = 0$$

or:

$$(\lambda - \tfrac{1}{2})(\lambda^3 + \tfrac{1}{2}\lambda^2 + \tfrac{1}{8}\lambda + \tfrac{1}{32}) = 0$$

which gives:

$$\lambda_1 = \tfrac{1}{2}$$

so that convergence towards the equilibrium genotypic structure is very rapid.

In this case, when all matings are between second cousins, the population does not tend towards homozygosity at all loci, despite the fact that all the matings are consanguineous. In the equilibrium state, which is quickly attained by the population, the excess of homozygotes over the panmictic proportion is only $\tfrac{1}{53}$.

For example, if there are two alleles, with frequencies p and q, the equilibrium genotype probabilities are:

$$\Pi_{11} = p^2 + \tfrac{1}{53}pq, \qquad \Pi_{12} = \tfrac{104}{53}pq, \qquad \Pi_{22} = q^2 + \tfrac{1}{53}pq.$$

2.7. Number of Ancestors
and the Approach Towards Homozygosity

As we have just seen, regular first-cousin marriage represents the limit beyond which regular systems of inbreeding no longer lead to complete homozygosity of the population. This result is due to Wright (1921).

This can be compared with the theorem of Malécot, which we derived in Section 1.4 of Chapter 8: "A population tends towards genetic homogeneity if its size increases at most as a linear function of the number of generations which have passed."

In the case of regular systems of inbreeding, we consider, not the future size of the population, but the number of ancestors an individuals has, at some particular time in the past.

As we have already noted, an individual has $n_i = 2^i$ ancestors, i generations ago, when no marriages between relatives occur; when all marriages are between first-cousins, the number is $2i$. In general, if all marriages are between k-th cousins, the number n_i of ancestors i generations back is such that:

$$n_i = 2 n_{i-1} - n_{i-k-1} \tag{24}$$

since all the ancestors in generation $i-1$ have two parents, but each pair of ancestors in generation $i-k-1$ have ancestors k generations earlier who are sibs. For some particular cases, this relation becomes:

Sib-matings (considered as matings between 0th cousins):

$$n_i = 2 n_{i-1} - n_{i-1} = n_{i-1}$$

i.e. the number of ancestors is constant.

First cousin marriages:

$$n_i = 2 n_{i-1} - n_{i-2}.$$

Hence:

$$n_i - n_{i-1} = n_{i-1} - n_{i-2}$$

so that the number of ancestors increases by a constant number in each generation.

For general values of k, Eq. (24) can be solved by the usual methods for difference equations, giving solutions of the form:

$$n_i = \lambda^i.$$

Hence:

$$\lambda^i = 2 \lambda^{i-1} - \lambda^{i-k-1}$$

or

$$f(\lambda) \equiv \lambda^{k+1} - 2 \lambda^k + 1 = 0. \tag{25}$$

If $k > 1$, the largest root of this equation is greater than 1. The derivative of the expression in Eq. (25) is:

$$f'(\lambda) = (k+1)\lambda^k - 2k\lambda^{k-1}$$

and this is equal to zero for $\lambda = 0$ and $\lambda = \dfrac{2k}{k+1}$. The curve of $f(\lambda)$ is therefore of the form shown in Fig. 9.9.

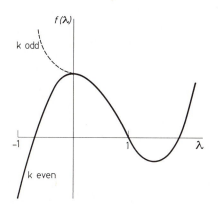

Fig. 9.9. The variation of $f(\lambda)$ with λ

This result shows that the number of ancestors increases faster than a linear function of the number of generations which have passed, when we consider more distant relationships than those between first-cousins. We have seen that in such cases, the population does not become homozygous.

This result can be considered as representing the same essential process as Malécot's result, but going backwards in time, instead of forwards.

2.8. Avoidance of or Preference for Certain Types of Marriage

The models which we have been studying so far have all been regular "systems of mating", such as could only be found in experimental stocks, and are a totally unrealistic picture of natural populations; we have studied them simply for their value as reference systems. In human populations, relationship affects mating choice in a far more subtle fashion (except perhaps for the near-universal prohibition of brother-sister marriage). Any system of mating choice, however complicated it may be, can be described by means of the simple model of Section 1.1.

Consider two individuals, A and B, belonging to a population with genic structure s; let the coefficient of kinship between A and B be Φ_{AB}. We saw in Section 1.2.1 (Eq. (7)) that the genotypic probability structure of their offspring is given by

$$\mathscr{S}_\Phi = (1 - \Phi) S_P + \Phi S_H.$$

Notice that $\mathscr{S}_\Phi$ is a linear function of Φ.

In a population in which marriages between some particular types of relative occur more or less frequently than would result from mating at random (e.g. sib matings prohibited, or marriages between first-cousins favoured, etc.), we can calculate a genotypic probability structure $\mathscr{S}_M$ by weighting each of the possible types of marriage by the appropriate coefficient of kinship, Φ, as well as by its frequency. Because $\mathscr{S}$, in Eq. (7) is a linear function of Φ, we will necessarily obtain by this procedure an equation of the form:

$$\mathscr{S}_M = (1 - \delta) S_P + \delta S_H,$$

where δ is a number between -1 and $+1$ (this is of the same form as Eq. (5)). It follows that *when choice of mates takes place solely according to the criterion of relatedness, the mating system can be characterised by a single parameter, the "coefficient of deviation from random mating", defined in Eq. (2).*

We shall next consider a few examples.

2.8.1. Prohibition of brother-sister matings. We saw in Section 4 of Chapter 8 that the expectation of the number of sibs of an individual chosen at random from a population is:

$$S = \frac{1}{\bar{x}} (V + \bar{x}^2 - \bar{x}),$$

where $\bar{x}$ is the mean and V the variance of number of "successful offspring" per mating. Consider a population of constant size N, i.e. $\bar{x}$ must be equal to 2. The above formula then gives:

$$S = \frac{V + 2}{2}.$$

Similarly, the expected number of first cousins in such a population is:

$$C = 2(V + 2).$$

Now suppose that in this population mating takes place at random, except that brother-sister matings are forbidden. What is the value of δ_s, the coefficient of deviation from random-mating?

Let X and Y be two individuals of opposite sex chosen at random from the population. Their *a priori* genic probability structure is the same as that of the population as a whole, s. Y may be either:

a sib of X, with probability $\dfrac{S/2}{N/2} = \dfrac{V+2}{2N}$. If this is the case, Y's genic probability structure is not independent of that of X; by Eq. (6) with $\Phi = \frac{1}{4}$, we get:

$$s_{Y|c} = \tfrac{1}{2} s_X + \tfrac{1}{2} s,$$

or else not a sib of X, which has the probability $1 - \dfrac{V+2}{2N}$; in this case, Y's genic probability structure is simply that of an individual who is a permitted mate for X. We shall call this s_M. We thus obtain the relation:

$$s_X = \frac{V+2}{2N} s_{Y|c} + \left(1 - \frac{V+2}{2N}\right) s_M,$$

which can be rearranged to give:

$$s_M = \frac{(4N - V - 2)s - (V+2)s_X}{2(2N - V - 2)}.$$

If when they reproduce, X contributes a gene A_i to the progeny, the genic probability structure of X is:

$$s_X = \tfrac{1}{2} i + \tfrac{1}{2} s.$$

Hence:

$$s_{M|A_i} = \frac{\left(4N - \dfrac{3V}{2} - 3\right) s - \left(\dfrac{V}{2} + 1\right) i}{2(2N - V - 2)}.$$

The probability that Y transmits allele A_j, different from A_i is then:

$$p_{j|i} = \frac{\left(4N - \dfrac{3V}{2} - 3\right)}{2(2N - V - 2)} p_j = (1 - \delta_s) p_j,$$

whence

$$\delta_s = -\frac{V+2}{4(2N - V - 2)}.$$

If we make the further assumption that family size follows a Poisson distribution (which is often true, to a good approximation), we have:

$$V = 2 \quad \text{and} \quad \delta_s = -\frac{1}{2(N-2)}. \, [1]$$

---

[1] Note that $p_i \geq \dfrac{1}{2N}$, so the condition for the lower limit of δ_s is not violated.

Prohibition of brother-sister marriage is therefore equivalent to a deviation from panmixia of $-\dfrac{1}{2(N-2)}$. This can be considered as a "negative inbreeding coefficient".

The genotypic structure of the next generation can be found using Eq. (5). This shows that, unless the population size N is extremely small, the genotype frequencies are essentially the same as in a panmictic population, except for alleles at low frequencies. For example, if $N = 10$, $\delta = -\frac{1}{16}$; for an allele with frequency 0.1, the expected frequency of homozygotes is 44×10^{-4}, compared with 100×10^{-4} in a random-mating population. If, however, N is 100, $\delta = -\frac{1}{196}$, and the effect on the genotypic structure is imperceptibly small.

2.8.2. Prohibition of both brother-sister mating and mating between first cousins. We can go through a similar argument here, as for the population considered above. The first step is to split up the population into the three following groups:

1. Sibs of X: these have a frequency of $\dfrac{V+2}{2N}$ in the population, and their genic probability structure is:

$$s_s = \frac{s_X + s}{2}.$$

2. Cousins of X: these form a fraction $2\dfrac{(V+2)}{N}$ of the population, and their genic probability structure is:

$$s_c = \tfrac{1}{8} s_X + \tfrac{7}{8} s.$$

3. Individuals who X is permitted to marry. These form $1 - \dfrac{5V+10}{2N}$ of the population, and their genic probability structure is equal to s_M, as defined in the previous section.

Hence, we obtain:

$$s = \left(\frac{V+2}{2N}\right)\left(\frac{s_X + s}{2}\right) + \frac{(V+2)(s_X + 7s)}{4N} + \left(1 - \frac{5V+10}{2N}\right) s_M,$$

which gives:

$$s_M = \frac{4(N - 2V - 4)s - 2(V+2)s_X}{2(2N - 5V - 10)}.$$

The probability that X's mate transmits allele A_j, given that X has transmitted A_i is therefore:

$$p_{j|i} = \frac{4N - 9V - 18}{2(2N - 5V - 10)} p_j$$

$$= (1 - \delta_c) p_j,$$

whence:

$$\delta_c = -\frac{V+2}{2(2N-5V-10)}.$$

Assuming a Poisson distribution for family size, we thus obtain:

$$\delta_c = -\frac{1}{N-10}.$$

2.8.3. Compensation for forbidden brother-sister matings by excess matings between first-cousins. In many human populations, incestuous brother-sister marriages are forbidden, whereas marriages between first-cousins are favoured. We shall next see to what extent these opposing tendencies cancel one another out, and result in an essentially random-mating population.

The first step is to define a parameter, λ, equal to the probability that an individual will mate with a first-cousin by choice, rather than because his randomly-chosen mate is, in fact, a first-cousin. We can then consider the population to be made up two component groups, one of size λN, in which all matings are between first-cousins, and the other of size $(1-\lambda)N$, in which mating takes place at random except that brother-sister mating is prohibited.

In the first group, the genic probability structure of an individual Y, given that of his mate X, is given by:

$$s_M^{(1)} = \tfrac{1}{8} s_X + \tfrac{7}{8} s.$$

Assuming that the second group is homogeneous with the whole population, the corresponding genotypic probability structure is:

$$s_M^{(2)} = \frac{[4(1-\lambda)N - V - 2]s - (V+2)s_X}{2[2(1-\lambda)N - V - 2]}.$$

In the whole population, it is:

$$\begin{aligned}
s_M &= \lambda s_M^{(1)} + (1-\lambda) s_M^{(2)} \\
&= \frac{\{2\lambda(1-\lambda)N - (4-3\lambda)(V+2)\}s_X}{8[2(1-\lambda)N - V - 2]} \\
&\quad + \frac{\{2(1-\lambda)(8-\lambda)N - (4-3\lambda)(V+2)\}s}{8[2(1-\lambda)N - V - 2]}.
\end{aligned}$$

If $s_X = s_M$, then for s_M to equal s, λ must satisfy

$$2N\lambda^2 - (2N + 3V + 6)\lambda + 4(V+2) = 0,$$

the genotype frequencies in the population will be the same as if the population were random-mating. Assuming a Poisson distribution of family size, the above equation reduces to:

$$N \lambda^2 - (N+6) \lambda + 8 = 0.$$

Provided that $N \geq 18$, this has two roots between 0 and 1:

$$\lambda = \frac{N+6 \pm \sqrt{N^2 - 20N + 36}}{2N},$$

or, if N is large:

$$\lambda_1 = \frac{8}{N}, \quad \lambda_2 = 1 - \frac{2}{N}.$$

Notice that the smaller of the two roots is equal to the frequency of matings between first-cousins under random-mating. Another way of looking at this is to say that the number of first-cousin matings must be double that in a random-mating population, if a population which prohibits sib-matings is to maintain random-mating proportions.

The larger root, $\lambda = 1 - \dfrac{2}{N}$, has no simple concrete interpretation. Its meaning is that the group within which, because of the prohibition of brother-sister matings, the members are forbidden to mate with one another, constitutes a fraction $1 - \lambda = 2/N$ of the population; this corresponds to $2 \times 2/N = 2$ individuals, who, of course, are sibs.

3. Assortative Mating

When individuals do not mate at random with respect to phenotype, there is said to be assortative mating.

Even if we only consider a character controlled by a single locus, it is clear that many different regimes of assortative mating are possible. For example, similar individuals may have a tendency to mate together (positive assortative mating); or they may avoid mating together (negative assortative mating); there may be a single preferred type for each individual, and all other types may be completely rejected (in which case the assortative mating is said to be total), or this tendency may be only partial; the choice might be made on the basis of the genotype, or, alternatively, the effects of dominance may be such as to render more than one genotype equivalent, so that there is only phenotypic assortative mating.

Unfortunately, each of the numerous different modes which have been studied has to be treated as a separate case, and it has been impossible to develop general methods for studying these problems. Usually the long-term results of a scheme of assortative mating have had to be studied by proceeding from generation to generation and seeing what happens. We shall examine a few simple cases here.

3.1. Total Positive Assortative Mating Based on Genotype

In this model, all matings are between two individuals who have the same genotype at the locus in question: an $(A_k A_l)$ individual will mate only with another $(A_k A_l)$ individual.

If the genotype of the pair of mating individuals is $(A_i A_i)$, all the offspring will also be homozygous $(A_i A_i)$.

If the mates are heterozygous $(A_i A_j)$, their offspring will be of the three genotypes $(A_i A_i)$, $(A_i A_j)$ and $(A_j A_j)$, in the proportions $\frac{1}{4}, \frac{1}{2}, \frac{1}{4}$.

If we assume that the population is of infinite size, we can equate frequencies and probabilities. We therefore have:

$$P_{ii}^{(g)} = P_{ii}^{(g-1)} + \frac{1}{4} \sum_{j \neq i} P_{ij}^{(g-1)}$$
$$= \frac{1}{2} P_{ii}^{(g-1)} + \frac{1}{2}\left(P_{ii}^{(g-1)} + \frac{1}{2}\sum_{j \neq i} P_{ij}^{(g-1)}\right)$$
$$= \frac{1}{2} P_{ii}^{(g-1)} + \frac{1}{2} p_i^{(g-1)}$$
$$P_{ij}^{(g)} = \frac{1}{2} P_{ij}^{(g-1)}$$

or, using the trimat notation:

$$S_g = \frac{1}{2} S_{g-1} + \frac{1}{2} S_H^{(g-1)}. \tag{26}$$

By taking the condensation of both sides of this equation, we obtain:

$$s_g = s_{g-1}.$$

Hence $S_H^{(g-1)} = S_H$, and Eq. (26) can be written in the form:

$$S_g - S_H = \frac{1}{2}(S_{g-1} - S_H) = (\tfrac{1}{2})^g (S_0 - S_H)$$

or:

$$S_g = S_H + (\tfrac{1}{2})^g (S_0 - S_H). \tag{27}$$

The population therefore tends towards the corresponding homozygous structure S_H; in other words, heterozygotes progressively disappear. This occurs very fast: Eq. (27) shows that the deviation from the equilibrium situation is halved in each generation.

3.2. Partial Positive Assortative Mating Based on Genotype

In this model, a fraction λ of matings are between individuals who both have the same genotype, while the rest of the population mates at random.

We shall assume that these two sections of the population are the same in all other respects, and that they have the same fecundity. For the first group of matings we have:

$$S_1^{(g)} = \tfrac{1}{2} S_{g-1} + \tfrac{1}{2} S_H$$

and, for the second group:

$$S_2^{(g)} = S_P.$$

Hence, for the population in general:

$$S_g = \lambda \left(\tfrac{1}{2} S_{g-1} + \tfrac{1}{2} S_H \right) + (1 - \lambda) S_P. \tag{28}$$

If there is an equilibrium state, we would have: $S_g = S_{g-1} = S_E$. From Eq. (28), therefore:

$$S_E = \frac{2(1-\lambda)}{2-\lambda} S_P + \frac{\lambda}{2-\lambda} S_H. \tag{29}$$

Therefore:

$$S_g - S_E = \frac{\lambda}{2} (S_{g-1} - S_E)$$

$$= \left(\frac{\lambda}{2} \right)^g (S_0 - S_E). \tag{30}$$

Thus the population tends towards an equilibrium state at a rate which depends on λ.

Eq. (29) shows that this equilibrium state is generally close to the panmictic structure. For example, if $\lambda = \tfrac{1}{2}$, i.e. if half the population mates assortatively with respect to genotype, we have:

$$S_E = \tfrac{2}{3} S_P + \tfrac{1}{3} S_H.$$

If $\lambda = 0.9$, we would have:

$$S_E = 0.18 S_P + 0.82 S_H.$$

This shows that a regime of only slightly less than total assortative mating results in an equilibrium population which contains a significant proportion of heterozygotes.

In the simple case when there are only two alleles, A_1 and A_2, the equilibrium structure is given by:

Homozygotes $(A_1 A_1)$: $\hat{D} = \dfrac{2(1-\lambda)}{2-\lambda} p^2 + \dfrac{\lambda}{2-\lambda} p$

Heterozygotes $(A_1 A_2)$: $\hat{H} = \dfrac{4(1-\lambda)}{2-\lambda} p q$

Homozygotes $(A_2 A_2)$: $\hat{R} = \dfrac{2(1-\lambda)}{2-\lambda} q^2 + \dfrac{\lambda}{2-\lambda} q.$

3.3. Total Positive Assortative Mating Based on Phenotype

Suppose that one of the alleles A_1 at a locus is dominant to all the others, so that all the genotypes $(A_1 A_1) \dots (A_1 A_i) \dots (A_1 A_n)$ have the same phenotype A_1. We shall assume that A_1 individuals mate only with one another, while the non-A_1 individuals mate with one another, but not with A_1 individuals.

Let the genotype frequencies in a given generation be $P_{11}, P_{12}, \dots,$ $P_{kl}, \dots, P_{nn}$. We will now find an expression for the frequency of genotypes containing the allele A_1 in the next generation.

We shall use the relations:

$$P'_{11} = \pi_1 \times \pi_{1|1}$$
$$P'_{1i} = \pi_1 \times \pi_{i|1} + \pi_i \times \pi_{1|i} \tag{31}$$

for the genotype frequencies in the next generation. Clearly, $\pi_1 = p_1$, the frequency of the dominant allele A_1, and $\pi_i = p_i$ for all the other alleles, but we also need to know the values of the conditional probabilities such as $\pi_{i|1}$.

a) $\pi_{1|1}$: We know that one mate has transmitted an A_1 gene. His genotype is therefore $(A_1 A_k)$. Now, whatever the allele A_k in the first mate, the probability that the genotype of the second mate is $(A_1 A_l)$ is:

$$\frac{P_{1l}}{\sum_i P_{1i}} = \frac{P_{1l}}{D}$$

where D is the total frequency of all the genotypes which contain A_1.

The probability that the second mate will transmit an A_1 gene is 1 if his genotype is $(A_1 A_1)$, and $\frac{1}{2}$ if his genotype is $(A_1 A_i)$ where $i \neq 1$, hence:

$$\pi_{1|1} = \frac{1}{D}\left(P_{11} + \tfrac{1}{2}\sum_{i \neq 1} P_{1i}\right) = \frac{p_1}{D}.$$

b) $\pi_{i|1}$: In this case we know that one mate has the genotype $(A_1 A_k)$, hence the probability that the second has the genotype $(A_1 A_l)$, where A_l is any allele, is P_{1l}/D; the second mate can transmit an A_i allele only if his genotype is $(A_1 A_i)$, hence:

$$\pi_{i|1} = \frac{1}{2}\frac{P_{1i}}{D}.$$

c) $\pi_{1|i}$: In order for the second mate to transmit an A_1 allele, his genotype must be $(A_1 A_k)$, and this happens only when the first mate also has the A_1 gene (because of our model of assortative mating); we know that the first mate transmitted an A_i gene, hence his genotype must be $(A_1 A_i)$.

Let $P(A_1 A_i | A_i)$ be the probability that an individual's genotype is $(A_1 A_i)$, given that he has transmitted an A_i gene. Using Bayes' theorem

(Chapter 2, Section 1.4) we obtain:

$$P(A_1 A_i \mid A_i) = \frac{P_{1i} \times \frac{1}{2}}{P_{1i} + \frac{1}{2} \sum\limits_{k \neq i} P_{ki}} = \frac{P_{1i}}{2p_i}.$$

Given that the genotype of the first mate is $(A_1 A_i)$, the probability that the genotype of the second mate is $(A_1 A_k)$ is P_{1k}/D; hence, finally:

$$\pi_{1 \mid i} = \frac{P_{1i}}{2p_i} \left(\frac{P_{11}}{D} + \frac{1}{2D} \sum_{k \neq i} P_{1k} \right) = \frac{P_{1i}}{2p_i} \cdot \frac{p_1}{D}.$$

Using the results we have just obtained in Eq. (31), we get:

$$P'_{11} = \frac{p_1^2}{D}, \qquad P'_{1i} = p_1 \frac{P_{1i}}{D}. \tag{32}$$

Therefore:

$$p'_1 = P'_{11} + \frac{1}{2} \sum_{i \neq 1} P'_{1i} = \frac{p_1}{D} \left(p_1 + \frac{1}{2} \sum_{i \neq 1} P_{1i} \right) = p_1$$

since

$$D = P_{11} + \sum_{i \neq 1} P_{1i} = P_{11} + \frac{1}{2} \sum_{i \neq 1} P_{1i} + \frac{1}{2} \sum_{i \neq 1} P_{1i}$$

$$= p_1 + \frac{1}{2} \sum_{i \neq 1} P_{1i}.$$

Therefore the frequency of the dominant allele will remain constant from one generation to the next.

Let H be the total frequency of heterozygotes for the allele A_1. Then in the following generation, this becomes:

$$H' = \sum_{i \neq 1} P'_{1i} = \frac{p_1 \sum\limits_{i \neq 1} P_{1i}}{D} = \frac{p_1 H}{p_1 + \frac{1}{2} H}.$$

Hence:

$$\frac{1}{H'} = \frac{1}{H} + \frac{1}{2p_1}. \tag{33}$$

Using this recurrence relation to go back from the g-th generation to the first generation, we obtain:

$$\frac{1}{H_g} = \frac{1}{H_0} + \frac{g}{2p_1}$$

or:

$$H_g = \frac{2p_1 H_0}{2p_1 + g H_0}. \tag{34}$$

These equations show that the frequency of heterozygotes for the A_1 gene decreases with the passage of the generations, while the frequency of homozygotes for this allele increases. The rate of decrease is, however, quite low, especially if the frequency p_1 of the allele is high.

For example, consider the case when $p_1 = 0.1$; then the initial value of H is (assuming panmixia) $H_0 = 0.18$. The table below gives some values for H_g in subsequent generations. For comparison, values for $p_1 = 0.5$ are also given.

p_1	$g=0$	1	2	3	4	5	10	∞
0.1	0.180	0.163	0.150	0.138	0.128	0.120 ...	0.090 ...	0
0.5	0.500	0.333	0.250	0.200	0.166	0.143 ...	0.083 ...	0

Finally, notice that if the initial population is panmictic, $H_0 = 2p_1(1-p_1)$, and Eq. (34) becomes:

$$H_g = \frac{H_0}{1+(1-p_1)g}.$$

Note. In the final equilibrium state, $H_E = 0$, hence:

$$D_E = P_E + H_E = P_E.$$

The first Eq. (32) then becomes: $P_E = p_1^2/P_E$, hence $P_E = p_1$. Thus the frequency of $(A_1 A_1)$ homozygotes at equilibrium is equal to the initial frequency of A_1, as expected.

3.4. Partial Positive Assortative Mating Based on Phenotype

Suppose that in each generation a fraction λ of the matings take place according to the same rule as in Section 3.3, while a fraction $1-\lambda$ of the population mates at random.

In the next generation, the frequency of heterozygotes $(A_1 A_i)$ in the first group will have changed from H to $H_1' = \dfrac{2p_1 H}{2p_1 + H}$; in the second group, this frequency is given by the Hardy-Weinberg formula: $H_2' = 2p_1(1-p_1)$. For the whole population, therefore, we have:

$$H' = \lambda \frac{2p_1 H}{2p_1 + H} + (1-\lambda) 2p_1(1-p_1).$$

At equilibrium, $H' = H = \hat{H}$, hence:

$$\hat{H} = \frac{2p_1 \lambda \hat{H}}{2p_1 + \hat{H}} + (1-\lambda) 2p_1(1-p_1)$$

giving:

$$\hat{H}^2 + 2p_1^2(1-\lambda)\hat{H} - 4p_1^2(1-p_1)(1-\lambda) = 0$$

which has the single positive root:

$$\hat{H} = (1 - \lambda) p_1^2 \left(\sqrt{1 + \frac{4(1 - p_1)}{(1 - \lambda) p_1^2}} - 1 \right). \tag{35}$$

This shows that the heterozygotes ($A_1 A_i$) are not eliminated from the population, but remain present at equilibrium at a frequency which decreases as the fraction λ of the population which mates assortatively increases.

For example, if the frequency of A_1 is $p_1 = \frac{1}{2}$, we obtain the following values of H, for different values of λ:

λ	0	0.25	0.50	0.75	0.90	1
$\hat{H}$	0.5	0.45	0.39	0.30	0.05	0

Note also that, if the initial population has a lower frequency of heterozygotes than $\hat{H}$, the equilibrium frequency, this system of mating will give rise to an increase in H.

This model corresponds quite well with the situation in a human population in which individuals with certain diseases caused by homozygosity for recessive genes are cared for together (e. g. in an institution), and may thus be more likely to marry one another than to marry unaffected individuals.

As we have already seen, the tendency for individuals of the same phenotype to marry one another has no effect on the gene frequency, but it leads to a decreased frequency of heterozygotes, and thus an increase in the proportion of homozygotes. The incidence of the disease in the population is therefore increased. For example, for a disease due to a recessive gene whose frequency is $q = 0.2$, the equilibrium frequencies of heterozygotes ($\hat{H}$), and of homozygotes $\left(\hat{R} = q - \frac{\hat{H}}{2} \right)$, are as follows, from Eq. (35):

λ	0	0.10	0.25	0.50	0.75	0.90	1
$\hat{H}$	0.320	0.314	0.302	0.278	0.232	0.163	0
$\hat{R}$	0.040	0.043	0.049	0.061	0.084	0.119	0.200

3.5. Total Negative Assortative Mating Based on Genotype

Suppose that matings between individuals of the same genotype never take place, while all other types of mating occur at random.

There are two ways in which an $(A_i A_j) \times (A_k A_l)$ mating can occur: either (1) the first mate is $(A_i A_j)$ (probability P_{ij}) and the second is $(A_k A_l)$, which, since we know that the second mate cannot have the same genotype as the first, has the probability $\dfrac{P_{kl}}{1-P_{ij}}$, or (2) the other way round. Hence:

$$P\{(A_i A_j) \times (A_k A_l)\} = P_{ij} P_{kl} \left(\frac{1}{1-P_{ij}} + \frac{1}{1-P_{kl}} \right) \tag{36}$$

for all types of mating, except that, by our initial assumption,

$$P\{(A_i A_j) \times (A_i A_j)\} = 0. \tag{37}$$

These relations enable us to determine the probability of any genotype in one generation as a function of the genotype frequencies in the generation before. These calculations become very laborious if there are many alleles, however, so we shall here deal only with the two allele case. We can write the three genotype frequencies as:

$$P_{11} = D, \quad P_{12} = H, \quad P_{22} = R.$$

The frequencies of the various possible mating types, and the genotypes that will be present among their offspring are as follows:

Mating type	Frequency	Offspring	
$(A_1 A_1) \times (A_1 A_2)$	$DH \left(\dfrac{1}{1-D} + \dfrac{1}{1-H} \right)$	$\frac{1}{2}(A_1 A_1)$	$\frac{1}{2}(A_1 A_2)$
$(A_1 A_1) \times (A_2 A_2)$	$DR \left(\dfrac{1}{1-D} + \dfrac{1}{1-R} \right)$	$(A_1 A_2)$	
$(A_1 A_2) \times (A_2 A_2)$	$HR \left(\dfrac{1}{1-H} + \dfrac{1}{1-R} \right)$	$\frac{1}{2}(A_1 A_2)$	$\frac{1}{2}(A_2 A_2)$

In the next generation, the genotype frequencies will therefore be:

$$D' = D \frac{H}{2} \left(\frac{1}{1-D} + \frac{1}{1-H} \right) = \frac{1}{2} \frac{DH(1+R)}{(1-D)(1-H)}$$

$$H' = DR \left(\frac{1}{1-D} + \frac{1}{1-R} \right) + \frac{H}{2} \left(1 + \frac{D}{1-D} + \frac{R}{1-R} \right)$$

$$= \frac{1}{2} + \frac{DR(1+H)}{2(1-D)(1-R)} \tag{38}$$

$$R' = R \frac{H}{2} \left(\frac{1}{1-H} + \frac{1}{1-R} \right) = \frac{1}{2} \frac{RH(1+D)}{(1-R)(1-H)}.$$

The equation for H' shows that $H' \geq \frac{1}{2}$, whatever the values of D, H and R. Also:

$$\frac{D'}{R'} = \frac{D}{R} \times \frac{1-R^2}{1-D^2} . \tag{39}$$

a) If $D < R$, for example, Eq. (39) implies that $\dfrac{D'}{R'} < \dfrac{D}{R}$. The series $\dfrac{D}{R}$ is thus a decreasing series with the limit 0 $\left(\text{because the hypothesis } \dfrac{D}{R} \to c, \right.$ where $0 < c < 1$, implies, by Eq. (39), that $\dfrac{1-D^2}{1-R^2} \to 1$, which contradicts the assumption we have just made$\left.\right)$. Therefore $D \to 0$, and the equilibrium situation is defined by:

$$\hat{D} = 0, \quad \hat{H} = \tfrac{1}{2}, \quad \hat{R} = \tfrac{1}{2}.$$

b) If, on the other hand, $D > R$, we find, by symmetry, that the population tends towards the equilibrium:

$$\hat{D} = \tfrac{1}{2}, \quad \hat{H} = \tfrac{1}{2}, \quad \hat{R} = 0.$$

c) Finally, if $D = R$, $H = 1 - 2D$, and the set of Eqs. (38) show that $D' = R'$ and $H' = 1 - 2D' = \frac{1}{2} + \dfrac{D^2}{1-D}$. The equilibrium is therefore defined by D, the root of the equation:

$$2D^2 - 5D + 1 = 0$$

which has only one root between 0 and 1. Therefore:

$$\hat{D} = \hat{R} = \frac{5 - \sqrt{17}}{4}, \quad \hat{H} = \frac{-3 + \sqrt{17}}{2}.$$

However, this is an unstable equilibrium, since a small deviation of D from equality with R will lead to the population moving towards one of the first two equilibria.

Note also that in the case where $D \to 0$, for example, and $R \to \frac{1}{2}$, the rate of convergence to the equilibrium state is given, according to Eq. (39), by the ratio $\dfrac{1-R^2}{1-D^2}$, which tends towards $\frac{3}{4}$. Thus the approach to equilibrium is very rapid, from points which are close to the equilibrium.

Finally, notice that in the case when no marriages between individuals of the same genotype take place, the population moves towards a state in which half of the individuals are heterozygous and half are homozygous; the homozygous type which was least common in the initial population disappears. The gene frequencies will therefore change during the approach to equilibrium: whatever the initial frequencies, p_0 and q_0, the frequencies at equilibrium will be (if $p_0 < q_0$): $p = \frac{1}{4}$, $q = \frac{3}{4}$.

The equilibrium with all three genotypes present is of no interest in practice, since it is unstable.

3.6. Partial Negative Assortative Mating Based on Genotype

It is possible to make many different models of these types of mating systems; for example, we can consider a population in which a fraction of individuals practice total negative assortative mating based on genotype, or a population in which certain types of mating between two individuals of the same genotypes do not occur.

3.6.1. Partial negative assortative mating based on genotype. Let λ be the fraction of the population which practises negative assortative mating, and $1 - \lambda$ be the fraction which is random-mating. In the notation of Section 3.5, the difference equations for the genotype frequencies are:

$$D' = \lambda \frac{DH(1+R)}{2(1-D)(1-H)} + (1-\lambda)\left(D + \frac{H}{2}\right)^2$$

$$H' = \lambda \frac{H/2 + DR(1+H/2)}{(1-D)(1-R)} + 2(1-\lambda)\left(D + \frac{H}{2}\right)\left(R + \frac{H}{2}\right) \quad (40)$$

$$R' = \lambda \frac{RH(1+D)}{2(1-R)(1-H)} + (1-\lambda)\left(R + \frac{H}{2}\right)^2.$$

In order to determine the equilibrium state, we write $D = D' = \hat{D}$, $H = H' = \hat{H}$ and $R = R' = \hat{R}$ and solve the resulting system of equations. Workman (1964) found that the equilibrium depends on the value of λ:

If $\lambda \leq 0.327$ there is a single stable equilibrium, with $\hat{D} = \hat{R}$.

If $\lambda > 0.327$ there are two equilibria: one, as in the previous case, where $\hat{D} = \hat{R}$, which is unstable, and a stable equilibrium with $\hat{D} \neq \hat{R}$.

Also, at all the equilibria, the frequency of heterozygotes is slightly greater than $\frac{1}{2}$. More specifically:

$$0.500 \leq \hat{H} \leq 0.523.$$

The extreme values arise when $\lambda = 0$ or 1 ($\hat{H} = 0.500$) and when $\lambda = 0.327$ ($\hat{H} = 0.523$).

In this type of population, therefore, the effects of assortative mating and of random mating work in opposite directions, and the net effect depends on the sizes of the two groups in the population, as follows:

1. If negative assortative mating is practised by less than about a third of the population ($\lambda < 0.327$), the two homozygous types are equally frequent at equilibrium; the effect of the assortative mating is thus to bring the gene frequencies to $p = q = \frac{1}{2}$.

2. If more than about $\frac{1}{3}$ of the population practises negative assortative mating ($\lambda > 0.327$), the only stable equilibrium is one with $\hat{D} \neq \hat{R}$. Since $\hat{H}$ is always greater than 0.5, it follows that one of the homozygotes is under-represented, and tends towards zero frequency as λ increases towards 1.

3.6.2. Suppression of certain types of mating between individuals of the same genotype.

Workman (1964) has classified the types of negative assortative mating based on genotype which are possible in the two-allele case as follows:

Type I: No $(A_1 A_1) \times (A_1 A_1)$, $(A_1 A_2) \times (A_1 A_2)$ or $(A_2 A_2) \times (A_2 A_2)$ matings. This is the case of total negative assortative mating.

Type II: No $(A_1 A_1) \times (A_1 A_1)$ or $(A_2 A_2) \times (A_2 A_2)$ matings, i.e. no matings between two homozygotes of the same type.

Type III: No $(A_1 A_1) \times (A_1 A_1)$ and $(A_1 A_2) \times (A_1 A_2)$ matings, i.e. no matings between two heterozygotes or between two individuals of one of the homozygous types.

Type IV: No $(A_1 A_2) \times (A_1 A_2)$ matings, i.e. no matings between two heterozygotes.

Type V: No $(A_1 A_1) \times (A_1 A_1)$ matings, i.e. no matings between two individuals of one of the homozygous types.

The properties of these models have been studied by Workman (1964), who examined the difference equations corresponding to each of them, and by Karlin and Feldman (1968), who traced the changes in populations of these sorts, starting from a variety of initial genetic structures.

These systems of mating usually lead to the elimination of one of the alleles, but the elimination is in general very slow, and alleles may be maintained in the long term in some cases.

For example, in Type II, there is a stable equilibrium defined by:

$$\hat{D} = \hat{R} = \frac{5 - \sqrt{17}}{4}, \qquad \hat{H} = \frac{-3 + \sqrt{17}}{2}.$$

This is identical to the equilibrium of Section 3.5c for total negative assortative mating based on genotype, except that the equilibrium was unstable.

As another example, for Type III there is a trivial equilibrium with $\hat{D} = \hat{H} = 0$, $\hat{R} = 1$, and also a stable equilibrium with $\hat{D} = \hat{H} = \frac{1}{2}$, $\hat{R} = 0$. Which of these equilibria is reached depends on the initial conditions.

These results are important: they show that an allele can be maintained in a population by an appropriate system of negative assortative mating, in the face of selective or stochastic pressures for its elimination. Thus negative assortative mating can help to maintain the genetic diversity of a population.

3.7. Total Negative Assortative Mating Based on Phenotype

Consider again the case of two alleles at a locus, with the genotype frequencies equal to D, H and R. Let us assume that A_1 is dominant to A_2. The only matings which occur are assumed to be between individuals who have the A_1 and A_2 phenotypes, in other words between the genotypes $(A_1 A_1)$ or $(A_1 A_2)$ and the genotype $(A_2 A_2)$. It is easy to show that the frequencies of these two types of mating, and the offspring which they produce, are as follows:

Mating type	Frequency	Offspring
$(A_1 A_1) \times (A_2 A_2)$	$\dfrac{D}{D+H}$	$(A_1 A_2)$
$(A_1 A_2) \times (A_2 A_2)$	$\dfrac{H}{D+H}$	$\frac{1}{2}(A_1 A_2) \quad \frac{1}{2}(A_2 A_2)$

The genotypic structure of the next generation is therefore:

$$D' = 0$$

$$H' = \frac{D + H/2}{D + H} \tag{41}$$

$$R' = \frac{H/2}{D + H}$$

and for the second generation:

$$D'' = 0$$

$$H'' = \frac{D' + H'/2}{D' + H'} = \tfrac{1}{2}$$

$$R'' = \frac{H'/2}{D' + H'} = \tfrac{1}{2}.$$

Thus the population attains, in the second generation, a stable equilibrium defined by $\hat{D} = 0$, $\hat{H} = \hat{R} = \tfrac{1}{2}$, hence $p = \tfrac{1}{4}$, $q = \tfrac{3}{4}$.

This system also (like certain regimes of negative assortative mating based on genotype) is capable of maintaining two alleles in the population in the long term, and thus can oppose the actions of selection or genetic drift tending to make the population homozygous.

Note. One could consider (Li, 1955) that sexual reproduction is a particular case of negative assortative mating based on phenotype, if we regard the Y chromosome as a dominant "male gene", and the X chromosome as a recessive "female gene". The genotype YY will not occur, and all matings would be between (XX) homozygotes and XY) heterozygotes, so that stable proportions of $\frac{1}{2}$ each of these "genotypes" will be maintained.

3.8. Partial Negative Assortative Mating Based on Phenotype

Now let us assume that a fraction λ of the population practises the type of negative assortative mating described in Section 3.7, while a fraction $1 - \lambda$ is random-mating. The difference equations for the genotype frequencies in any generation, in terms of the genotype frequencies in the generation before are:

$$D' = \lambda \times 0 + (1 - \lambda) \left(D + \frac{H}{2} \right)^2$$

$$H' = \lambda \frac{(D + H/2)}{D + H} + 2(1 - \lambda) \left(D + \frac{H}{2} \right) \left(R + \frac{H}{2} \right) \qquad (42)$$

$$R' = \lambda \frac{H/2}{D + H} + (1 - \lambda) \left(R + \frac{H}{2} \right)^2.$$

The frequency of the allele A_1 is therefore:

$$p' = D' + \frac{H'}{2} = \frac{\lambda}{2} \frac{p}{D + H} + (1 - \lambda) p (D + H + R)$$

$$= p \left\{ 1 - \lambda \left(1 - \frac{\frac{1}{2}}{D + H} \right) \right\}.$$

This equation shows that, if $D + H = \frac{1}{2}$, $p' = p$, whatever the value of λ. Thus there is an equilibrium at $D + H = \frac{1}{2}$ (which implies $R = \frac{1}{2}$ and $p \leq \frac{1}{2}$). This is a stable equilibrium (Li, 1955).

If we substitute in Eq. (42) the values $R = R' = D + H = \frac{1}{2}$, we obtain:

$$\hat{H} = \frac{-(1 + \lambda) + \sqrt{2(1 + \lambda)}}{1 - \lambda} \quad \text{and} \quad \hat{D} = \frac{3 + \lambda - 2\sqrt{2(1 + \lambda)}}{2(1 - \lambda)}.$$

This leads to:

$$\hat{p}=\hat{D}+\frac{\hat{H}}{2}=\frac{1-\sqrt{\dfrac{1+\lambda}{2}}}{1-\lambda}.$$

The final gene frequency is thus a function of λ only, and furthermore it is not very sensitive to changes in the value of λ; the genotype frequencies, however, are much more sensitive to changes in λ, as the following numerical results show:

λ	0	0.10	0.25	0.50	0.75	0.90	1
$\hat{p}$	0.293	0.287	0.278	0.268	0.258	0.255	0.250
$\hat{D}$	0.086	0.074	0.059	0.036	0.020	0.015	0
$\hat{H}$	0.414	0.425	0.441	0.464	0.476	0.480	0.500

4. The Offspring of Consanguineous Marriages

We have seen that the effect of consanguinity of parents is to increase the probability that their offspring will be homozygous, at any locus. They will therefore run an increased risk of manifesting a disease due to a recessive gene, and will also be on average less heterozygous than the offspring of a non-consanguineous mating.

It is tempting to think that the reason why most societies condemn marriages between close relatives as incestuous may be because such marriages have been observed to lead to harmful results. In fact, however, the bad effects of consanguineous unions are so slight that some people have even considered that the opposite was true. In the 19th century, a very bitter quarrel arose over this question, in which the two sides supported their opinions with unimpeachable, but mutually contradictory arguments. The "consanguinists" claimed that consanguinity of the parents had no bad effects on their offspring, or was even beneficial, while the "anticonsanguinists" believed the opposite.

The true nature of the situation could only be understood after the rediscovery of Mendel's work, at the beginning of the 20th century. This made possible, for the first time, a rigorous theory of the consequences of inbreeding; the offspring of consanguineous marriages will be less heterozygous than individuals whose parents are unrelated, and many observations suggest that this will normally have a deleterious effect. This, however, is only a prediction based on the Mendelian theory of inheritance, and on studies of the effects of homozygosity in experimental organisms. Whether inbreeding in human populations is indeed followed by harmful effects is still a matter for observation and study.

Unfortunately, this type of study is extremely laborious and complex; in order to obtain precise results, all sorts of precautions are necessary. It is necessary to use sophisticated statistical and demographic methods in order to ensure that the deviations observed are not due to causes other than those which one is studying.

4.1. The American Medical Association Study of 1856

In this early study of the question, the American Medical Association asked its member doctors to collect information about inbred and control families. The results were published by Bemiss in 1858, but no systematic use of the data was made until that of Sutter and Tabah, in 1951.

This study fortunately satisfied one of the essential conditions which must be met by any study of inbred families: only completed families should be considered, i.e. the mother should have reached the end of the reproductive period. This is necessary because the frequencies of many defects (e.g. mongolism) are highly correlated with the age of the mother when the child was born; in order to be able to make comparisons between groups, it is therefore important to remove this possible cause of differences in the mortality of the children.

The study included data on 819 consanguineous marriages (120 between second cousins, 600 between first cousins and 99 between closer relatives such as uncle and niece, double first cousins or other complicated types of relationship), and 156 non-consanguineous marriages. The number of children in each family was recorded, and also the number who died or who manifested certain defects (idiots, deaf-and-dumb children, epileptics, etc.). Sutter and Tabah (1951) give the following results based on this data (Table 9.2).

These results show that as the degree of relatedness between the parents rises, the following effects are observed:
1. A decrease in the mean number of children per family.
2. An increase in the number of children who died.
3. A very striking increase in the proportion of "defective" children.

Table 9.2. The effects of inbreeding. U.S.A., 1856

	Control families	Second-cousins	First-cousins	Closer relatives
Mean number of children per family	6.7	4.7	4.4	4.5
Percentage of children dying	16	19	24	32
Percentage of defective children	2	18	36	54

These conclusions must be accepted with reserve, since the data are very old, and it is not easy to be sure whether they are based on a really representative sample of the population. Also, the infant and juvenile mortality are given as overall figures, with no distinction between genetic causes and chance causes of death.

4.2. The Study in Morbihan and Loir-et-Cher of 1952. Definition of "Perinatal Mortality Rate"

It is clear from the last section that the 1856 study does not settle the question. The first modern study was done by Sutter and Tabah (1952) in the departments of Loir-et-Cher and Morbihan. In these areas, 264 and 262 inbred families were compared with 487 and 485 control non-inbred families. All the families studied were completed families; the marriages had taken place between 1919 and 1925 so that the mothers were all aged more than 45 at the time of the study.

The data were collected in such a way as to enable one to calculate the "perinatal mortality rate" for each group of families. We shall now define this measure.

Infant mortality statistics distinguish between "infants born dead" and "infants dying before the age of 1 year". These two categories are not clearly separated, and babies which die before a birth-certificate is made out are often classed as born dead.

Among the babies who die before their first birthday we can distinguish between those who die because of some external cause (e.g. infections, malnutrition, neglect), and those which are congenitally defective.

Bourgeois-Pichat (1951) has suggested a simple method of analysing infant mortality into the exogenous and endogenous components, to a very good approximation. He shows that for most countries, despite great heterogeneity in infant mortality rates, we can write:

$$m_{exo} = 1.25 \times m_{1-12}$$
$$m_{endo} = m_{0-12} - 1.25 \, m_{1-12}$$

where m_{1-12} is the mortality rate between 1 and 12 months, and m_{0-12} is the mortality rate between 0 and 12 months. The results obtained by these formulae agree very well with those given by the statistics on cause of death, whenever these are available.

We can assume that the cause of death for most infants who are born dead is some inborn defect. In addition to a very small number who die because of some difficulty in the birth, these babies are either dead before birth, or, if they survive the birth itself, die in the next few days.

Table 9.3. The effects of inbreeding. Loir-et-Cher and Morbihan, 1952

| | Loir-et-Cher | | | | Morbihan | | | |
| | Control families | Consanguineous families | | | Control families | Consanguineous families | | |
		First-cousins	Second-cousins	All couples		First-cousins	Second-cousins	All couples
Mean number of children per family	2.32	2.20	2.42	2.37	2.37	3.45	2.90	3.24
Percentage of sterile couples	6.0	12.5	7.1	10.6	5.6	5.2	10.4	8.4
Perinatal deaths (per 1000)	29	55	40	50	37	101	65	82
Proportion of abnormal children (not tubercular, per 1000)	16	163	46	104	19	132	75	113
Proportion of tubercular children (per 1000)	16	7	12	8	15	35	7	24

This shows that the difficulty of making distinctions between infants born dead and those who die in the first few weeks is not simply due to the arbitrariness of the procedure of making-out birth certificates; there is a real difficulty as well.

In order to estimate the mortality rate due to genetic causes, Bourgeois-Pichat has therefore proposed the use of the sum of the proportion of infants born dead and the rate of infant death due to endogenous causes. Although this is not a precise measure of the genetic constitution of individuals, it gives a useful approximate measure, which enables us to detect differences in genetic effects between different populations.

Table 9.3 summarises the data obtained by Sutter and Tabah (1952).

Bearing in mind the limited size of the sample of couples studied, we notice the following:

1. Family size appears to be independent of the relationship between the parents. However, the observed sizes show that the population must practise family-planning, since very few large families were found; the observed number of children in a family may therefore be mainly a reflection of the parents' attitude towards family-planning.

2. Except for the case of first-cousin marriages in Morbihan, the proportion of sterile couples was in general distinctly higher among consanguineous couples than non-consanguineous couples.

3. The rate of perinatal mortality is much higher among the children of consanguineous marriages. These differences are highly significant statistically (except for the first-cousin marriages in Loir-et-Cher).

4. The frequency of "defects" other than tuberculosis rises rapidly with inbreeding, while the proportion of offspring with tuberculosis does not rise.

These results therefore support the idea that the offspring of related couples will be subject to increased infant mortality and morbidity. However, we can make several objections to this data, in particular to the way in which the control families were chosen: these were chosen by a random draw, regardless of social factors; but we know that the probability of marrying a cousin varies with social class or religious affiliation, and these are different for different families and would be confounded with genetic differences between families. In this study, therefore, the control group of families may not be really comparable with the inbred group.

4.3. The Study in the Vosges in 1968

In order to remove this possible source of bias, Georges and Jacquard (1968) studied the offspring of consanguineous marriages in two valleys in the Vosges; in this study, the control families were a group of the families of brothers and sisters of the consanguineous marriage partners.

There were 189 consanguineous marriages and 646 control marriages, in each of which one member was a brother or sister of one of the 378 consanguineous marriage partners. The results are shown in Table 9.4.

As in the study in Loir-et-Cher and Morbihan, the number of children of the consanguineous marriages was slightly lower than the number of

Table 9.4. The effects of inbreeding. Vosges, 1968

	First-cousin marriages		Second-cousin marriages		All couples	
	Consan-guineous couples	Control couples	Consan-guineous couples	Control couples	Consan-guineous couples	Control couples
Mean number of children per family	4.2	5.1	4.8	4.8	4.5	4.8
Percentage of sterile couples	—	—	—	—	6.9	4.6
Perinatal deaths	11.1	9.0	8.0	7.9	8.9	8.5

children in the control families, and this was due mainly to an excess of sterile marriages in the consanguineous group.

The results for infant mortality, however, are quite surprising. The perinatal mortality among the children of first-cousin marriages (11.1 %) is certainly higher than that of their control group (9.0 %), but this is not statistically significant. For the second-cousin marriages, there is practically no difference from the control families. In the whole group of 189 consanguineous marriages, the perinatal mortality rate was 8.9 %, which is barely higher than the rate of 8.5 % for the 646 control marriages.

These results show that, in this population, consanguinity of the parents leads to at most very slight effects on their offspring, which cannot be detected even in a study of several hundred couples.

The only statistically significant result is the difference between the proportions of sterile couples (6.9 % for consanguineous couples, and 4.6 % of control couples), though it would still be desirable to have data on a larger sample. This difference is in agreement with other evidence that intra-uterine death in early pregnancy may play an important role in the elimination of genetically inferior offspring.

4.4. The Study in Japan in 1958–60

A very large study in the towns of Hiroshima and Nagasaki was carried out by a large team of doctors and biologists, and the results were published as a book: "The effects of inbreeding on Japanese children" (Schull and Neel, 1965). This gives details of the methods followed, the statistical techniques used, and the results obtained. The rate of perinatal mortality (as defined above) can be extracted from data presented in this book, for various groups of couples.

Table 9.5. The effects of consanguinity. Japan, 1965

	Hiroshima				Nagasaki			
	Consanguineous couples			Control couples	Consanguineous couples			Control couples
	First-cousins	Cousins of 5th, 7th etc. degree	Second-cousins		First-cousins	Cousins of 5th, 7th etc. degree	Second-cousins	
Number of births	951	356	390	1970	1542	466	600	2847
Perinatal mortality rate (%)	5.23	4.21	6.61	4.11	4.54	2.04	5.42	4.18

Table 9.5 shows that there is a clearer effect of consanguinity in Hiroshima than Nagasaki; in fact, there is a statistically significant difference between the inbred and the control children in Hiroshima, but not in Nagasaki.

Schull and Neel examine several hypotheses which could explain this result, without coming to any definite conclusion. We should perhaps note that this study was based on the registration of pregnancies, and not on completed families, so that there may be an effect of maternal age.

4.5. Sex-Linked Genes

We saw in Section 1.3 that the proportion of homozygotes for sex-linked genes among the offspring of consanguineous marriages depends on the sexes of the members of the pedigree they belong to. We might therefore expect that the frequency of offspring with diseases due to sex-linked genes will be different for the different types of consanguineous marriages whose members have the same autosomal coefficients of kinship Φ, but different coefficients of kinship for sex-linked genes, Φ_X. In general, for any character which is affected by sex-linked genes, we should find differences between individuals whose parents have different values of Φ_X.

Schull and Neel (1963) studied 853 girls whose parents were first-cousins for a set of ten anthropometric measures (height, head breadth etc.). They found no effect of the parents' Φ_X. The characters they studied are therefore not affected by sex-linked genes.

Georges (1969) in his study of inbreeding in the Vosges, also made comparisons of the offspring of the four different types of first-cousin marriages. The sex-ratios were the same for the four types. Mean family size, however, was clearly lower (2.3) in families of Types II and IV than in families of Types I and III (3.8). Unfortunately, the sample size was too small (82 couples, with a total of 254 children) for statistical tests of the significance of this difference to be possible.

4.6. Conclusions

The number of sufficiently large studies of the effects of consanguinity is still quite small. We have seen that different studies have given uncertain or even contradictory results. However, we must not underestimate the difficulty of doing this type of study. When we try to study the conditions which influence hereditary transmission we are working in a difficult area in which we can only have indirect knowledge. Our approach towards reality has to make alternate use of both theoretical models, which enable us to define the consequences of different hypotheses on the phenomena in question, and observational data, which we compare with the theoretical predictions.

The account of the studies of consanguinity which has been presented above shows how difficult this type of work is, and the extent to which it is subject to all kinds of bias, which can make interpretation impossible. Above all, therefore, we must not come to premature conclusions (especially when the results agree with our expectations) and must not forget that, in another population which is genetically different from the one in question, quite different results might have been obtained.

It therefore seems that we shall only have a clear idea of the effects of consanguineous marriage after studies have been done in many, diverse populations.

Further Reading

Barker, J. S. F.: The effect of partial exclusion of certain matings and restriction of their average family size on the genetic composition of a population. Ann. hum. Genet. **30**, 7–11 (1966).

Petit, C.: L'isolement sexuel; déterminisme et rôle dans la speciation. Ann. Biol. **6**, 271–285 (1967).

Scudo, F. M.: L'accopiamento assortativo basato sul fenotipo di parenti. Alcune consequenze en populazioni. R. C. scienze Instituto Lombardo, B. **101**, 435–455 (1967).

Watterson, G. A.: Non-random mating and its effect on the rate of approach to homozygosity. Ann. hum. Genet. **23**, 204–220 (1959).

Chapter 10
Selection

We saw in Chapter 3 that the demonstration of the Hardy-Weinberg equilibrium for a population required six assumptions about the conditions under which genes are transmitted from one generation to the next.

In the preceding three chapters, we have studied the effects of dropping conditions ② (non-overlapping generations), ⑥ ("infinitely" large population size) and ⑤ (random mating). We shall now study the consequences of dropping condition ④, which states that the number of offspring of an individual is independent of his genotype.

First we must define clearly what is meant by the word "offspring". When we assume that successive generations are distinct, it is possible to consider the population only at one particular stage in the life-cycle. In what follows, we shall define or characterise the state of the population, in particular its genetic structure, only in the adult stage, at the time of formation of mating couples. All other intermediate stages (conception, embryonic development, childhood, adolescence) are taken into account solely according to whether they result in the survival of an individual capable of reproduction. Starting with an individual in one generation, g, we are therefore interested only in the number of his offspring who reach adulthood in generation $g+1$; in Chapter 8, when we discussed sibship sizes, this number was referred to as the number of "successful offspring".

This number obviously depends on many factors, most of which are affected by the genetic make-up of the individuals concerned: the probability that a given gamete (ovum or spermatozoon) will be involved in the conception of a new individual may be a function of the genes which it carries; the probability that a foetus survives to be born, or that a child grows up to adulthood, depends on the genes he received from his parents. These probabilities are also functions of the environment in which the individuals in question live. In the face of such a complex situation, we can study only very simplified models, in which just one factor at a time is operating.

In this chapter, we shall first study some simple models of selection (selection acting only on the zygotes, with constant selection coefficients or with selection coefficients which are a function of genetic structure only). Then we shall study the consequences of these models of selection in terms of the population's mean selective value, and will describe some methods for measuring selection in populations.

1. Some Simple Models of Selection

1.1. Definition of Selective Values

We shall assume that selection acts only in the zygote stage. This is equivalent to saying that the probability of survival of each individual between the moment of conception and the adult stage is a function of his genotype. We are therefore assuming that every individual, regardless of genotype, has the same chance of mating, and that all gametes, regardless of the genes which they carry, have the same chance of being involved in the conception of a new individual.

We shall also assume that all the other conditions for Hardy-Weinberg equilibrium are met: ① (no migration), ② (non-overlapping generations), ③ (no mutation), ⑤ (random mating) and ⑥ (population size large enough to be considered infinite).

Consider a generation which has reached the adult stage. Let the set of frequencies of the different alleles present at the locus under consideration be: $s = (p_1, \ldots, p_n)$.

When the individuals mate, the probability, at the moment of conception, that the genotype of a zygote is $(A_i A_i)$ is $_c\Pi_{ii} = p_i^2$, and the probability that its genotype is $(A_i A_j)$ is $_c\Pi_{ij} = 2 p_i p_j$. In other words, the genotypic probability structure of the next generation, *at the moment of conception*, is:

$$\mathscr{S}_c = \vec{s}^2. \tag{1}$$

Under selection, the probability of survival to the adult stage, W_{ij}, of each individual depends on his genotype $(A_i A_j)$.

The *a priori* probability of survival to adulthood of an individual taken at random is:

$$\overline{W} = \sum_{ij} {_c\Pi_{ij}} W_{ij} \tag{2}$$

where the summation is over the set of all genotypes. From Eq. (1) and the general property of summations in Hardy-Weinberg populations which was noted in Section 2.4 of Chapter 3, we have:

$$\overline{W} = \sum_{ij} p_i p_j W_{ij}.$$

If we take an adult individual at random, the probability that his genotype is $(A_i A_j)$ is equal to the probability of having this genotype, given that he has survived to adulthood. By Bayes' theorem, this can be written:

$$\Pi_{ij} = \text{Prob}(A_i A_j | \text{survived})$$

$$= \frac{{}_c\Pi_{ij} \times W_{ij}}{\sum\limits_{ij} {}_c\Pi_{ij} \times W_{ij}} \tag{3}$$

$$= {}_c\Pi_{ij} \times \frac{W_{ij}}{\overline{W}}.$$

Equating probabilities and frequencies (which we can do since we are assuming condition ⑥ to hold), we can thus write the genotype frequencies in the new generation of adults as:

$$P'_{ii} = \frac{p_i^2 \times W_{ii}}{\overline{W}}, \qquad P'_{ij} = \frac{2 p_i p_j W_{ij}}{\overline{W}}. \tag{4}$$

Notice that if the mean number of offspring per couple, at the moment of conception, is $2n$, the population sizes in successive generations are related by:

$$N_{g+1} = n \overline{W} N_g.$$

However, in population genetics we are usually interested in changes in the proportions of the different genotypes, rather than in the total population size.

The Eqs. (4) show that, in order to study changes in the genotypic proportions, we do not need to know the probabilities of survival W_{ij} themselves, but only their relative values; we shall call a set of such relative values which are proportional to the W_{ij} values, the "*selective values*" of the genotypes, and we shall designate them by w_{ij}. We can adopt the convention of assigning a selective value of 1 to one of the genotypes.

We therefore have the following definition:

The selective value w_{ij} of a genotype is a number proportional to the probability of survival from conception to adulthood, of individuals of this genotype.

We can also define the "mean selective value of a gene" as follows:

The mean selective value of a gene is the mean of the selective values of the genotypes into which the gene enters, weighted by the frequencies of the other genes involved:

$$w_{i.} = \sum_j p_j w_{ij}.$$

Finally *the mean selective value of a population is defined as the mean of the selective values of the various genotypes, weighted by their frequencies at the time of conception:*

$$\bar{w} = \sum_{(ij)} {}_c\Pi_{ij} w_{ij} = \sum_{ij} p_i p_j w_{ij}.$$

It follows from these definitions that $\bar{w} = \sum_i w_{i.} p_i$; the mean selective value of the population is equal to the mean of the selective values of the genes, weighted by their frequencies.

It is important to notice that the mean selective values of the genes are functions of the gene frequencies p_i, and they therefore change when the gene frequencies change, even if the genotypic selective values w_{ij} themselves do not change.

1.2. Change in Gene Frequencies

The genotype frequencies among the adult members of a population are related to the gene frequencies in the adult population of the preceding generation by Eqs. (4), or by equivalent equations in which the probabilities of survival W_{ij} are replaced by the relative values w_{ij}:

$$P'_{ii} = \frac{p_i^2 w_{ii}}{\bar{w}}, \qquad P'_{ij} = \frac{2 p_i p_j w_{ij}}{\bar{w}}. \tag{5}$$

The frequency of the gene A_i in the new generation is therefore:

$$p_i = P'_{ii} + \tfrac{1}{2} \sum_{j \neq i} P'_{ij}$$

$$= \frac{p_i}{\bar{w}} \left(p_i w_{ii} + \sum_{i \neq j} p_j w_{ij} \right)$$

$$= \frac{w_{i.}}{\bar{w}} p_i.$$

The change $p'_i - p_i$ in gene frequency from one generation to the next is therefore:

$$\Delta p_i = \frac{p_i}{\bar{w}} (w_{i.} - \bar{w}). \tag{6}$$

This shows that the frequency of a gene will increase from one generation to the next if its mean selective value is greater than the mean selective value in the population; if it is lower, it will decrease.

The population is in equilibrium when all the Δp_i values are zero, in other words when for all genes either $p_i = 0$, or $w_{i.} = \bar{w}$. Thus, *in a population which is subject to zygotic selection, there is an equilibrium when the selective values of all the alleles at the locus in question are the same.*

Notice that when $p_i=1$, $p_j=0$ for all the alleles other than A_i, and $w_i. = w_{ii} = \overline{w}$; the selective values $w_j.$ of the other alleles are undefined.

There are therefore two types of equilibrium:

1. Only one allele present. This type of equilibrium is "trivial": it is obvious that a population which has reached this state cannot change.

2. More than one allele present; this type of equilibrium is called "polymorphic". The population has the same genic structure generation after generation, but we also have to find out whether the equilibrium is stable.

1.3. Loci with Two Alleles

Consider a locus with two alleles A_1 and A_2, with frequencies p and $q=1-p$, and with the following selective values:

$$(A_1 A_1) \quad (A_1 A_2) \quad (A_2 A_2)$$

$$w_{11} \qquad w_{12} \qquad w_{22}$$

In this notation we have:

$$w_1. = p\, w_{11} + q\, w_{12}$$

$$w_2. = p\, w_{12} + q\, w_{22}$$

$$\overline{w} = p^2\, w_{11} + 2\, p\, q\, w_{12} + q^2\, w_{22}$$

$$= p\, w_1. + q\, w_2..$$

Hence:

$$p' = p\, \frac{w_1.}{\overline{w}} = p\, \frac{p\, w_{11} + q\, w_{12}}{p^2\, w_{11} + 2\, p\, q\, w_{12} + q^2\, w_{22}}$$

and:

$$\Delta p = \frac{p}{\overline{w}}(w_1. - \overline{w})$$

$$= \frac{p\, q}{\overline{w}}(w_{11} - w_{12})\, p + (w_{12} - w_{22})\, q \tag{7}$$

$$= \frac{p\, q}{\overline{w}}(w_1. - w_2.).$$

This final result shows that in this particular case, just as in the general case, there are two types of equilibrium:

1. Trivial equilibria, when the population is fixed:

$$\hat{p}=0 \quad \text{or} \quad \hat{p}=1.$$

2. Polymorphic equilibria, with frequencies $\hat{p}$ such that:

$$(w_{11} - w_{12})\,\hat{p} = (w_{22} - w_{12})\,\hat{q}. \tag{8}$$

Since p and q are by definition positive numbers, this implies that $(w_{11} - w_{12})$ and $(w_{22} - w_{12})$ must be of the same sign, and consequently: *for there to be a polymorphic equilibrium at a locus with two alleles, the selective value of the heterozygotes cannot be between those of the homozygotes.*

1.3.1. Stability of equilibria. If the selective values which enter into Eq. (7) depend on p, the change in p from one generation to the next is also a function of p, say $\Phi(p)$:

$$\Delta p = \Phi(p). \tag{9}$$

Then the equilibrium states correspond to the roots of the equation:

$$\Phi(p) = 0. \tag{10}$$

The stability of the equilibria depends on the behaviour of the function $\Phi(p)$ (which we shall assume to be continuous and differentiable) in the neighbourhood of $\hat{p}$. Consider the graph of $\Phi(p)$ against p (Fig. 10.1).

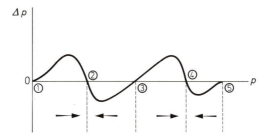

Fig. 10.1. The stability of equilibria

The roots of Eq. (10) correspond to the points where the curve crosses the p axis. At each of these points of intersection, the stability of the corresponding equilibrium depends on the derivative $\Phi'(\hat{p})$ of the function Φ.

Expanding Eq. (9) by Taylor's theorem, and remembering that $\Phi(\hat{p}) = 0$, we have:

$$\Delta p = (p - \hat{p})\,\Phi'(\hat{p}) + \frac{(p - \hat{p})^2}{2}\,\Phi''(\hat{p}) + \frac{(p - \hat{p})^3}{3!}\,\Phi'''(\hat{p}) + \cdots.$$

Hence:

$$p' - \hat{p} = (p - \hat{p})\,[1 + \Phi'(\hat{p})] + \frac{(p - \hat{p})^2}{2}\,\Phi''(\hat{p}) + \text{etc.} \tag{11}$$

From this equation, it is easy to obtain the following results:

1. $\Phi'(\hat{p})<0$:

a) $-2<\Phi'(\hat{p})$ (point 2 of Fig. 10.1): Neglecting terms of order $(p-\hat{p})^2$, we have:

$$|p'-\hat{p}|<|p-\hat{p}|$$

so that the deviation of p from $\hat{p}$ decreases towards zero.

b) $\Phi'(\hat{p})<-2$ (point 4 of Fig. 10.1): In this case:

$$|p'-\hat{p}|>|p-\hat{p}|$$

and $(p'-\hat{p})$ is of opposite sign from $(p-\hat{p})$. The value of p therefore oscillates from values greater than $\hat{p}$ to values less than $\hat{p}$, with ever-increasing amplitude. This type of equilibrium is therefore unstable.

c) $\Phi'(\hat{p})=-2$: From the first three terms of Eq. (11), we find that p oscillates about $\hat{p}$, with greater amplitude on one side of $\hat{p}$ than the other; if $\Phi''(\hat{p})\neq 0$ or the third derivative, $\Phi'''(\hat{p})$, is negative, the oscillations are damped, and gradually die away.

2. $\Phi'(\hat{p})>0$ (point 3 of Fig. 10.1): Here, $|p'-\hat{p}|>|p-\hat{p}|$, so that the equilibrium is unstable.

3. $\Phi'(\hat{p})=0$: In this case, the stability depends on the value of the second derivative. Neglecting terms of order $(p-\hat{p})^3$, Eq. (11) becomes:

$$p'-\hat{p}=(p-\hat{p})\left[1+\frac{(p-\hat{p})\,\Phi''(\hat{p})}{2}\right]$$

a) $\Phi''(\hat{p})<0$ (point 5 of Fig. 10.1):

$$\text{if } (p-\hat{p})<0, \quad |p'-\hat{p}|>|p-\hat{p}|,$$
$$\text{if } (p-\hat{p})>0, \quad |p'-\hat{p}|<|p-\hat{p}|.$$

Therefore this type of equilibrium is stable to perturbations to the right, but unstable to perturbations to the left of $\hat{p}$.

b) $\Phi''(\hat{p})>0$ (point 1 of Fig. 10.1): A similar argument to that just given shows that this type of equilibrium is stable at the left, and unstable at the right.

From Eq. (7) we have:

$$\Phi'(p)=\frac{q}{w}(w_{1.}-w_{2.})-\frac{p}{w}(w_{1.}-w_{2.})-\frac{pq}{w^2}(w_{1.}-w_{2.})\frac{d\bar{w}}{dp}$$

$$+\frac{pq}{w}\frac{d}{dp}(w_{1.}-w_{2.}).$$

Hence, for the trivial equilibria, with $p=0$ or 1:

$$\Phi'(0)=\frac{w_{1.}-w_{2.}}{\overline{w}}$$

$$\Phi'(1)=\frac{w_{2.}-w_{1.}}{\overline{w}}$$ (12)

and for the polymorphic equilibria, with $w_{1.}=w_{2.}$:

$$\Phi'(\hat{p})=\frac{\hat{p}\hat{q}}{\overline{w}}\frac{d}{dp}(w_{1.}-w_{2.}).$$ (13)

It is therefore relatively easy to determine the stability of particular cases of this type.

1.3.2. The approach towards stable equilibrium. Consider a population whose genic structure is very close to a stable equilibrium structure; it therefore lies in a zone in which $-2<\Phi'(p)<0$. The frequency of allele A_1 is p_0, close to $\hat{p}$. In the next generation, this frequency becomes:

$$p_1=p_0+\Delta p_0=p_0+\Phi(p_0).$$

But, by the theorem of the mean:

$$\Phi(p_0)=\Phi(\hat{p})+(p_0-\hat{p})\,\Phi'(p_{i\,0})$$

$$=(p_0-\hat{p})\,\Phi'(p_{i0})$$

where p_{i0} is between p_0 and $\hat{p}$. Hence:

$$p_1-\hat{p}=(p_0-\hat{p})\{1+\Phi'(p_{i0})\}$$

and, similarly:

$$p_2-\hat{p}=(p_1-\hat{p})\{1+\Phi'(p_{i1})\},\quad \text{etc.}$$

Since we are assuming that the population is close to its stable equilibrium, the values $p_1, p_2, ..., p_g$ must become closer and closer to $\hat{p}$. The p_{ig} values must therefore also approach $\hat{p}$, and we can write:

$$\frac{p_g-\hat{p}}{p_{g-1}-\hat{p}}\rightarrow 1+\Phi'(\hat{p}).$$ (14)

1.3.3. Changes in the mean selective value. By definition, the mean selective value is:

$$\overline{w}=p\,w_{1.}+q\,w_{2.}.$$

Its derivative with respect to p is therefore:

$$\frac{d\overline{w}}{dp}=w_{1.}-w_{2.}+p\frac{dw_{1.}}{dp}+q\frac{dw_{2.}}{dp}.$$

At equilibrium, when $w_{1.} = w_{2.}$, or p or q is zero, this derivative is not, in general, equal to zero. Therefore selection does not necessarily maximise the population's mean selective value.

The mean selective value is, however, maximised if the selective values are such that:

$$\hat{p}\left.\frac{dw_{1.}}{dp}\right|_{p=\hat{p}} + \hat{q}\left.\frac{dw_{2.}}{dp}\right|_{p=\hat{p}} = 0. \tag{15}$$

This is in particular the case when the selective values are constants. This can be proved as follows. From the definition of the selective value of a gene, we have, provided that the three genotypic selective values are constant:

$$\frac{dw_{1.}}{dp} = \frac{d}{dp}\{p\,w_{11} + (1-p)\,w_{12}\}$$

$$= w_{11} - w_{12}$$

and

$$\frac{dw_{2.}}{dp} = \frac{d}{dp}\{p\,w_{12} + (1-p)\,w_{22}\}$$

$$= w_{12} - w_{22}.$$

Substituting these values into Eq. (15), we see from Eq. (8) that the equality is satisfied. We shall next study this case.

1.4. Constant Selective Values

If in Eq. (7):

$$\Delta p = p\,q\,\frac{(w_{11} - w_{12})\,p + (w_{12} - w_{22})\,q}{w_{11}\,p^2 + 2\,w_{12}\,p\,q + w_{12}\,q^2}$$

we assume that w_{11}, w_{12} and w_{22} are constant, three solutions are formally possible:

a) $p=0, q=1$.

b) $p=1, q=0$.

c) $(w_{11} - w_{12})\,p + (w_{12} - w_{22})\,q = 0$, which gives $p = \dfrac{w_{22} - w_{12}}{w_{11} - 2\,w_{12} + w_{22}}$.

We shall write $\hat{p}$ for this value of p, in what follows.

But the only values of p which are biologically meaningful are those between 0 and 1. Solutions a) and b) are therefore always equilibrium solutions, but solution c) corresponds to an equilibrium only under certain conditions with respect to w_{11}, w_{12} and w_{22}. We shall now examine the various possible cases. There is no loss of generality if we assume that $w_{11} < w_{22}$ since the opposite case can be studied simply by altering the numbers attached to the alleles.

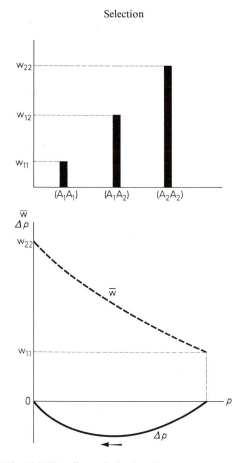

Fig. 10.2. The effects of selection when $w_{11} < w_{12} < w_{22}$

1.4.1. $w_{11} < w_{12} < w_{22}$. The heterozygote has a selective value between those of the homozygotes.

It is easy to see that:

$$\hat{p} = \frac{w_{22} - w_{12}}{(w_{22} - w_{12}) - (w_{12} - w_{11})}$$

is less than 0 if $w_{12} - w_{11} > w_{22} - w_{12}$ and greater than 1 if $w_{12} - w_{11} < w_{22} - w_{12}$. In these cases $\hat{p}$ cannot represent a gene frequency.

Eq. (7) shows that Δp is always negative in the interval 0–1, since the coefficients of p and q are both negative.

We can represent this situation as in Fig. 10.2.

Since Δp is always negative, p must always decrease. It follows that the equilibrium at $p=1$ is unstable: if the frequency is changed from

$p=1$ by some event (e.g. mutation or migration), it will not return to 1, but will steadily decrease towards zero, which is a stable equilibrium.

When p is sufficiently small for p^2 to be negligible compared with p, we can write:

$$\Delta p \approx p \frac{w_{12}-w_{22}}{w_{22}}$$

which shows that the change in p from one generation to the next is a constant fraction of p:

$$p'=p+\Delta p=p\left(1+\frac{w_{12}-w_{22}}{w_{22}}\right)=\frac{w_{12}}{w_{22}}p.$$

We can therefore write:

$$p_g=\left(\frac{w_{12}}{w_{22}}\right)^k p_{g-k}.$$

Thus, the population tends towards $p=0$ as $\left(\dfrac{w_{12}}{w_{22}}\right)^k$.

In other words, the speed with which a disfavoured allele disappears is a function of the selective value w_{12} of the heterozygote, and not of the selective value w_{11} of the homozygote for the allele; this is a consequence of the fact that when p is small the homozygotes have a lower frequency, p^2, than the heterozygotes, and the gene is eliminated mainly by selection against the heterozygotes.

Example [1]. Suppose $w_{11}=0.33$, $w_{12}=0.67$, $w_{22}=1.00$, and that $p_0=\frac{1}{2}$. Then we find the following values:

Generation (g)	Gene frequency p_g
0	0.500
1	0.375
2	0.271
3	0.190
4	0.132
5	0.090
6	0.061
⋮	⋮
∞	0

Thus the disfavoured gene is rapidly eliminated.

The special case when $w_{12}^2=w_{11}w_{22}$: The selective value of the heterozygote is equal to the geometric mean of the selective values of the homozygotes.

---

[1] The examples in this and the following sections are not intended to be realistic, but simply illustrate the application of the equations.

It can be shown that the mean selective value can be written in the form:

$$\bar{w} = (w_{11} - 2w_{12} + w_{22})(p - \hat{p})^2 + \frac{w_{11}w_{22} - w_{12}^2}{w_{11} - 2w_{12} + w_{22}}$$

with

$$\hat{p} = \frac{w_{22} - w_{12}}{w_{11} - 2w_{12} + w_{22}}.$$

In the case when $w_{12}^2 - w_{11}w_{22} = 0$, we therefore have:

$$\bar{w} = (w_{11} - 2w_{12} + w_{22})(p - \hat{p})^2$$

which gives:

$$\Delta p = \frac{pq}{p - \hat{p}}$$

or

$$p' = p + \frac{pq}{p - \hat{p}} = \frac{p\hat{q}}{p - \hat{p}}$$

and

$$q' = 1 - \frac{p\hat{q}}{p - \hat{p}} = -\frac{q\hat{p}}{p - \hat{p}}.$$

Hence:

$$\frac{p'}{q'} = -\frac{\hat{q}}{\hat{p}}\frac{p}{q}$$

and

$$\frac{p_g}{q_g} = \left(-\frac{\hat{q}}{\hat{p}}\right)^g \frac{p_0}{q_0}. \qquad (16)$$

This equation enables us to calculate p_g for any value of g. However, in this case $\hat{p}$ and $\hat{q}$ do not represent gene frequencies.

Example. Suppose we have $w_{11} = 0.25$, $w_{12} = 0.5$, $w_{22} = 1.0$, so that $w_{12}^2 = w_{11}w_{22}$. Then the values of $\hat{p}$ and $\hat{q}$ are:

$$\hat{p} = \frac{1.0 - 0.5}{0.25 - 1.0 + 1.0} = 2, \qquad \hat{q} = -1.$$

Therefore:

$$\frac{p_g}{q_g} = \left(\frac{1}{2}\right)^g \frac{p_0}{q_0}.$$

If $p_0 = q_0 = \frac{1}{2}$, p_g/q_g is therefore equal to $(\frac{1}{2})^g$, so that

$$p_g = \frac{(\frac{1}{2})^g}{1 + (\frac{1}{2})^g}.$$

Some values are as follows:

Generation (g)	Gene frequency (p_g)
0	0.500
1	0.333
2	0.200
3	0.111
4	0.058
5	0.030
6	0.015
7	0.0078
$\vdots$	$\vdots$
∞	0

As in the first example, the disfavoured allele A_1 is eliminated rapidly.

1.4.2. $w_{11} = w_{12} < w_{22}$. The heterozygote has the same selective value as the less favoured homozygote.

In this case, $\hat{p} = \dfrac{w_{22} - w_{11}}{w_{22} - w_{11}} = 1$. Therefore the third equilibrium coincides with the second.

The change in gene frequency from one generation to the next is:

$$\Delta p = \frac{p q^2}{\bar{w}} (w_{11} - w_{22}).$$

This is always negative, so the only stable equilibrium is when $p = 0$.

Near this equilibrium point, we have:

$$\Delta p \approx p \frac{w_{11} - w_{22}}{w_{22}}.$$

Hence:

$$p_g = \left(\frac{w_{11}}{w_{22}} \right)^k p_{g-k}.$$

Near the unstable equilibrium with $p = 1$, $q = 0$, however,

$$\Delta p \approx \frac{w_{22} - w_{11}}{w_{11}} q^2.$$

This shows that changes in gene frequency near this point will always be very small, since they are of the order of q^2. Thus, although the point $p = 1$, $q = 0$ is an unstable equilibrium, the population will move away from this point very slowly. In other words, a rare gene which is

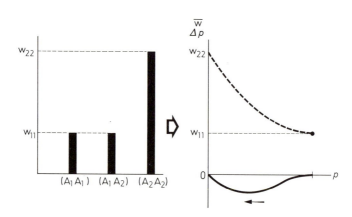

Fig. 10.3. The effects of selection when $w_{11} = w_{12} < w_{22}$

advantageous when homozygous will only spread in the population very slowly. These results are illustrated in Fig. 10.3.

1.4.3. $w_{12} < w_{11} < w_{22}$. The heterozygote has a lower selective value than the less favoured homozygote.

In this case

$$\hat{p} = \frac{w_{22} - w_{12}}{(w_{11} - w_{12}) + (w_{22} - w_{12})}$$

is between 0 and 1, since the two terms on the bottom and the term on the top of this equality are all positive. Therefore $\hat{p}$ corresponds to a real equilibrium state.

Notice that the value of Δp given by Eq. (7) can be written:

$$\Delta p = \frac{pq}{\bar{w}} \{(w_{11} - w_{12} + w_{22} - w_{12})\, p + w_{12} - w_{22}\}$$

$$= \frac{pq}{\bar{w}} (w_{11} - 2w_{12} + w_{22})(p - \hat{p}).$$

Since $(w_{11} - 2w_{12} + w_{22})$ is greater than 0, Δp has the same sign as $(p - \hat{p})$.

If $p > \hat{p}$, $\Delta p > 0$, hence p increases from one generation to the next, and tends towards the equilibrium with $p = 1$. If, however, $p < \hat{p}$, $\Delta p < 0$, and p decreases towards the equilibrium with $p = 0$.

There are therefore two stable equilibria (with $p = 0$ and with $p = 1$), and one unstable equilibrium (with $p = \hat{p}$).

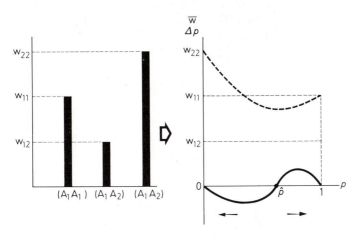

Fig. 10.4. The effects of selection when $w_{12} < w_{11} < w_{22}$

Near $p=0$, we have:

$$\Delta p = p \frac{w_{12} - w_{22}}{w_{22}}.$$

Hence:

$$p_g = \frac{w_{12}}{w_{22}} p_{g-1}$$

$$= \left(\frac{w_{12}}{w_{22}}\right)^k p_{g-k}.$$

Close to $p=1$, we have:

$$\Delta p = -\Delta q = q \frac{w_{11} - w_{12}}{w_{11}}$$

giving:

$$q_g = \frac{w_{12}}{w_{11}} q_{g-1}$$

$$= \left(\frac{w_{12}}{w_{11}}\right)^k q_{g-k}.$$

These results are illustrated in Fig. 10.4.

Notice that the equilibrium that the population will go to depends on the relative values of $\hat{p}$ and p_0, the initial gene frequency. If $p_0 < \hat{p}$, the equilibrium state will be $p=0$, whereas it will be $p=1$, if $p_0 > \hat{p}$.

Finally, if $w_{11} = w_{22}$, all the preceding arguments are still valid; the equality of the selective values of the homozygotes implies only that $\hat{p} = \frac{1}{2}$.

1.4.4. $w_{11} < w_{12} = w_{22}$. The selective value of the heterozygote is equal to that of the favoured homozygote.

In this case $\hat{p} = 0$: the third equilibrium coincides with the first. Then we have:

$$\Delta p = \frac{p^2 q}{\overline{w}} (w_{11} - w_{12}) < 0.$$

This shows that the frequency p decreases steadily each generation, towards the value $p = 0$, which is the only stable equilibrium.

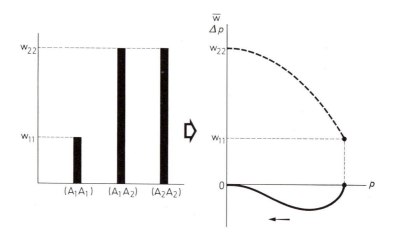

Fig. 10.5. The effects of selection when $w_{11} < w_{12} = w_{22}$

Near this equilibrium we have:

$$\Delta p \approx \frac{w_{11} - w_{22}}{w_{22}} p^2$$

giving:

$$p' = p + \Delta p = p \left[1 + \frac{w_{11} - w_{22}}{w_{22}} p \right].$$

The ratio p'/p therefore tends towards 1 as the equilibrium is approached; the gene frequency thus decreases more and more slowly. This means that a gene which is disadvantageous only when it is homozygous will be eliminated very slowly.

These results are illustrated in Fig. 10.5.

Example. Suppose that $w_{11} = 0.67$ and $w_{12} = w_{22} = 1.0$, and that the initial gene frequency is 0.5. We find the following values.

Generation (g)	Gene frequency (p_g)
0	0.500
1	0.450
2	0.407
3	0.390
4	0.355
5	0.325
6	0.300
7	0.278
8	0.258
9	0.241
10	0.226
$\vdots$	$\vdots$
∞	0

In this case, the rate of elimination of the disfavoured gene is lower than in the case (1.4.1) when its effect is expressed in the heterozygotes.

1.4.5. $w_{11} < w_{22} < w_{12}$. The selective value of the heterozygote is higher than those of both the homozygotes.

In this case, $\hat{p} = \dfrac{w_{12} - w_{22}}{(w_{12} - w_{11}) + (w_{12} - w_{22})}$ is between 0 and 1, as in case 1.4.3.

The change in p from one generation to the next is:

$$\Delta p = \frac{p\,q}{\overline{w}}\,(w_{11} - 2w_{12} + w_{22})(p - \hat{p})$$

but here $w_{11} - 2w_{12} + w_{22} = (w_{11} - w_{12}) + (w_{22} - w_{12})$ is negative, since each of the two terms in brackets is negative. Hence:

a) If $p > \hat{p}$, Δp is less than 0, and p decreases towards an equilibrium with $p = \hat{p}$.

b) If $p < \hat{p}$, Δp is greater than 0, and p increases towards an equilibrium with $p = \hat{p}$.

There are therefore two unstable equilibria (with $p = 0$ and $p = 1$) and one stable polymorphic equilibrium (with $p = \hat{p}$).

At the polymorphic equilibrium we have:

$$\overline{w} = w_{1.} = \hat{p}\,w_{11} + (1 - \hat{p})\,w_{12}$$

$$= \frac{w_{12}^2 - w_{11}\,w_{22}}{2\,w_{12} - w_{11} - w_{22}}.$$

Using formula (13), we then find:

$$\frac{d(\Delta p)}{dp} = -\frac{(w_{12} - w_{22})(w_{12} - w_{11})}{w_{12}^2 - w_{11}\,w_{22}}.$$

The convergence towards the equilibrium gene frequency, p, is therefore such that:

$$p' - \hat{p} = \left[1 - \frac{(w_{12} - w_{22})(w_{12} - w_{11})}{w_{12}^2 - w_{11} w_{22}} \right] (p - \hat{p}).$$

Thus, starting from generation k, the population tends towards the stable equilibrium as λ^{g-k} tends towards zero, where:

$$\lambda = \frac{w_{22}(w_{11} - w_{12}) + w_{11}(w_{22} - w_{12})}{w_{11} w_{22} - w_{12}^2}.$$

These results are illustrated in Fig. 10.6.

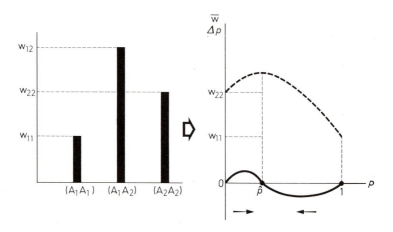

Fig. 10.6. The effects of selection when $w_{11} < w_{22} < w_{12}$

When the two homozygotes have the same selective value, these results are unaffected, and $\hat{p} = \frac{1}{2}$: at equilibrium, both alleles have the same frequency. Near this equilibrium we can write:

$$p' - \hat{p} = 2 \frac{w_{11}}{w_{11} + w_{12}} (p - \hat{p}).$$

Example. Suppose that $w_{11} = 0.40$, $w_{12} = 1.2$ and $w_{22} = 1.0$. The position of the stable equilibrium is given by:

$$\hat{p} = \frac{w_{12} - w_{22}}{-w_{11} + 2w_{12} - w_{22}} = \frac{1.2 - 1.0}{-0.4 + 2.4 - 1.0}$$

$$= 0.200.$$

The gene frequency will decrease towards this value if it is initially higher, or increase if it is initially lower. With $p_0 = \frac{1}{2}$, we find the values:

Generation (g)	Gene frequency (p_g)
0	0.500
1	0.421
2	0.366
3	0.328
4	0.300
5	0.280
6	0.264
7	0.252
8	0.242
⋮	⋮
∞	0.200

1.5. Some Particular Cases

1.5.1. Lethal genes. If the allele A_1 is lethal when homozygous, $w_{11} = 0$. The position of the equilibrium then depends on the selective value w_{12} of the heterozygotes.

Setting $w_{11} = 0$, the general Eq. (7) becomes:

$$\Delta p = \frac{pq\{p(w_{22} - 2w_{12}) + (w_{12} - w_{22})\}}{2pqw_{12} + q^2 w_{22}}$$

$$= \frac{p\{p(w_{22} - 2w_{12}) + (w_{12} - w_{22})\}}{(2w_{12} - w_{22})p + w_{22}}$$

which gives:

$$p' = p + \Delta p = \frac{w_{12}\,p}{(2w_{12} - w_{22})p + w_{22}}.$$

This can be written in the form:

$$\frac{1}{p'} = \frac{w_{22}}{w_{12}} \cdot \frac{1}{p} + \frac{2w_{12} - w_{22}}{w_{12}} \tag{17}$$

giving:

$$\frac{1}{p'} - \frac{2w_{12} - w_{22}}{w_{12} - w_{22}} = \frac{w_{22}}{w_{12}}\left(\frac{1}{p} - \frac{2w_{12} - w_{22}}{w_{12} - w_{22}}\right).$$

Hence:

$$\frac{1}{p_g} - \frac{2w_{12} - w_{22}}{w_{12} - w_{22}} = \left(\frac{w_{22}}{w_{12}}\right)^g \left(\frac{1}{p_0} - \frac{2w_{12} - w_{22}}{w_{12} - w_{22}}\right). \tag{18}$$

a) If $w_{12} < w_{22}$ (corresponding to case 1 above), $1/p$ increases indefinitely, i.e. p tends towards zero. Near $p=0$, we have:

$$\frac{p'}{p} \approx \frac{w_{12}}{w_{22}}.$$

b) If $w_{12} = w_{22}$ (corresponding to case 4 above), Eq. (17) becomes:

$$\frac{1}{p'} = \frac{1}{p} + 1.$$

Hence:

$$p_g = \frac{p_0}{1 + g\, p_0}.$$

Convergence towards the equilibrium is therefore very slow.

c) If $w_{12} > w_{22}$ (as in case 5 above) Eq. (18) shows that:

$$\frac{1}{p_g} - \frac{2w_{12} - w_{22}}{w_{12} - w_{22}}$$

tends towards zero as $\left(\dfrac{w_{22}}{w_{12}}\right)^g$, so that $p_g \to \dfrac{w_{12} - w_{22}}{2w_{12} - w_{22}}$. This shows that although the gene is lethal when homozygous, it remains in the population because of the advantage it confers on the heterozygous carriers.

Example. Suppose that $w_{11} = 0$, $w_{12} = 1.1$ and $w_{22} = 1.0$. The stable equilibrium value of p is given by:

$$\hat{p} = \frac{w_{12} - w_{22}}{2w_{12} - w_{22}} = \frac{0.1}{1.2} = 0.083.$$

In this case, we can use Eq. (18) to calculate p_g. We have:

$$\frac{1}{p_g} - \frac{1}{0.083} = \left(\frac{1}{1.1}\right)^g \left(\frac{1}{p_0} - \frac{1}{0.083}\right).$$

If $p_0 = \frac{1}{2}$, $p_g = \dfrac{1}{12 - 10\left(\dfrac{1}{1.1}\right)^g}$ which gives the following values:

Generation (g)	Gene frequency (p_g)
0	0.500
1	0.344
2	0.268
3	0.223
4	0.193
5	0.176
6	0.157
7	0.146
8	0.136
9	0.129
10	0.123
$\vdots$	$\vdots$
∞	0.083

1.5.2. The case when selective differences are small. Suppose that the selective values are such that we can write them as:

$$w_{11} = 1, \qquad w_{12} = 1 + h s, \qquad w_{22} = 1 + s$$

where s is a small, positive number (such that s^2 is negligible compared with s), and h can be regarded as a measure of the degree of dominance of allele A_2 over A_1.

In this notation, the change in the frequency p from one generation to the becomes, according to Eq. (7):

$$\Delta p = -p q s \frac{h p + (1-h) q}{1 + q s (2 h p + q)}$$

$$\approx -p q s \{ h p + (1-h) q \}. \tag{19}$$

a) If $h = 1$, i.e. if A_2 is fully dominant to A_1 and selection is effective only against the disfavoured homozygotes ($A_1 A_1$), we have:

$$\Delta p = -p^2 (1-p) s. \tag{20}$$

Since s is small, Δp must be small, and we can equate Δp and dp/dt, where t represents time, measured in generations. Then Eq. (20) becomes:

$$\frac{dp}{dt} = -p^2 (1-p) s$$

or

$$\frac{dp}{p^2 (1-p)} = -s \, dt.$$

Integrating this differential equation between generations $t=0$ and $t=g$, we obtain:

$$s g = \left[\frac{1}{p} + \log_e \frac{1-p}{p} \right]_{p_0}^{p_g} = \frac{p_0 - p_g}{p_0 \, p_g} + \log_e \frac{p_0 (1 - p_g)}{p_g (1 - p_0)}.$$

This equation enables us to calculate the number of generations it would take to go from a frequency of p_0 to p_g.

For example, to go from an initial frequency of $p_0 = 0.50$ (i.e. 25 % of the individuals homozygous recessive) to $p_g = 0.10$ (1 % of the individuals homozygous recessive), we have:

$$s g = \frac{0.5 - 0.1}{0.5 \times 0.1} + \log_e \frac{0.5 \times 0.9}{0.1 \times 0.5} = 10.2.$$

If $s = 0.01$, this change will therefore take 1020 generations. The process is even slower when the disfavoured gene is rare. Thus, to go from $p_0 = 0.10$ to $p_g = 0.01$ would take a number of generations such that:

$$s g = \frac{0.09}{0.001} + \log_e \frac{0.1 \times 0.99}{0.01 \times 0.9} = 92.4.$$

When s is 0.01, therefore, 9 204 generations are necessary.

b) If $h = \frac{1}{2}$ the heterozygote is exactly intermediate between the two homozygotes. Eq. (19) then becomes:

$$\Delta p = -\frac{s}{2} p(1-p).$$

Hence, in a similar way to the preceding case, we have:

$$\frac{dp}{p(1-p)} = -\frac{s}{2} dt.$$

Integrating between $t=0$ and $t=g$ we obtain:

$$\frac{s}{2} g = \left[\log_e \frac{1-p}{p} \right]_{p_0}^{p_g} = \log_e \frac{p_0 (1 - p_g)}{p_g (1 - p_0)}.$$

The disfavoured gene is thus eliminated much more rapidly in this case than in the case when only the homozygotes were selected against. For example, to go from $p_0 = 0.50$ to $p_g = 0.10$:

$$\frac{s}{2} g = \log_e \frac{0.5 \times 0.9}{0.1 \times 0.5} = 2.18$$

and to go from $p_0 = 0.10$ to $p_g = 0.01$:

$$\frac{s}{2} g = \log_e \frac{0.1 \times 0.99}{0.01 \times 0.9} = 2.39.$$

When $s = 0.01$, these correspond to 436 and 478 generations, respectively.

c) If $h = 0$, the heterozygote is selected against as strongly as the disfavoured $(A_1 A_1)$ homozygote. In this case:

$$\Delta p = -s p (1-p)^2$$

which gives:

$$\frac{dp}{p(1-p)^2} = -s dt.$$

Hence:

$$-s g = \left[\frac{1}{1-p} + \log_e \frac{p}{1-p} \right]_{p_0}^{p_g}$$

$$= \frac{p_g - p_0}{(1 - p_g)(1 - p_0)} + \log_e \frac{p_g(1 - p_0)}{p_0(1 - p_g)}.$$

For the same examples as before, we find that to go from $p_0 = 0.50$ to $p_g = 0.10$:

$$s g = \frac{0.4}{0.5 \times 0.9} - \log_e \frac{0.5 \times 0.1}{0.5 \times 0.9} = 3.07$$

and to go from $p_0 = 0.10$ to $p_g = 0.01$:

$$s g = \frac{0.09}{0.99 \times 0.9} - \log_e \frac{0.9 \times 0.01}{0.1 \times 0.99} = 2.49.$$

When $s = 0.01$ these correspond to 307 and 249 generations, respectively. These results were first derived by Haldane (1924).

1.6. Variable Selective Values

1.6.1. The Rhesus blood-group. This is a human blood-group system of the sort described in Section 5.1 of Chapter 4. A very much simplified account of this system is as follows. There are two alleles: Rh which is dominant, and a recessive allele rh. When a "rhesus negative", or "Rh −", woman (whose genotype is therefore (rh rh)) has a rhesus positive or "Rh +" baby (whose genotype must be (Rh rh)), the presence of the Rh positive foetus causes her to produce antibodies to rhesus positive red blood cells. If she later has another Rh positive baby, these antibodies may agglutinate its red blood cells. The baby can be made very ill by this process; this disease is called "haemolytic disease of the new-born" or "*erythroblastosis foetalis*" and is sometimes fatal.

There is therefore a selective difference between genotypes at this locus: children whose mother's genotype is (rh rh) have a lowered probability of survival if their genotype is (Rh rh). Let us represent this probability by $1-s$, while the probability of survival of babies of all other genotypes is 1. Let p be the frequency of the Rh allele in the population, and let $q=1-p$ be the frequency of the rh allele. If we assume that the genotype frequencies among adult individuals are given by the Hardy-Weinberg formula, and that mating is at random with respect to this locus, we can calculate the frequencies of the various types of marriages, and of the children they will produce, as follows.

Parents		Frequency of this type of mating	Offspring		
Mother	Father		(Rh Rh)	(Rh rh)	(rh rh)
(Rh Rh)	Any genotype	p^2	p	q	0
(Rh rh)	Any genotype	$2pq$	$p/2$	$\frac{1}{2}$	$q/2$
(rh rh)	(Rh Rh)	$p^2 q^2$	0	$1-s$	0
(rh rh)	(Rh rh)	$2pq^3$	0	$\frac{1}{2}(1-s)$	$\frac{1}{2}$
(rh rh)	(rh rh)	q^4	0	0	1

The genotypic proportions among the children will therefore be:

(Rh Rh): $\quad p^3 + p^2 q = p^2$

(Rh rh): $\quad p^2 q + p q + (1-s)(p^2 q^2 + p q^3) = 2pq - spq^2$

(rh rh): $\quad p q^2 + p q^3 + q^4 = q^2$

Total: $\quad 1 - spq^2$

Hence the frequency of the Rh gene among the children will be:

$$p' = p\,\frac{1 - s/2\,q^2}{1 - spq}.\tag{21}$$

The equilibria can be deduced from the equality:

$$\Delta p = spq^2\,\frac{p - \frac{1}{2}}{1 - spq^2} = 0$$

which leads directly to:

$$\hat{p}_1 = 0, \quad \hat{p}_2 = \tfrac{1}{2}, \quad \hat{p}_3 = 1$$

The first and last of these equilibria are stable, while the second is unstable. This can be seen as follows:

$$\Phi'(p) = s\,q\,\frac{3p\,q - q/2 - p^2 - s\,p^2 q^3}{(1 - s\,p\,q^2)^2}$$

which gives $\Phi'(0) = -\frac{1}{2}$, $\Phi'(\frac{1}{2}) = \dfrac{s/8}{1 - s/8}$, $\Phi'(1) = 0$, and $\Phi''(1)$ is found to be greater than zero.

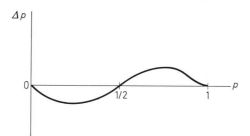

Fig. 10.7. The selective effect of the rhesus blood group system

With the assumptions we have made, we should expect that all human populations would become fixed for one or other of the rhesus alleles; if the initial frequency of the Rh allele was less than $\frac{1}{2}$, we would expect the final population to consists entirely of (rh rh) homozygotes; whereas if the initial frequency of Rh was greater than $\frac{1}{2}$, we would expect the Rh allele to be fixed.

This treatment is clearly an approximate one, and would not be strictly valid if the selection coefficients are large. Haldane (1942) has given an exact treatment, which leads to the same conclusions as the above approximate method.

However, when Rh frequencies in real populations are studied, most populations are found to be polymorphic at this locus, though with widely differing gene frequencies in different populations. In France, for example, it is found that:

$$p(\mathrm{Rh}) = 0.6, \quad q(\mathrm{rh}) = 0.4.$$

There are several possible explanations for this difference between the theoretical predictions and the observations.

1. The populations in which the frequencies were estimated may not have reached equilibrium. This is very likely, since convergence towards the equilibrium is slow. Also we know that human populations are subject to many influences which would perturb the gene frequencies from their equilibrium values, in particular migration.

2. The selective effect due to incompatibility between Rh negative mothers and their Rh positive children might be counterbalanced by mutation (though the mutation rate would have to be improbably high), or by increased fecundity of the couples which are affected by this selection.

For a review of this topic, see Cavalli-Sforza and Bodmer (1971).

1.6.2. Frequency-dependent selection. Up to now we have assumed that genotypic fitnesses are constant, or vary as a result of physiological interactions between mother and child (Rhesus). This assumption is reasonable when we are dealing with genes whose effects relate to general biological efficiency or to adaptations to the physical environment. In cases when a gene influences the reaction of its carriers to other members of the population, or to the members of a different species, it is likely that fitnesses will depend on the genotypic composition of the population. For instance, if the environment of a population is subdivided into different niches, a genotype which can exploit a niche which other members of the population cannot exploit will increase in numbers up to the point when this niche is fully exploited. The selective value of this genotype will thus be high when the genotype is rare, and decrease as it becomes more common and saturates the niche.

The subject of frequency-dependent selection has been considered by Wright (1949) and many later authors. We shall investigate some simple models which illustrate some of the more important properties of frequency-dependent selection.

a) *Dominant gene.* Suppose that the fitnesses of the genotypes are as follows:

Genotype	Frequency	Fitness
$A_1/-$	$1-q^2$	$1+s(1-q^2)$
A_2/A_2	q^2	$1+tq^2$

Then we have:

$$w_{1.} = 1+s(1-q^2)$$

$$w_{2.} = 1+tq^2$$

$$\overline{w} = 1+s(1-q^2)^2+tq^4.$$

The polymorphic equilibrium is given by:

$$w_{1.} = w_{2.}$$

i.e.:

$$\hat{q}^2(s+t)-s=0,$$

$$\hat{q} = \sqrt{\frac{s}{s+t}}, \quad \hat{p} = 1-\sqrt{\frac{s}{s+t}},$$

and

$$\hat{w}_{1.} = \hat{w}_{2.} = \hat{\bar{w}} = 1 + \frac{st}{s+t}.$$

There can be a biologically meaningful value of $\hat{q}$ only if s and $(s+t)$ are either both greater than zero, or both less than zero.

The stability of this equilibrium can be tested by the method introduced in Section 1.3. We have:

$$\Phi'(\hat{p}) = \frac{\hat{p}\hat{q}}{\hat{\bar{w}}} \frac{d}{dp}(w_{1.} - w_{2.})$$

$$= 2\frac{\hat{p}\hat{q}^2}{\hat{\bar{w}}}(s+t).$$

The equilibrium is therefore stable when $s+t<0$ and

$$|s+t| < \frac{\hat{\bar{w}}}{\hat{p}\hat{q}^2}.$$

This model of frequency-dependent selection therefore yields a single polymorphic equilibrium, which is stable only when s and $(s+t)$ are negative, i.e. when A_1 is at an advantage when rare. At equilibrium, all the genotypes have the same fitness, so that there is no question of heterozygous advantage here.

b) *Different fitness of all three genotypes.* Suppose that individuals are at an advantage, or disadvantage, depending on the frequency of other individuals of the same genotype, at the moment of conception. The selective values of the various genotypes are then of the form:

$$w_{11} = 1 + r p^2$$
$$w_{12} = 1 + 2 s p q$$
$$w_{22} = 1 + t q^2.$$

Hence the mean selective values are:

$$w_{1.} = 1 + p(r p^2 + 2 s q^2)$$
$$w_{2.} = 1 + q(2 s p^2 + t q^2)$$
$$\bar{w} = 1 + r p^4 + 4 s p^2 q^2 + t q^4.$$

The possible non-trivial equilibria are given by:

$$w_{1.} = w_{2.}$$

or:

$$p(r p^2 + 2 s q^2) = q(2 s p^2 + t q^2). \qquad (22)$$

If we now make the change of variable:

$$p = \tfrac{1}{2}(1+u), \qquad q = \tfrac{1}{2}(1-u)$$

Eq. (22) becomes:

$$(r+4s+t)\,u^3 + 3(r-t)\,u^2 + (3r-4s+3t)\,u + (r-t) = 0. \tag{23}$$

This equation has 0, 1, 2 or 3 roots between -1 and $+1$, depending on the values of r, s and t. These roots define the equilibria which are possible; by Eq. (13), the stability of these equilibria depends on the value of:

$$\Phi'(\hat{p}) = \frac{\hat{p}\,\hat{q}}{\hat{w}} \frac{d}{dp}(w_{1.} - w_{2.})$$

$$= \frac{(1-\hat{u}^2)}{16\hat{w}}\{3r - 4s + 3t + 6(r-t)\,\hat{u} + 3(r+4s+t)\,\hat{u}^2\}$$

in which

$$\hat{w} = 1 + \frac{r+4s+t}{16} + \frac{3}{16}(r-t)\,\hat{u} + \frac{3r-4s+3t}{16}\hat{u}^2 + \frac{r-t}{16}\hat{u}^3.$$

A particular case: $r = t$. When the coefficients r and t attached to the homozygotes are the same, the above equations become much simpler. Eq. (23) becomes:

$$2u\{(r+2s)\,u^2 + 3r - 2s\} = 0$$

which has the three roots:

$$u_1 = 0, \qquad u_2 = \sqrt{\frac{2s-3r}{2s+r}}, \qquad u_3 = -\sqrt{\frac{2s-3r}{2s+r}}$$

and the corresponding values of the derivative Φ' are:

$$\Phi'(u_1) = \frac{3r-4s}{8+r+2s},$$

$$\Phi'(u_2) = \Phi'(u_3) = \frac{r(2s-3r)}{2s(r+1)+r(1-r)}.$$

The question of whether the roots u_2 and u_3 correspond to gene frequencies, and the stability of these equilibria, depend on the signs of $r, s, 3r-4s, 2s-3r$ etc. We shall examine a few particular cases:

a) If $s < 0$ and $r < 0$, the roots u_2 and u_3 are real if $2s < 3r$, in other words if $|s| > \tfrac{3}{2}|r|$. The two corresponding equilibria are stable if:

$$2s(1+r) + r(1-r) < 0.$$

Thus, with: $r=-1$, $s=-2$, we have $u_2=\dfrac{1}{\sqrt{5}}$, $u_3=-\dfrac{1}{\sqrt{5}}$, and the equilibria are:

$$\hat{p}_1=\frac{1}{2}, \quad \hat{p}_2=\frac{1}{2}\left(1+\frac{1}{\sqrt{5}}\right)=0.723, \quad \hat{p}_3=\frac{1}{2}\left(1-\frac{1}{\sqrt{5}}\right)=0.277$$

for which:

$$\Phi'(\hat{p}_1)=\tfrac{5}{3}, \quad \Phi'(\hat{p}_2)=\Phi'(\hat{p}_3)=-\tfrac{1}{2}.$$

This shows that only the last two equilibria are stable. Also, the two trivial equilibria $p=0$ and $p=1$ can be shown to be unstable.

b) If $s>0$ and $r>0$, the roots u_2 and u_3 are real if $2s>3r$. The two corresponding equilibria are stable if $2s(1+r)+r(1-r)<0$, i.e. if $s<\dfrac{r(1-r)}{2(r+1)}$, but this is incompatible with the condition $s>\tfrac{3}{2}r$. Therefore the only stable polymorphic equilibrium in this case is $\hat{p}_1=\tfrac{1}{2}$.

It is also found that in this case the two trivial equilibria $\hat{p}=0$ and $\hat{p}=1$ are stable.

Frequency-dependent selection can thus yield stable polymorphisms when there is no heterozygous advantage, and can even maintain polymorphism against a net selective disadvantage to heterozygotes (Lewontin, 1958); with a particular selective situation, there may be more than one stable gene-frequency equilibrium. Recent evidence from experiments on *Drosophila* suggests that frequency-dependent selection may be a common phenomenon (Ehrman and Petit, 1968), so that these possibilities must be considered seriously when we attempt to explain polymorphisms.

1.7. Constant Selection for a Sex-Linked Gene

We shall now consider the case of selection acting on a sex-linked gene, i.e. a gene carried on the X chromosome. Women therefore will have two genes at this locus, but men will have only one.

With two alleles, there are five different possible genotypes: three in women and two in men. We shall write their selective values as follows:

	Women			Men	
Genotypes:	(A_1A_1)	(A_1A_2)	(A_2A_2)	(A_1)	(A_2)
Selective values:	w_{11}	w_{12}	w_{22}	w_1	w_2

Let p_f and p_m be the frequency of the A_1 allele in women and men respectively, among individuals of reproductive age. The genotype frequencies at the moment of conception of the next generation are, therefore, as follows:

$$(A_1 A_1): \quad p_f\, p_m$$
$$(A_1 A_2): \quad p_f\, q_m + p_m\, q_f$$
$$(A_2 A_2): \quad q_f\, q_m$$
$$(A_1): \quad p_f$$
$$(A_2): \quad q_f$$

When these new individuals reach adulthood, the frequencies of A_1 in the two sexes will become, under selection:

$$p'_f = \frac{w_{11}\, p_f\, p_m + \tfrac{1}{2} w_{12}(p_f\, q_m + p_m\, q_f)}{w_{11}\, p_f\, p_m + w_{12}(p_f\, q_m + p_m\, q_f) + w_{22}\, q_f\, q_m}.$$

$$p'_m = \frac{w_1\, p_f}{w_1\, p_f + w_2\, q_f}. \tag{24}$$

The population is in equilibrium if

$$p'_f = p_f = \hat{p}_f \quad \text{and} \quad p'_m = p_m = \hat{p}_m = \frac{w_1\, p_f}{w_1\, p_f + w_2\, q_f}.$$

The equilibrium is therefore defined by $\hat{p}_f$, which we shall represent as X. The first Eq. (24) then gives:

$$X = \frac{w_{11}\, w_1\, X^2 + w_{12}(w_1 + w_2)\, X(1-X)/2}{w_{11}\, w_1\, X^2 + w_{12}(w_1 + w_2)\, X(1-X) + w_{22}\, w_2(1-X)^2}$$

which has the solutions:

a) $X = 0$, i.e. $\hat{p}_f = \hat{p}_m = 0$,

b) $X = 1$, i.e. $\hat{p}_f = \hat{p}_m = 1$,

c) $X = \dfrac{w_{22}\, w_2 - w_{12}(w_1 + w_2)/2}{w_{11}\, w_1 - w_{12}(w_1 + w_2) + 2 w_{22}\, w_2}$

i.e.:

$$\hat{p}_f = \frac{2 w_{22}\, w_2 - w_{12}(w_1 + w_2)}{2 w_{11}\, w_1 - 2 w_{12}(w_1 + w_2) + 2 w_{22}\, w_2},$$

$$\hat{p}_m = \frac{w_1\, \{2 w_{22}\, w_2 - w_{12}(w_1 + w_2)\}}{2 w_1\, w_2(w_{11} + w_{22}) - w_{12}(w_1 + w_2)^2}.$$

In this case, the equilibrium frequencies can be different in men and women.

Whether the frequencies in the third case are real, and whether these three equilibria are stable, depends on the values of the five coefficients w_{11}, w_{12}, w_{22}, w_1 and w_2. We shall study the various possible cases, assuming that the hemizygote (A_1) is favoured compared with (A_2), whose selective value will be taken as unity; i.e. $w_2 = 1$, $w_1 = 1 + u$, where $u \geqq 0$.

a) *The equilibrium with* $\hat{p}_f = \hat{p}_m = 0$. Close to this equilibrium, we can represent p_f by a variable x, such that x^2 is negligible compared with x. Then we can write:

$$p'_m = \frac{w_1 x}{w_1 x + w_2 (1 - x)}$$

$$\approx \frac{w_1}{w_2} x = (1 + u) x,$$

$$p'_f = \frac{w_{12} \{ x(1 - x - ux) + (1 + u) x (1 - x) \}/2}{w_{22}(1 - x)(1 - x - u x)}$$

$$\approx \frac{w_{12}}{w_{22}} \left(1 + \frac{u}{2} \right) x$$

and

$$\Delta p = p'_f - p_f = \frac{w_{12} \left(1 + \dfrac{u}{2} \right) - w_{22}}{w_{22}} x.$$

The equilibrium is stable if $\Delta p_f < 0$, in other words if:

$$w_{12} \left(1 + \frac{u}{2} \right) < w_{22}. \tag{25}$$

If all the selective differences are small, we can write the selective values in the form:

$$w_{11} = 1 + r, \quad w_{12} = 1 + s, \quad w_{22} = 1 + t, \quad w_1 = 1 + u$$

where all the parameters, r, s, t and u are such that their squares and products with one another r^2, $r s$, etc. are negligible compared with r, s, t and u themselves. In this case, the inequality (25), neglecting second-order terms, becomes:

$$1 + s + \frac{u}{2} < 1 + t,$$

$$t - s > \frac{u}{2}.$$

The equilibrium with $\hat{p}_f = \hat{p}_m = 0$ (allele A_1 absent) is therefore stable, if the difference between the selection coefficients t and s of the homozygotes $(A_2 A_2)$ and the heterozygotes $(A_1 A_2)$ is greater than half the selective advantage u of the hemizygotes (A_1), even though A_1 is favoured in males.

b) *The equilibrium with $\hat{p}_f = \hat{p}_m = 1$.* By an analogous argument to the one for case a), we find that the equilibrium with allele A_2 absent is stable if

$$w_{12}\left(1+\frac{u}{2}\right) < w_{11}(1+u). \tag{26}$$

When the selective differences are small, this inequality takes the form:

$$s - r < \frac{u}{2}.$$

This shows that there is a stable equilibrium where the allele A_2 (which is disadvantageous in men) has zero frequency, provided that A_2 is not too advantageous in heterozygous women.

c) *The polymorphic equilibrium.* When the selective differences are small, the polymorphic equilibrium corresponds to:

$$\hat{p}_f \approx \hat{p}_m \approx \frac{t - s - u/2}{r - 2s + t},$$

$$\hat{q}_f \approx \hat{q}_m \approx \frac{r - s + u/2}{r - 2s + t}.$$

If $r - 2s + t > 0$, these can represent gene frequencies provided that:

$$t - s > \frac{u}{2} \quad \text{and} \quad s - r < \frac{u}{2}.$$

This means that the two trivial equilibria are stable, and therefore implies that the intermediate equilibrium is unstable. If however, $r - 2s + t < 0$, the above values can represent gene frequencies if:

$$t - s < \frac{u}{2} \quad \text{and} \quad s - r > \frac{u}{2}.$$

In this case, the trivial equilibria are unstable, which implies that the polymorphic equilibrium is stable.

If, therefore, $r - 2s + t$ is less than zero, a stable polymorphic equilibrium is possible. This is in particular the case when the heterozygote is favoured compared with both the homozygotes, i.e. when $s > r$ and $s < t$. This is the same type of equilibrium as we found for autosomal genes.

But in the case of sex-linked genes this condition can also be met when the selective value of the heterozygote is between those of the homozygotes, provided that the values are such that:

$$t - s < \frac{u}{2} < s - r.$$

This implies that $r < s < t$. This important result can therefore be summarised as follows:

If a sex-linked gene is favoured in the hemizygote form, slightly disfavoured in heterozygotes, and more disfavoured still in homozygotes (i.e. if the selective values are of the form $w_{11} = 1 + r, w_{12} = 1 + s, w_{22} = 1 + t, w_1 = 1 + u, w_2 = 1$, where $0 < r < s < t$ and $u > 0$), there is a stable polymorphic equilibrium if the selective values satisfy the condition:

$$t - s < \frac{u}{2} < s - r.$$

1.8. Selection in the Multi-Locus Case

In this chapter we have exclusively considered selection at a single locus. Clearly, this is unrealistic since we know that many loci are segregating simultaneously in populations, and some at least of these must be under selective control. Much work has been devoted to the complex problem of selection with more than one locus. To give an account of this important problem is beyond the scope of this book. An up-to-date review of this area is given by Kojima and Lewontin (1970).

2. The Consequences of Selection for the Mean Fitness of Populations

We have seen that selection can change the frequencies of the different alleles in a population towards an equilibrium state, which may be "trivial" (when only one allele is present) or "polymorphic" (when several alleles remain in the population). These changes in the genetic make-up of a population are interesting in themselves; but it is even more important to understand how they affect the adaptation of the population as a whole to its environment, and what effects they have on population's capacity to survive and increase in size.

We have defined the selective values of individuals as numbers proportional to their probability of survival from the moment of conception until reproductive age. The selective values therefore take into account all the factors which tend to eliminate individuals of one genotype rather than another. One could also broaden the idea of selective value to include variations in the fecundity of mating couples, by counting not only the conceptions which actually occur, but also the "potential conceptions" which would have been possible.

Note that when we assume that there is random mating, there will be Hardy-Weinberg proportions only at the stage of such "potential conceptions".

If the absolute probabilities of survival from conception to reproductive maturity for the various genotypes are constant in time, it follows from the discussion in Section 1.1 of this chapter that the rate of increase in size of the population is proportional to $\bar{w}$, the mean selective value at the stage of "potential conceptions". Any change in $\bar{w}$ caused by genetic factors will change the rate of growth of the population. It is reasonable to suppose that the absolute probabilities of survival are constant when the population is expanding into a new, previously uncolonised environment, where population size is not limited by finite environmental resources. The human species is at present in this situation, although this may be a very temporary state of affairs. If two species are competing in such a environment, then the species with the higher rate of growth will eventually predominate; we can therefore use the rate of growth of a population as a measure of its general level of adaptation. Changes in $\bar{w}$ caused by genetic changes in a population can thus be said to affect the general adaptedness of the population. We shall now study the way in which the mean selective value of a population is affected by selection.

2.1. Constant Selective Values

In the case of a single locus, with any number of alleles, it can be shown that when the selective values are constant, selection always leads to an increase in $\bar{w}$ from one generation to the next. This was first proved rigorously by Scheuer and Mandel (1959); Ewens (1968) gives a simpler proof. In this section we shall derive a version of this theorem which is valid in the neighbourhood of the equilibria, or globally if selection is weak. This result follows from Fisher's (1930) "Fundamental Theorem of Natural Selection", which states that *the rate of increase in the mean selective value of a population is approximately proportional to the additive genetic variance in selective value*. We shall therefore prove this theorem first.

2.1.1. Fisher's fundamental theorem. In multiple-allelic systems, we have, at the moment of conception:

$$\bar{w} = \sum_{ij} p_i p_j w_{ij}.$$

When these individuals reach reproductive age, the gene frequencies will have changed, under selection, to:

$$p_i' = \frac{w_{i.}}{\bar{w}} p_i = \sum_j \frac{p_j w_{ij}}{\bar{w}} p_i. \tag{27}$$

Assuming that the selective values w_{ij} do not change, the mean selective value in the next generation will be:

$$\overline{w}' = \sum_{ij} p'_i p'_j w_{ij}. \tag{28}$$

The selective values w_{ij} can be subjected to the type of analysis described in Chapter 5. In such an analysis, we can decompose each w_{ij} value into:

1. A general term $\overline{w}$, the general mean of the w_{ij} values.
2. Two terms corresponding to the "additive effects" of the A_i and A_j genes.
3. A term for the "dominance effect" of these two genes. We can thus write:

$$w_{ij} = \overline{w} + \alpha_i + \alpha_j + \delta_{ij}.$$

We have seen that the determination of the values of the α_i and δ_{ij} parameters involves making some arbitrary assumptions; the usual method employed is due to Fisher (1918) and involves minimising the value of the expression $\sum_{ij} p_i p_j \delta_{ij}^2$.

Using this "least squares method", the α_i are given by the following expression:

$$\alpha_i = \sum_j p_j(w_{ij} - \overline{w})$$

$$= \sum_j p_j w_{ij} - \overline{w} \sum_j p_j$$

$$= w_{i.} - \overline{w}.$$

Now we saw in Section 1.2 that:

$$\Delta p_i = \frac{p_i}{\overline{w}}(w_{i.} - \overline{w}).$$

Hence:

$$\alpha_i = \Delta p_i \frac{\overline{w}}{p_i}.$$

It is easy to prove that the α_i and δ_{ij} satisfy the relations:

$$\sum_i p_i \alpha_i = 0, \quad \sum_i p_i \delta_{ij} = 0. \tag{29}$$

Finally, with this method of analysis of the w_{ij} into components due to additive and dominance effects, the variance in w_{ij} is also partitioned into "additive genetic variance" and "dominance variance":

$$V_W = V_A + V_D$$

since $\mathrm{Cov}(AD) = 0$. Here, $V_A = 2 \sum p_i \alpha_i^2$.

If we substitute the components of fitness into Eq. (27), we obtain:

$$p_i' = \frac{\sum_j p_j(\overline{w} + \alpha_i + \alpha_j + \delta_{ij})}{\overline{w}} \, p_i$$

$$= \left(1 + \frac{\alpha_i}{\overline{w}}\right) p_i.$$

From Eq. (28), we find:

$$\overline{w}' = \sum_{ij} \left(\frac{\overline{w} + \alpha_i}{\overline{w}}\right) \left(\frac{\overline{w} + \alpha_j}{\overline{w}}\right) (\overline{w} + \alpha_i + \alpha_j + \delta_{ij}) \, p_i \, p_j.$$

There are 16 terms in the expansion of the expression after the summation sign; 12 of them are of the form

$$\sum_{ij} p_i p_j \alpha_i \quad or \quad \sum_{ij} p_i p_j \alpha_i \delta_{ij}$$

and these are equal to zero, from Eq. (29). The expression which remains is as follows:

$$\overline{w}' = \overline{w} + \frac{1}{\overline{w}} \sum_i p_i \alpha_i^2 + \frac{1}{\overline{w}} \sum_j p_j \alpha_j^2 + \frac{1}{\overline{w}^2} \sum_{ij} p_i p_j \alpha_i \alpha_j \delta_{ij}.$$

If all the $\delta_{ij} = 0$, we can therefore write:

$$\Delta \overline{w} = \overline{w}' - \overline{w}$$

$$= \frac{V_A}{\overline{w}}.$$

Again from Eq. (28) we have:

$$\overline{w}' = \sum_{ij} \left(1 + \frac{\alpha_i}{\overline{w}}\right) \left(1 + \frac{\alpha_j}{\overline{w}}\right) p_i p_j w_{ij}$$

$$= \sum_{ij} \left(1 + \frac{\alpha_i \alpha_j}{\overline{w}^2} + \frac{\alpha_i}{\overline{w}} + \frac{\alpha_j}{\overline{w}}\right) p_i p_j w_{ij}.$$

Now we have seen that:

$$\alpha_i = \Delta p_i \frac{\overline{w}}{p_i}.$$

Therefore

$$\frac{\alpha_i \alpha_j}{\overline{w}^2} = \frac{\Delta p_i \Delta p_j}{p_i p_j}$$

and this can be neglected in comparison to terms of the type:

$$\frac{\alpha_i}{\overline{w}} = \frac{\Delta p_i}{p_i}.$$

Hence:

$$\overline{w}' = \sum_{ij} p_i p_j w_{ij} + \frac{1}{\overline{w}} \sum_{ij} \alpha_i p_i p_j w_{ij} + \frac{1}{\overline{w}} \sum_{ij} \alpha_j p_i p_j w_{ij}$$

$$= \overline{w} + \frac{1}{\overline{w}} \sum_i p_i \alpha_i^2 + \frac{1}{\overline{w}} \sum_j p_j \alpha_j^2$$

and we have, approximately:

$$\Delta \overline{w} = \frac{V_A}{\overline{w}}. \tag{30}$$

Since a variance is necessarily positive, Eq. (30) shows that the mean selective value of a population always (under the conditions which we have assumed) increases from one generation to the next.

2.1.2. Two alleles. A more general result is easily derived for the two-allele case. In this case, we saw in Section 1.3 of this chapter that, with two alleles, A_1 and A_2, the rate of change in the frequency of A_1 is given by:

$$\Delta p = p q \frac{(w_{1.} - w_{2.})}{\overline{w}}$$

$$= \frac{p q \alpha}{\overline{w}}$$

where α is the average effect of the gene substitution on fitness. But:

$$\overline{w} = p^2 w_{11} + 2 p q w_{12} + q^2 w_{22}.$$

Hence:

$$\frac{d\overline{w}}{dp} = 2[p w_{11} + (q - p) w_{12} - q w_{22}]$$

$$= 2\alpha$$

and

$$\frac{d^2 \overline{w}}{dp^2} = 2(w_{11} + w_{22} - 2 w_{12}).$$

The derivatives of higher order are clearly zero. We therefore have:

$$\Delta p = \frac{p q}{2\overline{w}} \cdot \frac{d\overline{w}}{dp}.$$

This relation was first derived by Wright (1937).

To obtain the change in $\overline{w}$ resulting from a change Δp in the gene frequency p of allele A_1, we use Taylor's theorem to write the exact expression:

$$\Delta \overline{w} = \Delta p \, \frac{d\overline{w}}{dp} + \frac{\Delta p^2}{2} \, \frac{d^2 \overline{w}}{dp^2}$$

$$= 2 \, \frac{pq\alpha^2}{\overline{w}} + (w_{11} + w_{22} - 2w_{12}) \left(\frac{pq\alpha}{\overline{w}} \right)^2 \qquad (31)$$

$$= \frac{pq\alpha^2}{\overline{w}} \left[1 + \frac{p w_{22} + q w_{11}}{\overline{w}} \right].$$

This result is due to Li (1967).

It is easily seen from Eq. (31) that $\Delta \overline{w}$ is zero only at gene-frequency equilibrium (i.e. $\alpha = 0$ or $pq = 0$), and is otherwise positive. When the heterozygote is exactly intermediate between the two homozygotes (in other words, when there is no dominance), $w_{11} + w_{22} - 2w_{12}$ is zero, and the change in mean fitness is exactly equal to $2pq\alpha^2/\overline{w}$. From Eq. (20) of Chapter 5, the additive variance in fitness, in the two-allele case, is $V_A = 2pq\alpha^2$. Thus, when there is no dominance, we see that the change in mean fitness is equal to $V_A/\overline{w}$.

When we consider more than one locus, it can be shown that, even with constant selective values, the mean selective value is not necessarily maximised by selection (Kojima and Kelleher, 1961). This is also true when there is frequency-dependent selection, as we shall see below.

2.2. Variable Selective Values

Now let us consider systems in which the selective values are functions of the gene frequencies or genotype frequencies. As we have already seen, for the case of two alleles, the equilibria correspond to the solutions of the equation:

$$w_{1.} = w_{2.}.$$

But, by definition, $\overline{w} = p w_{1.} + q w_{2.}$. Hence,

$$\frac{d\overline{w}}{dp} = w_{1.} - w_{2.} + p \, \frac{dw_{1.}}{dp} + q \, \frac{dw_{2.}}{dp}.$$

The right hand side of this equation is zero at the equilibrium only if:

$$\hat{p} \, \frac{dw_{1.}}{dp} \bigg|_{\hat{p}} + \hat{q} \, \frac{dw_{2.}}{dp} \bigg|_{\hat{p}} = 0. \qquad (32)$$

It is easy to verify that this is the case for the examples of Section 1.6.2 in which the selective values were proportional to the genotype frequencies in the population.

However, one can imagine cases where Eq. (32) is not valid. L'Heritier (1959a) has studied the case when success in competition for mates is controlled by a single locus; $(A_1 A_1)$ homozygotes are assumed to have an advantage over $(A_1 A_2)$ heterozygotes, and these, in turn, are assumed to be more successful than the $(A_2 A_2)$ homozygotes. This advantage to carriers of the A_1 allele is, however, assumed to be counterbalanced by higher fecundity of the genotypes with the A_2 gene. If the fecundities of the three genotypes are assumed to be proportional to 2, 3 and 4, we can summarise the overall selective effects by means of the following selective values:

$$w_{11}=2-p^2, \quad w_{12}=3q, \quad w_{22}=2q^2$$

which give:

$$w_{1.}=p(2-p^2)+3q^2$$
$$=-p^3+3p^2-4p+3,$$
$$\overline{w}=p^2(2-p^2)+6pq^2+2q^4$$
$$=p^4-2p^3+2p^2-2p+2.$$

It is evident that:

$$w_{1.}-\overline{w}=(1-p)(p^3-p+1)$$

is always positive. Taking account of Eq. (6), Δp is therefore greater than 0, and allele A_1 will spread in the population. But $\overline{w}$ decreases as p increases, since $\dfrac{d\overline{w}}{dp}=4p^3-6p^2+4p-2$ is always less than 0.

At equilibrium, $p=1$ and $\overline{w}=1$, which is its minimum value. In this case, therefore, selection minimises the mean selective value.

3. Selection in Populations with Overlapping Generations

In our treatment of selection, we have so far assumed that generations are distinct. Clearly this is not a realistic representation of the facts, for many species, including man. In this section, we shall consider the problem of how to determine the effects of selection on a population of the type we studied in Chapter 7, where the generations overlap.

In order to deal with this complex problem in a simple manner, it is necessary to make several simplifying assumptions; this restricts the generality of the treatment, but makes the proof simpler to understand. However, it is possible to show that the result which we shall derive is valid in a wider context. We shall restrict our attention to the case of a population which mates at random with respect to both age and genotype. We shall consider the effect of selection on a single autosomal locus with two alleles, A_1 and A_2. Furthermore, we will assume that the differences in demographic parameters between males and females are sufficiently small that genotype frequencies in males and females can be assumed to be equal. This enables us to consider females only, in studying the effects of selection.

3.1. Demographic Parameters and Selective Differences

We shall employ the model introduced in Chapter 7, in which the continuum of ages was split up into a set of discrete age-classes 1, 2, ..., ω; the population was characterised demographically by $l(x)$, the probability of survival of a female up to the beginning of age-class x, and $f(x)$, her expected number of female offspring while she is in age-class x. The product $k(x)=l(x)f(x)$ will be used in what follows. When there are selective differences between the genotypes, there must be differences between the values of the $k(x)$ for the different genotypes. Let us write $k_{ij}(x)$ for the $k(x)$ function for individuals of genotype $(A_i A_j)$. These differences will lead to differences in the intrinsic rates of increase of the genotypes; we could therefore define $\hat{r}_{ij}$ as the real root (greater than -1) of the equation:

$$\sum_x (1+r)^{-x} k_{ij}(x)=1. \tag{33}$$

From the results of Chapter 7, we can consider $\hat{r}_{11}$ and $\hat{r}_{22}$ as measuring the asymptotic rates of growth of pure populations of $(A_1 A_1)$ and $(A_2 A_2)$, respectively; $\hat{r}_{12}$ has only a conceptual meaning, since in sexually-reproducing organisms there is no such thing as a pure population of heterozygotes, giving rise purely to heterozygotes.

Similarly, we can define net reproduction rates for the various genotypes as:

$$R_{ij}=\sum_x k_{ij}(x) \tag{34}$$

and genotypic generation times as:

$$T_{ij}=\frac{\sum_x x k_{ij}(x)}{\sum_x k_{ij}(x)}. \tag{35}$$

From Eq. (14) of Chapter 7, we see that, when the $\hat{r}_{ij}$ are all small:

$$\hat{r}_{ij}=\frac{R_{ij}-1}{R_{ij} T_{ij}}. \tag{36}$$

Now let us consider the evolution of the genetic structure of a population with overlapping generations, under the action of selection. As in Chapter 7, we write the frequency of the allele A_1 and A_2 among genes entering the population in age-class 1 during cycle t as $p(t)$ and $q(t)$ respectively, and the total number of individuals entering age-class 1 in cycle t as $B(t)$. We now introduce our first major assumption: *differences in the $k_{ij}(x)$ functions of different genotypes are small.*

One consequence of this assumption is that changes in gene-frequency within a cycle can be neglected, so that we can write the frequencies of individuals of the genotypes $(A_1 A_1), (A_1 A_2)$ and $(A_2 A_2)$ among individuals entering age-class 1 at time t as $p^2(t)$, $2p(t)q(t)$ and $q^2(t)$, respectively.

If we confine our attention to times t greater than ω, where ω is the last reproductive age-class, we need only consider contributions from individuals born after the initial generation. During cycle $(t-x)$, the number of females born is $cB(t-x)$, where c is the proportion of female births, and is assumed to be constant. The number of offspring produced by $(A_1 A_1)$ mothers in age-class x during cycle t is therefore:

$$c B(t-x) p^2(t-x) k_{11}(x)/c = B(t-x) p^2(t-x) k_{11}(x).$$

All these offspring must carry an A_1 gene; the other allele they carry at the locus in question depends on the genotype of the father.

A_1 genes are also contributed to new individuals entering age-class 1 during cycle t by $(A_1 A_2)$ mothers; the contribution due to mothers in age-class x is:

$$\tfrac{1}{2} c B(t-x) 2 p(t-x) q(t-x) k_{12}(x)/c = B(t-x) p(t-x) q(t-x) k_{12}(x).$$

Summing up these contributions over all reproductive age-classes, we therefore have:

$$p(t) = \sum_x B(t-x) p(t-x) \{p(t-x) k_{11}(x) + q(t-x) k_{12}(x)\}/B(t). \quad (37a)$$

Similarly, considering mothers of genotype $(A_1 A_2)$ and $(A_2 A_2)$ we obtain the following expression for $q(t)$:

$$q(t) = \sum_x B(t-x) q(t-x) \{p(t-x) k_{11}(x) + q(t-x) k_{22}(x)\}/B(t). \quad (37b)$$

These expressions can be simplified as follows. Let us write:

$$\Delta p(t) = p(t+1) - p(t)$$

and:

$$r(t) = \frac{B(t+1) - B(t)}{B(t)}.$$

Provided that changes in gene-frequency and the rate of population growth are small, we can neglect terms of the order of $(\Delta p)^2$ and $r(t)^2$ and can write, to a good approximation:

$$p(t-x) = p(t) - x \Delta p(t), \quad (38)$$

$$B(t-x) = B(t)[1 - x r(t)]. \quad (39)$$

Substituting these approximate equations into Eqs. (37) and neglecting small quantities, we obtain (dropping the argument t in $p(t)$ etc.):

$$p = p[p R_{11} + q R_{12}] - r p[p R_{11} T_{11} + q R_{12} T_{12}]$$
$$- \Delta p[2 p R_{11} T_{11} + (q - p) R_{12} T_{12}]$$
$$q = q[p R_{12} + q R_{22}] - r p[p R_{12} T_{12} + q R_{22} T_{22}]$$
$$+ \Delta p[2 q R_{22} T_{22} - (q - p) R_{12} T_{12}]$$

where R_{ij} and T_{ij} are defined in Eqs. (34) and (35).

From Eq. (36), these two equations reduce to:

$$p[p(\hat{r}_{11} - r) R_{11} T_{11} + q(\hat{r}_{12} - r) R_{12} T_{12}]$$
$$= \Delta p[2 p R_{11} T_{11} + (q - p) R_{12} T_{12}], \tag{40a}$$

$$q[p(\hat{r}_{12} - r) R_{12} T_{12} + q(\hat{r}_{22} - r) R_{22} T_{22}]$$
$$= -\Delta p[2 q R_{22} T_{22} - (q - p) R_{12} T_{12}]. \tag{40b}$$

We have assumed that selective differences are small. If we choose $(A_1 A_1)$, say, as a reference genotype, this means that we can neglect the squares of the differences of the demographic parameters of the other genotypes from those of $(A_1 A_1)$; we can further neglect their cross-products with Δp. We thus obtain from Eqs. (40) the following expressions:

$$p R T[p(\hat{r}_{11} - r) + q(\hat{r}_{12} - r)] = \Delta p R T$$
$$q R T[p(\hat{r}_{12} - r) + q(\hat{r}_{22} - r)] = \Delta p R T$$

where R and T are the values for the reference genotype and where

and
$$R_{12} = R + \varepsilon_{12}, \qquad R_{22} = R + \varepsilon_{22}$$
$$T_{12} = T + \delta_{12}, \qquad T_{22} = T + \delta_{22}$$

and the ε and δ values are small. Finally, therefore, we have:

$$\Delta p = p q(\hat{r}_{1.} - \hat{r}_{2.}) \tag{41}$$

where
$$\hat{r}_{1.} = p \hat{r}_{11} + q \hat{r}_{12}, \qquad \hat{r}_{2.} = p \hat{r}_{12} + q \hat{r}_{22}.$$

Comparing Eq. (41) with Eq. (7) in Section 1.3, where we dealt with selection in populations with non-overlapping generations, the similarity is obvious; $\hat{r}_{ij}$ enters into this equation where $w_{ij}/\bar{w}$ entered into Eq. (7). The only difference is that Δp in Eq. (41) refers to the change in gene-frequency in a "cycle", rather than a generation.

These considerations thus lead to the conclusion that, in a population with overlapping generations, the fitnesses of the different genotypes can be measured by their intrinsic rates of increase.

3.2. Some Examples of Selection in Human Populations

3.2.1. Characters causing excess deaths. Suppose that we know that a certain character C does not affect fecundity, but contributes a proportion C_x of the deaths in age-class x. Consider the value d_x, the probability that a member of the population who has survived up to the beginning of age-class x will die during this age-class. Some of these deaths are due to the character C, and others are due to other causes. Knowing C_x, the proportion of deaths in an age-class that are due to C, we can calculate the value $\hat{r}'$, of the intrinsic rate of increase which the population would have if character C did not occur in the population; comparing this with $\hat{r}$, the value given by the population's actual demographic parameters, we can get an idea of the effect of the character on the population.

Let d'_x be the death-rate in age-class x due to causes other than character C, and d^*_x be the death-rate in the population, in age-class x, due to the character, so that:

$$d^*_x = d_x\, C_x.$$

We therefore have:

$$1 - d_x = (1 - d'_x)(1 - d^*_x).$$

Thus l'_x, the probability of survival up to the beginning of age-class x, for the part of the population that does not have character C is given by:

$$l'_x = \prod_{i=1}^{x-1} (1 - d'_i)$$

$$= \frac{\displaystyle\prod_{i=1}^{x-1} (1 - d_x)}{\displaystyle\prod_{i=1}^{x-1} (1 - d^*_x)}$$

$$= \frac{l_x}{\displaystyle\prod_{i=1}^{x-1} (1 - d^*_x)}$$

$$= \frac{l_x}{\displaystyle\prod_{i=1}^{x-1} (1 - d_x\, C_x)}.$$

Assuming that no fecundity difference is associated with the character C, we can use the fecundity table for the whole population to calculate both $\hat{r}$, the value for the whole population, and $\hat{r}'$, the intrinsic rate of increase the population would have if C were absent.

Example. We can use this example to study the effect of the presence of individuals with duodenal ulcer on the intrinsic rate of increase of a population. The data for this example are taken from Cavalli-Sforza and Bodmer (1971) and refer to an Italian population.

Table 10.1. The effect of duodenal ulcer on the intrinsic rate of increase of a population

Age	$C_x \times 10^3$	$_5d_x \times 10^3$	$_5d_x^* = {_5d_x}\, C_x\; l_x$ $(\times 10^6)$		f_x	l_x'
15–19	2.30	2.85	6.56	0.937	0.0172	0.937000
20–24	2.65	3.95	10.47	0.934	0.192	0.934006
25–29	1.96	5.20	10.19	0.930	0.340	0.930016
30–34	2.36	6.55	15.46	0.925	0.379	0.925025
35–39	2.25	9.00	20.25	0.918	0.147	0.918039
40–44	3.38	11.80	39.98	0.909	0.0867	0.909057
45–49	4.40	17.70	77.88	0.896	0.0122	0.896092

The proportions of deaths due to duodenal ulcer in various age-groups in this population are given in the second column of Table 10.1. If we use these proportions as the values of C_x, we obtain the values of d_x^* given in Table 10.1, which also shows the steps in the calculation of the intrinsic rate of increase. Using the approximate formula of Chapter 7, the values found were as follows:

Intrinsic rate of increase per year for whole population: 0.0025997
(including people with ulcers)

For hypothetical population with no ulcer sufferers: 0.0026005

3.2.2. Characters which affect only fecundity. Now let us examine the case of a character C whose sole effect is to alter the fecundity of the individuals who show the character; C's effect might be due to an effect on the proportion of individuals who remain unmarried, or on the age of marriage, or on the actual fertility of the couples. An example of such a character might be a certain physical appearance which, without having any effect on the death-rate of the individuals concerned, would lower their chances of marrying; or else a disease which manifests itself during the reproductive years, so that individuals with the character would marry and reproduce like normal people up to the age of the appearance of the disease.

The intrinsic rate of increase for the whole population is given approximately by (see Chapter 7):

$$\hat{r} = \frac{\sum_x l_x f_x - 1}{\sum_x x \, l_x f_x}$$

and, for the group of individuals showing character C:

$$\hat{r}^* = \frac{\sum_x l_x f_x^* - 1}{\sum_x x \, l_x f_x^*} .$$

Examples. a) Suppose that people with polydactyly (six fingers instead of five) have a decreased chance of marrying while they are young, but that this effect wears off as they become older. Then, in each age-group, the proportion of polydactylous women who are married will be lower than the proportion of normal women who are married, but the difference will decrease with age. The fecundities f_x^* will therefore also be reduced in the same way. As an example (which does not claim to represent reality at all) let us suppose that the fecundities of the affected women in the different age-classes are:

Age-class (x)	Fecundity (f_x^*)
15–19	$\frac{1}{2} f_{15}$
20–24	$\frac{3}{5} f_{20}$
25–29	$\frac{4}{5} f_{25}$
30–34	$\frac{9}{10} f_{30}$
35–39	$\frac{9}{10} f_{35}$
40–44	$\frac{9}{10} f_{40}$

Using the demographic parameters of the 1830 cohort of French women, we find:

$$\sum_x l_x f_x^* = 0.80$$

$$\sum_x x \, l_x f_x^* = 24.8 .$$

Hence:

$$\hat{r}^* = \frac{-0.20}{24.8} = -0.0081 .$$

The value of $\hat{r}$ for the 1830 cohort of women is, as we saw earlier $+0.0008$; these two values quantify the selective disadvantage of polydactyly.

b) Now consider the case of a disease which develops late in life, such as Huntington's chorea. This disease is caused by a dominant gene, but does not manifest itself until after the age of about 30–35. It is a reasonable assumption that people who carry this gene suffer no increased mortality or decreased fertility until the disease appears, but that after this time their fertility is nearly zero (this is mainly due to their being in hospital). Reed and Chandler (1958) give the results of observations on Michigan families in which one member had the symptoms of Huntington's chorea. Their data enable us to calculate the probability P_x that the disease will appear by the age of x in a person who has the gene. Hence we deduce the probability $\tilde{P}_x$ that the disease does not appear by age x as:

$$\tilde{P}_x = (1 - P_{15}) \times (1 - P_{20}) \times \cdots \times (1 - P_{x-1}).$$

The values of these probabilities (from the data of Reed and Chandler, for the year 1940) are as follows:

Age of appearance of Huntington's chorea in women

Age	P_x (%)	$\tilde{P}_x$ (%)
15–19	5	100
20–24	11	95
25–29	8	84
30–34	17	77
35–39	18	64
40–44	23	53
45–	18	41

With the assumptions which we have made, the fecundity of a woman who has this gene is, at age x:

$$f_x^* = \tilde{P}_x \, f_x.$$

Using the 1939–1940 U.S. census data as the source of the l_x values, and data for the state of Michigan in 1940 as the source of the fecundity table, we can calculate the following intrinsic rates of increase (in years):

for the whole population: $r = -0.00172$,

for the women with Huntington's chorea: $r = -0.00796$.

These figures again demonstrate the selective disadvantage of this disease. The difference between the two intrinsic rates of increase is equal to 0.00624. Multiplying by the estimated generation time, we obtain the difference over a period of one generation, 0.17. Thus the selection coefficient against this genotype is approximately equal to 0.17, in discrete-generation terms.

3.2.3. Characters affecting both mortality and fecundity.

Galton (1869), in his book "Hereditary Genius", drew attention to the fact that families of high social class tend to have relatively few offspring. Many other authors, notably R. A. Fisher, have considered this question, and have concluded that human society runs the risk of regressing intellectually, because of the high fertility of the least gifted individuals, in the advanced nations.

We shall not now enter into any discussion of the doubtful nature of the assumptions behind this theory, namely that there are strong correlations between social level and intellectual level, and between intellectual level and genetic constitution. We shall confine ourselves to showing that there is little solid evidence for the claimed differential fertility in favour of the less gifted, in general.

Bajema (1963) gives the results of a study of 1 144 white Americans born in 1916 and 1917. He grouped the individuals into five classes according to their intelligence quotient, and he also determined the demographic parameters of each group l_x (proportion of survivors to age x) and $_5F_x$ (number of births to females aged between x and $x+5$). From these data, he calculated the intrinsic rate of natural increase $\hat{r}$, by solving Lotka's equation; we can use this to estimate the selection coefficient of each group, relative to the group with the highest I.Q. The results are shown in Table 10.2.

Contrary to the opinion which is often held, the individuals with the highest intelligence quotient had the highest fecundity in this population. They therefore had the highest selective value. The value for individuals whose I.Q. was between 80 and 95 was also high, but the others were all lower. The very low value for individuals whose I.Q. was below 80 is probably partly due to social factors which act to prevent the marriage of intellectually subnormal people.

Table 10.2. Intelligence quotient and selective value. Data from Bajema (1963)

I.Q.	$\hat{r}$ $(\times 10^3)$	T (years)	R_0	Selection coefficient
≥ 120	8.9	29.4	1.29	0
105–119	3.9	28.9	1.12	-0.14
95–104	0.3	28.4	1.01	-0.25
80– 95	7.5	28.0	1.23	-0.04
≤ 80	-10.0	28.8	0.75	-0.56
Mean	3.9	28.5		

Of course, these results must be interpreted with caution. They refer to a particular population, living in certain defined conditions. A different population might have given quite different results. Nevertheless, these results show that we cannot accept the widespread idea that differential fertility of the less gifted will cause human populations to regress intellectually, until we have much more data on the subject, and this is both difficult and expensive to obtain.

4. The Study of Selection in Human Populations

4.1. Difficulties in Detecting Selective Effects

If there is selection against one or more genotypes, some individuals will be eliminated because of their genetic constitution, and the genotype frequencies among the survivors will not agree with Mendel's law of segregation (for the offspring of a mating of individuals of known genotype) or the Hardy-Weinberg proportions (if we are considering a population). It ought, therefore, to be possible to study the effects of selection by studying genotype frequencies in groups of individuals of different ages. However, we cannot get statistically significant results from this type of study unless the selective differences are very high, or the numbers of individuals studied are extremely large.

4.1.1. The numbers of observations required to detect deviations from a multinomial or binomial distribution. When we want to test the significance of the deviation of a set of observations from a theoretical distribution, we usually do a χ^2 test. For example, suppose we are looking for evidence of abnormal segregation ratios among the offspring of a certain type of mating. Suppose that Mendel's law predicts that the proportions of the k classes of offspring should be $(p_1, \ldots, p_i, \ldots, p_k)$, whereas if selection occurred the proportions would be $(p'_1, \ldots, p'_i, \ldots, p'_k)$; these would be different from the values predicted by Mendel's law, due to the selective elimination of some genotypes. To carry out the study, we examine n offspring from the type of mating in question, and find their distribution $(n_1, \ldots, n_i, \ldots, n_k)$ among the k classes. We want to ask the question: what sets of observed numbers will leads us to reject the hypothesis H_0 "agreement with Mendel's law", when hypothesis H_1 "selection occurs" is true?

The method of testing the hypothesis H_0 is to calculate a parameter δ, defined by:

$$\delta = \sum_i \frac{(n_i - n p_i)^2}{n p_i} = \frac{1}{n} \sum_i \frac{n_i^2}{p_i} - n$$

and to reject H_0 if $\delta > \chi^2_{(k-1, \alpha)}$, where α is the risk of a Type I error (rejection of H_0 when it is true) which we will accept (we usually choose $\alpha = 5\%$ or $\alpha = 1\%$).

If hypothesis H_1 is true, the expectations of the numbers n_i are: $E(n_i) = n\, p_i'$; from the properties of the multinomial distribution, the expectation of the squares of the n_i values are:

$$E(n_i^2) = n^2\, p_i'^2 + n\, p_i'(1 - p_i').$$

Hence the expectation of δ is:

$$E(\delta) = \frac{1}{n} \sum_i \frac{n^2\, p_i'^2 + n\, p_i' - n\, p_i'^2}{p_i} - n$$

$$= n \left(\sum_i \frac{p_i'^2}{p_i} - 1 \right) + \sum_i \frac{p_i'(1 - p_i')}{p_i}. \tag{42}$$

The probability that we will reject the incorrect hypothesis H_0 increases as δ increases; Eq. (42) shows that as p_i' and p_i become closer to one another, n has to be higher to give the same expectation of δ. A rigorous proof would require a study of the variance of δ, and this involves some very laborious calculations (see Kendall and Stuart, 3rd ed., 1960, vol. II, p. 227).

To find the order of magnitude of n which is required, we shall simply find the condition for the expectation of δ to be above the significant value of $\chi^2_{(k-1, \alpha)}$. We thus obtain:

$$n > \frac{\chi^2_{(k-1, \alpha)} - \sum_i \frac{p_i'(1 - p_i')}{p_i}}{\sum_i \frac{p_i'^2}{p_i} - 1}$$

This number will be very large, if p_i' is close to p_i.

For example, when we would expect two classes of individuals with the same frequency ($p_1 = p_2 = \frac{1}{2}$), and when there is selection, which acts with intensity s against the second class, such that $p_1' = \dfrac{1}{2-s}$, $p_2' = \dfrac{1-s}{2-s}$, we obtain:

$$n > \frac{\chi^2_{(1, \alpha)} - 4\dfrac{1-s}{(2-s)^2}}{s^2/(2-s)^2} = \frac{\chi^2_{(1, \alpha)}(2-s)^2 - 4(1-s)}{s^2}$$

or, if s is sufficiently small:

$$n > \frac{4(\chi^2_{(1, \alpha)} - 1)}{s^2}.$$

If we will accept a risk of 5 % of a Type I error, we have $\chi^2 = 3.8$, which gives $n > \dfrac{11.2}{s^2}$. Thus, to detect a selective effect of $s = 0.01$, we must study more than 112 000 individuals in order for the expectation of δ to be above the chosen threshold level. If $s = 0.1$, we need more than 1 100 individuals.

These numbers show that it is practically impossible to detect weak selection by this method. Also, the arguments we have used here assume that the character in question is controlled exclusively by genetic factors, and did not take into account the possibility of environmental variation due to the age of the parents, the family's socio-economic level, or the part of the world the individuals live in, etc. It therefore seems doubtful that one could ever collect a sufficiently large and homogeneous set of data for this method of detecting selection to be used.

4.2. Direct Evidence for Selective Differences Associated with Human Polymorphisms

There is a large literature concerned with selective effects associated with some of the human blood groups. We have already discussed the problem raised by the selection against rhesus-positive (Rh/rh) children of rhesus-negative mothers (see Section 1.6.1 of this chapter). Mother-child incompatibility is also known in the AB0 system, and in this case also there is a certain loss of heterozygotes. Despite much work on this problem, the mechanism of the compensating selection which must be maintaining this widespread polymorphism, has not yet been discovered (see, for example, Matsunaga and Itoh, 1958; Morton, Krieger and Mi, 1966; Hiraizumi, 1964, 1964a; Reed et al., 1964).

Some of these studies also involved other polymorphic systems, but no clear and repeatable evidence of selection associated with them has been obtained. For example, Morton and Chung (1959) found an excess of heterozygotes for the two alleles of the MN system, among the offspring MN × MN matings, but later work has failed to confirm this. Matsunaga (1959) found large differences in the numbers of abortions between AB0-compatible and incompatible matings (10.3 % of all pregnancies, and 15.3 %, respectively), and in the proportions of sterile matings (9.8 % and 18.1 % respectively). These results, also, have not been confirmed.

Another approach has been to study associations between different phenotypes or genotypes of the polymorphic systems, with certain diseases. For reviews of this work, see Clarke (1959) and Cavalli-Sforza and Bodmer (1971).

The only human polymorphism for which there is strong evidence of heterozygous advantage is the case of sickle-cell anaemia. This is due to a gene which codes for a haemoglobin molecule with one amino acid residue different from the normal residue for this position in the polypeptide (a glutamic acid residue is replaced by valine). The SS homozygotes suffer from a severe anaemia. It is therefore a problem to explain the high frequency of the gene in many African populations.

Following the work of Beet (1946), Allison (1954) showed that there is a correlation between the presence of the gene for haemoglobin S, and malaria. He found that the heterozygotes, who show the "sickle cell trait" but not the anaemia, are most common in areas where malaria is frequent (up to 40 % in the most malarial areas). The simplest explanation for this is that these heterozygotes are somehow protected against malaria, compared with normal homozygotes. There is now considerable evidence to support this idea (Allison, 1964).

Assuming that mortality is the only component of fitness, the average fitness of normal homozygotes in malarious areas has been estimated as 0.85, compared with a value of 1 for heterozygotes for haemoglobin S (Rucknagel and Neel, 1961).

A number of other polymorphisms are also found principally in malarial parts of the world. For instance β-thalassaemia (a defect in the synthesis of the haemoglobin β-chain) shows a correlation with the incidence of malaria, and Haldane (1949) suggested that this polymorphism was maintained by heterozygote advantage of a similar kind to that maintaining the sickle-cell anaemia polymorphism. Correlations between the distribution of malaria and the haemoglin variants C and E (Allison, 1961) have also been reported, and Siniscalco et al. (1961) found a very striking correlations between the frequency of deficiency of the enzyme glucose-6-phosphate-dehydrogenase and malaria, in a set of Sardinian villages. Unlike the haemoglobin S polymorphism, however, no direct evidence has been obtained for selective differences associated with these polymorphisms.

4.3. Indirect Evidence for Selection

We saw in Section 6 of Chapter 8 that the frequency of a gene in a set of subdivisions of a single, ancestral population will tend, under the action of genetic drift, to diverge progressively from their initial uniform value, and that simultaneously the mean inbreeding coefficients within the groups increase in value. These two consequences of the finite size of the sub-populations are related, in quantitative terms, by the following equation:

$$E\left[\frac{V_i^{(g)}}{p_i(1-p_i)}\right] = -E\left[\frac{\text{Cov}_{ij}^{(g)}}{p_i p_j}\right] = \bar{\alpha}_g$$

in which $\bar{\alpha}_g$ is the mean of the inbreeding coefficients of the groups, in generation g, p_i is the initial frequency of A_i, $V_i^{(g)}$ is the variance in frequency of A_i between groups, in generation, and $\mathrm{Cov}_{ij}^{(g)}$ is the covariance of the frequencies of alleles A_i and A_j, in generation g.

The value of $\bar{\alpha}_g$ depends on the genealogy of the population in question, and is thus independent of which allele we are considering. The equation given above shows that the standardised variances and covariances:

$$\frac{V_i}{p_i(1-p_i)} \quad \text{and} \quad -\frac{\mathrm{Cov}_{ij}}{p_i\, p_j}$$

have the same expectation regardless of which gene is being studied.

If we collect data, and find that these values differ significantly for different loci, we would be forced to conclude that some of the assumptions on which the theory is based are not valid for the population studied; in particular, this kind of data can provide evidence for selection. Two recent reports of studies of this kind will now be discussed.

Lewontin and Krakauer (1973) base their argument on Eq. (56) of Chapter 8:

$$V(\hat{\alpha}_j) = \frac{k}{n}\,(\bar{\alpha})^2$$

where $\hat{\alpha}_j$ is the standardised variance of a gene in a sample of populations, $\bar{\alpha}$ is the mean of the $\hat{\alpha}_j$, n is the number of populations studied, and k is a coefficient which is close to 2 if the distribution of a gene frequency between populations is either a binomial or a normal distribution.

Using data on gene frequencies for a number of blood group systems in different human populations, they obtained the following numerical results:

$$\bar{\hat{\alpha}} = 0.148$$

$$V(\hat{\alpha}_j) = 0.00741, \quad \text{with } n = 60$$

whence:

$$k = 20.3\,.$$

A value of k as high as this proves that the assumption that the genes are neutral, and change in frequency purely as a result of genetic drift, is not a reasonable assumption for the systems studied, Lewontin and Krakauer therefore conclude that selective forces have been acting on at least some of the genes.

Thoma (1971) calculated the standardised variances and covariances of the frequencies p, q and r of the alleles at the locus controlling the A, B and 0 antigens of human red blood cells. He used data on three sets of populations: a set of 51 samples from Germanic populations; a set of 25 samples from European countries; and a set of 42 samples from different Chinese provinces.

The standardised variances and covariances are shown in Table 10.3. This table shows that the estimate of $\bar{\hat{\alpha}}$ derived from the covariance between q and r is the highest of all the values, in all three population samples; the value based on the variance of q is next in rank, and that based on the variance of r is next, in all three populations. Thoma shows that these puzzling regularities in the pattern of the values support the idea of a selective advantage of the AB, A0 and B0 heterozygotes, and also maternal-foetal incompatibility affecting non-0 children of 0 mothers.

Table 10.3. Values of $\bar{\hat{\alpha}}$ estimated from the AB0 locus, for three population sets (all values $\times 10^3$). Data from Thoma (1971)

Estimate of $\bar{\hat{\alpha}}$ based on	Germanic races		European countries		Chinese provinces	
	Value	Rank	Value	Rank	Value	Rank
$V(p)$	0.5	5	2.4	5	2.1	6
$V(q)$	2.9	2	20.0	2	9.0	2
$V(r)$	1.3	3	8.8	3	8.2	3
Cov pq	0.7	4	6.2	4	2.9	5
Cov qr	3.9	1	26.2	1	13.1	1
Cov pr	0.4	6	1.7	6	3.6	4

4.4. The Index of the Opportunity for Selection

4.4.1. Definition. We have seen that Fisher's fundamental theorem of natural selection shows that the change in the mean selective value of a population is proportional to the additive variance of the selective values of the individuals in the population:

$$\frac{\Delta \bar{w}}{\bar{w}} = \frac{V_A}{\bar{w}^2}$$

where $\bar{w}$ is the mean selective value, and V_A is the variance of the additive portion of the selective values w_{ij} of the genotypes.

This relative change in $\bar{w}$ characterises the speed at which the population changes under selection; in other words, it is a measure of the speed of evolution. It would thus be interesting to estimate this parameter; unfortunately it is very difficult to determine the additive genetic portion of the variance in the character, without making unrealistic assumptions.

We have seen that in a human population, the selective value of a group of individuals can be estimated from their "number of successful offspring" (see definitions on pages 171 and 269).

To simplify the argument, we shall assume that the "number of successful offspring" of an individual is a function only of his genetic constitution, and that the variance V_N of this number is equal to the additive variance of this inherited character (we are assuming, therefore, that there is no dominance variance or interaction variance). With these assumptions, the relative change in the mean selective value, $\dfrac{\Delta \overline{w}}{\overline{w}}$ is equal to:

$$I = \frac{V_N}{\overline{N}^2}$$

where $\overline{N}$ is the mean number of offspring, and I is an index which characterises the rate of evolution of the population.

This index measures the change in selective value only insofar as this is determined by the genetic constitution. In reality, of course, other factors are involved, so that the population will not change as fast as the index I indicates. I therefore measures the maximum change in selective value which the population's mortality and fecundity characteristics could entail. Crow (1958) has therefore suggested that I should be called the "index of the opportunity for selection".

4.4.2. The analysis of I. The time interval between two generations, g and $g+1$, can be divided into two periods:

1. The period from the birth of the individuals of generation g until they become adult, i.e. from 0 to 15 years, in man, by our convention (Section 1, Chapter 7).

2. The interval between the time when these individuals become adult, and the time when their offspring are born.

Parallel with this division, we can analyse the variance V_N of the number of successful offspring into:

1. A portion corresponding to the events which influence this number during the first period, in other words, corresponding to death between the ages of 0 and 15 years.

2. A portion corresponding to events in the second period, i.e. to differences in fecundity between women who survive to the age of 15.

Let l_a be the proportion of girls who survive to 15, and $n_0, \dots, n_i, \dots, n_\omega$ be the proportions of women who survive to 15 years and have $0, \dots, i, \dots, \omega$ daughters, respectively. Then we obviously have:

$$\sum_0^\omega n_i = l_a.$$

The mean and variance of the numbers of daughters are therefore:

$$\overline{N} = \sum_i i \, n_i$$

and

$$V_N = (1 - l_a)\,\overline{N}^2 + \sum_i n_i (i - \overline{N})^2.$$

Then the index of opportunity for selection, I, can be written:

$$I = 1 - l_a + \sum_i n_i \left(\frac{i - \overline{N}}{\overline{N}}\right)^2.$$

If all women had the same fecundity, in other words if all the women who do not die before the age of 15 had exactly the same number of offspring $\overline{N}_a$ (with, obviously, $\overline{N}_a = \overline{N}/l_a$), the variance V_N would be due solely to mortality, and would be equal to:

$$V_m = (1 - l_a)\,\overline{N}^2 + l_a \left(\frac{\overline{N}}{l_a} - \overline{N}\right)^2$$

which gives:

$$V_m = \frac{1 - l_a}{l_a}\,\overline{N}^2.$$

This enables us to define an "index of selection due to mortality" as:

$$I_m = \frac{V_m}{\overline{N}^2} = \frac{1 - l_a}{l_a}.$$

If, on the other hand, we consider the group of women who live to adulthood, the variance in their numbers of offspring is due solely to fecundity differences. This variance is:

$$V_f = \frac{1}{l_a} \sum_i n_i \left(i - \frac{\overline{N}}{l_a}\right)^2$$

$$= \frac{1}{l_a} \sum_i n_i \left\{(i - \overline{N}) - \overline{N}\left(\frac{1 - l_a}{l_a}\right)\right\}^2$$

$$= \frac{1}{l_a} \sum_i n_i (i - \overline{N})^2 - \overline{N}^2 \left(\frac{1 - l_a}{l_a}\right)^2$$

and so we can define an "index of selection due to fecundity" as:

$$I_f = \frac{V_f}{\overline{N}_a^2} = l_a \sum_i n_i \left(\frac{i - \overline{N}}{\overline{N}}\right)^2 - (1 - l_a)^2.$$

The relation between the three indices I, I_m and I_f is as follows:

$$I = I_m + \frac{I_f}{l_a}.$$

This anylsis is obviously an arbitrary one. In fact, of course, we know that the variance in fecundity depends on deaths of women during the reproductive years, so the index I_f is also a function of the death-rate. However, it is reasonable to assume that it measures mainly fecundity differences, since mortality is very low in this period.

4.4.3. Some examples of the indices of selection. To calculate I and the two components I_m and I_f, we need to know not only the mortality and fecundity tables, but also the distribution of offspring number among the surviving women (more precisely, the numbers of daughters, but since the selection indices are calculated from ratios of a variance to the square of a mean, it is equivalent to use total number of offspring, regardless of sex). This distribution has only rarely been calculated.

We can estimate it by a simulation of human reproductive behaviour, by "Monte-Carlo methods"; this consists of calculating by computer the number of offspring of a woman, assuming that the probability distributions of the various random events which determine this number are given by the mortality and fecundity tables of the population in question. Then, if we do the calculations with a large number of women, we can find the distribution of offspring number, and calculate its variance.

We can also use the fecundity statistics of a group of women to calculate the probabilities of increase in size of their families; these are the ratios of the number of women who have $i+1$ offspring to the number who have had i offspring.

a) *The 1830 and 1900 cohorts of French women.* The probabilities of increase in family size have been calculated by Henry (1953) as follows:

Probabilities of increase in family size ($\times 10^3$)

	Cohort	
	1830[a]	1900[b]
a_0	886	814
a_1	836	690
a_2	744	590
a_3	725	580
a_4	712	600
a_5	714	630
a_6	694	630
a_7	681	630
a_8	657	630
a_9	647	630

[a] Henry (1953).
[b] Calculated from data of Henry (1953).

These figures are based on 1 000 women who survived to adulthood. We see that, for the cohort of 1830, 886 women had at least one child, thus $1000 - 886 = 114$ had none; $886 \times \frac{836}{1000} = 741$ had at least two children, so that $886 - 741$ had just one child. In this way, we find that the distribution of offspring number among the women is as follows (making some simple assumptions for women with more than nine children):

Distribution of offspring number among 1000 women

Number of offspring	Cohort	
	1830	1900
0	114	186
1	145	255
2	190	232
3	152	137
4	115	77
5	82	41
6	62	27
7	45	20
8	33	9
9	22	6
10	14	3
11	13	3
12	10	1

From these figures, we obtain the mean numbers of offspring of women who survive to adulthood:

$$\overline{N}_a(1830) = 3.442 \quad \text{and} \quad \overline{N}_a(1900) = 2.151.$$

These numbers are close to those found using the crude birth rates (3.475 and 2.110, respectively, see p. 145), which shows that the two methods of calculation are consistent. We also obtain the variances of these numbers:

$$V_f(1830) = 7.56 \quad \text{and} \quad V_f(1900) = 3.90.$$

We also found (Table 7.1) that the proportions of women who survived to the age of 15 were:

$$l_a(1830) = 0.672, \quad l_a(1900) = 0.796.$$

These enable us to calculate the three selection indices given in Table 10.4.

Although we cannot rely on the exact numerical values in this table, we can regard them as giving an indication of the magnitude of the change which occurred in the 70 year period between the years of birth of the two cohorts. Mainly because of advances in medicine, mortality has far less effect on the 1900 cohort than on the 1830 cohort: I_m is only about half the 1830 value. However, we know that deaths due to infectious diseases have decreased far more than deaths due to genetic causes, so that the rate of change in the population's genetic make-up due to deaths between the ages of 0 and 15 has probably decreased less than the decrease in I_m would indicate, although we cannot measure this effect. This shows the reason for the term "index of *opportunity* for selection", rather than "index of selection".

Table 10.4. Selection indices for two groups of French women

	1830 cohort	1900 cohort
l_a	0.672	0.796
$I_m = \dfrac{1-l_a}{l_a}$	0.49	0.26
$\overline{N}_a$	3.44	2.15
V_f	7.56	3.90
$I_f = \dfrac{V_f}{\overline{N}_a^2}$	0.64	0.84
$I = I_m + \dfrac{I_f}{l_a}$	1.44	1.32

The index I_f, on the other hand, has increased slightly. This seems paradoxical, in view of the decreased heterogeneity in family size. The explanation is that the mean number of children has decreased even faster than the standard deviation of this number, because of the introduction of contraception.

The changes in I_m and I_f approximately cancel one another out, and I is nearly the same for both cohorts.

b) *The 1870 and 1930 cohorts of white American women.* Pressat (1966) calculated the probabilities of increase in family size for several cohorts, using the age-specific fecundity statistics. His results for the two extreme cohorts, the 1870 and 1930 cohorts, give the following distribution of women according to offspring number.

The distribution of 1000 white American women, according to their number of offspring

Number of children	Cohort	
	1870	1930
0	170	100
1	164	90
2	140	203
3	121	243
4	97	182
5	68	82
6	59	44
7	45	70
8	40	15
9	36	10
10	25	6
11	15	4
12	10	1
13	10	0

The means and variances of offspring number derived from these distributions are:

$$\overline{N}_a(1870)=3.5, \qquad \overline{N}_a(1930)=3.1,$$
$$V_N(1870)=9.97, \qquad V_N(1930)=4.05.$$

Also, the proportions of women surviving to the age of 15 are:

$$l_a(1870)=0.75, \qquad l_a(1930)=0.94.$$

These give the three selection indices shown in Table 10.5.

Table 10.5. Selection indices in two groups of white American women

	1870 cohort	1930 cohort
l_a	0.75	0.94
$I_m=\dfrac{1-l_a}{l_a}$	0.33	0.06
$\overline{N}_a$	3.5	3.1
V_f	10.0	4.1
$I_f=\dfrac{V_f}{\overline{N}_a^2}$	0.81	0.42
$I=I_m+\dfrac{I_f}{l_a}$	1.41	0.51

There is a more marked difference between the two cohorts of American women than between the two French cohorts. Medical progress has brought early mortality down to such a low level for the 1930 cohort that I_m is only one-fifth of the value 60 years before. Also, family-planning resulted in a profoundly changed distribution of family sizes, which is much narrower in 1930 than in 1870, so that the variance for the 1930 cohort is half the 1870 value. The mean number of offspring, however, decreased only very slightly, unlike the French case. This resulted in a decrease in I_f to about half its 1870 value.

The total index I of opportunity for selection for the 1930 cohort is reduced to only a third of the value for the 1870 cohort.

c) *The Hutterite population*. This population's founders emigrated to the United States in the 19th century. They live as a closed community, and forbid any kind of contraceptive practice. Eaton and Mayer (1954) give some demographic parameters for this population, from which the three parameters we need for the calculation of the selection indices can be found:

$$l_a = 0.94, \quad \overline{N}_a = 8.9, \quad V_N = 13.9$$

which gives:

$$I_m = 0.06, \quad I_f = 0.16, \quad I = 0.23$$

In this population I_m is very low because of good health care, and I_f is also low, since the number of offspring is extremely high.

This population is a good example of a population in which fecundity is uncontrolled. This example shows that differences in fertility will permit only slow changes in the population's genetic make-up.

However, the low overall value of I among the Hutterites is due to the fact that they have maintained uncontrolled fecundity while adopting modern health care, so that their mortality is characteristic of an advanced society. For a primitive population, we would expect that I_f would be similar to the value for the Hutterites, but the proportion of individuals surviving to the age of 15 would be much lower, of the order of 65–70%, which would give: $I_m \approx 0.5$ and $I \approx 0.7$. This shows that in such a "natural" situation, the portion of selection which is due to mortality would be by far the larger portion.

d) *The Bourgeoisie of Geneva*. The data which we have been reviewing deals simply with total numbers of offspring. However, the really significant variable which we would like to be able to deal with, is the number of "successful offspring". The results which we obtained in the preceding pages are valid only if we assume that differences in total number of offspring between women will result in different numbers of successful offspring. Only a full study of the historical demography of a population can provide better data. Henry (1956) has carried out such a study on the bourgeoisie of Geneva.

Choisy (1947) compiled an extremely detailed account of the demographic events (births, marriages and deaths) that had taken place since the 16th century, in nineteen families belonging to the bourgeoisie in Geneva, and whose members had played prominent roles in the city's affairs. Henry (1956) used these data to show that the demographic history of these families is characterised by a sharp decline in the number of births at the beginning of the 18th century; for the women who were married before reaching the age of 24, and who remained married until the end of their child-bearing years (in other words, for completed families), the average numbers of offspring were:

9.6 for marriages in which the husband was born between 1600 and 1750
6.2 for marriages in which the husband was born between 1650 and 1700
3.4 for marriages in which the husband was born between 1700 and 1750

This striking change must reflect the rapid adoption of family planning. This population therefore provides exceptionally favourable material for studying the effect of family planning on the distribution of family size.

Table 10.6 shows the distributions of both total number of offspring and also of number of successful offspring. The data are given according to the husband's year of birth, and the value of the overall index of the opportunity for selection:

$$I = \frac{V_M}{M^2}$$

where V_M and M are the variance and mean of number of successful offspring.

Table 10.6 shows that the considerable demographic change produced only a transient effect on the value of the index I. The decreased infant mortality was almost exactly compensated for by the adoption of family

Table 10.6. Mean family size and variance in family size in the Geneva data. (From Henry, 1956)

Husband's year of birth	Total number of children born		Number of successful offspring		I
	Mean ($\bar{N}$)	Variance (V_N)	Mean (M)	Variance (V_M)	
Before 1600	4.5	12.7	2.4	4.9	0.85
1600–1650	5.5	20.1	3.1	7.0	0.73
1650–1700	3.6	10.3	2.4	5.0	0.87
1700–1750	2.9	5.9	2.0	2.7	0.69
1750–1800	2.8	4.3	2.3	2.6	0.49
1800–1850	2.7	4.9	2.4	3.4	0.59
1850–1900	2.0	3.2	1.9	2.9	0.80

planning: while the number of offspring born has decreased by at least half over the course of the three centuries, the number of successful offspring has shown no clearly-marked downward trend.

In summary, we find that the two demographic revolutions, brought about by the success of medical science in decreasing infant mortality, and by the adoption of family size limitation, may together result in no more than a small change in the potential for natural selection to act.

Further Reading

Neel, J.V.: The study of natural selection in primitive and civilised human populations. Hum. Biol. **30**, 43–72 (1958).

Scudo, F.M.: Selection on both haplo and diplophase. Genetics **56**, 693–704 (1967).

Workman, P.L.: Gene flow and the search for natural selection in man. Hum. Biol. **40**, 260–279 (1968).

Chapter 11

Mutation

Genes are not perfectly unchanging entities: with a low, but not zero, frequency, a gene can change to another state, as a result of irradiation, chemical treatment, or simply because of an error in the process of chromosome replication. If such a change occurs in a line of cells which will give rise to gametes, then the new form of the gene can be transmitted to the offspring of the individuals in whom it arose. Thus an individual who received the alleles A_i and A_j at a locus at the time of his conception, can transmit a different allele, A_k, to his offspring.

Although this gene is new with respect to the individual in which the mutation happened, it can be the same as an allele which is already present in the population. Or it may be an entirely new allele, unknown until the mutation occurred; if such an allele spreads in the population, and increases in frequency from generation to generation, the population will become more and more differentiated from other populations of the same species. Such mutations are the only source of new genetic information, and the slow process of evolution is dependent on their occurrence.

There are therefore two distinct problems to be studied. First, how can a single mutant gene be maintained in a population and even gradually spread through the population, eliminating the other alleles and transforming the population's genetic structure? And second, how does the process of successive mutation of alleles to other alleles affect the genetic structure of a population in the long term?

1. The Probability of Survival of a Mutant Gene

Suppose that at a locus A with $n-1$ different alleles $A_1, \ldots, A_{n-1}$ a gene in an individual's germ line mutates to a new allele A_n, which is new to the population. Half his offspring will receive this allele. If A_n is dominant, the mutant character will therefore appear in the next generation (provided, of course, that A_n is in fact transmitted to one or more offspring); if, however, A_n is recessive, the character will not be manifest, and we would not be aware that a mutation had happened. In this case, the character would appear only some generations later, when two descendants of the mutant individual, both carrying the A_n gene, mate; $\frac{1}{4}$ of their offspring would be $(A_n A_n)$ homozygotes, and would manifest the character which they had inherited from their ancestor.

It is obvious that a new mutant gene has a strong probability of being lost from the population soon after it first arises, so that the new character would never be expressed. We shall first examine the case of a neutral gene, with no selective effect.

1.1. Elimination of a Neutral Allele

If a mutant individual has r offspring, the probability E_r that the mutant gene A_n is not transmitted to any of them is:

$$E_r = (\tfrac{1}{2})^r.$$

We shall assume that the distribution of number of offspring among couples follows a Poisson distribution, with parameter m, the mean number of offspring per couple. The probability of having r offspring, P_r, is therefore given by:

$$P_r = e^{-m} \frac{m^r}{r!}.$$

The probability E that a new gene will be lost without being transmitted to any members of the next generation is therefore:

$$E = \sum_r P_r E_r = \sum_r e^{-m} \frac{m^r}{r!} \left(\frac{1}{2}\right)^r$$

$$= e^{-m/2} \sum_r e^{-m/2} \frac{\left(\dfrac{m}{2}\right)^r}{r!}$$

$$= e^{-m/2}$$

since the summed terms are simply the probabilities corresponding to the terms of the Poisson distribution with parameter $m/2$, and they therefore sum to 1.

Thus the probability $E(1)$ that a neutral gene is eliminated by chance in the first generation is $e^{-m/2}$.

In what follows, we shall assume that the population size is constant, so that $m = 2$. Hence:

$$E(1) = e^{-1} = 0.368.$$

Thus the chance that the gene escapes elimination in the first generation is 0.632. What happens to the genes which survive?

Suppose that the mutant gene is transmitted to j offspring of the original mutant individual. The probability of this is $P_j = e^{-1} \dfrac{1}{j!}$.[1] Each gene that is transmitted to the next generation (generation 2) has a probability of $j/2N$ of being a copy of the mutant gene A_n, where N is the population size. The probability that no A_n gene is transmitted is therefore:

$$E(2|j) = \left(1 - \frac{j}{2N}\right)^{2N} = e^{-j}$$

since N is assumed to be very large.

The probability that no A_n gene survives to generation 2, whatever the number j of these genes in generation 1, can therefore be written as:

$$E(2) = \sum_j P_j E(2|j) = \sum_j e^{-1} \frac{1}{j!} e^{-j}$$
$$= e^{-1} \sum_j \frac{(e^{-1})^j}{j!} = e^{-1} e^{(e^{-1})} \tag{1}$$
$$= e^{E(1)-1}.$$

Hence:

$$E(2) = e^{0.368-1} = e^{-0.632} = 0.532.$$

Thus the probability that a newly-arisen neutral mutation will remain in the population for two generations is less than $\frac{1}{2}$.

Fisher (1930) showed that Eq. (1) can be generalised to later generations E_g, E_{g+1} etc. We have:

$$E(g) = e^{E(g-1)-1}. \tag{2}$$

This enables us to calculate the successive probabilities of elimination; we can also show that the probability of elimination tends towards 1: E_g increases with g, and its limit E is such that:

$$\hat{E} = e^{\hat{E}-1}$$

which has the obvious solution $\hat{E} = 1$. Thus the probability that a new mutant gene with no selective advantage will be maintained in the population in the long term is zero.

---

[1] In order to transmit this gene j times, the mutant individual must have $u \geq j$ offspring, which has probability $e^{-2} \dfrac{2^u}{u!}$, and he must have transmitted the mutant gene to j of his u offspring, which has probability $\dfrac{u!}{j!(u-j)!} \left(\dfrac{1}{2}\right)^u$; hence:

$$P_j = \sum_{u=j}^{u=\infty} e^{-2} \frac{2^u}{u!} \frac{u!}{j!(u-j)!} \left(\frac{1}{2}\right)^u = \frac{e^{-2}}{j!} \sum_{u=j}^{u=\infty} \frac{1}{(u-j)!} = \frac{e^{-1}}{j!}.$$

1.2. Survival of a Neutral Mutant Gene in a Finite Population

Up to now, we have assumed that population size is effectively infinite; we have seen that a neutral mutant gene will eventually be lost from the population with probability one. If, however, the population is finite, with N breeding individuals in each generation, then it can be shown that there is a probability of $1/2N$ that the mutant gene will survive and eventually become fixed in the population. The proof is as follows. In any generation, there are $2N$ genes at a given locus. We saw in Chapter 8 that, for a future generation, g, the probability that all the genes at a given locus are descended from a single gene in the population in generation 0 tends to 1 as g tends towards infinity. If one of the $2N$ genes present in generation 0 is a new, neutral mutation, the chance that it will be the ancestor of all the genes in the population in some distant future generation is therefore $1/2N$. Clearly, if N is very large, this probability is extremely small, although finite.

Kimura has proposed that the majority of gene substitutions that have taken place in evolution are due to this process of random fixation of neutral mutations. We shall return to this topic in Section 4 of this chapter.

1.3. The Probability that an Advantageous New Mutant Gene Will be Maintained in the Population

It is obvious that a gene which gives its carriers a selective disadvantage (i.e. a gene which increases the risk of dying at any age, or decreases fecundity) will be lost from a population even more quickly than a neutral gene, and will consequently have even less chance of remaining present in the long term.

However, a gene which gives a selective advantage to its carriers can be maintained even if the advantage is very slight.

To define precisely what we mean by "selective advantage" we proceed as follows:

1. Let the population in a given generation consist of N individuals, among whom the mutant allele is present in j copies.

2. If A_n is a neutral allele, the probability that a gene that is transmitted to the next generation is A_n is equal to $j/2N$.

3. If A_n has a selective advantage, the probability that a gene that is transmitted to the next generation is an A_n allele is:

$$(1+s)\frac{j}{2N} > \frac{j}{2N}$$

where s is called the "selective advantage" of A_n.

Following through a similar argument to that for the case of a neutral mutation, it can be shown, assuming large population size, that the probability $E(g)$ of elimination of a gene with selective advantage s during the course of g generations after its occurrence is:

$$E(g) = e^{-(1+s)\{1-E(g-1)\}}. \tag{3}$$

This enables us to calculate $E(1)$, $E(2)$ etc. The limit $\hat{E}$ of $E(g)$ as g increases is given by the solution of the equation:

$$\hat{E} = e^{-(1+s)(1-\hat{E})}.$$

When s is sufficiently small, and $\hat{E}$ sufficiently close to 1, we may reasonably use the approximation:

$$\hat{E} = 1 - (1+s)(1-\hat{E}) + \frac{(1+s)^2}{2}(1-\hat{E})^2$$

which has the solution

$$\hat{E} = 1 - \frac{2s}{(1+s)^2}.$$

This equation gives the following values of $\hat{E}$, for various values of s:

$$s=1\%: \quad \hat{E}=0.980$$
$$s=2\%: \quad \hat{E}=0.962$$
$$s=3\%: \quad \hat{E}=0.943$$

In other words, out of every 1000 new mutant genes, twenty are expected to survive in the long term if they confer a selective advantage of 1% on their carriers; if the selective advantages is 2%, 38 will, on average, survive, and if the advantage is 3%, 57 will remain in the population.

In a small population, the elimination of a gene with a slight selective disadvantage is no longer a certainty. Kimura (1962) has shown that, if the effective population size is N_e, the probability of elimination is given by the following expression:

$$\hat{E} = \frac{e^{-2s} - e^{-4N_e s}}{1 - e^{-4N_e s}}.$$

We shall return to the study of mutation and selection in finite populations, in Chapter 13.

2. Recurrent Mutations

We will now consider the case, not of entirely new mutant alleles, but of mutation to alleles which are already present in the population.

Such mutation will obviously have very little effect on the overall genetic structure of the population unless it happens many times or, more precisely, unless it happens regularly in every generation, with a low, but not zero, frequency.

A mutation is, by its very nature, an unpredictable event. When we want to study the progressive changes in the genetic structure of populations that are caused by recurrent mutation, it is essential to use probabilistic reasoning. The only information which we need to know about a mutation, for these purposes, is the probability that a gene A_i which an individual receives will have mutated to A_j when it is transmitted to his offspring.

2.1. Change in Genic Structure due to Recurrent Mutation

Let $v_{ij}^{(g)}$ be the probability that an A_i gene received by an individual in generation g mutates to A_j before being transmitted to a new individual in generation $g+1$, i.e. the probability that a mutation $A_i \to A_j$ occurs in generation g.

The probability that an A_i gene does not mutate, u_i, is therefore:

$$v_{ii}^{(g)} = 1 - \sum_{j \neq i} v_{ij}^{(g)}.$$

The genic structure in generation g has been defined as the set of frequencies $p_i^{(g)}$ of the various alleles $A_1, \ldots, A_n$ in this generation. It can be written as a vector:

$$s_g = \{p_1^{(g)}, \ldots, p_n^{(g)}\}.$$

We want to find s_{g+1} given s_g, assuming that all the conditions for Hardy-Weinberg equilibrium are satisfied, except that mutations can occur.

The A_i genes in generation $g+1$ are copies of A_i genes in generation g, which have not mutated, and of A_j genes, which have undergone the mutation $A_j \to A_i$. The probability of encountering an A_i gene in generation $g+1$ (in other words, since population size is assumed to be infinite, the frequency of A_i) is therefore:

$$p_i^{(g+1)} = p_i^{(g)} - \sum_{j \neq i} v_{ij}^{(g)} p_i^{(g)} + \sum_{j \neq i} v_{ji}^{(g)} p_j^{(g)}$$
$$= v_{ii}^{(g)} p_i^{(g)} + \sum_{j \neq i} v_{ji}^{(g)} p_j^{(g)}$$
$$= \sum_j p_j^{(g)} v_{ji}^{(g)}.$$

This shows that the vectors s_{g+1} and s_g are related by the equation:

$$s_{g+1} = s_g V_g \tag{4}$$

where V_g is the matrix of v_{ij} values with the i, j-th element equal to the probability of a mutation from A_i to A_j. Notice that the sum of the elements in each row of this matrix is equal to 1, so that V_g is a "stochastic matrix" (see Appendix B).

Eq. (4) shows that:

$$S_{g+1} = S_{g-1} V_{g-1} V_g.$$

Hence, going back generation by generation:

$$S_{g+1} = S_0 \times V_0 \times V_1 \times \cdots \times V_g. \tag{5}$$

If the probabilities of the various possible types of mutation remain the same from generation to generation, we can write $V_g = V$, and Eq. (5) becomes:

$$S_g = S_0 V^g. \tag{6}$$

In this case, therefore, the problem of the changes in genic structure with generation number is solved when we know the successive powers of the matrix V.

The elementary results of matrix algebra which are given in Appendix B show that:

1. As g increases without limit, the matrix V^g tends towards a matrix P_1 whose rows are all identical.

2. The speed of convergence towards this limit is measured by the largest root, other than 1, of the equation in λ:

$$\det (V - \lambda I) = 0.$$

3. The vector s_g tends towards a vector $\hat{s}$ which is identical to the rows of the matrix P_1, and is therefore independent of the initial vector s_0.

In other words, the genic structure of a panmictic population which is subjected to mutation, but not to any other evolutionary force, tends towards an equilibrium which is determined only by the probabilities of mutations between the various alleles.

2.2. The Case of a Locus with Two Alleles

When the locus in question has only two alleles, A_1 and A_2, the matrix of mutation frequencies is determined by only two parameters:

1. the probability u that an A_1 gene will mutate to an A_2 allele; and
2. the probability v that an A_2 allele will mutate to A_1:

$$V = \begin{bmatrix} 1-u & u \\ v & 1-v \end{bmatrix}.$$

These two parameters are called the mutation rates of A_1 to A_2 and of A_2 to A_1, respectively.

If the genic structure s in a given generation is characterised by the frequencies p and q, the structure in the next generation will be:

$$s' = (p, q) \begin{bmatrix} 1-u & u \\ v & 1-v \end{bmatrix}$$

$$= (p - pu + qv, \, pu + q - qv).$$

For example, the frequency p of allele A_1 will change to:

$$p' = p - pu + qv = v + p(1 - u - v). \tag{7}$$

If $p' = p$, the population is in equilibrium. From Eq. (7), the value, $\hat{p}$, of p at the equilibrium is such that:

$$\hat{p} u = \hat{q} v$$

which gives:

$$\hat{p} = \frac{v}{u+v}, \qquad \hat{q} = \frac{u}{u+v}.$$

Eq. (7) can therefore be written in the form:

$$p' - \hat{p} = p(1 - u - v) + v - \hat{p}$$

$$= p(1 - u - v) + \hat{p}(u + v) - \hat{p}$$

$$= (p - \hat{p})(1 - u - v).$$

Going backwards to generation 0, we therefore have:

$$p_g - \hat{p} = (1 - u - v)^g \, (p_0 - \hat{p}). \tag{8}$$

Since mutation rates are very low, of the order of 10^{-6}, this equation shows that the population will approach equilibrium with respect to mutations very slowly. For example, a given deviation from the equilibrium frequency will be halved in m generations such that:

$$(1 - u - v)^m = \tfrac{1}{2}$$

which gives:

$$m = \frac{-\log_e 2}{\log_e (1 - u - v)} \approx \frac{\log_e 2}{u + v} = \frac{0.7}{u + v}.$$

If $u = 3 \times 10^{-6}$ and $v = 2 \times 10^{-6}$, $\hat{p} = 0.4$ and $m = \dfrac{0.7}{5} \times 10^6 = 140\,000$. For a human population, with three or four generations per hundred years, it would therefore take many thousands of years to progress noticeably towards equilibrium. This conclusion has scarcely any meaning, however, since the mutation rates would not stay constant for such a long time, nor is it reasonable to assume that no other evolutionary forces would have any importance during such a long period.

In summary, therefore, mutation is extremely important because it is the origin of all hereditary variation, but it affects the genetic make-up of populations only very slowly. Over a small number of generations, the effect of mutations will be insignificant compared with the effects of selection.

3. The Resultant Effect of Selection and Mutation at a Locus with Two Alleles

In the last chapter, we studied in some detail the effects of selection on a locus with two alleles, A_1 and A_2. We shall now study the joint effect of selection and mutation, and here too we shall study the two-allele case.

3.1. The Equilibrium between Mutation and Selection

If a population is exposed to selection causing a change in gene frequency, $\Delta_s p$, of allele A_1 between one generation and the next, and also mutation causes a change $\Delta_m p$, and if we can assume that these two causes of change are independent, then we have:

$$\Delta p = \Delta_s p + \Delta_m p.$$

But we saw earlier (Chapter 10, Section 1.4) that, if the selective values w_{ij} of the different genotypes remain constant:

$$\Delta_s p = pq \frac{(w_{11} - w_{12})p + (w_{12} - w_{22})q}{w_{11} p^2 + 2w_{12} pq + w_{22} q^2}. \tag{9}$$

Also, Eq. (7) gives:

$$\Delta_m p = -up + vq = -(u+v)p + v. \tag{10}$$

The net change in frequency of A_1 is therefore:

$$\Delta p = pq \frac{(w_{11} - w_{12})p + (w_{12} - w_{22})q}{w_{11} p^2 + 2w_{12} pq + w_{22} q^2} - up + vq.$$

It is simple to find the equilibrium points corresponding to $\Delta p = 0$, graphically. We have already studied, for the various possible cases, the curves of $\Delta_s p$ against p. When there is recurrent mutation, the equilibrium points no longer correspond to the points of intersection of these curves with the p axis (these points satisfy the relation $\Delta_s p = 0$), but are found from the points of intersection of the curves with the line:

$$y = -\Delta_m p = (u+v)p - v$$

(these points satisfy the relation $\Delta_s p = -\Delta_m p$, and therefore $\Delta p = 0$).

3.2. Constant Selective Values

We shall now consider the various possible relative values of the selective values of the three genotypes.

3.2.1. $w_{11} \leqq w_{12} < w_{22}$. The heterozygotes have a selective value between those of the homozygotes, or equal to that of the less-favoured homozygote (i.e. the disfavoured gene is dominant).

In the absence of mutation, we know that $\hat{p}=0$. If mutations occur this equilibrium is slightly displaced to a value of $\hat{p}$ which is greater than 0. We saw earlier that, in the neighbourhood of $p=0$, we can write:

$$\Delta_s p \approx \frac{(w_{12}-w_{22})}{w_{22}} p.$$

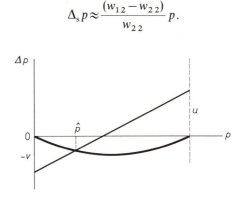

Fig. 11.1. The equilibrium between mutation and selection when $w_{11} \leqq w_{12} < w_{22}$

At equilibrium, therefore:

$$\frac{(w_{12}-w_{22})}{w_{22}}\hat{p} = -\Delta_m p$$

$$= (u+v)\hat{p} - v$$

which gives:

$$\hat{p} = \frac{v}{1+u+v-\dfrac{w_{12}}{w_{22}}}.$$

But, since u and v will in general be negligible compared with the ratio of the selective values, we can write:

$$\hat{p} = \frac{v}{1-\dfrac{w_{12}}{w_{22}}} \tag{11}$$

and since v is small, it follows that $\hat{p}$ is close to 0.

3.2.2. $w_{11} < w_{12} = w_{22}$. The heterozygotes have the same selective value as the favoured homozygote (i.e. total dominance of the favourable gene).

In the absence of mutation, the only stable equilibrium is when $\hat{p} = 0$; close to this equilibrium we can write:

$$\Delta_s p = \frac{(w_{11} - w_{22})}{w_{22}} p^2.$$

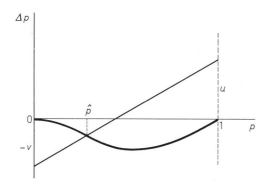

Fig. 11.2. The equilibrium between mutation and selection when $w_{11} < w_{12} = w_{22}$

At equilibrium with mutation, we therefore have:

$$\frac{(w_{11} - w_{22})}{w_{22}} \hat{p}^2 = -\Delta_m p = (u + v)\hat{p} - v.$$

But, since $\hat{p}$ is small, $(u + v)\hat{p}$ is negligible compared with v, so that we can write:

$$\frac{(w_{22} - w_{11})}{w_{22}} \hat{p}^2 = v. \tag{12}$$

The relative difference between the selective values, of the two homozygotes $\frac{w_{22} - w_{11}}{w_{22}}$, is often designated by s. With this notation, Eq. (12) becomes:

$$\hat{p} = \sqrt{\frac{v}{s}}. \tag{13}$$

3.2.3. $w_{12} < w_{11} < w_{22}$. The heterozygote has a lower selective value than either homozygote.

We have seen that in this case there are two stable equilibria, defined by $\hat{p} = 0$ and $\hat{p} = 1$.

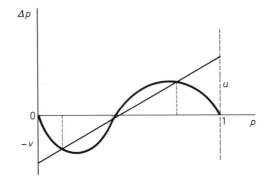

Fig. 11.3. The equilibrium between mutation and selection when $w_{12} < w_{11} < w_{22}$

Close to $\hat{p} = 0$ we have:

$$\Delta_s p \approx \frac{(w_{12} - w_{22})}{w_{22}} p.$$

Hence, taking account of mutation:

$$\frac{(w_{12} - w_{22})}{w_{22}} \hat{p} = (u + v)\hat{p} - v$$

which gives, neglecting $u + v$ compared with $1 - \dfrac{w_{12}}{w_{22}}$,

$$\hat{p} \approx \frac{v}{1 - \dfrac{w_{12}}{w_{22}}}.$$

Similarly, in the neighbourhood of $p = 1$, we have

$$\Delta_s p \approx (1 - p)\frac{(w_{11} - w_{12})}{w_{12}}.$$

Taking account of mutation, therefore:

$$(1 - \hat{p})\frac{(w_{11} - w_{12})}{w_{12}} = u\hat{p} - v(1 - \hat{p}).$$

Hence:

$$1 - \hat{p} = \hat{q} \approx \frac{u}{1 - \dfrac{w_{12}}{w_{11}}}.$$

Therefore the two equilibria are slightly displaced because of the occurrence of mutations.

3.2.4. $w_{11} < w_{22} < w_{12}$. The selective value of heterozygotes is higher than those of the homozygotes.

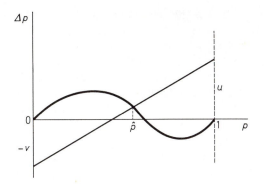

Fig. 11.4. The equilibrium between mutation and selection when $w_{11} < w_{22} < w_{12}$

The stable equilibrium defined by $\hat{p}_s = \dfrac{w_{12} - w_{22}}{(w_{12} - w_{11}) + (w_{12} - w_{22})}$ is shifted when mutations occur, to a lower value of $\hat{p}$ if $\dfrac{u}{u+v} < \hat{p}_s$, or to a higher value if $\dfrac{u}{u+v} > \hat{p}_s$. This can be shown as follows:

We saw that, in the neighbourhood of the equilibrium point under selection alone, we could write:

$$\Delta_s p = (p - \hat{p}_s) \frac{(w_{22} - w_{12})(w_{11} - w_{12})}{w_{11} w_{22} - w_{12}^2}.$$

Taking account of mutations, we therefore have, at equilibrium:

$$(\hat{p} - \hat{p}_s) \frac{(w_{22} - w_{12})(w_{11} - w_{12})}{w_{11} w_{22} - w_{12}^2} = (u+v)\hat{p} - u$$
$$= (u+v)(\hat{p} - \hat{p}_s) - u + (u+v)\hat{p}_s.$$

Hence:

$$\hat{p} - \hat{p}_s = [-u + (u+v)\hat{p}_s] \left[\frac{(w_{22} - w_{12})(w_{11} - w_{12})}{w_{11} w_{22} - w_{12}^2} - u - v \right]^{-1}.$$

The second bracketed term on the right-hand side is clearly always negative.

4. The Human Mutation Rate

As we have already mentioned, a mutation to a recessive gene cannot be detected immediately, but only after the passage of some generations. On the other hand, a new dominant gene would be expected to manifest its effect in the first generation after it occurred. However, things are not always so simple: many dominant genes are "incompletely pene-trant", and do not always give rise to a clearly detectable character difference. For this reason, a child may appear to carry a new mutation when in fact one of his parents already carried the gene, but did not show its effect. Also, mistaken attribution of paternity could make it seem that a mutation had occurred, when this was not so at all.

These factors make precise measurements of human mutation rates very difficult. Few studies have been done in this area, and the inter-pretation of the data obtained depends on making several unrealistic assumptions (for a review, see Neel, 1962).

There are two methods of estimating the mutation rate to dominant genes:

1. A large number of births, say N, are observed, and the number of children who show the character, but whose parents are normal, are noted. Suppose this number is n. Then, if the sample is free from bias, and if the gene is fully penetrant, the mutation rate is estimated by $n/2N$. This is called the "direct method".

2. The selective value of affected individuals is estimated. Then, assuming that the population is in equilibrium, the mutation rate can be calculated using the appropriate formula from Section 3.2 of this chapter. This is called the "indirect method"; the idea of using equilibrium gene frequencies and selective values to estimate mutation rates was first proposed by Haldane (1935) for the case of haemophilia.

Mørch (1941) studied the mutation rate to chondrodystrophic dwarf-ism (which is caused by an autosomal dominant gene) in Denmark, by these two methods. Out of 94075 births registered at the Rigshospital in Copenhagen during a period of 30 years, 10 children with this form of dwarfism were born, of whom two were offspring of dwarf parents. The frequency with which a gene for dwarfism is transmitted in a gamete from a normal parent, which is an estimate of the mutation rate, is therefore:

$$\frac{8}{94075 \times 2} = 4.25 \times 10^{-5}.$$

Mørch also found that chondrodystrophic dwarfs have on average five times fewer children than their normal brothers and sisters.

In Section 3.2.1. of this chapter, we saw that the equilibrium fre-quency p of a disfavoured dominant gene is given in terms of the mutation rate v to this gene, and the selective values w_{12} and w_{22} of affected and normal individuals, by the approximate relation:

$$\hat{p} = \frac{v}{1 - \dfrac{w_{12}}{w_{22}}}. \tag{14}$$

Here, we have $\hat{p}=\frac{1}{2}\frac{10}{94\,075}=0.53\times10^{-4}$ (since 10 children out of 94075 were found to have the gene for dwarfism, and this was one of their two genes). Also:

$$\frac{w_{12}}{w_{22}}=\frac{1}{5}.$$

Hence:

$$v=0.53\times10^{-4}\times\tfrac{4}{5}=4.2\times10^{-5}.$$

The remarkably close agreement between the two estimates of the mutation rate is probably due to a lucky coincidence, and should not be taken as showing that the rates have been precisely evaluated.

Reed and Neel's (1959) study of Huntington's chorea in Michigan shows how difficult it is to estimate even the order of magnitude of a mutation rate in man. For this disease, the late and variable time of onset make it difficult even to detect which individuals have the gene, and it is never possible to be certain that the parents of an individual did not have the gene.

Reed and Neel studied 231 choreic people in Michigan, of whom at most 25 had parents who did not have the gene. The frequency of choreics in the population was 1.01×10^{-4}, hence a maximum estimate of the mutation rate by the direct method is:

$$v=\tfrac{1}{2}\times\tfrac{25}{231}\times1.01\times10^{-4}=5.4\times10^{-6}.$$

In Section 3 of Chapter 10, we discussed the estimation of the selection coefficient associated with Huntington's chorea; we obtained the value 0.17, based on Reed and Chandler's (1958) data. Using this value in Eq. (14), we obtain the following indirect estimate of the mutation rate for this gene:

$$v=\tfrac{1}{2}\times1.01\times10^{-4}\times0.17=8.6\times10^{-6}.$$

These two estimates differ quite widely; this is no doubt related to the fact that some of the assumptions we have made must be invalid. This is in particular likely to apply to the assumption that the population is in equilibrium.

Finally, we shall examine the study of Neel and Falls (1952) on retinoblastoma in Michigan. Out of 1.05 million children born to normal parents, 49 had retinoblastoma, which is caused by an autosomal dominant gene. This gives an estimated mutation rate of 2.3×10^{-5}. However, the penetrance of the gene appears to be low (20%, according to the estimate of Robert, 1966) so that little reliance can be placed on this result.

Further notes on mutation rates. 1. The estimates of mutation rates for different characters are very variable. Robert (1966) quotes values ranging from 5×10^{-7} for the "Landouzy-Degerine" form of muscular dystrophy, to 10^{-4} for the "Duchenne" form of muscular dystrophy. Cavalli-Sforza and Bodmer (1971) give the value of the median of known mutation rates in man as 3.6×10^{-6}. They note, however, that this figure suffers from an important source of bias.

Suppose that we knew the mutation rates for all loci, and could classify them into classes with different rates. Then it is reasonable to assume that the probability that we will observe mutations at a given locus will be higher, the higher its mutation rate (this is a similar argument to the one we used in Chapter 8, when we calculated the number of sibs of an individual taken at random from the population, and found that the proportion of individuals belonging to large sibships is higher than the proportion of such sibships in the population). In order to correct for this bias, we need to know the frequency distribution of mutation rates; assuming a log-normal distribution, Cavalli-Sforza and Bodmer arrive at a corrected estimate of 3.1×10^{-7} per locus per generation. This is less than one tenth of the value of the uncorrected estimate.

We may also note another reason why the methods we have described in this chapter do not give us an estimate of the mutation rate per locus. The methods we have described are based on the study of mutation to specific disease conditions, e.g. Huntington's chorea. Now we know that a locus consists of a length of DNA many bases long; if we take the average length of a protein to be 300 amino acids, then the average locus coding for a protein would be 900 bases long. We do not know what proportion of base-changes at the locus (or loci) which causes Huntington's chorea lead to this disease, and what proportion have some other, perhaps negligible, effect. Consider the locus coding for the β-chain of haemoglobin A; this protein consists of 146 amino acids in a particular sequence. A change from valine to glutamic acid at the 6th position from the end of the polypeptide chain causes the disease sickle-cell anaemia. If we estimate the mutation rate for this condition, it will be considerably lower than the mutation rate for the β-chain locus as a whole. It is possible, therefore, that mutation rates for specific disease conditions may seriously underestimate the per-locus mutation rate in man.

2. When we study mutation rates for specific diseases, these are taken to indicate changes in the genetic information carried in the chromosomes, which become evident to us after translation of the DNA sequences into protein sequences, and after the phenotypic effects of changes in the proteins has become manifest. We can, however, study protein sequences directly.

If we compare the sequence of homologous proteins from two species, we generally find that there are sequences which are identical in both, and residues which are different in the two species. For example, there are 72 positions in the haemoglobin α-chain of the carp which are the same as in the α-chain of man, and 68 residues which differ in these two species.

Kimura and Ohta (1971) have made the following argument, based on this kind of data:

1. Let k be the probability that, at a given site, an amino-acid will be substituted for another one, during a period of one year.

2. Let T be the number of years which have passed since the separation of the two lines of descent leading to the carp and to man.

3. Then the probability that a site has remained unchanged throughout this period is:

$$P = (1-k)^{2T}$$

since the period of time since the separation of the two lines is T for the carp, and T for the human species. Hence:

$$\log_e P = 2T \log_e (1-k) \approx -2Tk.$$

It seems reasonable to assume that the common ancestor of the carp and man lived in the Devonian, i.e. about 400 million years ago. We also know that 72 out of 140 amino acids have remained the same in the two species. Therefore:

$$k = -\frac{\log_e P}{2T} = \frac{-1}{8 \times 10^8} \log_e \frac{72}{140} \approx 8 \times 10^{-10}.$$

The same calculation can be done with other pairs of species for which we can estimate the time since their lines of descent diverged, and for which we know the sequences of homologous proteins. Curiously enough, the annual rates of substitution, k, which we obtain by this method are very similar; the rate appears to be the same for different species-pairs used, regardless of the generation-times involved (the mean of the values which have been estimated is of the order of 1.6×10^{-9} per amino-acid site per year).

In order to determine the rate of substitution per locus per generation, we may assume that the average protein consists of 300 amino acids. In man, the generation time can be taken as 25 years, so that the substitution rate per locus per generation is:

$$1.6 \times 10^{-9} \times 300 \times 25 = 1.2 \times 10^{-5}.$$

Now the number of nucleotides in the human genome has been estimated to be about 4×10^9. A rate of substitution of 1.6×10^{-9} per amino acid site per year therefore leads to the conclusion that 6 or 7 nucleotide substitutions occur each year, in man, or about 100 or more per generation.

Assuming that this estimate of the rate of substitution is correct, we may ask the question: is this rate consistent with our ideas about the evolutionary process? Kimura and Ohta (1971) argue that such a high rate of substitution cannot be due to natural selection causing the spread and fixation of favourable new mutations, but must reflect the occurrence of neutral mutations, and their ultimate fixation due to chance events in finite populations. Their two chief arguments are as follows.

First, they assume that most substitutions of new mutations under natural selection must occur because of changes in the environments of the populations concerned. We can therefore apply the classical theory of the genetic load due to the substitution of genes. This theory was developed by Haldane (1957), and is discussed in detail by Kimura and Ohta (1971); these authors conclude that the load caused by the rate of substitution per genome based on data on protein sequences in different species, is too high for populations to survive. The substitutions could only have occurred at this rate if the load they caused was much lower, in other words if they were neutral substitutions.

A second argument is based on the striking uniformity of substitution rates in different lines of descent. It is not reasonable to assume that the rate of environmental change (and thus the rate of selective changes in the genetic composition of populations) is constant over different lines of descent. A constant rate of substitution is, however, perfectly consistent with substitution due to mutation to neutral alleles, occurring at an approximately constant rate per year, in all the species in the different lines of descent.

Both these lines of argument involve assumptions that may be questioned, and this is a field which is the subject of much dispute at present.

Finally, if we accept the arguments in favour of the idea that the bulk of amino-acid changes in proteins reflect neutral substitutions, we can show that the rate of substitution is equal to the neutral mutation rate. The proof is as follows. Consider a population with N members, in other words there are $2N$ genes at each locus. Now suppose that each year a fraction u of the genes mutate to new, neutral alleles, which have never before occurred in the population; in each year, therefore the population will acquire $2Nu$ new alleles. We saw in Section 1.2 that the probability of eventual fixation of a given one of the genes present in the population at a particular time is $1/2N$. The probability that the gene which will eventually be fixed at this locus will be one of the new mutant alleles rather than the original allele, is $2Nu/2N = u$. If the population has reached a state of equilibrium in which the rate at which it acquires new alleles is equal to the rate at which alleles are lost by chance, it is clear that the number of alleles which complete the process of substitution each year will be equal to the number of alleles arising each year which will eventually become fixed. In other words, the rate of substitution (k) is equal to the rate of neutral mutation (u), and so the calculations given above are equivalent to an estimate of the neutral mutation rate.

5. The Spread of a Mutation: Congenital Dislocation of the Hip

In order to illustrate the difficulties of using the mathematical techniques which we have developed, in actual studies of human populations, we shall discuss a study of congenital dislocation of the hip (Sutter, 1972).

The mode of inheritance of this condition is not fully understood, although it is certainly a hereditary condition. It is probably due to a dominant major gene, with penetrance depending on sex (girls are affected six times more frequently than boys); there may also be a number of modifying genes with smaller effects.

Congenital dislocation of the hip is especially common in the Bigouden region in the southern part of the department of Finistère, in France; in this area, 3 % of all girls are found to be affected. This, however, appears to be a recent phenomenon, as there was no mention of it until the 19th century. Whether the gene originated locally in this area, by mutation, or was introduced into the area by migration, it has certainly spread rapidly. A geographical study of the distribution of affected individuals shows that the centre from which the gene must have spread is the small town of Pont l'Abbé. Fig. 11.5 shows the distribution of cases in the different cantons of Finistère.

Sutter's study is confined to children of school age, so that comparable samples from the different cantons are assured. The total school population was 132000, of whom 128500 were examined. Among these, 356 affected individuals were found. An attempt was made to elucidate the genetic basis of the disease, by family studies of the cases which were found, but only 106 families could be studied (a total of 5100 people).

In such studies it is obviously important to carry out the study on the largest possible scale; even so, the results are not always conclusive. In the present case, the mode of inheritance of the condition could not be decided with any certainty, nor could the reason for the rapid spread of the condition be determined. In particular, there was no evidence for heterosis of the gene. We may, however, note one important observation: the frequencies of health problems and of various abnormalities were found to be the same among the affected children as among their normal

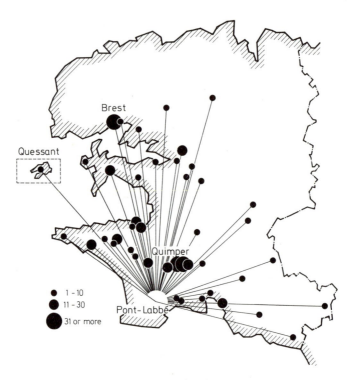

Fig. 11.5. The distribution of congenital dislocation of the hip in Finistère

sibs, and among the children of entirely unaffected families. Also, their growth rates were the same as those of normal children. It therefore seems that the gene for congenital dislocation of the hip has no disadvantageous effect, other than that of causing the hip abnormality itself.

Further Reading

Popham, R. E.: The calculation of reproductive fitness and the mutation rate of the gene for chondrodystrophy. Amer. J. Hum. Genet. **5**, 73–75 (1953).

Chapter 12

Migration

by Daniel Courgeau

Totally isolated human populations are very rare. In Chapter 16, we shall study several cases of human populations which have remained separated from other groups for some time (for example, the Kel Kummer Tuareg, or the Bedik tribesmen of Senegal). Apart from a few such extreme cases, however, most human and animal populations communicate with other populations of the same species by migration. The amount of migration that will take place, and the exact type of migration that occurs, will depend on geographical, sociological and economic factors.

If a population receives a number of migrants from other communities, its genetic structure may become altered. This will happen if the frequencies of the various alleles at a locus are different in the group of immigrants from those in the host population. Thus migration is a cause of changes in the genetic constitution of populations. In this chapter, we shall study such genetic changes due to migration.

First of all, we shall assume that all other conditions for Hardy-Weinberg equilibrium are satisfied, and we will consider the case of a number of groups, each of infinite size, which exchange members by migration. Such a model is clearly unrealistic; we shall therefore next study finite populations and shall establish "stochastic models", which enable us to study the variance in allele frequencies between groups when there is migration. Thirdly, we shall try to show which of the assumptions used in these models of migration need to be altered in order to agree more closely with reality; this will be done by considering some studies of human migration.

1. Deterministic Models with Migration

Consider an infinitely large population divided into m groups, each of "infinite" size. Suppose that in each generation migration between groups occurs; the proportion of members of the k-th group who are migrants from the r-th group, in generation g, will be written as $l_{kr}^{(g)}$. We are assuming that all migrations take place in the period between birth and the start of the reproductive period; during the reproductive period itself, we assume that each group is panmictic. The proportion of individuals in the r-th group who are native to that group will be written as $l_{rr}^{(g)}$; hence we have:

$$\sum_r l_{kr}^{(g)} = 1.$$

It will be useful to write the set of values of $l_{kr}^{(g)}$ as a matrix:

$$L_g = (l_{kr}^{(g)}),$$

L_g is a square stochastic matrix of order m; the first subscript k represents the row number, and the second subscript r is the column number.

The matrix L_g represents the exchanges of migrants between all the m groups in the g-th generation. Clearly, L_g will change in each generation, since the numbers of migrants exchanged between particular populations will vary; however, in order to simplify matters, we shall assume from now onwards that the terms l_{kr} remain constant during the period in question, so that the exchanges between groups can be characterised by a single matrix L.

1.1. Changes in Genic Structure

Consider a locus with n alleles $A_1, \dots, A_i, \dots, A_n$. We shall write $s_k^{(g)}$ for the genic structure of the k-th group in generation g, i.e.:

$$s_k^{(g)} = (p_{k1}^{(g)}, \dots, p_{ki}^{(g)}, \dots, p_{kn}^{(g)})$$

where the p_{ki} are the frequencies of the various alleles in the group.

The genic structures of the m groups are therefore represented by m vectors, each with n elements. To simplify the notation, we shall write Ω_g for the matrix of order $m \times n$ whose rows are the vectors $s_k^{(g)}$, i.e.:

$$\Omega_g = (p_{ki}^{(g)}), \quad \text{with} \quad \sum_i p_{ki}^{(g)} = 1.$$

Assuming that the allele frequencies are the same in the migrants from a group as in the whole of the group they came from, we obtain:

$$p_{ki}^{(g)} = \sum_r l_{kr} \, p_{ri}^{(g-1)}$$

which can be expressed simply, in matrix notation, as:

$$\Omega_g = L\Omega_{g-1}.$$

Going back to the initial generation, we therefore have:

$$\Omega_g = L^g \Omega_0. \tag{1}$$

Thus, if the matrix L^g tends towards a limit, as $g \to \infty$, the genic structure will also tend towards a limit. In order to study this limit, it is convenient to express the matrix L^g as a function of the eigenvalues and eigenvectors of L. It is shown in Appendix B that, since L is a stochastic matrix, all its eigenvalues are less than or equal to 1 in absolute magnitude, and that at least one of them is equal to 1. Also, if all the elements of the principal diagonal are non-zero, the eigenvalue $\lambda = 1$ is the only eigenvalue whose absolute value is equal to 1. This condition is, in general, satisfied for human populations: it is equivalent to saying that there are no communities which, in every generation, send all their members to other groups, so that none remain in the original group.

The matrix L can then be written in the form:

$$L = USU^{-1}.$$

If we assume that the eigenvalues of L are distinct [1], then the matrix S is a diagonal matrix whose non-zero elements are the eigenvalues $\lambda_1, \lambda_2, \ldots, \lambda_m$ of L (see Appendix B):

$$S \equiv \begin{bmatrix} \lambda_1 & 0 & \ldots & 0 \\ 0 & \lambda_2 & \ldots & 0 \\ \vdots & & \ddots & \vdots \\ 0 & \ldots & \ldots & \lambda_m \end{bmatrix}.$$

Using the results for stochastic matrices developed in Appendix B, we see that the product $U S^g U^{-1}$ $(= L^g)$ tends towards a matrix in which all the rows are identical, and equal to the column eigenvector associated with the eigenvalue unity.

The product of this matrix with the matrix Ω_0 will therefore give a matrix Ω in which all the rows are identical. This shows that in the limit, the genic structures of all the groups will become the same. This limiting genic structure, s, depends on the initial structure Ω_0 of the set of populations, and on the migration matrix L.

The speed with which the difference $\Omega_g - \Omega$ tends to zero depends only on the eigenvalue with the largest modulus, other than $\lambda = 1$.

1.1.1. Some particular cases. The simplest case is that of two groups, of which only one receives migrants from the other; the migration matrix

---

[1] The result derived here does not depend on this assumption, but the proof for the general case is too long to give here.

in this case is:

$$L = \begin{bmatrix} 1-m & m \\ 0 & 1 \end{bmatrix}$$

where m is the proportion of individuals in group 1 who came from group 2. It is easy to see that:

$$L^g = \begin{bmatrix} 1 & 1 \\ 0 & 1 \end{bmatrix} \begin{bmatrix} (1-m)^g & 0 \\ 0 & 1 \end{bmatrix} \begin{bmatrix} 1 & -1 \\ 0 & 1 \end{bmatrix}$$

$$= \begin{bmatrix} (1-m)^g & 1-(1-m)^g \\ 0 & 1 \end{bmatrix}.$$

As $g \to \infty$, $L^g \to \begin{bmatrix} 0 & 1 \\ 0 & 1 \end{bmatrix}$, and the genic structure of the first group, which is the only one of the two structures which will change, tends towards that of the second group. The genic structure of the first group will be essentially equal to that of the second group after the passage of g generations, where g is of the order of $1/m$.

In the slightly more complex case of reciprocal exchanges, the migration matrix is of the form:

$$L = \begin{bmatrix} 1-m_1 & m_1 \\ m_2 & 1-m_2 \end{bmatrix}.$$

This gives the equilibrium:

$$L^g_{g \to \infty} = \begin{bmatrix} \dfrac{m_2}{m_1+m_2} & \dfrac{m_1}{m_1+m_2} \\[2ex] \dfrac{m_2}{m_1+m_2} & \dfrac{m_1}{m_1+m_2} \end{bmatrix}. \tag{2}$$

If the initial genic structure of the set of two populations was:

$$\Omega_0 = \begin{bmatrix} p_1 & 1-p_1 \\ p_2 & 1-p_2 \end{bmatrix}$$

the limit of the genic structure of the population would be:

$$\Omega = \begin{bmatrix} \dfrac{p_1 m_2 + p_2 m_1}{m_1 + m_2} & 1 - \dfrac{p_1 m_2 + p_2 m_1}{m_1 + m_2} \\[2ex] \dfrac{p_1 m_2 + p_2 m_1}{m_1 + m_2} & 1 - \dfrac{p_1 m_2 + p_2 m_1}{m_1 + m_2} \end{bmatrix}.$$

It is obvious that this structure depends both on the initial structure and on the migration matrix.

1.2. Changes in Genotypic Structure

The genotypic structures of the different groups will tend, like the genic structures, towards a common limit, as $g \to \infty$.

However, until the homogenisation of the populations is complete, the populations will have different genotypic structures. This state will last for a period of time which is greater the lower the migration rates. The heterogeneity which remains at any time can be measured by the deviation of the actual genotypic structure S of the whole population, from the theoretical structure of the population in Hardy-Weinberg equilibrium, with the same genic structure as the population in question, i.e. the corresponding Hardy-Weinberg structure.

Suppose that the size of the k-th group (which is assumed to be very large) is N_k; we shall write $N = \sum_k N_k$. Then the genotypic structure of the set of groups can be written as:

$$S = \frac{\sum_k N_k S_k}{N} = \frac{\sum_k N_k \vec{s}_k^2}{N}.$$

Hence, the overall genic structure is:

$$s = \frac{\sum_k N_k s_k}{N}$$

and so the corresponding Hardy-Weinberg structure is:

$$S_P = \vec{s}^2.$$

The deviation of the actual genotypic structure from the corresponding Hardy-Weinberg structure is therefore:

$$S - S_P = \frac{\sum_k N_k \vec{s}_k^2}{N} - \vec{s}^2 = \frac{\sum_k N_k (\vec{s}_k - \vec{s})^2}{N}. \tag{3}$$

The i-th diagonal element of this trimat is:

$$\sum_k \frac{N_k (p_{ik} - p_i)^2}{N} = V_{ii}$$

where V_{ii} is the variance in gene frequency of the i-th allele between groups. The element in the i-th row and the j-th column is:

$$\frac{2 \sum_k N_k (p_{ik} - p_i)(p_{jk} - p_j)}{N} = 2 V_{ij}$$

where V_{ij} is the covariance in frequency of the i-th and j-th alleles. The elements of the trimat $S - S_P$ will thus all be equal to zero at the equilibrium state, when all the variances and covariances are zero.

1.2.1. The two allele case. In the particular case when there are only two alleles, the difference between the actual genotypic structure and the Hardy-Weinberg structure can be characterised by a single coefficient. This can be shown as follows. Write

$$s = (p, q) \quad \text{with} \quad p + q = 1.$$

Then we have:

$$\frac{\sum\limits_k N_k (p_k - p)^2}{N} = \frac{\sum\limits_k N_k (q_k - q)^2}{N} = V$$

$$\frac{2 \sum\limits_k N_k p_k q_k}{N} - 2pq = -2V.$$

We can therefore write:

$$S - S_P = \begin{vmatrix} V & \\ -2V & V \end{vmatrix} = V \begin{vmatrix} 1 & \\ -2 & 1 \end{vmatrix}.$$

If we define f by the relation:

$$V = fpq$$

we can write:

$$S = S_P + f \begin{vmatrix} pq & \\ -2pq & pq \end{vmatrix} = S_P + f \begin{vmatrix} p & \\ 0 & q \end{vmatrix} - f \begin{vmatrix} p^2 & \\ 2pq & q^2 \end{vmatrix}$$

or

$$S = (1 - f) S_P + f S_H \tag{4}$$

where S_H is the corresponding homozygous structure of the population with the same gene frequencies as the population in question.

This corresponds to the classical formulae of Wahlund (1928) and Wright (1943). If we compare Eq. (4) above with Eq. (24 b) of Chapter 8, it is clear that they are formally identical, and that f here plays the same role as the coefficient of kinship α in the earlier equation. However, in the general case, with any number of alleles, we do not have a single coefficient which measures the inbreeding effect of subdivision of a population; when there are more than two alleles, the trimat $S - S_P$ depends on more than one coefficient. It is also important to notice the difference between α and f. The coefficient α is a probability, while f is the ratio of the variance in frequency between the groups to the product pq of the mean frequency over all the groups.

1.3. Applications to Actual Populations

This model can be applied in a number of ways, according to the kind of data that is available. Very few genic structures of different human populations are known for more than short periods of time. However, a number of cases are known where several sub-populations have been formed by migration from a large population and have colonised new areas. Assuming that the genic structure of the population of origin has remained unchanged, and that the groups of migrants were representative of the population they came from, we know the initial structures of the new sub-populations; Ω_0 is thus assumed to be the same as the present-day structure of the population from which the new populations originated. If we know the number of generations g which have elapsed since the colonisation occurred, and the present genic structure Ω_g of a colony, we can deduce the migration matrix L, using the relation:

$$\Omega_g = L^g \, \Omega_0.$$

1.3.1. The black and white populations of the United States. This method has been used by Glass (1955), Glass and Li (1953) and Roberts and Hiorns (1962) to estimate the rates of migration between the black and white populations of the U.S.A.

Glass and Li (1953) give data for 4 loci (14 alleles). They assume that the gene frequencies in the white population have remained constant, and that 10 generations have passed since the two populations first came into contact. This corresponds to the case we gave above where:

$$L = \begin{bmatrix} 1-m & m \\ 0 & 1 \end{bmatrix}.$$

These authors obtained estimates of m which were all of the same order of magnitude:

$$0.028 \leq m \leq 0.056.$$

If the present migration rate is maintained (mean $m=0.036$) the equilibrium frequency of the R° gene (rhesus blood group cDe), for example, would not be attained until 60.7 generations had passed; this corresponds to 1670 years, if we take 27.5 years as the generation time in man. Also a migration rate of 0.03 would mean that about 30% of the genes of the present "black" population must have originated from white ancestors.

Roberts and Hiorns (1962) basing their calculations on more up-to-date views about the original populations from which the black population was derived, obtained estimates of m between 0.02 and 0.025.

1.3.2. The Sudanese Nilotes. In this case, data are available on inter-marriages between different tribes. Knowing the genic structure at any time, we can use these to find the structure at a given time in the past or future, on the assumption that the migration rates are constant in time.

The data show that marriages between members of different groups are in general rare[2]. For example, Roberts and Hiorns (1962) give the following matrix of frequencies of marriage between three Sudanese tribes, the Nuer, Dinka and Shilluk:

$$L = \begin{bmatrix} 0.9850 & 0.0125 & 0.0025 \\ 0.0138 & 0.9775 & 0.0087 \\ 0.0000 & 0.0098 & 0.9902 \end{bmatrix}.$$

With these migration rates, if the genic structures of these populations are different, it would take a large number of generations for these populations to reach an equilibrium. It is because of low migration rates between groups that we can talk of human "races". The results of this section show that, if the population sizes of the different races are effectively infinite, genetic differences between the races must steadily decrease, and eventually will disappear altogether. In Section 2, however, we shall see that in the case when the sizes of the groups are finite, heterogeneity between the groups may be maintained by random fluctuations in allele frequencies.

1.4. Deterministic Models of Migration when Other Forces for Change are Acting

In the following sections, we shall continue to assume that the sub-groups are all of infinite size, and that the matrix of migration rates does not change with time.

1.4.1. The continuous-time model of migration[3]. If we assume that migration, reproduction and death occur continuously in time, as is reasonable for a human population, we can write the genetic structure of a set of m populations as the matrix $\Omega(t)$, which is a function of the continuous variable t. Let us define a matrix M such that in a small interval of time dt we have:

$$\Omega(t+dt) = \Omega(t) + M\,\Omega(t)\,dt + \text{terms of order } (dt^2)$$

---

[2] These rates are, however, much higher than mutation rates, which, as we saw in Chapter 11, are of the order of 10^{-5}.

[3] This model has been studied by Roberts and Hiorns (1962); they made the implicit assumption that individuals are capable of reproducing as soon as they are born. Courgeau (1971) has studied the case when there is a lag between birth and the beginning of reproductive life.

so that:

$$\frac{d\Omega(t)}{dt} = M\,\Omega(t). \tag{5}$$

Note that M is not a stochastic matrix, but that it must satisfy the relation $\sum_j m_{ij}=0$. The matrix $M+I$, where I is the unit matrix, is therefore a stochastic matrix.

If we assume that all the eigenvalues μ_j of M are distinct, as we have done previously, it follows from the theory of differential equations that Eq. (5) has the solution:

$$\Omega(t) = U\,S^t\,U^{-1}\Omega(0)$$

where S is a diagonal matrix whose j-th term is equal to e^{μ_j}.

As before, if S^t tends to a limit as $t \to \infty$, then the genic structure of the population tends towards a limiting structure. We have seen that the matrix $M+I$ is a stochastic matrix; it follows that all its eigenvalues are of absolute magnitude less than or equal to 1. Now, the eigenvalues of this matrix are equal to $1+\mu_j$, where μ_j are the eigenvalues of M. It therefore follows that:

$$|1+\mu_j|\leq 1 \qquad \text{for all values of } j.$$

If μ_j is real, we therefore have $\mu_j \leq 0$, hence $e^{\mu_j t}$ will tend towards 0 (if $\mu_j < 0$) or 1 (if $\mu_j = 0$), as $t \to \infty$.

If μ_j is a complex number, we can write $\mu_j = r_j\,(\cos\Theta + i \sin\Theta)$. The preceding condition gives $r_j \cos\Theta \leq 0$. Hence:

$$e^{\mu_j t} = e^{r_j t \cos\Theta}\,e^{r_j i t \sin\Theta}.$$

If $\cos\Theta < 0$, the first term on the right hand side of this equation tends to 0 as $t \to \infty$; the second term can be written

$$\cos(r_j\,t \sin\Theta) + i \sin(r_j\,t \sin\Theta)$$

and this is always finite. In this case, therefore, $e^{\mu_j t} \to 0$ as $t \to \infty$. On the other hand, if $\cos\Theta = 0$, we have $r_j = 0$, hence $e^{\mu_j t} = 1$.

Thus, as $t \to \infty$, S^t tends towards a particular matrix, and $\Omega(t)$ has a limit. Therefore, as $t \to \infty$, the genic structures of the different groups become identical, as in the discrete-time case.

If there are just two groups, and the migration matrix is:

$$M = \begin{bmatrix} -m_1 & m_1 \\ m_2 & -m_2 \end{bmatrix}$$

we obtain:

$$S = \begin{bmatrix} 1 & 0 \\ 0 & e^{-(m_1+m_2)} \end{bmatrix}.$$

Hence:

$$\Omega(t) = \begin{bmatrix} 1 & -m_1 \\ 1 & m_2 \end{bmatrix} \begin{bmatrix} 1 & 0 \\ 0 & e^{-t(m_1+m_2)} \end{bmatrix} \begin{bmatrix} \dfrac{m_2}{m_1+m_2} & \dfrac{m_1}{m_1+m_2} \\ \dfrac{1}{m_1+m_2} & \dfrac{1}{m_1+m_2} \end{bmatrix} \Omega(0)$$

$$= \begin{bmatrix} \dfrac{m_2+m_1 e^{-t(m_1+m_2)}}{m_1+m_2} & \dfrac{m_1(1-e^{-t(m_1+m_2)})}{m_1+m_2} \\ \dfrac{m_2(1-e^{-t(m_1+m_2)})}{m_1+m_2} & \dfrac{m_1+m_2 e^{-t(m_1+m_2)}}{m_1+m_2} \end{bmatrix} \Omega(0).$$

As $t \to \infty$, therefore:

$$\Omega(t) \to \begin{bmatrix} \dfrac{m_2}{m_1+m_2} & \dfrac{m_1}{m_1+m_2} \\ \dfrac{m_2}{m_1+m_2} & \dfrac{m_1}{m_1+m_2} \end{bmatrix} \Omega(0).$$

This continuous-time model of migration is of interest when the population is studied for only a small number of years. However, the study of the limits shows that these are identical in both the continuous and discontinuous cases, as can be seen by referring back to Eq. (2); the only difference is in the rate at which the population approaches the equilibrium state.

In the discontinuous case, the rate of convergence towards the equilibrium can be obtained from:

$$\Omega_g = \begin{bmatrix} 1 & -m_1 \\ 1 & m_2 \end{bmatrix} \begin{bmatrix} 1 & 0 \\ 0 & \{1-(m_1-m_2)\}^g \end{bmatrix} \begin{bmatrix} \dfrac{m_2}{m_1+m_2} & \dfrac{m_1}{m_1+m_2} \\ \dfrac{-1}{m_1+m_2} & \dfrac{1}{m_1+m_2} \end{bmatrix} \Omega_0$$

$$= \begin{bmatrix} \dfrac{m_2+m_1\{1-(m_1+m_2)\}^g}{m_1+m_2} & \dfrac{m_2(1-\{1-(m_1+m_2)\}^g)}{m_1+m_2} \\ \dfrac{m_2(1-\{1-(m_1+m_2)\}^g)}{m_1+m_2} & \dfrac{m_1+m_2\{1-(m_1+m_2)\}^g}{m_1+m_2} \end{bmatrix} \Omega_0.$$

Thus when the migration is discontinuous in time, $(\Omega_g - \Omega_\infty)$ tends to zero as $(1-a)^g$, where $a = m_1 + m_2$. In the continuous-time case, however, this difference tends towards zero as e^{-ag}, when the unit of time is the generation time. The discontinuous model therefore predicts a more rapid approach to the limit than the continuous one. The difference is greater the larger a is, i.e. the greater the amount of migration which takes place.

1.4.2. Migration and mutation. We shall now examine the case when the genes are subject to mutation. The probability (or rate) of a mutation from one allele to another is assumed to be the same for all the groups of populations. Let v_{xy} be the probability of mutation from allele A_x to A_y (so that $v_{xx} = 1 - \sum_{y \neq x} v_{xy}$).

The frequency $p_{ki}^{(g)*}$ of the i-th allele in the k-th group before mutation, but after migration, is given by:

$$p_{ki}^{(g)*} = \sum_r l_{kr} \, p_{ri}^{(g-1)}.$$

After mutation, this frequency will have changed to:

$$p_{ki}^{(g)} = p_{ki}^{(g)*} - \sum_{y \neq i} v_{iy} \, p_{ki}^{(g)*} + \sum_{x \neq i} v_{xi} \, p_{kx}^{(g)*}$$

$$= p_{ki}^{(g)*} \left[1 - \sum_{y \neq i} v_{iy} \right] + \sum_{x \neq i} v_{xi} \, p_{kx}^{(g)*}$$

$$= \sum_x v_{xi} \, p_{kx}^{(g)*}.$$

Writing V for the stochastic matrix of order $n \times n$, whose terms are the v_{xy}, this relation can be written:

$$\Omega_g = L \Omega_{g-1} V = L^g \Omega_0 V^g.$$

We have seen that when the eigenvalues of the matrix L are all distinct, L^g tends towards a matrix Ω whose rows are all identical. With the same conditions, V^g tends towards a matrix Y whose rows are all identical. Therefore, under these conditions, the product $L^g \Omega_0 V^g$ also tends towards a matrix whose rows are all identical, and it is easy to show that this matrix is Y. In this case, therefore, the limiting structure is independent of the initial structure of the population and also of the migration matrix L. It depends only on the matrix of mutation frequencies.

Malécot (1948) has treated the particular case of two alleles. In this case, writing u for the mutation rate v_{12} and v for v_{21}, the matrix Y is equal to:

$$Y = \begin{bmatrix} \dfrac{v}{u+v} & \dfrac{u}{u+v} \\[2mm] \dfrac{v}{u+v} & \dfrac{u}{u+v} \end{bmatrix}.$$

Thus, for example, the frequency

$$p_{k1}^{(g)} \to \frac{v}{u+v} \qquad \text{as } g \to \infty.$$

The speed with which the gene frequencies $p_{k1}^{(g)}$ tend towards their limits depends only on $(1-u-v)^g$. The limiting value is therefore essentially reached when g is greater than approximately $\dfrac{1}{u+v}$. Now, we have seen that mutation rates are generally of the order of 10^{-5}. Therefore the time necessary for the gene frequencies to be equalised is very high, of the order of 10^5 generations.

However, migration rates are usually much higher than mutation rates. Hence the set of populations will at first behave as if mutation did not occur, and will tend towards an equilibrium structure Ω, which is a function of the initial structure and of the migration matrix L, as we showed above. Later, this structure will slowly move towards the equilibrium structure Y under mutation, which depends only on the matrix of mutation rates. In the first stage, we would find inbreeding effects due to the splitting-up of the population, but in the second stage Hardy-Weinberg frequencies would be satisfied. It is, however, important to realise that these conclusions are based on a quite unrealistic model; in particular, we have had to assume that the mutation and migration rates remain constant throughout all these stages.

2. Stochastic Models with Migration

We shall now consider a finite population divided into m groups of sizes N_k between which migration can occur. First of all, we shall assume that all the other conditions for Hardy-Weinberg equilibrium are satisfied. We shall study the limit to which the distribution of gene frequencies in the different groups tends, and the values of the first two moments of this distribution; these provide a measure of the differences between the groups, at equilibrium[4]. We shall then study some cases in which some of the other conditions for Hardy-Weinberg equilibrium are dropped.

In order to define the conditions under which $2N$ "successful gametes" are drawn in generation g, we have to make several assumptions. First, we assume that the number of gametes produced by members of generation $g-1$ is very large. From this assumption, it follows that the gametes which will go to form the new individuals of generation g are drawn from an essentially infinite pool of gametes, among which the genes have the proportions $p_{ki}^{(g-1)}$. We also have to assume that the sample of these gametes which will go to form males has an identical composition with the sample of gametes which will form the females of

---

[4] See footnote to p. 364.

generation g. We are thus implicitly assuming that the sampling is with replacement.

With these assumptions, the number of successful gametes $2N_k p_{ki}^{(g)}$ which carry the i-th allele is a binomial variate; the probability that it takes the value n_{ki} (which is an integer such that $0 \leq n_{ki} \leq 2N_k$) is:

$$P(n_{ki}) = \binom{2N_k}{n_{ki}} \{p_{ki}^{(g-1)}\}^{n_{ki}} \{1 - p_{ki}^{(g-1)}\}^{2N_k - n_{ki}}.$$

As we saw in Chapter 8 (Eq. (35)) the mean and variance of $p_{ki}^{(g)}$, given the value of $p_{ki}^{(g-1)}$, are:

$$E_{g-1}\{p_{ki}^{(g)}\} = p_{ki}^{(g-1)},$$

$$V_{g-1}\{p_{ki}^{(g)}\} = \frac{p_{ki}^{(g-1)}\{1 - p_{ki}^{(g-1)}\}}{2N_k}.$$

We shall now examine how the introduction of migration modifies these equations.

2.1. Migration

We shall use the same notation as in Section 1 for the migration matrix L, which is assumed to be constant from generation to generation.

Also, consideration of Eq. (1) of Section 1.1 shows that we can consider the changes in the frequency of one allele in isolation from the others. The frequencies of the i-th allele in the m groups constitute a column vector:

$$p_i^{(g)} \equiv \begin{bmatrix} p_{1i}^{(g)} \\ p_{2i}^{(g)} \\ \vdots \\ p_{mi}^{(g)} \end{bmatrix}$$

We shall leave out the index i in what follows.

Finally, we assume that migration takes place before the start of the reproductive period, and that individuals remain in the same group throughout their reproductive life. This is in particular the case when we consider the case of matings between individuals who belong to different groups, e. g. different races or tribes, in man.

2.1.1. The expectations of gene frequencies. Since migration is assumed to occur before reproduction, the genic structure from which the gametes which will go to form generation g are drawn is:

$$p^{(g)*} = L p^{(g-1)}.$$

After the random draw, the genic structure obtained can be written:

$$p^{(g)} = L p^{(g-1)} + E^{(g)}$$

where $E^{(g)}$ is a column vector whose k-th element $e_k^{(g)}$ represents the chance variation in the frequency of the allele in question, in the k-th group. The expectation and variance of $e_k^{(g)}$, given the genic structure in generation $g-1$ are:

$$E_{g-1}\{e_k^{(g)}\}=0,$$

$$V_{g-1}\{e_k^{(g)}\}=\frac{p_k^{(g)*}\{1-p_k^{(g)*}\}}{2N_k}.$$

The expectation of $p^{(g)}$ is therefore:

$$E\{p^{(g)}\}=E[E_{g-1}\{p^{(g)}\}]=LE\{p^{(g-1)}\}.$$

If the genic structure of the initial generation is assumed to be known, the solution of this equation gives:

$$E\{p^{(g)}\}=L^g\,p^{(0)}.$$

Thus the expectation is the same as for the deterministic case treated in Section 1. As $g\to\infty$, $E\{p^{(g)}\}\to p$, which is a vector whose elements are all identical. This vector depends in general on the migration matrix and on the genic structures of the initial sub-populations.

In the particular case when one group sends migrants to all the other groups, but does not itself receive migrants from other groups, this result shows that the expected frequency of the allele in question will become the same as the initial frequency of the "colonising" group.

2.1.2. The second moments[5]. We shall write the covariances and variance between the different groups as:

$$u_{jk}^{(g)}=\text{Cov}\{p_i^{(g)},p_k^{(g)}\}=u_{kj}^{(g)},$$

$$u_{jj}^{(g)}=V\{p_j^{(g)}\}.$$

We shall also need the conditional variance and covariances, given the genic structure of generation $g-1$, which we shall denote by $u_{jk}^{(g|g-1)}$.

We want to find a difference equation for the $u_{jk}^{(g)}$ in terms of the $u_{jk}^{(g-1)}$, and to see whether this leads to a limit for the set of $u_{jk}^{(g)}$, as $g\to\infty$.

First let us examine the conditional covariances (and variances). If the gene frequencies in generation $g-1$ are assumed to be known, we

---

[5] The variances and covariances are found as follows. We suppose that we can make repeated "trials", each time starting the ensemble of populations in the same initial state, and that we observe the outcomes in a particular population (for the variances) or pair of populations (for the covariances). The variances and covariances are calculated over the different "trials". When the set of populations reaches a steady state (in cases where this can occur), these variances and covariances are equivalent to the variances and covariances between the different populations in the set.

obtain:

$$u_{jk}^{(g|g-1)} = \{\sum_z l_{jz}[p_z^{(g-1)} - E\{p_z^{(g-1)}\}]\}$$
$$\times \{\sum_w l_{kw}[p_w^{(g-1)} - E\{p_w^{(g-1)}\}]\} + E_{g-1}(e_j e_k).$$

Assuming that the gametes in the j-th and k-th "pools" are sampled independently, we have:

$$\text{if } j \neq k: E_{g-1}(e_j e_k) = 0,$$

$$\text{if } j = k: E_{g-1}(e_j e_k) = \frac{p_j^{(g)*}\{1 - p_j^{(g)*}\}}{2 N_j}.$$

The expectation of the conditional covariance gives us the *a priori* covariance:

$$u_{jk}^{(g)} = \sum_z \sum_w l_{jz} l_{kw} u_{zw}^{(g-1)} + \delta_{jk} \frac{E\{p_j^{(g)*}\} - E[\{p_j^{(g)*}\}^2]}{2 N_j}$$

where $\delta_{jk} = 0$ if $j \neq k$ and $\delta_{jj} = 1$. Now we know that:

$$E\{p_j^{(g)*}\} = \sum_k l_{jk} E\{p_k^{(g-1)}\}$$

$$E[\{p_j^{(g)*}\}^2] = [E\{p_j^{(g)*}\}]^2 + E[(p_j^{(g)*} - E\{p_j^{(g)*}\})^2]$$
$$= [\sum_k l_{jk} E\{p_k^{(g-1)}\}]^2 + \sum_z \sum_w l_{jz} l_{kw} u_{zw}^{(g-1)}.$$

Thus the *a priori* covariances in generation g can be expressed in terms of the covariances and expectations in generation $g - 1$.

We saw in the last section that $E\{p_j^{(g)}\}$ tends towards a limit p: this limit will be effectively reached when g exceeds a certain number which is a function solely of the eigenvalue of L with the largest modulus other than 1. We shall assume in what follows that this limit has been reached. Then the equation for $u_{jk}^{(g)}$ given above takes the simplified form:

$$u_{jk}^{(g)} = \sum_z \sum_w l_{jz} l_{kw} \left(1 - \frac{\delta_{jk}}{2 N_j}\right) u_{zw}^{(g-1)} + \delta_{jk} \frac{p - p^2}{2 N_j}.$$

We can write the covariances and expectations as column vectors with m^2 elements[6]:

---

[6] We could write them as vectors with $\dfrac{m(m-1)}{2}$ elements, since $u_{jk} = u_{kj}$, but the form we have used is easier to handle in the formation of the matrix A, which we use below; for doing calculations, however, the vectors with the lower number of elements would be preferable.

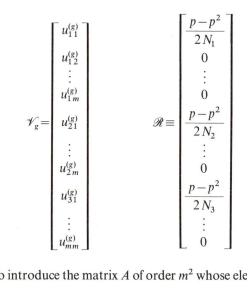

$$
\mathcal{V}_g = \begin{bmatrix} u_{11}^{(g)} \\ u_{12}^{(g)} \\ \vdots \\ u_{1m}^{(g)} \\ u_{21}^{(g)} \\ \vdots \\ u_{2m}^{(g)} \\ u_{31}^{(g)} \\ \vdots \\ u_{mm}^{(g)} \end{bmatrix}
\qquad
\mathcal{R} \equiv \begin{bmatrix} \dfrac{p-p^2}{2N_1} \\ 0 \\ \vdots \\ 0 \\ \dfrac{p-p^2}{2N_2} \\ \vdots \\ 0 \\ \dfrac{p-p^2}{2N_3} \\ \vdots \\ 0 \end{bmatrix}
$$

We shall also introduce the matrix A of order m^2 whose elements are:

$$
a_{rs} = l_{jz}\, l_{kw} \left(1 - \frac{\delta_{jk}}{2N_j} \right)
$$

(where the two subscripts j and k determine the row number r, and the pair of subscripts z and w determine the column number s).

In this notation, the relation between the variances in generations g and $g-1$ becomes:

$$
\mathcal{V}_g = A \mathcal{V}_{g-1} + \mathcal{R}. \tag{6}
$$

If we again assume that $E\{p_j^{(g)}\}$ is effectively equal to its limiting value from generation g onwards, we have:

$$
\mathcal{V}_{g+n} = A^{n+1} \mathcal{V}_{g-1} + \sum_{j=0}^{n} A^j \mathcal{R}. \tag{7}
$$

For rows for which $j \neq k$, the elements of A sum to 1; when $j = k$, the sum of these elements is $1 - \dfrac{1}{2N_j}$. Now it is known that for a non-singular matrix $A = (a_{ij})$ whose elements are all greater than zero, the following property holds:

$$
\text{minimum } s_i \leq \lambda \leq \text{maximum } s_i
$$

$$
\text{where } s_i = \sum_{j=1}^{n} a_{ij}
$$

and λ is the eigenvalue with the largest modulus. Furthermore, the two relations can only be equalities when all the "row-sums" $s_1, \ldots, s_n$ are equal [7], and we have seen above that this is not the case for the matrix A which we are considering. It follows that all the eigenvalues of A are less than 1 in value.

If we assume, as before, that the eigenvalues of the matrix A are all distinct [8], A can be written as:

$$A = U S^{-1} U$$

where S is a diagonal matrix whose non-zero elements are the eigenvalues of A.

Since we have just shown that all the eigenvalues of A are less than 1 in absolute value, it follows that, as $g \to \infty$, A^g (i.e. $U S^g U^{-1}$) tends towards a matrix all of whose elements are zero. Eq. (7) thus tends towards the expression:

$$\mathscr{V}_{g+n} = \sum_{j=0}^{n} A^j \mathscr{R}.$$

To determine $\mathscr{V}_{g+n}$, it thus remains to determine the limit of

$$\sum_{j=0}^{g} A^j = U \left(\sum_{j=0}^{g} S^j \right) U^{-1}$$

as $g \to \infty$.

The matrix $\sum_{j=0}^{g} S^j$ consists of terms of the form:

$$1 + \lambda + \lambda^2 + \cdots + \lambda^g$$

where λ is an eigenvalue of A. This is a geometric series. Since $|\lambda| < 1$, the series tends, as $g \to \infty$, towards the limit $\dfrac{1}{1-\lambda}$. The matrix $\sum_{j=0}^{g} A^j$ therefore tends to a finite limit:

$$A^* \equiv \begin{bmatrix} \dfrac{1}{1-\lambda_1} & 0 & \cdots & 0 \\ 0 & \dfrac{1}{1-\lambda_2} & \cdots & 0 \\ \vdots & & & \vdots \\ 0 & \cdots\cdots\cdots & & \dfrac{1}{1-\lambda_{m2}} \end{bmatrix}$$

----

[7] For a proof of these results, see Gantmacher (1959), Vol. II, p. 63, remark 2.

[8] The result remains valid when this assumption is dropped, but the proof is long and tedious.

and $\mathscr{V}_{g+n}$ tends towards the limit:

$$\mathscr{V} = A^* \mathscr{R}.$$

$\mathscr{V}$ therefore depends on the set of eigenvalues of A, and on the vector $\mathscr{R}$. Another method of finding $\mathscr{V}$ is to write the equilibrium condition $\mathscr{V}_g = \mathscr{V}_{g-1} = \mathscr{V}$, which, from Eq. (6), gives $\mathscr{V} = A\mathscr{V} + \mathscr{R}$. This matrix equation can be solved to give $\mathscr{V}$.

It is also possible to show that the speed with which the variances approach their asymptotic values is a function both of the speed with which the mean approaches its limit (which is a function of the largest eigenvalue of L that is less than 1), and the speed with which $\sum\limits_{j=0}^{g} A^j \mathscr{R}$ tends towards its limit. This is a function of the largest eigenvalue of A.

The following case is of interest. Suppose that the k-th group is infinite in size and that migrants from this group go to all the other groups, but that group k itself receives no migrants. We have seen that the expected genic structures of the other groups tend towards that of the k-th group. Let us now examine whether the variance also tends towards a limit. Since the "colonising" group k is of infinite size, $\dfrac{p-p^2}{2N_k} = 0$; also, since the k-th group does not receive any migrants, we have $V_{kk}^{(g)} = 0$ and $V_{kj}^{(g)} = 0$. The relation between the second moments in generations g and $g-1$ can therefore be written, excluding this k-th group:

$$\mathscr{V}_g' = A' \mathscr{V}_{g-1}' + \mathscr{R}'$$

where the vectors $\mathscr{V}'$ and $\mathscr{R}'$ are of order $m^2 - m = m(m-1)$ and the square matrix A' is of order $m(m-1)$. Furthermore, since the k-th group sends migrants to every other group, we have:

$$\sum_{s=m+1}^{m^2} a_{rs} < 1.$$

Therefore the vector $\mathscr{V}_{g+n}'$ tends towards some constant vector, as $n \to \infty$.

Consider the case when there are just two groups, one of which is infinite; the other group is of size N, of which a proportion m are migrants from the first case. This is a particular case of Wright's (1943) "island model". It is easy to prove that the variance u_g of the group of size N is not zero, since it must satisfy the relation:

$$u_g = (1-m)^2 \left(1 - \frac{1}{2N}\right) u_{g-1} + \frac{p(1-p)}{2N}$$

where p is the frequency of the allele in question in the infinite population. The limit u of u_g as g tends to infinity satisfies the relation:

$$u\left[1-(1-m)^2\left(1-\frac{1}{2N}\right)\right]=\frac{p(1-p)}{2N}.$$

Hence:

$$u=\frac{p(1-p)}{1+4Nm-2m-2Nm^2+m^2}.$$

When m is small, we can neglect terms in m, Nm^2 and m^2, and then:

$$u=\frac{p(1-p)}{1+4Nm}. \tag{8}$$

Now let us find the rate at which u_g will approach u. We have seen that this depends on the value of the eigenvalue of L which has the largest modulus less than 1; in this case, this is m. The quantity $E\{p^{(g)}\}-p$ can be considered as essentially equal to zero when g is greater than approximately $1/m$. Assuming, therefore, that this quantity is equal to zero, we can write:

$$u_{g+n}-u=(1-m)^2\left(1-\frac{1}{2N}\right)(u_{g+n-1}-u)$$

$$=(1-m)^{2n}\left(1-\frac{1}{2N}\right)^n(u_g-u)$$

$$\approx\left(1-2m-\frac{1}{2N}\right)^n(u_g-u).$$

This shows that when n exceeds the order of magnitude of $2m+\dfrac{1}{2N}$, the equilibrium state will have been reached. Thus this state is reached after approximately $\dfrac{1}{m}+\dfrac{2N}{1+4Nm}$ generations. Notice that this is the maximum number of generations that must elapse before equilibrium is attained.

Bodmer and Cavalli-Sforza (1968) studied the two-allele case when the gene frequencies remain close to $\frac{1}{2}$. Using the transformation: $p_k^{(g)}=\sin^2\Theta_k^{(g)}$, where Θ is in radians, they obtained the following expression for the conditional variance:

$$V_{g-1}\{\Theta_k^{(g)}\}=\frac{1}{8N_k}+O\left(\frac{1}{N_k}\right)^2 \tag{9}$$

which is independent of the gene frequencies in generation g.

In this case, it is not necessary to assume that the limit of $E\{p_k^{(g)}\}$ has been attained, and we can study the change in the variance simply from the first generation onwards. If we write:

$$\mathscr{R}' \equiv \begin{bmatrix} \dfrac{1}{8\,N_1} \\ 0 \\ \vdots \\ 0 \\ \dfrac{1}{8\,N_2} \\ \vdots \\ 0 \end{bmatrix}$$

we have:

$$\mathscr{V}_g = A^g \mathscr{V}_0 + \sum_{j=0}^{g} A^j \mathscr{R}' = \sum_{j=0}^{g} A^j \mathscr{R}'.$$

The results obtained by Bodmer and Cavalli-Sforza for the case when there are groups which send out migrants to other groups, but which do not themselves receive migrants, show that the approximation expressed in Eq. (9) remains satisfactory so long as the gene frequencies are not close to 0 or 1. The variance increases rapidly with time, and reaches its equilibrium value after a number of generations which is of the order of $1/\alpha$, where α is the smallest proportion of migrants from another group that any group receives.

2.1.3. Alternative methods of investigating the effects of migration. Another way of studying the second moments of the distribution of gene frequencies (which is due to Malécot, 1948) consists of considering the coefficient of kinship between individuals in two sub-populations, k and j; this coefficient of kinship Φ_{kj} is defined as the probability that a gene taken at random from the gamete pool in group k is identical with a gene taken at random from the gamete pool in group j. It can be shown that the *a priori* variance and covariances between groups can be expressed in terms of the coefficients of kinship, as follows.

We introduce the random variables X_{ky}; these take the value 0 or 1 according to whether the y-th gamete in the k-th colony (out of the $2\,N_k$ gametes in this colony) is the allele in question or not. We can then write:

$$p_k^{(g)} = \frac{\sum\limits_{y} X_{ky}^{(g)}}{2\,N_k}, \qquad p_j^{(g)} = \frac{\sum\limits_{z} X_{jz}^{(g)}}{2\,N_j}.$$

We know that once equilibrium is reached $E\{p_k^{(g)}\}=E\{p_j^{(g)}\}=p$. The covariance in gene frequencies between colonies is found by the formula:

$$\text{Cov}\{p_k^{(g)},p_j^{(g)}\}=\frac{1}{4N_kN_j}\sum_{yz}E[(X_{ky}^{(g)}-p)(X_{jz}^{(g)}-p)].$$

We have to consider two possible cases; either two gametes drawn one from the k-th colony and one from the j-th colony are identical by descent – the probability of this is $\Phi_{kj}^{(g)}$, hence we can write, remembering that $X_{ky}^2=X_{ky}$:

$$E\{(X_{ky}^{(g)}-p)^2\}=E\{(X_{ky}^{(g)})^2\}-p^2=p-p^2.$$

The other possibility is that the two genes drawn are not identical; this has the probability $1-\Phi_{kj}^{(g)}$, and in this case we have the following relation:

$$E\{(X_{ky}^{(g)}-p)(X_{jz}^{(g)}-p)\}=0$$

because the two draws are independent. We therefore have:

$$\text{Cov}\{p_k^{(g)},p_j^{(g)}\}=\frac{1}{4N_kN_j}\times 4N_kN_j(p-p^2)\,\Phi_{kj}^{(g)}=(p-p^2)\,\Phi_{kj}^{(g)}.$$

In a similar way, we find that:

$$V\{p_k^{(g)}\}=(p-p^2)\,\Phi_{kk}^{(g)}+\frac{p-p^2}{2N_k}(1-\Phi_{kk}^{(g)}).$$

If $\dfrac{1}{2N_k}$ can be neglected in comparison with $\Phi_{kk}^{(g)}$, we therefore obtain:

$$V\{p_k^{(g)}\}=(p-p^2)\,\Phi_{kk}^{(g)}.$$

These expressions enable us to study the moments of order 2 in terms of the coefficients of kinship.

2.2. Stochastic Models of Migration with Other Evolutionary Forces also Acting

In this section we shall study the effects of migration together with mutation and linearised selection pressure. We shall consider only the stationary state which the population will reach, and we shall assume that the dispersion of gene frequencies around the point of equilibrium which they would reach under selection alone is small, so that selection can be linearised.

2.2.1. Migration and mutation. We now introduce the possibility of mutations, whose frequencies are given by values v_{xy}, as in Section 1.4.2.

We assume that migration occurs first, and that then mutation occurs at the time of production of an infinite number of gametes by the members of the group; finally the "successful gametes", which will go to form the members of the next generation, are assumed to be drawn at random. We will have to consider all the alleles at the locus in question, and cannot consider just one of them, in isolation; we shall therefore use the matrix Ω of gene-frequencies again.

We have seen that, before the drawing of the successful gametes:

$$p_{ki}^{(g)*} = \sum_{xr} v_{xi} l_{kr} p_{rx}^{(g-1)}$$

where p_{ki} is the frequency of the i-th allele in the k-th group. Consequently:

$$E(\Omega_g) = LE(\Omega_{g-1}) V = L^g \Omega_0 V^g.$$

Thus the change in the expected gene frequency is the same as for the deterministic model: as $g \to \infty$, $E(\Omega_g)$ tends towards a matrix Y whose rows are all identical. Let the expected frequency of allele A_x be p_x. The covariances and variances will be written:

$$\text{Cov}(p_{ki}, p_{lj}) = u_{klij} \quad \text{and} \quad V(p_{ki}) = u_{kkii}.$$

The corresponding conditional covariances and variances will be denoted by u^*_{klij} and u^*_{kkii}.

We can write the following expression for the conditional covariances:

$$u_{klij}^{*(g-1)} = \sum_{xr} v_{xi} l_{kr} (p_{rx}^{(g-1)} - p_x) \sum_{ys} v_{yj} l_{ls} (p_{sy}^{(g-1)} - p_y) + E_{g-1}(e_{ki} e_{lj}).$$

The *a priori* covariance is therefore:

$$u_{klij}^{(g)} = \sum_{xrys} v_{xi} v_{yj} l_{kr} l_{ls} u_{rsxy}^{(g-1)} - \frac{\delta_{klij} E\{p_{ki}^* p_{kj}^*\}}{2N_k} + \frac{\delta^*_{klij} E\{p_{ki}^*(1-p_{ki}^*)\}}{2N_k}$$

$$= \sum_{xrys} v_{xi} v_{yj} l_{kr} l_{ls} \left(1 - \frac{\delta_{kl}}{2N_k}\right) u_{rsxy}^{(g-1)} - \delta_{klij} \frac{p_i p_j}{2N_k} + \delta^*_{klij} \frac{(p_i - p_i^2)}{2N_k}$$

where:

$$\delta_{klij} = 0, \quad \text{if } k \neq l \text{ or if } k = l \text{ and } i = j$$
$$\delta_{kkij} = 1, \quad \text{if } i \neq j$$
$$\delta^*_{klij} = 0, \quad \text{if } k \neq l \text{ or if } i \neq j$$
$$\delta^*_{kkii} = 1$$
$$\delta_{kl} = 0, \quad \text{if } k \neq l$$
$$\delta_{kk} = 1.$$

This can be written more simply in matrix notation if we assume that the means have reached their stationary state by generation g. The pair of indices k and l which refer to the sub-populations will be used to define the row number, and the pair of indices i and j, which denote the alleles in question, will be used to define the column number. We then define the following matrices:

A: a square matrix of order m^2 whose terms are of the form

$$l_{kr} l_{ls} \left(1 - \frac{\delta_{kl}}{2 N_k} \right).$$

W: a square matrix of order n^2, whose general term is $v_{xi} v_{yj}$.

U_g: a matrix of order $m^2 \times n^2$, whose general term is u_{rsxy}.

R: a matrix of order $m^2 \times n^2$ in which the only non-zero terms are r_{kkij}, and these are equal to:

$$-\frac{p_i p_j}{2 N_k} \quad \text{if } i \neq j \qquad \text{and} \qquad \frac{p_i - p_i^2}{2 N_k} \quad \text{if } i = j.$$

In this notation, the expression for the covariance above becomes:

$$U_g = A U_{g-1} W + R.$$

This gives:

$$U_{g+n} = A^n U_{g-1} W^n + \sum_{j=0}^{n} A^j R W^j.$$

Now we know that A has no eigenvalues equal to 1, so that $A^n \to 0$, and the limit of U_{g+n} as $n \to \infty$ is equal to the limit of the sum:

$$U = \sum_{j=0}^{\infty} A^j R W^j.$$

This limit can be found by solving the matrix equation:

$$U = A U W + R.$$

2.2.2. Migration, mutation and linearised selection (the two allele case).

We make similar assumptions to those in the preceding section: migration is assumed to take place first, then mutation and linearised zygotic selection, and random sampling of the gametes is assumed to occur last of all. It is not difficult to obtain useful expressions for this type of model, for the case of a locus with just two alleles. We shall therefore drop the subscript for the allele in what follows, and will confine our attention to one allele.

If the frequency of the allele in question is $p_k^{(g)}$ in the k-th group, it will be changed, after migration has occurred, to:

$$p_{k1}^{(g)} = \sum_r l_{kr}\, p_r^{(g)}.$$

Mutation and linearised selection can be taken into account by a coefficient h of approach towards the equilibrium position, at which the value of p is $\hat{p}$. The gene frequency after this second stage will then be:

$$p_{k2}^{(g)} = p_{k1}^{(g)} - h\,(p_{k1}^{(g)} - \hat{p}).$$

Finally, the effect of chance during the random draw of the $2N_k$ gametes can be expressed as follows:

$$p_k^{(g+1)} = p_{k2}^{(g)} + e_k^{(g)}.$$

As before, we have

$$E_g\{e_k^{(g)}\} = 0, \qquad V_g\{e_k^{(g)}\} = \frac{p_{k2}^{(g)}\{1 - p_{k2}^{(g)}\}}{2N_k}.$$

The passage from one generation, g, to the next, $g+1$, can be expressed by the matrix equation:

$$p^{(g+1)} - p = (1-h)\, L\{p^{(g)} - p\} + e^{(g)}$$

where p is the column vector whose elements are all equal to $\hat{p}$.

It is simple to derive the following expression for the mean gene frequencies:

$$E\{p^{(g+1)} - p\} = (1-h)\, LE\{p^{(g)} - p\}.$$

This shows that $E\{p^{(g)} - p\}$ tends to zero, as $g \to \infty$; hence, in the limit, $E\{p^{(g)}\}$ tends towards p. The expectation of the gene frequency in a group therefore tends to a limit which is independent of the initial gene frequency in the group.

We shall now study the special case of a simple migration matrix, such that the groups are arranged in a linear order, and each is of constant size N and exchanges a fixed proportion m of migrants with the two neighbouring groups which lie on either side of it (Fig. 12.1). We assume that no other type of migration occurs. This model is called the "stepping-stone" model of migration.

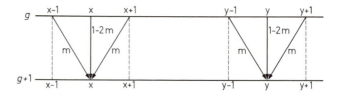

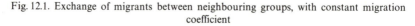

Fig. 12.1. Exchange of migrants between neighbouring groups, with constant migration coefficient

The asymptotic stationary state is defined by relations of the following form:

$$u_{kl}=(1-h)^2 \{m^2 (u_{k-1,l-1}+u_{k-1,l+1}+u_{k+1,l-1}+u_{k+1,l+1})$$
$$+m(1-2m)(u_{k-1,l}+u_{k+1,l}+u_{k,l-1}+u_{k,l+1})+(1-2m)^2 u_{kl}\}$$
$$\times\left(1-\frac{\delta_{kl}}{2N}\right)+\delta_{kl}\frac{(\hat{p}-\hat{p}^2)}{2N}.$$

It turns out that the covariance between colonies depends only on the distance between them. If we write d for the distance between two groups, k and l, and neglect terms in m^2, we have the following expression for the covariance between populations situated at a distance d apart:

$$\text{Cov}(d)=(1-h)^2 \{2m[\text{Cov}(d+1)+\text{Cov}(d-1)]+(1-4m)\,\text{Cov}(d)\}$$
$$+\delta_{kl}\frac{(\hat{p}-\hat{p}^2)}{2N}-\delta_{kl}\frac{(1-h)^2}{2N}[(1-4m)\,\text{Cov}(0)+4m\,\text{Cov}(1)]. \tag{10}$$

In the case when $d\neq0$, $\text{Cov}(d)$ is therefore defined by a second order difference equation; the solutions of this type of equation are known (see Appendix A) to be of the form μ^d, where μ^d satisfies the relation:

$$\mu^d=(1-h)^2 \{2m\,\mu^{d+1}+2m\,\mu^{d-1}+(1-4m)\,\mu^d\}.$$

Neglecting terms in h^2 and in $h\,m$, this equation simplifies to:

$$\mu=(1-4m-2h)\,\mu+2m+2m\,\mu^2$$

which gives:

$$\mu^2-\left(2+\frac{h}{m}\right)\mu+1=0,$$

$$\mu=1+\frac{h}{2m}\pm\sqrt{\frac{h^2}{4m^2}+\frac{h}{m}}.$$

As $d\to\infty$, the solution $\text{Cov}(d)$ remains bounded; it is therefore only necessary to consider the root which is less than 1. We then have:

$$\text{Cov}(d)=\lambda\left\{1+\frac{h}{2m}-\sqrt{\frac{h^2}{4m^2}+\frac{h}{m}}\right\}^d \tag{11}$$

where λ is a constant which can be determined as a function of $\text{Cov}(0)$, as follows.

From Eq. (10), we have the following relation for $d=1$:

$$\text{Cov}(1)=(1-4m+2h)\,\text{Cov}(1)+2m\,\text{Cov}(0)+2m\,\text{Cov}(2).$$

Replacing $\text{Cov}(1)$ and $\text{Cov}(2)$ by the values given by Eq. (11), we obtain:

$$2m\,\lambda\,\mu^2-(4m+2h)\,\lambda\,\mu+2m\,\text{Cov}(0)=0.$$

Hence:

$$\lambda = \text{Cov}(0).$$

Now let us calculate the variance in gene frequency between colonies, $V(=\text{Cov}(0))$; this comes from Eq. (10):

$$V = (1 - 4m - 2h) V + 4m \, \text{Cov}(1) + \frac{\hat{p} - \hat{p}^2 - (1 - 4m - 2h) V - 4m \, \text{Cov}(1)}{2N}.$$

Hence:

$$(4m + 2h) V - 4m \, \text{Cov}(1) = \frac{\hat{p} - \hat{p}^2 - V}{2N - 1}$$

$$2(\sqrt{4mh + h^2}) V = \frac{\hat{p} - \hat{p}^2 - V}{2N - 1}.$$

If N is large, $2N - 1 \approx 2N$, so we obtain, finally:

$$V = \frac{\hat{p} - \hat{p}^2}{1 + 4N \sqrt{4mh + h^2}}. \tag{12}$$

This enables us to find the general expression:

$$\text{Cov}(d) = \frac{\hat{p} - \hat{p}^2}{1 + 4N \sqrt{4mh + h^2}} \left\{ 1 + \frac{h}{2m} - \sqrt{\frac{h^2}{4m^2} + \frac{h}{m}} \right\}^d. \tag{13}$$

When the changes in gene frequency due to migration are high, compared with those due to mutation and selection, h can be neglected, in comparison to m, and Eq. (13) simplifies to:

$$\text{Cov}(d) = \frac{\hat{p} - \hat{p}^2}{1 + 8N \sqrt{mh}} \left\{ 1 - \sqrt{\frac{h}{m}} \right\}^d.$$

This can be approximated by the exponential formula:

$$\text{Cov}(d) = \frac{\hat{p} - \hat{p}^2}{1 + 8N \sqrt{mh}} e^{-d \sqrt{h/m}}. \tag{14}$$

This shows that the correlation coefficient $\dfrac{\text{Cov}(d)}{V}$ decreases exponentially with distance, and also that the correlation is independent of N, the size of the sub-populations.

These results are due to Malécot (1966).

2.2.3. Migration and mutation in a spatially continuous population.
A rather different type of model of the interaction of stochastic factors and migration has been studied by Wright (1943, 1946) and Malécot

(1948, 1969). In this model, the population is assumed to be continuously distributed in space, over an infinite line or plane. Although the total population size is infinite in this model, as in the discontinuous "stepping-stone" model discussed above, differences in gene frequency between different points on the line or plane can arise, because the parents of an individual born at a given point are themselves likely to therefore have been born nearby, and they therefore have a finite chance of being related. Wright has reviewed his approach in his book (Wright, 1969), so we shall here confine ourselves to a consideration of Malécot's method, for the simple case of a linear continuum.

It is convenient to treat this problem in terms of the coefficient of kinship, rather than in terms of the variances and covariances of gene frequencies. As we saw in Section 2.1.3, results obtained in terms of the coefficient of kinship can be translated into variance terms. We shall consider how to determine the coefficient of kinship, $\Phi(x)$, between two individuals who were born at α and β, which are a distance x apart (see Fig. 12.2). The limit, $\Phi(0)$, of $\Phi(x)$ as x tends to zero is the inbreeding coefficient of an individual in such a population, since two individuals born a distance 0 apart are, in fact, the same individual.

In the case of migration along a continuous line, we have to characterise the migration process by a continuous function, the "migration distribution", which specifies the probability density $g(y)$ that a parent of a given individual was born at a distance y from the place where the individual himself was born (we shall arbitrarily assign a negative sign to displacements in the left-hand direction in Fig. 12.2, and a positive sign to displacements to the right, with respect to the birth-place of the individual in question). We shall assume that the probability density $g(y)$ is independent of time and also of the position of the individual in question, i.e. that the migration is homogeneous in both space and time. Clearly, we have:

$$\int_{-\infty}^{\infty} g(y)\,dy = 1.$$

We will also make the simplifying assumption that the number of individuals per unit length, ρ, is independent of position. The number of individuals in an element of length dx is thus $\rho\,dx$.

Finally, we shall assume that the probability that a given allele mutates to some other allele is the same for all alleles, i.e. that $\sum_{j \neq i} v_{ij}$ (see Section 2.1 of Chapter 11) is the same for all i. We shall denote this probability by v.[9]

---

[9] Note that it follows from this assumption that the mean frequencies of all the alleles, at equilibrium, are the same.

Referring to Fig. 12.2, we want to calculate the coefficient of kinship $\Phi(x)$ between individuals I and J, who were born a distance x apart. We shall assume that the population has reached a steady state with respect to the distribution of allele frequencies, so that $\Phi(x)$ remains constant from generation to generation. Let us consider the relations between two genes, one drawn at random from I and one from J. Suppose that the first comes from the parent I′ of I, and the second from the parent J′ of J. The probability that I′ was born in a neighbourhood of length dy around a point γ, a distance y from α is $g(y)\,dy$; the probability that J′ was born in a neighbourhood of length dz around δ, a distance z from β is $g(z)\,dz$. The probability that I′ and J′ both come from a neighbourhood of length dz around the point δ, a distance z from β (and

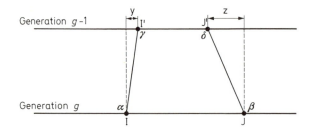

Fig. 12.2. Distances between parents and offspring in a spatially continuous population

therefore, from Fig. 12.2, a distance $x+z$ from α) is $g(z)g(x+z)\,dz$. In this case, there is a probability of $\dfrac{1}{\rho\,dz}$ that I′ and J′ are the same individual. If this is the case, the probability that the two genes are identical by descent is $\dfrac{1+\Phi(0)}{2}$, given that neither of them has mutated.

If the two genes are descended from different individuals in generation $g-1$, who were born a distance $(x-y+z)$ apart, their probability of identity by descent is $\Phi(x-y+z)$.

Taking all the possibilities into account, and noting that the chance that neither gene has mutated is $(1-v)^2$, we obtain:

$$\Phi(x)=(1-v)^2\left\{\int_{-\infty}^{\infty}\int_{-\infty}^{\infty}\Phi(x-y+z)g(y)g(z)\,dy\,dz\right.$$
$$\left.+\frac{1}{\rho}\int_{-\infty}^{\infty}\left[\frac{1+\Phi(0)}{2}-\Phi(0)\right]g(x+z)g(z)\,dz\right\}. \tag{15}$$

$\Phi(0)$ is subtracted from $\dfrac{1+\Phi(0)}{2}$ in the right-hand integral, to correct for the fact that the left-hand integral should include no contribution from the cases when the two genes come from the same individual.

This integral equation can be transformed into a linear differential equation with constant coefficients by replacing $\Phi(x-y+z)$ in the double integral above by its Taylor series:

$$\Phi(x-y+z)=\Phi(x)+(z-y)\Phi'(x)+\frac{(z-y)^2}{2!}\Phi''(x)+\cdots.$$

If we neglect terms in v^2, and those containing the differential coefficients $\Phi'(x)$, $\Phi''(x)$ etc. multiplied by v, we get:

$$2v\Phi(x)-\Phi'(x)\int_{-\infty}^{\infty}\int_{-\infty}^{\infty}(z-y)\,g(y)\,g(z)\,dz$$

$$-\frac{\Phi''(x)}{2}\int_{-\infty}^{\infty}\int_{-\infty}^{\infty}(z-y)^2\,g(y)\,g(z)\,dy\,dz+\cdots$$

$$=\frac{1-\Phi(0)}{2\rho}\int_{-\infty}^{\infty}g(x+z)\,g(z)\,dz.$$

The left-hand side of this equation includes the moments of the dispersal distribution. If we assume that this distribution is symmetrical, so that odd moments are equal to zero, then, neglecting moments of higher order than 2, we obtain the expression:

$$2v\Phi(x)-\sigma^2\Phi''(x)=\frac{1-\Phi(0)}{2\rho}\int_{-\infty}^{\infty}g(x+z)\,g(z)\,dz. \tag{16}$$

Since $g(x+z)$ tends towards zero with increasing x, it must be negligible for large x. For large x, therefore, $\Phi(x)$ is given by the solution of the differential equation:

$$\Phi''(x)=\frac{2v}{\sigma^2}\Phi(x)$$

which gives the exponential form for $\Phi(x)$:

$$\Phi(x)\propto e^{-\sqrt{2v}|x|/\sigma}. \tag{17}$$

To determine $\Phi(0)$, the inbreeding coefficient, we require the solution of the fundamental Eq. (16). It can be shown that this equation has the solution:

$$\Phi(x)=\frac{1-\Phi(0)}{4\pi\rho}\int_{-\infty}^{\infty}\frac{G^2(t)}{2v+\sigma^2 t}e^{-itx}\,dt$$

where $G(t)$ is the Fourier transform of $g(x)$:

$$G(t) = \int_{-\infty}^{\infty} e^{itx} g(x)\,dx$$

(with t a real number), and $i = \sqrt{-1}$.

For the case of a normal dispersal distribution such that:

$$g(x) = \frac{1}{\sqrt{2\pi}\,\sigma}\, e^{-\frac{x^2}{2\sigma^2}}$$

we have:

$$G(t) = e^{-\frac{\sigma^2 t^2}{2}}$$

so that:

$$\Phi(0) = \frac{1}{1 + 4\sigma\rho\sqrt{2v}}. \tag{18}$$

The even moments of order 2 and higher of a normal distribution are all powers of σ. It is therefore reasonable to neglect moments of higher order than 2, provided that the unit of distance with which we are working (i.e. one over which there is a significant decline in genetic relationship) is large compared with the standard deviation of the migration distribution.

These results can be compared with those obtained with the very similar discontinuous model of migration which we studied in Section 2.2.2. If we assume that only two alleles are present at the locus in question, it follows from the discussion of Section 2.1.3, together with the assumption of equal mutation rates for all the alleles (so that $p = 1 - p = \frac{1}{2}$), that $\dfrac{\Phi(0)}{4}$ measures the variance in gene frequency between regions, and $\dfrac{\Phi(x)}{4}$ measures the correlation coefficient of gene frequency between populations separated by a distance x. Eq. (18) is clearly analogous to the corresponding expression of Section 2.2.2, Eq. (12), if we equate $2m$ with σ^2 (which we can do if we take the distance between adjacent groups in the model of Section 2.2.2 as unity). Eq. (17) is analogous to Eq. (13).

A similar model can be set up for the case of dispersion over a two-dimensional plane. If migration is assumed to follow a normal distribution, with the same standard deviation, σ, in all directions, then the following expressions for $\Phi(0)$ and $\Phi(x)$ are obtained:

$$\Phi(0) = \frac{1}{1 - 8\pi\rho\sigma^2\,(1/\log 2v)}$$

and when the distance x is large:

$$\Phi(x) \propto \frac{1}{\sqrt{x}}\, e^{-\sqrt{2v}\,|x|/\sigma}.$$

The effect of migration is clearly very sensitive to dimension: $\Phi(0)$ is smaller in the two-dimensional case than the one-dimensional case, and $\Phi(x)$ falls off more quickly.

3. Data on Migration in Human Populations

In all the models which we have discussed so far we assumed that the matrix of all the migration rates between different populations are known. These were assumed to remain constant during the whole period of approach to the stationary state. We also made further assumptions during the development of several of the models; in particular we had to assume that all migration occurred before the reproductive period.

We shall now compare these hypotheses with some observational data on human migration, from various countries.

We shall first discuss some of the many models that have been proposed which would afford a precise description of migration and its evolutionary consequences.

3.1. Models of the Migration Process

First, we need to define the term "migration" more clearly. As population geneticists, we are obviously not interested in short-term movements of individuals, but only in permanent migrations from one population to another. In particular, we are concerned with "matrimonial migration", which can be measured in one of two ways.

1. By comparing the places of birth of individuals with the places where their offspring are born (we would therefore have to take the migrations of both parents into account).

2. By comparing the birthplaces of men and their wives.

3.1.1. The principal types of model. Many studies have shown that the number Y_{ab} of migrants from one community a to another b depends on the distance between the two communities, and on the size of both of them. The most commonly used model gives Y in terms of these parameters, according to the equation:

$$Y_{ab} = k\, \frac{\delta(a)\,\delta(b)}{r^{\alpha}}\, ds(a)\, ds(b)$$

where $\delta(x)$ is the density of the population at x, r is the distance between the two places, k and α are constants and $ds(a)$ and $ds(b)$ are the areas of the two places in question. This is called the Pareto model.

Stouffer (1940) has tried to substitute a measure of "social distance" for the measure of geographic distance in this equation; social distance is defined in terms of the total number of migrants from community a who are found in all the intervening communities between a and b. It is obviously very highly correlated with geographic distance, and only empirical data can tell us which model best fits a given population.

Hägerstrand (1957) considered a different model of migration. He assumed that the migration rate at any time is closely correlated with the rates at earlier periods. One of the parameters in this model is therefore the number of migrants from a to b, n years before the time we are considering; another parameter, which implicitly introduces the distance between the two communities, depends on the number (assumed to be small) of individuals who move to b but are not attracted there by earlier migrants.

Other, more complex, models introduce a variety of socio-economic variables. Olsson (1965a, b), for example, showed that the geographic distance that a migrant moves depended on eight parameters which could be used to characterise the migrants. The distance moved thus sums up information of many different sorts about the migrant.

Models of matrimonial migration are usually analogous to the first two types of model.

3.1.2. Some applications to real populations. The Pareto model has been applied to population data from several countries (including France, America, Sweden and Japan), and generally gives satisfactory agreement with observed migration data. The coefficient α, which was intitially thought to be a constant, has, however, proved to vary from one region to another, and also in time.

For Swedish populations, for example, Hägerstrand (1957) found values of α from 0.4 to 3.3; the large values mostly corresponded to migration in rural communities, and the small values to towns. α was generally found to decrease with time.

Courgeau (1969) studied migration in French populations, between 1896 and 1962. He found that α could be taken as equal to 2 for the whole of this period, by introducing a correction term l, which is a function of time. The expression for Y then becomes:

$$Y_{ab} = \delta(a)\,\delta(b)\left(\frac{k}{r^2} + 1\right) ds(a)\,ds(b)$$

which gives good agreement with observation.

The finding that migration rates change with time means that it is not strictly valid to apply the genetic theory we have developed above to human populations, since this theory was based on the assumption of constant migration rates.

A difficulty of the Pareto model is that it is not applicable over the whole of the interval $(0, \infty)$, since the number of non-migrants $(r=0)$ is undefined. This can be overcome, by introducing a constant β, and writing the equation for Y as:

$$Y_{ab}=k\,\frac{\delta(a)\,\delta(b)}{(\beta+r)^{\alpha}}\,ds(a)\,ds(b).$$

In order to find the frequency distribution of migration starting at a point a, we must know the form of the territory in which the migration occurs, and the population density at all points. Two particular cases give simple results. First, we can assume that the territory is uni-dimensional (e.g. a valley), and that the density is constant. The probability distribution is then of the form:

$$f(r)=\frac{(\alpha-1)\,\beta^{\alpha-1}}{(\beta+r)^{\alpha}}\qquad\text{for } r\geqq 0.$$

Alternatively, we could consider the territory to be an infinite surface, and assume constant density δ. We then have the following probability distribution:

$$f(r)=\frac{(\alpha-1)(\alpha-2)\,\beta^{\alpha-2}\,r}{(\beta+r)^{\alpha}}.$$

Cavalli Sforza (1962) used a model similar to this one, in a study of matrimonial migration in a valley in Parma:

$$f(r)=\frac{k^{\alpha}}{\Gamma(\alpha)}\,e^{-kr}\,r^{\alpha-1}$$

where

$$\Gamma(\alpha)=\int_{0}^{\infty}e^{-x}\,x^{\alpha-1}\,dx.$$

This model using the Γ function of Pearson is easier to fit to empirical data than the one given above.

The Pareto model has been compared with Stouffer's model, for American and Swedish population data, and both proved almost equally satisfactory. However, Hägerstrand's model was shown, in a Swedish population, to be greatly superior to the other two. It is also capable, in a modified form, of fitting data on migration in French populations, and their changes between 1896 and 1962. Unfortunately, this model has not yet been applied to population genetics.

3.2. Comparison of the Genetic Models with the Models of Migration

The models of migration in human populations and of the genetic consequences of migration were developed independently. We would therefore like to know whether the assumptions on which the population genetic theory is based agree with the observations of anthropologists and demographers.

3.2.1. Migration independent of time. In the theoretical sections of this chapter, we generally assumed that migration rates were constant in time. This assumption is not necessary for the study of deterministic models, since it is possible to write L_g for the migration matrix in generation g, and we then have:

$$\Omega_g = L_g \Omega_{g-1} = L_g L_{g-1} \dots L_1 \Omega_0.$$

If migration rates are constant, or only change gradually, this equation allows us to study the changes in the genetic composition of a population, and even to predict future changes.

However, when we dealt with the stochastic model of migration we assumed that the population had reached its stationary state. If migration rates are changing in time, this assumption is clearly invalid.

Studies in Sweden have shown that migration rates were effectively constant from 1785 to 1870, but have since changed considerably. This also appears to be the case in France, since 1896. Thus it is probable that European populations were in a stationary state before 1870, and could have been studied using our stochastic model, but that these populations are at present changing, and are far from the new stationary state which they may one day attain.

There may, of course, be populations which have not undergone this process of change, so that we could use the stochastic model. It is, however, important to establish over a long period, that the population is really stationary. Cavalli Sforza (1962), for example, states that migration rates in the province of Parma have remained the same for three centuries.

These considerations show that it would be desirable to have a treatment, not only of the stationary state, but also of the changes in the variance before this state is reached. Bodmer and Cavalli Sforza (1968) have studied the changes in variance from generation to generation, with various migration matrices. However, they did not consider the possibility of the migration matrices changing with time.

3.2.2. The nature of the migration matrices. Most genetical models of migration involve matrices of migration rates. Non-genetic studies of migration, however, consider migration as a function of distance. We have described one genetic model which has this property, and others have also been developed by Wright and Malécot. However, the probability distributions for migration to various distances which these models use (usually the normal distribution) do not correspond to reality at all. It would be desirable to study models like this, but incorporating migration distributions which agree better with observations on migration, and to see what changes this introduces.

Another important point is that most genetic models of migration assume constant population density over the whole of the area in question. This is of course quite unlike the situation for human populations, which vary greatly in density, in particular if we compare country areas and towns.

Finally, the probability distribution for migration is usually assumed to be the same for all points of origin, whereas studies on real populations show that there can be large differences between different populations.

For these reasons, it would seem that there is as yet no satisfactory treatment of the genetic consequences of migration which is continuous in space.

However, as a first attempt to study this problem, we can use the matrix model. We can divide the population into a larger or smaller number of sub-populations and calculate the rates of migration between the sub-populations, whose sizes and gene frequencies are known. Then we can predict the future genetic state of the population, provided that the migration matrix can be assumed to be changing slowly.

3.2.3. Stability of marriages. In a discrete-generation treatment of migration, we assume that mates remain together, once a couple has formed. This is not the case in human populations. In order to allow more than one migration during a lifetime, we would have to make a model with overlapping generations. Many studies have been done of the probability that an individual who has migrated once will do so again during the next n years. It is often assumed that second migrations follow the same frequency distribution of migration distances as the first ones; it would be desirable to see whether this is a reasonable approximation to the real situation.

Although this factor could thus be introduced into the genetic theory, it seems probable *a priori* that the effect would be slight.

4. Conclusions

Observations on real populations show that several of the assumptions behind our theories of the genetic effects of migration are of doubtful validity, when applied to human populations. The stochastic models developed by Malécot (1948, 1969) are possibly applicable to populations of the snail *Cepaea nemoralis* (Lamotte, 1951), but cannot validly be used for man, since migration rates in man change considerably with time. Since we do not know the limit towards which human migration rates are moving, we can only predict future changes in the short or medium term. For further understanding, we need a theory of the changes during the approach to equilibrium, as well as of the stationary state. This will require precise knowledge of the rates of matrimonial migration between groups. Sutter (1958) and Sutter and Thanh (1962) give data which show that these are very similar to the rates of migration between groups, so that we can probably use this type of data in genetic studies.

It should perhaps also be mentioned that our assumption, in Section 2.2 that mutation rates and selection coefficients are the same in all groups, may not be valid. Malécot (1966) has developed a model in which these values depend on the group in question. Under this model, we can no longer estimate the migration rates from the variance between groups, unless we also know the mutation rates and selection parameters for all the groups; conversely if the migration rates are known, we could estimate the parameter which characterises the mutation rate and selection coefficient. However, this model has only been developed for the stationary state.

Finally, it is important to remember that all the models we have discussed assumed that migration occurred between the gamete pools (assumed to be of infinite size) of the different groups, and not at the level of individuals. In this way, we could consider the frequency of a gamete in a group after migration as deterministic, and the only random stage was the stage of drawing the useful gametes.

In fact, of course, migration takes place because individuals move. This introduces another random process into the system, so it is only the expected frequency of a gene in the group who migrate which is equal to the frequency in the group they come from.

If we also consider the number of migrants exchanged by two groups to be random variable, instead of deterministic, we have a third random stage. It is very difficult to study such a model, with several random stages, except by "Monte Carlo" methods. These will be described in the next chapter.

Further Reading

Alström, C. H.: First-cousin marriages in Sweden 1750–1844 and a study of the population movement in some Swedish populations from the genetic-statistical point of view. Acta Genet. (Basel) **8**, 295–369 (1958).

Azevedo, E., Morton, N. E., Miki, C., Yee, S.: Distance and kinship in Northeastern Brazil. Amer. J. Hum. Genet. **21**, 1–22 (1969).

Cavalli-Sforza, L. L.: Genetic drift in an Italian population. Sci. Amer. **221**, 30–37 (1969).

Cavalli-Sforza, L. L., Zonta, L. A., Nuzzo, F., Bernini, L., Jong, W. W. W. de, Meera Khan, P., Ray, A. K., Went, L. N., Siniscalco, M., Nijenhuis, L. E., Loghem, E. van, Modiano, G.: Studies on African pygmies. I. A pilot investigation of Babinga pygmies in the Central African Republic (with an analysis of genetic distances). Amer. J. Hum. Genet. **21**, 252–274 (1969).

Courgeau, D.: Mutations, migrations et structures géniques. Population **24**, 935–940 (1969).

Haldane, J. B. S.: The theory of a cline. J. Genet. **48**, 277–284 (1948).

Hanson, W. D.: Effects of partial isolation (distance), migration, and different fitness requirements among environmental pockets upon steady-state gene frequencies. Biometrics **22**, 453–468 (1966).

Hiorns, R. W., Harrison, G. A., Boyce, A. J., Kuchemann, C. F.: A mathematical analysis of the effects of movement on the relatedness between populations. Ann. Hum. Genet. **32**, 237–250 (1969).

Imaizumi, Y.: Variation of inbreeding coefficient in Japan. Hum. Heredity **21**, 216–230 (1971).

Imaizumi, Y., Morton, N. E., Harris, D. E.: Isolation by distance in an artificial population. Genetics **66**, 569–582 (1970).

Katz, A., Hill, R.: Residential propinquity and marital selection: a review of theory, method and fact. In: Les deplacements humains, Entretiens de Monaco en sciences humaines. Paris: Hachette 1962.

Kimura, M., Weiss, G. H.: The stepping stone model of population structure and the decrease of genetic correlation with distance. Genetics **49**, 461–576 (1964).

Morton, N. E., Harris, D. E., Yee, S., Lew, R.: Pingelap and Mokil atolls: migration. Amer. J. Hum. Genet. **23**, 339–349 (1971).

Roberts, D. F.: Genetic fitness in a colonising human population. Hum. Biol. **40**, 494–507 (1968).

Saldanha, P. H.: Gene flow from white into negro populations in Brazil. Amer. J. Hum. Genet. **9**, 299–309 (1957).

Smith, C. A. B.: Local fluctuations in gene frequencies. Ann. Hum. Genet. **32**, 251–260 (1969).

Ward, R. H., Neel, J. V.: Gene frequencies and microdifferentiation among the Makiritare Indians. IV. A comparison of a genetic network with ethnohistory and migration; a new index of genetic isolation. Amer. J. Hum. Genet. **22**, 538–561 (1970).

Workman, P. L., Niswander, J. D.: Population studies on southwestern Indian tribes. II. Local genetic differentiation in the Papago. Amer. J. Hum. Genet. **22**, 24–49 (1970).

Wright, S.: The genetical structure of populations. Ann. Eugen. (Lond.) **15**, 325–354 (1951).

Yanase, T.: A note on the patterns of migration in isolated populations. Jap. J. Hum. Genet. **9**, 136–152 (1964).

Zipf, G. K.: The $P_1 P_2/D$ hypothesis: on the intercity movement of persons. Amer. Sociol. Rev. **11**, 677–686 (1946).

Chapter 13

The Combined Effects
of Different Evolutionary Forces

In the earlier chapters of this book we have examined in turn the consequences of dropping each of the six conditions for the establishment of Hardy-Weinberg equilibrium. We have thus approached a little closer to the real conditions under which a population transmits its genes to the next generation. However, in any real population, it is clear that many different factors will all act at the same time: the generations will often overlap, rather than being separate, migration may occur, so that each group is in communication with other groups, new genes will arise by mutation, selection will occur, and may be very strong in favour of certain genes, mating will not really take place at random, and, finally, the population will not be of infinite size, so that gene frequencies will be subject to random fluctuations.

This last factor is different from the other ones: there is no clearly-defined evolutionary "force" acting, but gene frequencies are changed through a purely random effect, which must depend on intangible deterministic factors, whose overall effect can be considered to be "random".

The various models which have been proposed to include the combined effects of these different factors are often classified according to whether they include such random effects. Models which do not include random elements are said to be "deterministic"; such models assume that the population size is infinite, so that the frequency of a gene at the time of conception is equal to the probability of encountering this gene among the gametes formed by the members of the previous generation. Models in which finite population size is assumed, and in which the gametes which enter into newly conceived zygotes are only a sample of those produced by the parental generation, are called "stochastic models".

In Chapter 12, when we studied the consequences of migration, we saw how difficult it is to study models in which more than one evolutionary force acts at the same time. Even with many simplifying assumptions, so that the models become quite unrealistic, we became involved in some very complex equations. In such a situation we may well ask whether our mathematical formulations can do what we want of them, i.e. to throw light on a situation where the results are not obvious. In this chapter, we shall not try to describe all the complex models that have been developed, but will be mostly concerned to the study of the simplest of these, the original model of Wright (1931). In Section 2, we shall give a brief introduction to the ideas of computer simulation of complex systems.

1. Wright's Model

The assumptions made in this model are as follows:

1. We consider a locus with two alleles whose frequencies are p and q.

2. Mutations occur at constant rates, $u(A_1 \rightarrow A_2)$ and $v(A_2 \rightarrow A_1)$.

3. Selection occurs; the selective values are constant and the values for different genotypes are close to one another.

4. The effective population size N_e is finite and constant.

5. Mating is not at random, but can be characterised by a coefficient of deviation from panmixia, δ, which is assumed to remain constant from one generation to another.

6. A constant proportion m of the population consists of migrants from another population, whose genic structure remains constant throughout, with gene frequencies $\pi(A_1)$ and $1 - \pi(A_2)$.

Let the selective values be $w_{11} = 1$, $w_{12} = 1 + h s$, $w_{22} = 1 + s$, where s is a small number, such that s^2 can be neglected compared with s. Also, let the generation number be denoted by g. Of the various parameters listed above, only p and q are functions of g.

1.1. Change in Gene Frequency from One Generation to the Next

The contributions to the change in gene frequency due to each of the factors involved are as follows:

1. Mutation. It is obvious that:

$$\Delta p_m = -u\,p + v(1 - p).$$

2. Migration. The change is given by:

$$\Delta p_M = (1 - m)\,p + m\,\pi - p$$
$$= -m(1 - \pi)\,p + m\,\pi(1 - p).$$

3. Selection. Taking non-random mating into account, the genotype frequencies at the moment of conception are:

$$
\begin{array}{ccc}
(A_1 A_1) & (A_1 A_2) & (A_2 A_2) \\
p^2 + \delta p q & 2(1-\delta) p q & q^2 + \delta p q.
\end{array}
$$

It is easy to see that the change in gene frequency due to selection is:

$$
\begin{aligned}
\Delta p_s = p q & \frac{(p + \delta q)(w_{11} - w_{12}) + (q + \delta q)(w_{12} - w_{22})}{w_{11}(p^2 + \delta p q) + 2 w_{12}(1-\delta) p q + w_{22}(q^2 + \delta p q)} \\
= s p q & \frac{(h-1)(q + \delta p) - h(p + \delta q)}{1 + 2 h s (1-\delta) p q + s q (q + \delta p)}.
\end{aligned}
\tag{1}
$$

If the selection coefficient s is very small, we can approximate Δp_s by the first term of the expansion of Eq. (1) as a series, and write:

$$
\Delta p_s = p(1-p)(t + w p)
$$

where $t = s\{h(1-\delta) - 1\}$ and $w = s(1-\delta)(1-2h)$.

It is important to realise that we can "linearise" the effect of selection in this way only when s is very small, so that this is a highly restrictive assumption.

4. Chance fluctuations resulting from finite population size. We saw in Chapter 8 that the change δ_p in gene frequency is such that the frequency in the next generation is a binomial variate with expectation p and variance $V = \dfrac{p(1-p)}{2N_e}$.

We now want to bring together all the different factors changing gene frequency, to see what the overall effect will be. In order to do this, we have to make assumptions about the order in which the different processes take place. This is because the gene frequency which results after the action of one of these forces, e.g. migration, depends on the gene frequencies in the populations before migration occurred, and similarly for mutation and selection. The equation giving the overall effect of the different factors will therefore be very complicated.

A simple equation can, however, be obtained if we assume that none of the systematic forces changing gene frequency has a large effect. Then we can calculate the effect of each force using the initial gene frequency among the zygotes in the generation in question, before any of the changes have taken place, and can assume that the overall effect is equal to the sum of the effects due to the separate forces.

For example, we can combine the effects of mutation and migration as follows, writing:

$$
u' = m(1 - \pi), \qquad v' = m \pi
$$

we obtain:
$$\Delta p_M = -u' p + v'(1-p).$$

This equation is of the same form as the equation for the effect of mutations. If we now write:
$$\mu = u + m(1-\pi) \quad \text{and} \quad \lambda = v + m\pi$$

we can combine the effects of these two forces, in the following approximate equation:
$$\Delta p_{m+M} = -\mu p + \lambda(1-p).$$

If the systematic effects change gene frequency only slightly (by Δp, say), we can also assume that the variance due to random factors is equal to that calculated using the gene frequency, p, in the zygotes, before these factors have any effect. This can be shown as follows. Suppose that the gene frequency p at the moment of conception changes to $p^* = p + \Delta p$ because of selection, migration and mutation. Then, as we have seen, the frequency in the next generation, at the moment of conception, is a binomial variable with variance:

$$V = \frac{p^*(1-p^*)}{2N_e}$$
$$= \frac{(p+\Delta p)(1-p-\Delta p)}{2N_e}$$
$$= \frac{p(1-p)+\Delta p(1-2p)-\Delta p^2}{2N_e}.$$

If Δp is very small, we can neglect $\Delta p(1-2p)$ and Δp^2, compared with the term $p(1-p)$, so that the variance is approximately equal to $\frac{p(1-p)}{2N_e}$; this is the value of the variance which would be obtained, using the gene frequency at conception, before it was altered by selection, migration and mutation.

Bringing together all these different changes in gene frequency to give the overall effect, we therefore have:

$$\Delta p + \delta_p = -\mu p + \lambda(1-p) + p(1-p)(t+wp) + \delta_p. \tag{2}$$

1.2. The Fundamental Equation

In Eq. (2) the elements represented by Δp are fully deterministic when p is known, but the last term, δ_p, is a random variable. Because of this term, the frequency p can have any value between 0 and 1, in any generation, whatever the value in the previous generation. This means that we cannot calculate the frequency of the A_1 gene in a particular generation g, either by working from generation to generation, or by an *a priori* calculation. The most we can do is to calculate the probabilities of different frequencies, in a given generation.

Since the population size N is finite, there is a finite number of possible gene frequencies: $0, \dfrac{1}{2N}, \dfrac{2}{2N}, ..., 1$. We shall assume that N is sufficiently large for p to be considered to be a continuous variable, and we shall write $\Phi(p, g)\,dp$ for the probability that the frequency of gene A_1 is between p and $p+dp$, in generation g.

We want to determine the form of the function $\Phi(p, g)$.

If the frequency p_g of A_1 in generation g is known, Eq. (2) shows that the frequency p_{g+1} follows a binomial distribution, with expectation:

$$E_g = p_g - \mu\,p_g + \lambda(1-p_g) + p_g(1-p_g)(t + w\,p_g)$$
$$= p_g + \Delta p_g$$

and variance:

$$V_g = \frac{p_g(1-p_g)}{2N_e}.$$

Therefore the probability that the frequency of A_1 in generation $g+1$ is between p and $p+dp$ follows a binomial distribution.

If we assume that the parameters μ, λ, t, w and N_e are constants, we see that the frequencies p_g in successive generations are the elements of a series of random variables, each of which is related to the preceding member of the series by the same transition law. In other words, this is a Markov process. Now it can be shown (Section 5.2, Appendix B) that the *a priori* distribution $\Phi(p, g)$ tends, as g increases without limit, towards an asymptotic distribution $\Phi(p)$, which is independent of the initial distribution. Therefore the probabilities of different gene frequencies tend towards limits which depend on the parameters μ, λ, t, w and N_e, but do not depend on the initial frequency.

If we do not know p_g, but only the *a priori* probability distribution $\Phi(p, g)$ of the gene frequency in generation g, we can deduce the frequency distribution in generation $g+1$:

$$\Phi(p, g+1) = \int_0^1 \Phi(x, g)\,f(x, p)\,dx \tag{3}$$

where $f(x, p)$ is the probability that a gene frequency of x will change to the value p between one generation and the next. As we have seen, $f(x, p)$ is a binomial distribution with mean $x_g + \Delta x_g$ and variance equal to $x_g(1-x_g)/2N_e$.

Since we are assuming that the change in gene frequency in one generation is very small, we can approximate this discrete-time process by a continuous-time formulation. If it is assumed that the moments of $f(x, p)$ of higher order than 2 can be neglected, this leads to the following relation:

$$\frac{\partial}{\partial g}\left[\int_0^p \Phi(x, g)\,dx\right] = \frac{1}{2}\frac{\partial}{\partial p}[V(p)\,\Phi(p, g)] - \Delta p\,\Phi(p, g). \qquad (4)$$

(In the case of a binomial distribution, the above assumption is a reasonable one, provided that N is large, since the moments of order 3, 4 etc. are of the order of $1/(2N_e)^2$, $1/(2N_e)^3$ etc.) This equation is a fundamental one, which will enable us to define the asymptotic probability distribution $\Phi(p)$. Eq. (4) is of the same form as the equation used in physics to study the process of diffusion, and is therefore known as a "diffusion equation". For the derivation of this equation, see Crow and Kimura (1970).

1.3. The Asymptotic Probability Distribution

The function $\Phi(p, g)$ has reached its equilibrium state when it ceases to change from one generation to the next, i.e. when we have:

$$\frac{\partial}{\partial g}\Phi(p, g) = 0.$$

Eq. (4) then shows that the asymptotic distribution $\hat{\Phi}$ is such that:

$$\frac{1}{2}\frac{\partial}{\partial p}[V(p)\,\hat{\Phi}(p)] = \Delta p\,\hat{\Phi}(p).$$

This gives:

$$\frac{1}{\hat{\Phi}}\frac{\partial\hat{\Phi}}{\partial p} = \frac{1}{V}\left[2\Delta p - \frac{\partial V}{\partial p}\right]$$

$$= \frac{4N_e}{p(1-p)}\left\{-\mu p + \lambda(1-p) + p(1-p)(t+w\,p)\right\} - \frac{1}{V}\frac{\partial V}{\partial p}$$

$$= 4N_e\left[-\frac{\mu}{1-p} + \frac{\lambda}{p} + t + w\,p\right] - \frac{1}{V}\frac{\partial V}{\partial p}.$$

Hence:

$$\frac{\partial}{\partial p}\{\log_e \hat{\Phi}\} = 4N_e\frac{\partial}{\partial p}\left[\mu\log_e(1-p) + \lambda\log_e p + t\,p + w\frac{p^2}{2}\right] - \frac{\partial}{\partial p}\{\log_e V\}.$$

Integrating, we therefore have:

$$\hat{\Phi}(p) = K(1-p)^{4N_e\mu-1} p^{4N_e\lambda-1} e^{2N_e(wp^2+2tp)} \tag{5}$$

where K is a constant such that $\int_0^1 \hat{\Phi}(p)\,dp = 1$.

This equation defines the final probability distribution of the frequency of a gene which is subject to constant evolutionary forces for a long period. Unfortunately, to derive this equation, we have had to make a large number of assumptions, so that we cannot expect it to be an accurate representation of any real situation. In particular, it is important to realise clearly that we have assumed that N is large and that all the systematic forces changing the gene frequency are small and constant in their action. Despite these limitations, we can use Eq. (5) to study the behaviour of gene frequencies in certain particular cases.

1.3.1. Some Particular Cases

1.3.1.1. No selection. If the selection parameters t and w are zero, Eq. (5) becomes:

$$\hat{\Phi}(p) = K(1-p)^{4N_e\mu-1} p^{4N_e\lambda-1}. \tag{6}$$

The asymptotic distribution is therefore a beta distribution with parameters $4N_e\mu$ and $4N_e\lambda$. It is known that the mean of such a distribution is:

$$\bar{p} = \frac{4N_e\lambda}{4N_e\mu + 4N_e\lambda} = \frac{\lambda}{\mu+\lambda},$$

its variance is:

$$\sigma^2 = \frac{\bar{p}(1-\bar{p})}{4N_e(\mu+\lambda)+1},$$

and that the constant K is given by:

$$K = \frac{1}{B(4N_e\mu, 4N_e\lambda)} = \frac{\Gamma(4N_e\mu + 4N_e\lambda)}{\Gamma(4N_e\mu)\Gamma(4N_e\lambda)}.$$

a) If $4N_e\mu$ and $4N_e\lambda$ are both greater than 1, in other words if the mutation and migration rates are sufficiently high, $\hat{\Phi}(p)$ becomes zero for $p=0$ and $p=1$, and is a maximum for a frequency $\tilde{p}$ such that:

$$\frac{\partial\hat{\Phi}}{\partial\tilde{p}} = 0.$$

Hence:

$$\tilde{p} = \frac{4N_e\lambda - 1}{4N_e(\mu+\lambda)-2}.$$

The formula for the variance shows that this will decrease as the values of $4N_e\lambda$ and $4N_e\mu$ increase. In the limit, when the population size is infinite, the variance becomes zero, and the gene frequency at the equilibrium is $\bar{p}$; this is the result we derived in Chapter 11, when we studied mutation in a panmictic population (Fig. 13.1).

b) If $4N_e\mu$ and $4N_e\lambda$ are both equal to 1, Eq. (6) shows that $\hat{\Phi}(p)=K$. In this case, therefore, all frequencies are equiprobable (Fig. 13.1).

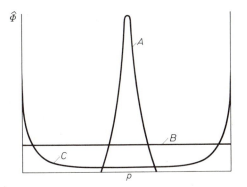

Fig. 13.1. The asymptotic distribution of gene frequency with no selection
$$A: 4N_e(\mu+\lambda)=200$$
$$B: 4N_e(\mu+\lambda)=2$$
$$C: 4N_e(\mu+\lambda)=0.2$$

c) Finally, if $4N_e\mu$ and $4N_e\lambda$ are less than 1, the curve representing $\hat{\Phi}(p)$ is U-shaped, and becomes more and more concentrated in the regions of $p=1$ and $p=0$, as $4N_e\mu$ and $4N_e\lambda$ decrease. We saw in Chapter 8 that only one allele will survive in the long term, in a closed, finite population. The present result is clearly equivalent to this earlier result (Fig. 13.1).

These considerations give us a clearer idea of what we mean by the terms "large" or "small" population size.

1. If N_e is such that $4N_e\mu<1$ and $4N_e\lambda<1$, genetic drift plays an important role in determining the population's genic structure; mutation and migration will only slightly hinder the achievement of total genetic uniformity. Such a population is "small".

2. If N_e is such that $4N_e\mu$ and $4N_e\lambda$ are definitely greater than 1, drift is less important, and the gene frequency will approach an equilibrium which depends on the amount of mutation and migration. Such a population is "large".

3. If N_e is such that $4N_e\mu$ and $4N_e\lambda$ are close to 1, drift will be as important as mutation and migration. The population will not tend

towards a particular equilibrium gene frequency. Such a population is of "intermediate size".

1.3.1.2. Small mutation and migration rates. If the effects of mutation and migration are negligible compared with selection, Eq. (5) becomes:

$$\hat{\Phi}(p) = K \frac{e^{2N_e(w p^2 + 2 t p)}}{p(1-p)}.$$

The curve of $\hat{\Phi}(p)$ has two asymptotes, $p=0$ and $p=1$. In between these values, its form depends on the parameters t and w. The derivative of $\hat{\Phi}(p)$ is:

$$\hat{\Phi}'(p) = \frac{\hat{\Phi}(p)}{p(1-p)} \{4N_e p(1-p)(w p + t) - (1-p) + p\}.$$

The term between the brackets is a polynomial of degree 3 in p; its value is -1 for $p=0$ and $+1$ for $p=1$, so it must be equal to zero for some odd number of values of p between these values.

If $\hat{\Phi}'(p)$ is zero for just one value of p between 0 and 1, this must be a minimum, and the curve is of the form shown in Fig. 13.2. If there are three such values, two will be minima, and one will be a maximum, so that $\Phi(p)$ will be of the form shown in Fig. 13.3.

This second case is particularly important when the selective value of the heterozygote is higher than those of the homozygotes. For example, suppose that we have $\delta = 0$, $w_{11} = w_{22} = 1$ and $w_{12} = 1 + c$. Then it follows that $t = c$, $w = -2c$, hence:

$$\hat{\Phi}(p) = K \frac{e^{4N_e c p(1-p)}}{p(1-p)}.$$

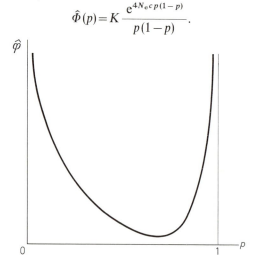

Fig. 13.2. The asymptotic distribution of gene frequency with small mutation rates, and with one minimum

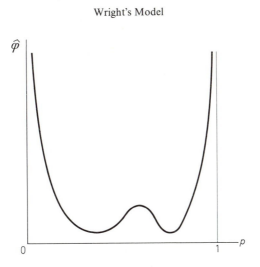

Fig. 13.3. The asymptotic distribution of gene frequency with small mutation and migration rates, and with two minima

The derivative of this is:

$$\hat{\Phi}'(p) = \frac{\hat{\Phi}(p)}{p(1-p)} (1-2p)\{4N_e c\, p(1-p)-1\}$$

which shows that $\hat{\Phi}(p)$ is a maximum when $p = \frac{1}{2}$, and that there are two minima $\left(\text{when } p = \frac{1}{2} \pm \frac{1}{2}\sqrt{1 - \dfrac{1}{N_e c}}\right)$.

This shows that, despite the superiority of the heterozygotes, the locus has only a limited probability of remaining polymorphic, and that (unless N_e is very high) genetic drift will often outweigh the effect of heterosis, and bring about total homozygosity of the population. However, we must remember that this conclusion is only valid under the assumptions on which our model was based, and these include the assumption that selective effects are very small.

To summarise, this model has enabled us to derive again some of our earlier results, and to understand the combined results of some kinds of evolutionary forces acting together. However, the generality of these results is obviously limited by the assumptions that the migration rate and the selection coefficients are very small; to be rigorous, these must be of the order of $1/N_e$, the inverse of the effective population size. It has been possible to extend some of the conclusions to less restricted cases (Coursol, 1969, for example has treated a model where these effects are of the order of $1/\sqrt{N_e}$), but highly restrictive assumptions are still necessary.

It should be noted that, unless migration or mutation occur, the asymptotic probability distribution will be a trivial one, with every population fixed for one or other of the alleles. This is because a population which has become fixed for A_1 or A_2 can only return to a state of segregation for the two alleles when mutation or migration introduce new alleles into the population. In determining asymptotic distributions, this should be borne in mind. We can, of course, study the interaction of finite size with selection as the major deterministic force, by assuming that the coefficients μ and λ in Eq. (5) are negligible in relation to the selection parameters and to $1/N_e$.

1.4. Some Further Results on Selection and Mutation in Finite Population

More modern work on stochastic models in population genetics, which is mainly due to Kimura, is reviewed in detail by Crow and Kimura (1970) and Kimura and Ohta (1971). We shall now describe some of the more important results.

1.4.1. The probability of survival of a mutant gene in a finite population.
In Chapter 11, we studied the probability of survival of a mutant gene, assuming that population size was infinite. We found that, in this case, only selectively favourable genes can survive and spread. If population size is finite, the elimination of an unfavourable mutation no longer has a probability of one, and the probability that a favourable mutation will remain in the population is increased, compared with its value in an infinite population. We shall now study the case of a finite population in more detail. Mating is assumed to take place at random, and the effects of recurrent mutation and migration are ignored.

Suppose the population size is N. Then the frequency of a newly-arisen mutant allele in generation 0 is $p_0 = \dfrac{1}{2N}$. This frequency will change in each subsequent generation, until it becomes equal to 0 (extinction of the mutant gene) or to 1 (fixation of the mutant gene).

Let us assume that the mutant gene confers a selective advantage of s when heterozygous, and $2s$ when homozygous (with $s < 0$ if the gene is unfavourable). It was shown by Kimura (1962) that the probability of fixation of a gene of this sort, when it has attained a frequency of p, is given by the expression:

$$P_f = \frac{1 - e^{-4N_e s p}}{1 - e^{-4N_e s}}.$$

(The probability that the gene will be eliminated is therefore $P_e = 1 - P_f$. Notice that this expression is meaningful only when $s \neq 0$; if $s = 0$, $P_f = \dfrac{1}{2N}$, as we saw in Section 1.2 of Chapter 11.)

If the gene we are considering is a newly-arisen mutant we have:

$$p = \frac{1}{2N}$$

and therefore:

$$P_f = \frac{1 - e^{-2N_e \, s/N}}{1 - e^{-4N_e s}}.$$

Now if N and N_e are both very large, and not very different from one another, this leads to:
$$P_f \approx 2s \quad (s > 0)$$

which is the same as the result derived in Section 1.3 of Chapter 11 for an infinite population. This result shows that, in an infinite population, only advantageous genes can be maintained.

If, however, N is small, P_f is a positive number even for a disfavoured gene. For example, with $N = N_e = 20$, the probabilities of fixation corresponding to different values of s are as follows:

s	P_f
−0.03	0.0062
−0.02	0.0104
−0.01	0.0166
0	0.0250
+0.01	0.0359
+0.02	0.0492
+0.03	0.0641

An unfavourable mutation thus has a non-negligible chance of replacing the other alleles in the population, provided that the population size is small. This result may appear paradoxical. However it is not really so, but merely shows that the effects of chance, in a sufficiently small population, can sometimes outweigh the effects of systematic factors, such as selection pressure.

1.4.2. The mean time to fixation or extinction for a neutral mutation.
Consider a new mutant gene. Its ultimate fate is either elimination or fixation in the population.

Kimura and Ohta (1969 a, b) have shown that the number of generations which pass before a neutral mutant gene is eliminated from the population is a random variable with expectation:

$$\bar{t}_e \approx \frac{2N_e \log_e(2N)}{N},$$

and variance:

$$V_e \approx \frac{16N_e^2}{N}.$$

They also showed that the number of generations until fixation of such a mutant is a variable with expectation:

$$\bar{t}_f \approx 4N_e,$$

and variance:

$$V_f \approx 4.58N_e^2.$$

These formulae show that, whatever the effective population size, $\bar{t}_e$ is much smaller than $\bar{t}_f$. For example, if $N = N_e = 1\,000$, we find $\bar{t}_e = 7$, $\bar{t}_f = 400$. Thus, although it is true that a proportion $\frac{1}{2N}$ of new mutant genes ultimately become fixed, this process is extremely slow; elimination, however, is the fate of a fraction $1 - \frac{1}{2N}$ of such genes, and occurs in only a few generations, on average.

2. Simulation

Any natural phenomenon is far too complex for us to understand it completely. When we describe such a complex phenomenon, and especially when we try to explain the mechanism behind it, we are bound to simplify it, and pay attention only to the features which seem to be essential – in other words, we make a "model" of the system.

Model-building is thus an essential part of the process by which we understand a real situation. The model is the connection between "the two horses drawing the carriage of human knowledge: empirical evidence and theoretical analysis" (Wold, 1964).

The more complex the phenomenon we are trying to explain, the more necessary it is to have a model which will help us to understand the significance of the observations we make, especially when these are incomplete or derived from different sources.

The genetic constitution of a population is a complex phenomenon of this sort; also, our knowledge of the genetic parameters of a population is usually imperfect.

In theoretical population genetics, we usually want to try to establish relationships between the different parameters, or to establish a probability distribution, in terms of one or more of the parameters. For example, earlier in this book we found expressions for the frequency of an allele, or the probability distribution of this frequency, as a function of mutation rates, selective values, the population size etc.

However, although some of these formulations appear very complex, these techniques are capable of treating only highly simplified situations; to obtain useful expressions, we have to ignore practically all the factors which affect the phenomenon in question. There is therefore a real risk that the model will become so detached from reality that the results are of little significance.

There is another way of approaching these questions. This is to simulate models of the system, in particular probabilistic models; with this method, we can continue to take into account all the parameters we think are necessary.

We have seen that many of the events which determine changes in the genic structures of populations can be described only in terms of probability distributions. The passage from one generation to the next can thus be viewed as the net result of many processes, which are not fully deterministic, but have an element of chance. Therefore, to simulate this type of process, we must be able to introduce chance factors, and this is what "Monte Carlo" methods allow us to do.

2.1. The Principles of Monte Carlo Methods

The steps involved in these methods are as follows:

1. The process must be analysed into elementary events which follow one another in a sequence which may be complex.

2. The probability distribution for the outcome of each event must be specified. The distribution can be as complicated as desired, and can depend on any number of parameters.

3. The outcome of each random event is decided according to the result of a draw which is arranged so as to have the same probability distribution as the event in question. By following through the whole sequence of events in this way, the outcome of the entire process which we are simulating is obtained.

4. We repeat the simulated process many times and thus obtain, to the desired accuracy, the probability distribution of the variable we are studying.

For example, suppose that we want to study the distribution of women at a particular time, according to their numbers of offspring, and to see how family planning will affect the distribution. We can analyse the "fertile period" of a woman (from 15 to 45 years) into stages of one month's duration; in each stage, the woman is subject to several "risks", as follows:

1. Death.
2. If she is still alive, she may marry.
3. If married, she may become pregnant.

4. If married, she may be widowed.

5. If married, she may get divorced.

6. If a widow, or divorced, she may re-marry.

The "risk" of becoming pregnant is affected by the woman's preferred number of children, and by her views about the spacing of births in her family, and it also depends on the success of the contraceptive methods (if any) which she uses. Also, this risk is zero during pregnancy, and for a variable period after the birth of a child.

In order to make the transition from conception of a child to birth of a child, we also have to take intra-uterine death into account.

These events can be represented by a diagram (Fig. 13.4), which shows how the outcomes of each of the events determine whether or not the woman is subject to the other risks.

The next step is to specify a probability distribution for each of the "risks". For example, we might assume that a married woman aged 30, who is not practising contraception, runs a 25% risk of becoming pregnant, in each month; or that an unmarried 25-year-old woman has a probability of 1.36% of getting married, each month.

Once these distributions have been specified, it is easy to simulate the reproductive life of a woman; it is simply necessary to find the outcome of the series of risks, for each of the 360 months of the years 15–45; this is done by a series of "draws", to decide the outcome of each risk as it arises. For example, to decide whether an unmarried woman aged 25 gets married in a given month, we draw a random number between 1 and 10000; if the number drawn is less than 137 (which occurs with the probability 1.36%), the woman is considered to be married, and she is then subject to the risks of pregnancy, divorce or widowhood. Similarly, to decide whether a married woman of 30, who is not using contraception, will become pregnant during the month, we draw a number between 1 and 100 at random, and the woman will be considered to have become pregnant if the number is below 26.

To simulate the fertile life of one woman according to this scheme requires about 1500 draws of random numbers, each of which is then compared with the numbers defining the probability distributions of the elementary events. But the result obtained for just one woman is of no interest on its own; we want to find the distribution of offspring number, so we have to simulate a large number of women, for example 1000. This will involve about 1.5 million draws.

These considerations show that we can only do such simulations on high-speed computers. These can be made to generate large numbers of random numbers, or, more precisely, "pseudo-random numbers", which we can use to compare with the numbers expressing the probabilities of the elementary random events. Thus we can simulate

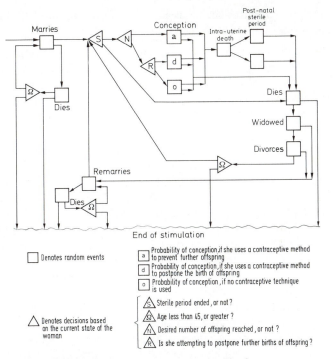

Fig. 13.4. Simulation of the reproductive life of a woman

processes, using a computer, and can also calculate the parameters characterising the variable in question.

In the example which we have been considering, the results of the simulation correspond to observations on a cohort of woman, at the end of their 30-year reproductive period. On a modern computer, a cohort of 1 000 women can be simulated in less than five minutes.

2.2. The Use of Monte Carlo Methods

There are four main types of use for simulation of populations.

a) Fitting parameters. Often the information at our disposal does not permit us to assign values to the parameters which characterise the process under study. In such a situation, we can attempt to fit values of a parameter by simulating the process, using several different possible values, and trying to obtain a good fit to the observational data.

Clearly, such an "inductive" approach cannot prove anything about the validity of the assumptions we have made; at most, we can demonstrate that a particular set of assumptions leads to results which are not inconsistent with our observations.

b) The effect of changes in a parameter. When we have obtained a set of values which give good agreement with observation, we can answer with reasonable exactitude the question: how will a given characteristic of the population (e.g. the number of offspring per female, or the variance of this number) be affected by a given change in one of the factors (e.g. contraception).

c) The analysis of the variance of a variable. The important property of Monte Carlo methods is that they include chance events: each event which forms part of the process in question is assumed to follow a probability distribution which is specified as one of the basic assumptions of the model. The result of the simulation process is therefore itself a random variable (e.g., in our example, the women are all considered to be the same at the start of reproductive life, and yet they end up with different numbers of children). It is clear that the dispersion in the final results is a function of the dispersion implicit in the probability distributions assumed for the elementary events. However, the events are combined into the final result in such a complicated way that it is usually impossible to calculate the resulting variance as an *a priori* function of the variances of these elementary events.

We can also use simulation to find how a particular factor contributes to the resulting distribution; we simply suppress the random element for that factor, determine the distribution which results, and compare it with the previous one.

It is also possible to repeat a simulation, changing certain of the random events to deterministic ones. This gives us a way of finding out the joint effects of these stages on the final variance. In other words, we can use simulation to carry out an "analysis of variance" of the characteristic in question.

d) Intrinsic biological differences between individuals. The dispersion in the values of a characteristic (e.g. number of offspring per woman) in a population results both from heterogeneity of the individuals, with respect to the biological factors which govern the characteristic (e.g. fertility), and also from the existence of certain random factors (e.g. age at marriage). Since we can estimate, by simulation, the variance that is due to the random factors, we can use this technique to estimate the portion of the overall variance due to biological heterogeneity.

As an example of this, simulation of the fecundity of Hutterite women, based on the demographic parameters of this population, yields the result that the variance of offspring number in a uniform cohort of women would be 5.8, whereas the actual variance is found to be 13.9.

Although we cannot place too great reliance on these numbers, it seems reasonable to assume that these variances are different because the women differ in fertility. It is known that fertility has indeed remained very variable in this population, despite the fact that it has remained an effectively closed isolate for several centuries. It appears that this causes a dispersion of family size of at least the same magnitude as that due to random factors (marriage, death, conception etc.) which can affect family size.

2.3. Simulation of the Genetic Structure of a Population

It is easy to see how we could use Monte Carlo simulation methods to study changes in the genetic make-up of a population subjected to a given set of conditions.

Starting with a population of known genotypic structure, we can formulate the process of transition to the next generation, taking into account all the factors that could change gene frequencies, for example:

1. The formation of mating couples can be simulated, taking account of all kinds of mating preference or avoidance.

2. The number of children of a couple can be determined, taking into account selective factors, which could be constant, or could depend on gene frequencies.

3. The genes which the parents transmit to their offspring can be decided by a random draw, and we could also take gametic selection into account.

4. The survival of a child to adulthood, or his death during childhood, can be simulated using life tables of the different genotypes.

We can use as complicated assumptions as we like, in this type of simulation. All that is necessary is to specify the probability distribution of each "node" in the series of random events. The only difficulty with nodes which involve very complex and realistic assumptions is that these will inevitably increase the number of parameters so that it will become harder and harder to interpret the results.

Simulation is an extremely useful tool in population genetics. It allows us to carry out experiments on imaginary populations. The scientist is no longer constrained by the necessity to obtain simple mathematical expressions which he can handle, but can freely use his imagination to develop new hypotheses and test their effects against observational data.

Simulation of genetic systems on computers has been used to study several different types of population genetic problems. For example, Kimura and Ohta (1969 a, b) and Ohta and Kimura (1971) have used this method to check the validity of diffusion equations derived for several different situations. Lewontin (1967), and Franklin and Lewontin (1970) have simulated multi-locus systems in both finite and infinite populations. Fraser and Burnell (1970) describe the method used for this type of simulation. A number of systems with special properties have also been simulated (for example, see Lewontin and Dunn, 1960). Robertson (1970) used simulation in a study of artificial selection.

Simulation has also been used to investigate some problems in human population genetics. For example, MacCluer (1967) has used computer simulation to study inbreeding in populations with over-lapping generations; Gilbert and Hammel (1966) have simulated a set of populations with a particular mating pattern. Levin (1967) has used simulation to study the problem of the maintenance of the Rhesus polymorphism. Several studies of human reproduction patterns have been done using computer methods (e.g. Perrin and Sheps, 1964).

3. Maintenance of Polymorphisms. Genetic Load

One of the problems which confronts population genetics is to explain the phenomenon of polymorphism. We have seen that genetic drift causes a kind of "erosion" of genetic diversity, leading, in the long term, to a population in which all the individuals have the same genes, and are all homozygous at all loci.

There is a great deal of evidence that populations are not genetically homogeneous. Many characters are known for which individuals show genetic diversity. In particular, it has been found that about one third of a sample of loci which control the amino-acid sequence of human proteins and enzymes are polymorphic. This data is especially important, because the loci can reasonably be assumed to be a random sample (for a review, see Harris and Hopkinson, 1972). It therefore seems to be true that the genic structure of most populations shows a high degree of polymorphism: many loci have more than one allele (often a large number) and an individual is heterozygous at, on average, about 12 % of all loci.

How can we explain these facts? In a closed population, there are two factors which oppose the tendency of drift to make populations homozygous: mutation and selection. Either of these evolutionary forces may lead to a polymorphic equilibrium situation.

3.1. The Equilibrium under Mutation and Selection

If, in each generation, a certain fraction of the A_1 alleles in a population mutate to A_2 alleles, and if individuals who carry A_2 are disfavoured, the population will come to an equilibrium when the number of A_2 genes eliminated by selection in each generation is equal to the number of mutations to A_2.

As we saw in Chapters 10 and 11, the mathematical formulation of this type of model is particularly simple when there are just two alleles. Following the classical notation, we shall write:

p and $q=1-p$ for the frequencies of the A_1 and A_2 alleles,
1, $1-hs$ and $1-s$ for the selective values of the three genotypes $(A_1 A_1)$, $(A_1 A_2)$ and $(A_2 A_2)$, respectively,
u for the mutation rate from A_1 to A_2.

We shall further assume that the genotype frequencies agree with the Hardy-Weinberg formula, in other words that they are: $p^2, 2pq, q^2$.

In this notation, the change Δp_s in the frequency of A_1 between one generation and the next, due to selection, is:

$$\Delta p_s = p q s \{h p + (1-h) q\}.$$

This assumes that s is small, so that s^2 can be neglected compared with s. The change Δp_m in the frequency of A_1 due to mutation of A_1 to A_2 is given by:

$$\Delta p_m = -u p.$$

The equilibrium is given by $\Delta p_s + \Delta p_m = 0$. The value p of the frequency of A_1 at the equilibrium is therefore such that:

$$\hat{q} s \{h \hat{p} + (1-h) \hat{q}\} = u$$

which gives:

$$s \hat{q} \{(1-2h) \hat{q} + h\} = u. \tag{7}$$

We now have to consider the possible types of dominance relations of A_1 and A_2; these can be expressed in terms of the parameter h.

1. If $h=0$, in other words if the allele A_2 is recessive to A_1, so that only the $(A_2 A_2)$ homozygotes are disfavoured, the solution of Eq. (1) is:

$$\hat{q} = \sqrt{\frac{u}{s}}. \tag{8}$$

2. If h is not negligible compared with the mutation rate u (which we know is always very low), Eq. (7) shows that q will always be small. Neglecting q^2 compared with q we therefore obtain:

$$\hat{q} \approx \frac{u}{h s}. \tag{9}$$

This is precisely true when $h = \frac{1}{2}$.

3.2. Maintenance of Variability by Neutral Mutation

Kimura and Crow (1964) have proposed that a large amount of variability can be maintained in a population by mutation to selectively neutral alleles. This theory is discussed in detail by Kimura and Ohta (1971). The argument is based on the fact that genes consist of lengths of DNA whose average size is about 300–600 nucleotide bases. Since there are four different nucleotides, the number of possible allelic forms of a gene, i.e. the number of nucleotide base sequences, is very large. Mutation rates at the individual nucleotide level are known to be very low (10^{-8} or less), so that in a population of reasonable size a mutation from one nucleotide sequence to some particular other sequence is a virtually unique event. We can denote the frequency of mutation from an allele to another, selectively equivalent allele by u. This is assumed to be the same for all alleles, and is called the "neutral mutation rate". In every generation, a fraction u of the genes at a locus change to different, but selectively equivalent forms. The problem is to determine the effect of this process on the amount of variability at the locus in question in a population of finite size (if the population were infinite, any neutral mutation would have zero chance of survival, as we saw in Chapter 11).

Let us consider a population of constant effective size N_e. Let the chance that two genes in uniting gametes in generation g are the same allele be f_g. Consider an individual A taken at random in generation g. Referring to Fig. 8.1, we see that A's two genes can be descended:

1. From the same gene of one ancestor in generation $g-2$. The probability that this will be the case is $1/2N_e$. These two genes will be the same allele if neither has mutated, which has the probability $(1-u)^4$.

2. From the two different genes of one ancestor in generation $g-2$, with probability $1/2N_e$; they will be the same allele if this ancestor was a homozygote (probability f_{g-2}), and if neither gene has mutated (probability $[1-u]^4$).

3. From two different individuals in generation $g-2$, with probability $1-1/N_e$; in this case, the probability that they are the same allele is $f_{g-1}(1-u)^2$. Hence, finally, we have:

$$f_g = (1-u)^2 \left(1 - \frac{1}{N_e}\right) f_{g-1} + (1-u)^4 \frac{1}{2N_e}(1+f_{g-2}).$$

At equilibrium, when chance loss of alleles is balanced by new mutation, $f_g = f_{g-1} = f_{g-2} = \hat{f}$, say, hence:

$$\hat{f} = (1-u)^4 \bigg/ \left[1 - (1-u)^2\left(1 - \frac{1}{2N_e}\right) - (1-u)\frac{1}{2N_e}\right] 2N_e.$$

Neglecting u compared with 1, and u^2 compared with u, we therefore have the approximate expression for $\hat{f}$:

$$\hat{f} \approx \frac{1}{4N_e u + 1}.$$

The probability H that an individual is heterozygous is $1 - \hat{f}$, so that:

$$H = \frac{4N_e u}{4N_e u + 1}.$$

Clearly, if $4N_e u \ll 1$, H is negligible: if $4N_e u \approx 1$, H is approximately equal to 0.5, whereas if $4N_e u > 1$, $H \approx 1$. Large population size and high mutation rate lead to a high level of variability maintained in this way. Empirical estimates of H in human populations have been obtained by the study of electrophoretic variants of proteins, and the values obtained are in the region of 0.12 (Harris and Hopkinson, 1972); this corresponds to a value of $4N_e u$ of 0.14. If the mutation rate were 10^{-6}, N_e would be about 3.5×10^4. Clearly this is very much lower than the present-day human population size, but it is probable that past population sizes were much smaller than present ones. Since gene frequencies change very slowly under genetic drift, the present degree of polymorphism could simply be a reflection of the past.

Note that in this model the allele frequencies are not constant at "equilibrium", but are changing slowly under genetic drift, though the rate of change may be too low to be detectable. The word "equilibrium" merely refers to the fact that the amount of variability in the population becomes constant.

3.3. Heterotic Equilibrium

In our discussion of selection, in Chapter 10, we saw that in the case of a locus with two alleles, and when the selective values are constant, there can be a stable polymorphic equilibrium; for such an equilibrium to be possible, it is necessary only that the selective value of the homozygotes are less than that of the heterozygote (heterosis).

If we take the selective value of the heterozygotes as 1, and write $w_{11} = 1 - s$, $w_{22} = 1 - t$, it is easy to see from the results of Section 1.4.5 of Chapter 10 that the equilibrium genetic structure is defined by:

$$\hat{p} = \frac{t}{s+t}, \qquad \hat{q} = \frac{s}{s+t}. \tag{10}$$

Such an equilibrium would remain stable unless the selective values change, in other words unless the population's environment is altered.

A case of an environmetal change, which changes the selective values of the alleles, has been studied in man. We saw in Chapter 10 that the polymorphism for haemoglobin S in some African populations is probably maintained by heterozygous advantage due to the increased capacity of Hb A/S heterozygotes to withstand malaria infections, compared with Hb A/A individuals, while the Hb S/S individuals have lowered fitness due to sickling of their red blood cells in the small blood vessels. If we assume that 90% of homozygotes for haemoglobin S die without reproducing, and that 25% of homozygotes for haemoglobin A will die of malaria before adulthood, we can write:

$$s = 0.25 \quad \text{and} \quad t = 0.9$$

we then obtain:

$$\hat{p}_A = 0.78, \quad \hat{q}_S = 0.22.$$

The American negro population is descended from West Africans who lived in malarial areas, and these ancestors probably had a high frequency of the gene for haemoglobin S. In America, however, they were not subject to malaria infections, so that the advantage to the Hb A/S heterozygotes disappeared. Under these conditions, we can put $t = 0$, so that the population would be expected to move towards $q_S = 0$. The present frequency of this gene in American negroes is about 2–6%, whereas the frequency among their African ancestors is estimated to have been about 8 to 14% (Allison, 1964). There has been much interbreeding with Americans of non-African descent, so it is difficult to tell whether this drop in frequency is really due to the relaxation of selection. Comparison with the changes in other gene frequencies (Workman, Blumberg, and Cooper, 1963), however, suggest that the change in Hb S frequency is too great to be due to inter-racial marriage alone.

3.4. The Genetic Load of a Locus

The existence of unfavourable genes or combinations of genes decreases the overall selective value of populations. We saw in Chapter 10 that this value is related to the population's rate of increase in size. The concept of "genetic load" has been introduced, as a way of studying the effect of genes on the mean selective value. Unfortunately, this term is not always given a clear definition. Following Crow (1958), we shall adopt the following definition:

The genetic load L of a population, with respect to a given locus, is equal to the relative difference between the population's mean selective value and the value, w_{max}, which the population would have if all its members had the most favoured genotype at this locus:

$$L = \frac{w_{max} - \overline{w}}{w_{max}}.$$

For the equilibrium under mutation and selection, and the equilibrium with heterozygous advantage, which we have just been considering, it is easy to find expressions for the genetic load in terms of the parameters which characterise the equilibria.

a) The equilibrium of selection and mutation. In the notation of Section 3.1, we have:

$$L_m = 2pqhs + q^2 s = qs(2ph+q).$$

When the population has reached equilibrium the values of p and q are given by Eq. (7).

If the disfavoured allele A_2 is fully recessive, i.e. if $h=0$, the value of q at equilibrium is $\hat{q} = \sqrt{u/s}$; hence:

$$\hat{L}_m = u. \tag{11}$$

If the disfavoured allele is fully or partially dominant (i.e. $0 < h \leq 1$), the equilibrium value of q is $\hat{q} \approx u/hs$; therefore, neglecting q^2 compared with q:

$$\hat{L}_m \approx 2u. \tag{12}$$

Thus, when the population is in a state of equilibrium between recurrent mutation and selection against a deleterious gene, the genetic load is between u and $2u$; in other words, the load is of the same order of magnitude as the mutation rate, regardless of the selective values of the different genotypes.

For example, the genetic load due to chondrodystrophic dwarfism, which is controlled by a dominant gene, is equal to twice the mutation rate to this gene, i.e. 8×10^{-5}; it is independent of the selective disadvantage of the affected individuals.

As we saw earlier, this disadvantage is such that the carriers of this gene have, on average, only $\frac{1}{5}$ of the number of offspring of normal people; this corresponds to a value of $s = \frac{4}{5}$. The equilibrium situation corresponding to this mutation rate and selection coefficient is therefore:

$$\hat{q} = \frac{u}{s} \quad \frac{4 \times 10^{-5}}{0.8} = 5 \times 10^{-5}.$$

If, due to some medical advance, the fecundity of chondrodystrophic dwarfs was doubled, s would be equal to $\frac{3}{5}$, giving

$$\hat{q} = \frac{4 \times 10^{-5}}{0.6} = 6.6 \times 10^{-5}.$$

The proportion of carriers of the gene would thus be 32% higher than before, but the genetic load would remain the same, since the handicap of the carriers is less.

b) Heterotic equilibrium. When the equilibrium is maintained by selective advantage of the heterozygotes, the results of Section 3.3 show that:

$$L_h = p^2 s + q^2 t.$$

At equilibrium, the allele frequencies are:

$$\hat{p} = \frac{t}{s+t}, \qquad \hat{q} = \frac{s}{s+t}.$$

Hence:

$$\hat{L}_h = \frac{s\,t}{s+t}. \tag{13}$$

For example, the genetic load due to sickle-cell anaemia in a population which is exposed to malaria can be calculated, assuming the values:

$$s = 25\%, \qquad t = 90\%.$$

We obtain:

$$\hat{L}_h = \frac{0.225}{1.15} = 0.19.$$

Since we anticipate that selection coefficients for heterotic genes would in general be much higher than mutation rates, we expect the genetic load contributed by a heterotic locus to be greater than the load from a locus where variability is maintained by recurrent mutation to harmful alleles. The example of sickle-cell anaemia is an extreme case, because the selection coefficients are very high. However, we know that, in order for a stable equilibrium to be maintained, selection must be strong enough to overcome the effects of drift; we saw in Section 1.4c) of this chapter that, in order for this to be possible, the selection coefficients must be of the order of magnitude of the reciprocal of the effective population size. For example, if $s = t = 1\%$, we obtain $\hat{L}_h = 5 \times 10^{-3}$, which is a hundred times higher than the kind of value we would get for a locus in mutational equilibrium.

3.5. The Total Genetic Load

The concept of the genetic load due to a single locus appears to be a simple one. There has, however, been some confusion in this area. In particular, it must be realised that L_m and L_h are not strictly comparable with one another. The reference selective value for calculating L_m is that of the favoured homozygote, while for L_h it is the selective value of the heterozygote. Furthermore, in the case of a heterotic gene it is highly artificial to take the mean selective value of a population consisting entirely of heterozygotes, as a point of reference, since such a population is clearly not an equilibrium one.

Finally, it is clear that what matters for a population is the genetic load due to all the loci that are segregating, and not the load at any particular locus.

The total genetic load of a population can be defined in a similar way to the genetic load due to a single locus, as the difference between the selective value of the optimum genotype and the mean selective value of the population, divided by the selective value of the optimum genotype. When we are considering mutational load, we can take the optimum genotype to be the homozygote for the wild-type allele, at all the loci which are capable of mutating to deleterious alleles; for calculating the heterotic load, the optimum genotype is the genotype which is heterozygous at all the heterotic loci.

If L_i is the genetic load at the i-th locus, and if we assume that the effects of different loci on survival and fertility are multiplicative with one another, we can write the following expression for the total genetic load, L:

$$1 - L = (1 - L_1) \times (1 - L_2) \times \cdots \times (1 - L_i) \times \cdots.$$

If all the L_i are small, we have:

$$L \approx 1 - e^{\sum_i L_i}$$

and if $\sum_i L_i$ is small, this is further approximated by:

$$L \approx \sum_i L_i. \tag{14}$$

The total genetic load can be approximately analysed into two components, L_m due to loci subject to recurrent mutation to deleterious alleles, and L_h due to loci in heterotic equilibrium.

For the first type of loci, we have already seen that L_i is equal to $k_i u_i$, where k_i is a number between 1 and 2, depending on the degree of dominance of the disfavoured allele, and u_i is the rate of mutation to this allele. Summing over all loci, we obtain:

$$L_m = 1 - e^{-\bar{k} n \bar{u}} \tag{15}$$

where $\bar{k}$ is a number between 1 and 2, n is the number of loci maintained in equilibrium by the balance between mutation and selection, and $\bar{u}$ is the average mutation rate.

If we adopt values of $\bar{k} = 1.5$ and $\bar{u} = 2 \times 10^{-5}$, the number of loci that could be maintained in a population in this way, with a genetic load of 30%, is such that:

$$n = \frac{\log_e 0.7}{1.5 \times 2 \times 10^{-5}} = 10\,800.$$

This means that the occurrence of deleterious mutations at $11\,000$ loci would, with the values of the parameters which we have assumed, cause a decline in the population's growth rate of about one third, relative to the rate of growth of a population in which these deleterious mutations did not occur.

For heterotic alleles, using the results for single loci, we have:

$$L_h = 1 - e^{\sum_i \left(\frac{s_i t_i}{s_i + t_i}\right)}. \tag{16}$$

If we assume that $s_i = t_i = 1\%$, so that $L_i = 0.5\%$, the number of loci that could be maintained with a genetic load of 30% is:

$$n = \frac{\log_e 0.7}{0.5 \times 10^{-2}} = 72.$$

This is a small number compared with the large number of polymorphic loci which must exist in man. As mentioned earlier, it has been shown that about 30% of randomly-chosen loci controlling the structure of enzymes are polymorphic. If we assume that this figure is typical of the genome in general, and take $100\,000$ as a conservative estimate of the total number of loci in man, we would have about $30\,000$ polymorphic loci. If there is a heterotic load of 0.1% at each of these loci, then the total resulting load would be:

$$1 - e^{-3 \times 10^4 \times 10^{-3}} = 1 - e^{-30} \approx 1 - 10^{-13}.$$

Even such a moderate selection intensity as we have assumed (a load of 0.1% at each locus) would thus reduce the mean selective value of the population to nearly zero, relative to the selective value of completely heterozygous individuals, if every polymorphic locus were maintained by this form of selection. This implies either that the population's rate of growth would be reduced to such a low level that extinction is inevitable, or else that the multiple heterozygote must have an enormously high fitness, which is highly improbable, on physiological grounds.

There are several possible solutions to the paradox of the observation of a high frequency of polymorphic loci, and the generation of an unacceptably high genetic load if they are maintained by heterosis. First of all, we have assumed that the actions of different loci on an individual's overall selective value are independent. It seems probable that any change in an individual's genetic constitution will have effects on all kinds of characters, and there is much evidence that genes affecting the same character do not act independently, but interact, often in complex ways. Models which take interaction into account have been studied theoretically. Sved, Reed and Bodmer (1967) suggested that the advantage of heterozygosity might fall off as the number of loci for which an individual is heterozygous increases – a sort of biological law of diminishing returns. This has the effect of greatly reducing the total variation in fitness in the population, while leaving the selection coefficients at individual loci virtually unaffected. King (1967) and Milkman (1967) have proposed a model in which there is a "threshold" level of heterozygosity: above this threshold, individuals survive and reproduce, whereas below it they fail to contribute to the next generation. With this model, over a thousand loci could be maintained in heterotic equilibrium with selection coefficients of 1 %, without imposing too great a genetic load on the population.

Secondly, polymorphisms could be maintained without heterosis, by frequency-dependent selection. This possibility has been emphasized by Kojima (1971). Our calculations of genetic load will not apply to this case. Kimura and Ohta (1971) discuss the problem of defining genetic load with frequency-dependent selection.

Finally, the maintenance of variability by mutation to neutral alleles creates no genetic load at all: provided that $N_e u$ is of the right order of magnitude, any degree of polymorphism can be explained, without any problem of too great genetic load.

It is one of the major concerns of population genetics to discriminate experimentally between these different explanations of the very high degree of variability observed in populations. A first step towards this goal is to try to estimate the different types of genetic loads in populations. Morton, Crow and Muller (1956) proposed a method for doing this, using data on the effects of inbreeding on selective value. Unfortunately, as we shall see, this method is subject to several practical and theoretical difficulties, and has not yielded a conclusive answer.

3.6. The Effect of Inbreeding on Selective Value

All our arguments up to now have assumed that the population is in Hardy-Weinberg equilibrium. If we isolate a group of consanguineous matings (e.g. a set of first-cousin marriages) in such a population, the genotypic proportions among the offspring will be:

$$p^2 + \Phi p q, \quad 2 p q (1 - \Phi), \quad q^2 + \Phi p q$$

where Φ is the coefficient of kinship of the couples (1/16 for first-cousin marriages).

Consider first loci where variation is maintained by mutation to deleterious alleles. What effect will such loci have on the selective value of the offspring of such consanguineous matings? For a single such locus, their mean selective value, $\overline{w}(\Phi)$, relative to the fitness of mutation-free individuals, w_{max}, is:

$$\frac{\bar{w}(\Phi)}{w_{max}} = 1 - 2(1-\Phi)\,\hat{p}\,\hat{q}\,h\,s - (\hat{q}^2 + \Phi\hat{p}\hat{q})\,s$$

$$= 1 - \hat{L}_m\left[1 + \frac{(1-2h)}{2\hat{p}h+\hat{q}}\,\hat{p}\,\Phi\right]$$

where $\hat{L}_m$ is the load in the whole population. Assuming independent gene action for all such loci, we get:

$$\frac{\bar{w}(\Phi)}{w_{max}} \approx e^{-\sum_i \hat{L}_{im}\left[1 + \frac{(1-2h_i)}{2\hat{p}_i h_i + \hat{q}_i}\hat{p}_i\Phi\right]}.$$

For a heterotic locus, we have:

$$\frac{\bar{w}(\Phi)}{w_{max}} = 1 - (\hat{p}^2 + \Phi\hat{p}\hat{q})\,s - (\hat{q}^2 + \Phi\hat{p}\hat{q})\,t$$

$$= 1 - \hat{L}_h(1+\Phi).$$

For all heterotic loci, therefore:

$$\frac{\bar{w}(\Phi)}{w_{max}} \approx e^{-\sum_i \hat{L}_{ih}(1+\Phi)}.$$

We can define σ, the sensitivity of fitness to inbreeding, as:

$$\sigma = \frac{\log_e \hat{\bar{w}} - \log_e \bar{w}(\Phi)}{\Phi \log_e \hat{\bar{w}}}$$

where $\hat{\bar{w}}$ is the mean selective value of non-inbred individuals.

Taking both types of loci into account, this becomes (when $w_{max} = i$):

$$\sigma = \frac{\sum_i L_{im}\dfrac{(1-2h_i)\,\hat{p}_i}{2\hat{p}_i h_i + \hat{q}_i} + \sum_i L_{ih}}{\sum_i L_{im} + \sum_i L_{ih}}.$$

The factor $\dfrac{(1-2h_i)\,\hat{p}_i}{2\hat{p}_i h_i + \hat{q}_i}$ in the term for mutational load, in the numerator, is generally much greater than 1. For example, if $u_i = 10^{-5}$, $s_i = 0.1$, $h_i = 0$, it is equal to 100; if $u_i = 10^{-5}$, $s_i = 0.01$ and $h_i = 1/40$, its value is 11. If there were no load due to heterotic loci, σ would therefore be much greater than 1. If all the load were due to heterotic loci, each with two alleles, the value of σ would be 1. This suggests that measurement of σ should tell us whether mutation to deleterious genes or heterosis contributes most to the genetic load. Taking the numerical example used earlier, if there are 30000 heterotic loci, each with a genetic load of 0.1%, $\sum_i L_{ih} = 30$. If $\sum_i L_{im} = 1$, and the multiplying

Table 13.1. The sensitivity of perinatal mortality to inbreeding

Region	Perinatal mortality rate (per 1000)			$\sigma = \dfrac{m_c - m}{\Phi m}$	
	Control couples (m)	Consanguineous couples (m_c)			
		First-cousins	Second-cousins	First-cousins	Second-cousins
Loir-et-Cher	29	55	40	14.3	24.3
Morbihan	37	101	65	27.5	48.4
Vosges	90	111	—	3.7	—
Vosges	79	—	80	—	0.8
Hiroshima	41	52	66	4.3	39.0
Nagasaki	42	45	54	1.1	18.3

factor for mutational load is 20, we obtain:

$$\sigma = \frac{20 + 30}{31} \approx 1.7.$$

The results of Section 4 of Chapter 9 are suitable for this type of analysis, if we assume that perinatal survival is a measure of selective value. The data, and the values of σ for the offspring of first- and second-cousin marriages are shown in Table 13.1.

The results of the Loir-et-Cher and Morbihan studies agree reasonably well. They show that selective value is highly sensitive to inbreeding, which would lead to the conclusion that the great majority of loci are maintained in a state of equilibrium by mutation. However, the Japanese results differ considerably from these French data, and from one another, while the study in the Vosges indicates that selective value is not at all sensitive to inbreeding.

Apart from these practical difficulties, this technique is open to several theoretical objections. For example, if more than two alleles are maintained at a locus by heterotic selection, its contribution to σ will be higher than with just two alleles, so that high values of σ up to about 4, could be due to multiple allelism. Such problems are further discussed by Schull and Neel (1965).

3.7. Conclusion: "Neo-Darwinian" Versus "Non-Darwinian" Evolution

Lamarck explained evolution by means of the inheritance of acquired characters. This explanation initially seemed a natural one, and appeared to be in agreement with common sense. Darwin's theory proposed that evolution occurs by means of "spontaneous variation" together with the "struggle for existence", in other words, by natural selection. The "neo-

Darwinian" theory of evolution is the result of a re-examination of Darwin's theory, in the light of discoveries about the physical basis of inheritance. This theory has for some time been accepted as a satisfactory explanation of the evolutionary process. In this modern theory, the counterpart of Darwin's "spontaneous variation" is mutation: with a low frequency, random changes in genes can occur, altering their phenotypic effects. The carriers of a mutant gene in single dose, if it is dominant, or double dose if it is a recessive mutation, will show some new character. If the new character is advantageous, for example if it confers some degree of increased resistance to disease, or increases fecundity, the mutant individuals will tend to have more descendants than the average member of the population, so that the mutant gene will spread. If, however, the new character is harmful, the mutant gene will be eliminated from the population. The mean selective value of the population will therefore increase: in other words, adaptation to the environment takes place. Thus, the discovery of the mechanism of inheritance has led to fuller development and confirmation of Darwin's original theory.

This apparently satisfactory theory is nevertheless incapable of explaining all known observations. As we have seen in this chapter, and in Section 4 of Chapter 11, it leads to the paradox that the observed rates of molecular evolution and the amount of variability in protein structure in natural populations are apparently too great to be accounted for on the theory of natural selection. The theory of evolution by the random fixation of selectively neutral genes enables us to resolve these paradoxes, but is contrary to the "neo-Darwinian" theory; King and Jukes (1969) have even used the term "non-Darwinian evolution" for this process. According to this theory, the central factor in evolution is not selection bringing populations into ever greater adaptedness towards their environments, but *chance*, random changes with no aim or directedness, but constantly producing new diversity.

In fact, both processes must be acting in nature, and it is one of the chief aims of population genetics to determine their relative importance.

Further Reading

Cavalli-Sforza, L. L., Zei, G.: Experiments with an artificial population. Proc. III Int. Congress Hum. Genet., p. 473–478. Baltimore: Johns Hopkins Press 1966.

Crow, J. F., Maruyama, T.: The number of neutral alleles maintained in a finite geographically structured population. Theoret. Pop. Biol. **2**, 437–453 (1971).

Li, C. C.: The way the load ratio works. Amer. J. Hum. Genet. **15**, 316–321 (1963).

Morton, N. E.: The mutational load due to detrimental genes in man. Amer. J. Hum. Genet. **12**, 348–364 (1960).

Sanghvi, L. D.: The concept of genetic load: a critique. Amer. J. Hum. Genet. **15**, 298–309 (1963).

PART 4

The Study of Human Population Structure

In the earlier sections of this book, we have studied models of the genetic structures of populations, and we have seen what conditions will lead to changes in these structures, but we have so far paid very little attention to the study of data from real populations. We would like to be able to trace the long-term effects on human populations of environmental factors, and to understand how the complex and variable behaviour of the populations' members will affect their genetic make-up.

It is difficult to obtain answers to these questions. One approach which we can use is to study the "genetic distances" between different populations, and between members of the same population. This should tell us what changes have taken place in the genetic composition of the populations in question. In the next three chapters, we shall discuss this type of study, laying particular emphasis on the basic concepts which are involved, and on the difficulties encountered.

Chapter 14

Genetic Distance.
I. Basic Concepts and Methods

1. The Idea of Distance

When we wish to characterise an object, we usually do this by means of a list of properties, which may be qualitative or quantitative. For example, we can characterise individuals by their intellectual ability, blood groups or anthropometric measurements; a population can be characterised by the number of individuals of which it consists, by the gene frequencies at certain loci, or by the frequencies of certain phenotypes among the population's members.

Whenever we characterise an individual object or a set of objects in this way, we can express the classification as a set of numbers; each quantitative characteristic we observe will furnish a measure, and the qualitative characters can be transformed into numbers by an appropriate convention (e.g. $0 \equiv$ absence, $1 \equiv$ presence). This set of numbers can be considered as a vector x_i characterising the i-th object:

$$x_i \equiv \begin{bmatrix} x_{i1} \\ x_{i2} \\ \vdots \\ x_{ip} \end{bmatrix}.$$

A comparison of two objects of the same sort, or of two populations, can be carried out by calculating the differences between the numbers characterising them, in other words by finding the value of Δ_{ij}, whose elements are equal to the difference between the elements of vectors characterising the i-th and j-th objects:

$$\Delta_{ij} = x_i - x_j = \begin{bmatrix} x_{i1} - x_{j1} \\ \vdots \quad \vdots \\ x_{ip} - x_{jp} \end{bmatrix}.$$

If we are dealing with a classification involving many different characteristics, it is difficult to assimilate the result of such a calculation, and to answer the question "is object i closer to object j or to object k?". In practice, with vectors consisting of more than four or five elements, it is impossible to form a global assessment of the set of differences which make up the Δ_{ij} vector, and it is therefore necessary to try to find a method of condensing the information on the differences between two objects into a single number, the "distance" between them.

The apparently simple word "distance", however, must be used with care. It suggests to us an idea which is part of ordinary, everyday life, namely "geographic distance", measured in Euclidean space, and in which Pythagoras' theorem holds (i.e. the distance d_{ij} between two points, i and j, is given by:

$$d_{ij} = (x_i - x_j)^2 + (y_i - y_j)^2 + (z_i - z_j)^2 \tag{1}$$

where x_i, y_i and z_i are the co-ordinates of a point in a system of normal, orthogonal axes in three dimensions). If we consider the set of the three co-ordinates of a point i as a column vector c_i, this relation can be expressed in the form:

$$d_{ij} = (c_i - c_j)'(c_i - c_j) \tag{2}$$

where c_i' is the transpose of the vector c_i (i.e. a row vector consisting of the same elements as the column vector c_i).

Clearly, a distance between two objects i and j which are each characterised by p numbers could be defined by a formula analogous to (1), but extending the sum over all the p numbers; this is called "distance in p-dimensional space". The difficulties in this are, however, immediately obvious: we have no justification for the procedure of equating the numbers in the vector characterising an object with co-ordinates in orthonormal space. Some examples will illustrate the problems:

1. If the classification is in terms of anthropometric measures, there are often correlations between the different measures, so that knowledge of one constitutes information with regard to the others.

2. If the classification deals with the frequencies of the alleles at a locus, these are related by the obvious relation:

$$p_1 + p_2 + \cdots + p_p = 1.$$

3. If the classification is based on qualitative differences, the numbers are assigned quite arbitrarily.

It is obvious, finally, that any method by which n numbers are condensed into a single index will be to some extent arbitrary. A "good" measure of distance will simply be one which gives results which agree with our intuitive views about the objects in question, or else which has suitable properties in relation to the use which we wish to make of the measure of "closeness" and "distance".

First of all, we must recall the definition of distance.

1.1. The Definition of Distance

Consider a set E; a distance defined on the set E is an operation on $E \times E$ and on the set of real, positive numbers: a real, positive number d_{ij} is associated with each pair of elements i and j of E, in such a way that:

$$
\begin{aligned}
d_{ij} &> 0 \quad \text{if } i \neq j \\
d_{ii} &= 0 \\
d_{ij} &= d_{ji} \\
d_{ij} &\leq d_{ik} + d_{kj}.
\end{aligned}
\tag{3}
$$

The first three of these properties express our intuitive view that the distance between two objects is a positive number, which is independent of the direction (whether i to j or j to i), and which is zero when the two objects are in the same place. The last property corresponds to the result in two dimensions that one side of a triangle is always shorter than the sum of the lengths of the two other sides. This property ensures the desirable property for a measure of distance, that two elements which are both close to a third element will also not be far away from one another.

Clearly, we could construct as many measures of distance as we like, given a set of elements; each measure would have its own properties, which would be better or worse suited to the problem under study, and would correspond to a greater or lesser degree to the aim of the study. In what follows, we shall try to find a measure of distance which is appropriate for studying distances between individuals or populations, defined on the basis of information about their genetic make-up.

The special features of hereditary transmission from one generation to the next has made it essential to develop specific measures of distance for these situations. First, however, it will be useful to recall some measures of distance which are widely used; we shall see how to proceed when the objects are defined by sets of measurements, and when they are defined by sets of qualitative characters.

1.2. Distance between Objects Characterised by Measurements

The "classical", Euclidean measure of distance is given by the formula:

$$
\begin{aligned}
d_{ij}^2 &= \sum_{r=1}^{p} (x_{ir} - x_{jr})^2 \\
&= (x_i - x_j)' (x_i - x_j).
\end{aligned}
$$

The different measures $X_1, ..., X_r, ..., X_p$ may, however, be non-independent. For example, when we are dealing with anthropometric measures, these are often correlated to some extent: the height of an individual is correlated with his arm-length, though this correlation is not absolute. The relations between the different measures can be expressed in general in the form of a matrix (S) of variances and covariances, defined by:

$$S \equiv E[(x_i - \bar{x})(x_i - \bar{x})']$$

where E is the expected value operator, and $\bar{x}$ is the vector of the means of the x_{ir} values for the characters.

In this type of situation, if we use a measure of distance which takes independent account of each of the different measures, e.g. height and arm-length, we would to some extent be using the same information twice.

Furthermore, if the variability of the characters measured differs between characters, the same absolute difference in the measures of two individuals for two characters may have quite different significance for the "distance" between them. For example, a difference of 1 cm in the length of a finger corresponds to a much greater distance than a difference of 1 cm in height, since height varies by much greater amounts, in absolute terms, than finger-length.

1.2.1. Mahalanobis' distance. We wish to define distances so that the space thus defined obeys Pythagoras' theorem, i.e. so that the distance between the points corresponding to the vectors of measurements c_1 and c_2 agrees with Eq. (2). To do this, new measures, $Y_1, ..., Y_p$ are usually chosen, such that:

1. They are linear functions of the original measures $X_1, ..., X_p$.
2. They are uncorrelated with one another.
3. They are standardised, so that their variances are equal to 1.

In other words, we wish to find a matrix A, such that the measures Y, defined by $y = Ax$ have a variance-covariance matrix which is a unit matrix:

$$E[(y_i - \bar{y})(y_i - \bar{y})'] = I. \tag{4}$$

It can be shown that, if the variance-covariance matrix of x is S, that of $y = Ax$ is ASA'; property (4) can therefore be written:

$$ASA' = I$$

or, multiplying on the left by A^{-1}, and on the right by $(A')^{-1}$:

$$S = A^{-1}(A')^{-1} = (A'A)^{-1} \quad \text{or} \quad A'A = S^{-1}. \tag{5}$$

Because of the way in which the measures Y are calculated, they define a space in which, as in geographical space defined by orthogonal axes, the distance between two points can be found using Pythagoras' theorem, i.e. by the relation:

$$D^2 = (y_1 - y_2)' (y_1 - y_2)$$

or

$$D^2 = (x_1 - x_2)' A' A (x_1 - x_2) \tag{6}$$

or, from Eq. (5)

$$D^2 = (x_1 - x_2)' S^{-1} (x_1 - x_2).$$

This measure of distance is called "Mahalanobis' distance" (Mahalanobis, 1930, 1936).

When the measures X_i are uncorrelated, their variance-covariance matrix S is a diagonal matrix of the form:

$$S = \begin{bmatrix} V_1 & \dots & 0 & \dots & 0 \\ \vdots & & \vdots & & \vdots \\ 0 & \dots & V_r & \dots & 0 \\ \vdots & & \vdots & & \vdots \\ 0 & \dots & 0 & \dots & V_p \end{bmatrix}.$$

In this case, it is obvious that:

$$S^{-1} = \begin{bmatrix} 1/V_1 & \dots & 0 & \dots & 0 \\ \vdots & & \vdots & & \vdots \\ 0 & \dots & 1/V_r & \dots & 0 \\ \vdots & & \vdots & & \vdots \\ 0 & \dots & 0 & \dots & 1/V_p \end{bmatrix}$$

and Mahalanobis' distance reduces to:

$$D^2 = \sum_r \frac{(x_{1r} - x_{2r})^2}{V_r}. \tag{7}$$

In this special case, but only in this case, the distance D^2 is equivalent to the measure proposed by Pearson in 1921 ("coefficient of racial likeness") and by Penrose in 1954 (C_H^2); these measures are often used, as the calculations involved are easy to do, but the use of them is often unjustifiable, because it is frequently invalid to assume non-correlated values of the different characters.

1.2.2. The case when the measures in question are frequencies. A case which is often encountered is when the measures used to characterise the objects are frequencies, or proportions, or probabilities. This is the case, for example, when the objects in question are populations characterised by allele frequencies at a locus.

In general, let us consider a set of populations P_i, each of which is characterised by p positive or zero numbers p_{ir} such that:

$$\sum_{r=1}^{p} p_{ir} = 1.\tag{8}$$

The arc and chord measures of Cavalli-Sforza and Edwards (1967). The set of vectors $p_i \equiv (p_{i1}, \ldots, p_{ip})$ correspond to points in p-dimensional space, in a space bounded by the conditions:

$$\left.\begin{array}{l} p_{ir} \geq 0 \quad \text{for all } r \\ \sum_r p_{ir} = 1 \end{array}\right\} \text{ for all } i.$$

In the case when $p = 2$, the space is a segment of the straight line AB situated in the first quadrant (Fig. 14.1).

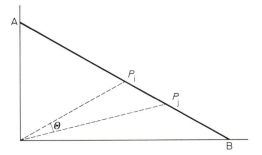

Fig. 14.1. The angle between two vectors of gene frequencies

One possible measure of the difference between the vectors p_i and p_j that suggests itself is the angle Θ between them; this is classically defined by the relation:

$$\cos \Theta = \frac{\displaystyle\sum_r p_{ir} p_{jr}}{\left(\sum_r p_{ir}^2 \sum_r p_{jr}^2\right)^{\frac{1}{2}}}.\tag{9}$$

This measure, however, has the disadvantage that it changes at a different rate depending on whether the points p_i and p_j are close to one of the extremities of the line AB; a given linear distance between p_i and p_j, along the line AB, corresponds to a smaller angle Θ when p_i and p_j are close to A or B than when they are close to the middle of AB.

In order to overcome this difficulty, Cavalli-Sforza and Edwards proposed the following change of variable: the frequencies p_{ir} are replaced by their square roots, $\pi_{ir} = \sqrt{p_{ir}}$, so that Eq. (8) becomes:

$$\sum_r \pi_{ir}^2 = 1, \quad \text{for all } i;$$

the points defined by the co-ordinates π_{ir} will lie on a segment of the hypersphere in p-dimensional space, with radius 1, in the zone in which all the co-ordinates are positive (see Fig. 14.2).

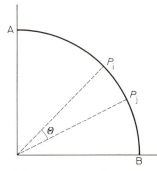

Fig. 14.2. The angle between vectors of square roots of gene frequencies

If we use the angle between the two vectors π_i and π_j to measure the distance between them, the problem of variable scale does not arise. Also, this angle is given by the appropriate modification of Eq. (9):

$$\cos \Theta = \sum_r \pi_{ir} \pi_{jr} = \sum_r (p_{ir} p_{jr})^{\frac{1}{2}}.$$

The angle Θ varies between 0 and $\pi/2$. Cavalli-Sforza and Edwards considered it preferable to work with a parameter varying between 0 and 1; they therefore adopted the following measure of distance:

$$\operatorname{arc}(i, j) = \frac{2}{\pi} \operatorname{arc} \cos \sum_r (p_{ir} p_{jr})^{\frac{1}{2}}. \tag{10}$$

In order to facilitate the study of certain types of problem, the same authors also proposed, as a measure of distance, the length of the chord subtending the arc $p_i p_j$. It is easy to derive the relation:

$$\operatorname{chord} p_i p_j = [\sin^2 \Theta + (1 - \cos \Theta)^2]^{\frac{1}{2}}$$
$$= [2(1 - \cos \Theta)]^{\frac{1}{2}}.$$

To obtain a parameter with values between 0 and 1, they again multiply by $2/\pi$, and finally define the measure of distance:

$$\operatorname{chord}(i, j) = \frac{2\sqrt{2}}{\pi} \sqrt{1 - \sum_r \sqrt{p_{ir} p_{jr}}} \tag{11}$$

Kidd and Sgaramella-Zonta (1971) note that the factor of $2/\pi$ is not particularly appropriate here, since the chord measure of distance takes values between 0 and $\dfrac{2\sqrt{2}}{\pi} = 0.9003$.

The χ^2 measure of distance. Another method of measuring the distance between populations characterised by a set of p frequencies or probabilities p_{ir} is based on the two following considerations about the closeness or distance of two objects:

1. We should give greater weight to a difference $(p_{ir} - p_{jr})$ between the frequencies of the r-th character in two populations, if the character in question is rare, in other words if its mean frequency $\bar{p}_r = \dfrac{\sum\limits_i n_i\, p_{ir}}{\sum\limits_i n_i}$ is small.

2. If two characters, C_r and C_s have frequencies that are in the same proportion to one another in all the populations, in other words if, for all i, p_{ir}/p_{is} is a constant a, we can replace the two characters r and s by a single character m, such that:

$$p_{im} = p_{ir} + p_{is}.$$

This second condition corresponds to the fact that in this case the information provided by character s is entirely contained in that provided by character r; it should therefore be possible to condense them into a single character without changing the distances between the populations.

It is easy to show that classical Euclidean distance defined by:

$$d_{ij}^2 = \sum_r (p_{ir} - p_{js})^2$$

does not satisfy this condition. A "χ^2 distance", however, defined by the relation:

$$\chi_{ij}^2 = \sum_r \frac{(p_{ir} - p_{jr})^2}{\bar{p}_r} \tag{12}$$

satisfies condition (2), and has the additional advantage that the observations are weighted according to consideration (1), above. We can see this as follows:

If

$$p_{ir} = a\, p_{is}$$

then

$$\bar{p}_r = a\, \bar{p}_s$$

and

$$p_{im} = p_{ir} + p_{is} = (1 + a)\, p_{is}.$$

Hence

$$\frac{(p_{ir}-p_{jr})^2}{\bar{p}_r}+\frac{(p_{is}-p_{js})^2}{\bar{p}_s}=\frac{(p_{is}-p_{js})^2\,(1+a)}{\bar{p}_s}$$

$$=\frac{[(1+a)\,p_{is}-(1+a)\,p_{js}]^2}{(1+a)\,\bar{p}_s}$$

$$=\frac{(p_{im}-p_{jm})^2}{\bar{p}_m}.$$

In order to see that χ^2 is a "distance", it is sufficient to note that the change of variable:

$$q_{ir}=\frac{p_{ir}}{\sqrt{\bar{p}_r}}$$

leads to:

$$\chi^2_{ij}=\sum_r (q_{ir}-q_{jr})^2.$$

χ^2 is thus a classical Euclidean distance in the space defined by this new variable, q.

A generalisation. The frequencies of several series of mutually exclusive characters (for example allele frequencies at several loci) can be considered as measures characterising populations. Assuming that the different characters are independent, we can define a global distance in the space of these m measures by a simple application of Pythagoras' theorem.

Cavalli-Sforza and Edwards (1967) propose a distance G_{ij} for such a case, calculated by the formula:

$$G^2_{ij}=\sum_s [\mathrm{chord}_s(i,j)^2]$$

where the summation is over the set of all the sets of frequencies, and $\mathrm{chord}_s(i,j)$ is the distance between i and j calculated according to Eq. (11) for the series of frequencies s.

Similarly, we can generalise the χ^2 measure of distance to the case of a set of frequencies, by using the formula:

$$\chi^2=\sum_s \chi^2_s$$

where the summation is over all the sets.

1.3. Distance between Objects Characterised by Qualitative Attributes

Now let us consider objects characterised in qualitative, rather than quantitative terms. For example, we might be concerned with individuals characterised by their phenotypes at a given locus, or by the presence of certain genes in their genome.

We could measure distances in such cases by assigning arbitrary numbers for the presence or absence of the characters in question (e. g. 1 for presence and 0 for absence), and we might then use the methods described above. This, however, is clearly an artificial procedure, since the true qualitative nature of the data is made to appear to be quantitative, by a purely formal procedure. It would be better to develop a method specifically suitable for qualitative data.

1.3.1. Similarity indices. Suppose that each object is classified according to whether each of N characters is present or absent. Then to compare two objects i and j, we can classify the characters into four classes:

I: present in both i and j. Let the number of these be n_1.
II: present in i but not j. Let their number be n_2.
III: present in j but not i. Let their number be n_3.
IV: present in neither i nor j. Let their number be n_4.

This can be represented as in Table 14.1.

Table 14.1. Comparison of two objects according to N qualitative criteria

Object 1 / Object 2	Present	Absent
Present	I (n_1)	III (n_3)
Absent	II (n_2)	IV (n_4)

Two objects are similar to one another when the characters in class I and class IV form a large proportion of the characters studied. Many different authors have proposed "indices of similarity" calculated from the four numbers $n_1, \ldots, n_4$. Practically every conceivable combination of these numbers that could reasonably be used as a measure of resemblance has been proposed. We may mention the following:

$$i = \frac{n_1}{n_1 + n_2 + n_3} \qquad \text{Jaccard (1908)}$$

$$i = \frac{n_1}{n_2 + n_3} \quad \text{or} \quad i = \frac{n_1}{2}\left(\frac{1}{n_1 + n_2} + \frac{1}{n_1 + n_3}\right) \qquad \text{Kulzynski (1927)}$$

$$i = \frac{2n_1}{2n_1 + n_2 + n_3} \qquad \text{Dice (1945)}$$

$$i = \frac{n_1}{n_1 + 2(n_2 + n_3)} \qquad \text{Sokal and Sneath (1963)}$$

This profusion of indices bears witness to the difficulty of finding an objective measure of resemblance. It is obvious that we would conclude sometimes that object i was more similar to object k than to object l, and sometimes the opposite, depending on which index we used.

Furthermore, all these indices suffer from a difficulty, which may be serious in practice: the only comparisons between objects that they permit are between objects all of which are defined by the same series of characters; there would be no meaning to an attempt to compare the distances of three inviduals, one characterised by the genes for the AB0 and Gm blood systems, another by the genes of the Gm and HL-A systems while a third is characterised by the genes of the HL-A and AB0 systems.

In practice, however, one often has incomplete data, in which certain characters were not studied in certain of the objects. In such a situation, one is either forced to eliminate from the study the objects for which the data is incomplete, or else to eliminate the characters for which some individuals were not studied, or both of these things simultaneously. This means losing some of the information which has been collected, often at great expense.

It therefore seems necessary to try to develop a measure of the difference between objects which allows us to use all the available information. The concept of the "probability of the symmetric difference of two sets" allows us to do this.

1.3.2. The probability of the symmetric difference of two sets. The symmetric difference of two sets A and B is the set of elements which either belong to A but not to B, or else belong to B but not to A; in conventional set notation, this can be written:

$$A \triangle B = (A \cap \tilde{B}) \cup (\tilde{A} \cap B).$$

When the sets $A, B, C, \ldots$ in question are events, and their elements e_i are the results of a probabilistic trial, in other words when each e_i has a corresponding probability p_i, the probability $P(A)$ of the set A is defined

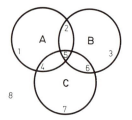

Fig. 14.3. The symmetric differences of sets

by $P(A) = \sum\limits_{e_i \in A} p_i$; we can show that, whatever the nature of the events A, B, C, we have:

$$P(A \vartriangle B) \leq P(A \vartriangle C) + P(C \vartriangle B). \tag{13}$$

The proof is as follows. Consider Fig. 14.3 above. The results e_i of a trial can fall into one of eight classes $E\,1$ to $E\,8$, according to which of sets A, B and C they belong to; thus, class 1 consists of the elements which belong to A, but not to B or C. We see that:

$$P(A \vartriangle B) = \sum_{e_i \in E\,1} p_i + \sum_{e_i \in E\,3} p_i + \sum_{e_i \in E\,4} p_i + \sum_{e_i \in E\,6} p_i$$

$$P(A \vartriangle C) = \sum_{e_i \in E\,1} p_i + \sum_{e_i \in E\,2} p_i + \sum_{e_i \in E\,6} p_i + \sum_{e_i \in E\,7} p_i$$

$$P(C \vartriangle B) = \sum_{e_i \in E\,2} p_i + \sum_{e_i \in E\,3} p_i + \sum_{e_i \in E\,4} p_i + \sum_{e_i \in E\,7} p_i$$

and the inequality (13) follows immediately.

It is also obvious that $P(A \vartriangle B) = 0$ if, and only if, $A = B$ and $P(A \vartriangle B) = P(B \vartriangle A)$; it therefore follows that the probability of the symmetric difference of two sets has all the fundamental properties of a measure of distance.

The introduction of the concept of probability also enables us to define a measure of distance which makes use of the whole of the available information.

To see this, consider the example of a set of individuals belonging to the same population, and suppose that their blood groups have been determined. However, for practical reasons, suppose that certain blood groups have not been determined for certain of the individuals, e.g. MN for individual 1. It is incorrect to say that we have no information about the MN type of this individual, since we know that he comes from a particular population, and the proportions of M, MN and N individuals in a sample of this population are known. Using this information, we can therefore calculate the probabilities that individual 1 has the phenotype M, MN or N; for example, these probabilities could be taken as equal to the phenotype frequencies in the group of individuals whose MN blood group phenotypes have been determined; one might also take into account individual 1's relationships with individuals of known MN blood groups.

Thus lack of direct knowledge is not the same as complete lack of knowledge; it is equivalent to a certain indeterminacy, or to a partial set of information, and can be expressed objectively as a probability.

However slight our knowledge of the system, we can always assign numbers to the possible outcomes so as to represent the information we possess.

From this point of view, we can regard our data as complete for all the objects, and all the characters in question; some of the data will be in the form of code numbers (1 for presence of the character, and 0 for its absence), and some will consist of probabilities. We can thus adopt the following definition of the distance between two objects:

In a set of objects characterised by a number of characters, the distance between two objects i and j is the probability that a character chosen at random will be present in i, but not in j, or vice versa.

If we give the same "weight" to each character, in other words if each has the same probability of being chosen at random, the distance d_{ij} is given by the formula:

$$d_{ij} = \frac{1}{p} \sum_{r=1}^{p} [p_{ri}(1-p_{rj}) + p_{rj}(1-p_{ri})] \tag{14}$$

where the summation is over all the p characters, and where p_{ri} is the probability that the i-th object has the r-th character.

If, however, we wish to attach weights w_r to the different characters, the definition of distance becomes:

$$d_{ij} = \frac{\sum\limits_{r=1}^{p} w_r [p_{ir}(1-p_{jr}) + p_{jr}(1-p_{ir})]}{\sum\limits_{r=1}^{p} w_r}. \tag{15}$$

The weights might, for example, be chosen to take account of the amount of information provided by each character; this is greater the smaller the character's variance, so that we would choose the definition:

$$w_r = \frac{1}{V_r} = \frac{1}{\bar{p}_r(1-\bar{p}_r)}.$$

In practice, the characters in question can be divided into nine groups, depending on whether they are present, absent or not studied in each of the two objects i and j, which are to be compared. The nine possible groups are shown in Table 14.2. This is a generalisation of the table we drew up earlier, for the case when all the characters were determined in each object. If we give all the characters studied the same weight, formula (15) becomes:

$$d_{ij} = \frac{1}{p} \left\{ n_2 + n_3 + \sum_{r \in V}(1-p_{rj}) + \sum_{r \in VI}(1-p_{ri}) + \sum_{r \in VII} p_{rj} \right.$$
$$\left. + \sum_{r \in VIII} p_{ri} + \sum_{r \in IX}[p_{ri}(1-p_{rj}) + p_{rj}(1-p_{ri})] \right\}. \tag{16}$$

Table 14.2. Comparison of two objects according to N qualitative criteria, not all determined for both objects

Object 1 Object 2	Present	Absent	Not determined
Present	I (n_1)	III (n_3)	VI (n_6)
Absent	II (n_2)	IV (n_4)	VIII (n_8)
Not determined	V (n_5)	VII (n_7)	IX (n_9)

Although this is a cumbersome equation, the calculation of this measure of distance is very easy, especially in cases when we can assume simple formulae for the probabilities p_{ir}. For example, we can often treat the objects for which we have not determined certain characters as randomly-chosen objects from the population, so that the objects whose characters have been studied are a representative sample from this population. The p_{ri} terms in Eqs. (15) and (16) are then given by the mean $\bar{p}_r$ of the character, based on the objects which have been studied for this character.

2. Distance between Individuals of Known Ancestry

We saw in Chapter 6 that we can measure the relationship between two individuals of known ancestry, by means of their coefficient of kinship Φ, or by their coefficients of identity δ_i or Δ_i. We shall now show how this leads to a measure of "distance" between individuals.

2.1. Inadequacy of the Coefficient of Kinship

The coefficient of kinship Φ is an index whose value decreases as the relationship between the individuals in question becomes more remote; for example $\Phi = \frac{1}{4}$ for sibs, $\frac{1}{16}$ for first cousins, $\frac{1}{64}$ for second cousins, etc. It would therefore seem natural to try to define distance d between individuals as a decreasing, monotonic function of Φ; d would thus increase as the relationship became more distant.

Unfortunately, this procedure will not work, because of a special property of Φ, which we have encountered before. The coefficient of kinship of an individual with himself is given by the formula:

$$\Phi_{AA} = \tfrac{1}{2}(1 + \Phi_{FM})$$

where Φ_{FM} is the coefficient of kinship of his father and mother. If we know nothing about the relationship between F and M, we have $\Phi_{FM}=0$, and $\Phi_{AA}=\frac{1}{2}$. But it is perfectly possible for two individuals to have a coefficient of kinship greater than $\frac{1}{2}$; for example a brother and sister whose parents, grandparents and greatgrandparents were each brother and sister would have a coefficient of kinship $\Phi_{AB}=\frac{19}{32}$.

It follows from these considerations that for any monotonically decreasing function of Φ such that $d_{AA}=0$, d_{AB} would be less than zero for $\Phi_{AB}>\frac{1}{2}$, and this is incompatible with the definition of a distance.

Although it is useful when we wish to predict the nature of the offspring of a mating, we thus find that the coefficient of kinship does not help us to define a distance between individuals of known ancestry. We are therefore forced to try to develop some other measure to deal with this problem, either by using the coefficients of identity δ_i or Δ_i, or else by introducing other concepts in addition to that of identity by descent.

2.2 Genotypic Distance between Relatives

2.2.1. Definition and method of calculation. At a given locus, two individuals are characterised by their genotypes. We can say that they are close to one another when there is a high probability that their genotypes are identical. We can thus attempt to base a measure of distance on the probability of the "symmetric difference" between them, i.e. on the probability that one of the two individuals has a given genotype while the other does not.

Consider individuals characterised by their genotypes at a given locus. We will assume that there are n alleles $A_1, \ldots, A_n$ at this locus, and that they are all co-dominant (so that the difference between the genotype and phenotype is eliminated); also, we will denote the allele frequencies in the population by $p_1, \ldots, p_n$, and will assume that the population is in Hardy-Weinberg equilibrium, so that the genotype frequencies are p_i^2 for the homozygotes $(A_i A_i)$, and $2p_i p_j$ for the heterozygotes $(A_i A_j)$. The relationship between two individuals, A and B, can be measured in terms of the nine coefficients of identity $\Delta_1, \ldots, \Delta_9$ (see Chapter 6).

Taking account of this information about the genes at this locus, and about the relationship between A and B, how can we measure the distance between them?

We shall adopt the following definition:

The genotypic distance $D(AB)$ between two individuals A and B, for a given locus, is the probability that A's genotype is not the same as B's.

We can see that this probability is a distance, as follows. Consider three individuals, A, B and C. Taking two of these individuals at a time, their genotypes may or may not be the same ($G_A = G_B$ or $G_A \neq G_B$). There are five possible cases:

$$I: \quad G_A = G_B = G_C$$
$$II: \quad G_A = G_B \neq G_C$$
$$III: \quad G_A = G_C \neq G_B$$
$$IV: \quad G_B = G_C \neq G_A$$
$$V: \quad G_A \neq G_B \neq G_C \neq G_A.$$

By definition, we have:

$$D(AB) = \text{Prob}(III) + \text{Prob}(IV) + \text{Prob}(V)$$
$$D(BC) = \text{Prob}(II) \ + \text{Prob}(III) + \text{Prob}(V)$$
$$D(CA) = \text{Prob}(II) \ + \text{Prob}(IV) + \text{Prob}(V)$$

so that the triangle inequality is immediately shown to be valid, i.e.:

$$D(AB) \leq D(AC) + D(CB).$$

In order to relate this definition of distance to the coefficients of identity Δ_i, which measure the relatedness of A and B, we must consider each of the nine modes of identity:

1. Mode $S'1$: both of A's genes are identical with both of B's genes. A and B therefore have the same genotype, so that $D(AB)|S'1 = 0$. Similarly in identity mode $S'7$, each of A's genes is identical with one of B's genes, and *vice versa*, so that A and B have the same genotype, and $D(AB)|S'7 = 0$.

2. With identity mode $S'2$, A's two genes are identical with one another, and so are B's two genes, but they are not identical with A's genes. Both A and B have a probability of p_i of being of the genotype $(A_i A_i)$. The probability that they are both of the same genotype is therefore $\sum_i p_i^2$, and the distance between them is $D(AB)|S'2 = 1 - \sum_i p_i^2$. The same result is reached for identity modes $S'3$, $S'5$ and $S'8$.

3. With identity mode $S'4$, A's two genes are identical, so that the probability that A's genotype is $(A_i A_i)$ is p_i. In order for B to have the same genotype, his two genes must each be allele A_i which has the probability p_i^2. The probability that A's and B's genotypes are the same is therefore $\sum_i p_i^3$, and the distance between them is $D(AB)|S'4 = 1 - \sum_i p_i^3$. The same result is found for identity mode $S'6$.

4. With identity mode $S'9$, none of the genes of A and B is identical with any of the others. In order for A and B to have the same genotype, one of the two following situations must occur:

a) The two paternal genes are the same allele, and so are the two maternal genes. This has the probability:

$$\sum_{ij} p_i^2 p_j^2 = \left(\sum_i p_i^2\right)^2.$$

b) The paternal gene of A is the same allele as the maternal gene of B, and A's maternal gene is the same allele as B's paternal gene. The probability that this is the case is $\left(\sum_i p_i^2\right)^2$.

We saw in Chapter 2 that

$$\text{Prob}(a \text{ or } b) = \text{Prob}(a) + \text{Prob}(b) - \text{Prob}(a \text{ and } b).$$

In the present case, the event $(a \text{ and } b)$ is the event: the four genes are all the same allele; the probability of this is $\sum_i p_i^4$. Finally, therefore, the probability that A and B have the same genotype is equal to:

$$2\left(\sum_i p_i^2\right)^2 - \sum_i p_i^4$$

and the distance between them is

$$D(AB)|S'9 = 1 + \sum_i p_i^4 - 2\left(\sum_i p_i^2\right)^2.$$

Gathering up the terms for each of the identity modes, and taking account of their probabilities Δ_i, we obtain the following overall probability that A and B have different genotypes, in other words, the "distance" between them:

$$D(AB) = (\Delta_2 + \Delta_3 + \Delta_5 + \Delta_8)\left(1 - \sum_i p_i^2\right) + (\Delta_4 + \Delta_6)\left(1 - \sum_i p_i^3\right)$$
$$+ \Delta_9\left[1 + \sum_i p_i^4 - 2\left(\sum_i p_i^2\right)^2\right]. \tag{17}$$

The genotypic distance between two individuals is thus a function not only of the coefficients of identity, but also depends on the frequencies of the alleles at the locus in question, which enter into the formula for the distance in the form of terms such as $\sum_i p_i^2$, $\sum_i p_i^4$. A given relationship can therefore correspond to different distances, depending on the locus and the population under study.

The special case of a locus with two alleles. When there are only two alleles at the locus being studied, and their frequencies are p and $q = 1 - p$, the sums $\sum_i p_i^2$ etc. can be expressed in terms of the product pq, as follows:

$$p^2 + q^2 = (p+q)^2 - 2pq = 1 - 2pq$$
$$p^3 + q^3 = (p+q)^3 - 3p^2q - 3pq^2 = 1 - 3pq, \quad \text{etc.}$$

Table 14.3. Genotypic distances between relatives for a locus with two alleles

Relationship	Distance
Parent-offspring	$2pq$
Half-sibs	$3pq(1-pq)$
Full sibs	$pq(2-\frac{3}{2}pq)$
Uncle-nephew	$3pq(1-pq)$
First-cousins	$\frac{pq}{2}(7-9pq)$
Double first-cousins	$pq(3-\frac{27}{8}pq)$
Second-cousins	$\frac{pq}{8}(31-45pq)$
Sibs whose parents are:	
Sibs	$\frac{pq}{8}(13-3pq)$
First-cousins	$\frac{pq}{32}(61-39pq)$

Eq. (17) then becomes:

$$D(AB)=pq[2(\Delta_2+\Delta_3+\Delta_5+\Delta_8)+3(\Delta_4+\Delta_6)+2(2-3pq)\Delta_9].$$

For the most frequently studied relationships, we thus obtain the distances given in Table 14.3.

2.2.2. The expectation of genotypic distance. If we do not know the frequencies p_i of the alleles at the locus under study, we clearly cannot calculate genotypic distances. We can, however, attempt to determine the expectation of this distance, by making assumptions about the distribution of the allele frequencies. Alternatively, we may wish to determine the "mean" distance corresponding to the frequencies at a large number of loci.

a) *Assuming equal frequencies of all alleles.* The simplest hypothesis, which allows us to solve these problems without difficulty, is that the frequencies of all the alleles are the same. If there are n alleles at the locus, we will then have:

$$p_i=\frac{1}{n}, \quad \sum_i p_i^2=\frac{1}{n}, \quad \sum_i p_i^3=\frac{1}{n^2}$$

which gives:

$$D(AB)=(\Delta_2+\Delta_3+\Delta_5+\Delta_8)\left(1-\frac{1}{n}\right)+(\Delta_4+\Delta_6)\left(1-\frac{1}{n^2}\right)$$

$$+\Delta_9\left(1-\frac{2}{n^2}+\frac{1}{n^3}\right). \tag{18}$$

When the number n is very large, we therefore have:

$$D(AB) \approx 1 - \Delta_1 - \Delta_7.$$

The assumption of equal allele frequencies is one that is often made, because of its simplicity. However, it is totally unrealistic. It would seem that, when we do not have any knowledge of the genic structure at the locus in question, we should try to formulate a set of assumptions which agree more closely with what we know about the frequencies of alleles at typical loci.

b) Bayesian models. When we know that a number is between the limits α and β, it is often reasonable to consider it as a random variable with uniform probability density in the interval $(\alpha - \beta)$; this is called Bayes' postulate.

In the present case, we have not one but n variables, the frequencies p_i of the various alleles A_i; these are related by:

$$\sum_{i=1}^{n} p_i = 1.$$

We shall here consider the application of two different "Bayesian models" to this type of problem.

(i) *First model.* We assume that the genic structure $s = (p_1, \ldots, p_n)$ at the locus in question follows a uniform distribution; in other words, we are assuming that the point s is a random variable in p-dimensional space, on the surface defined by:

$$\sum_i p_i = 1, \quad p_i \geq 0$$

and that the probability density of s on this domain is constant.

In this model, the probability densities for the different alleles are all the same. It is easy to see that if one frequency, p_k say, is known, the genic structure can be represented by a point in $n - 2$ dimensions. The density function $\Phi(p_k)$ is therefore proportional to $(1 - p_k)^{n-2}$, and the condition:

$$\int_0^1 \Phi(p_k)\, dp_k = 1$$

leads to:

$$\Phi(p_k) = (n-1)(1 - p_k)^{n-2}.$$

The moments of order 1–4 for each allele can be found as follows:

$$E(p_k) = (n-1) \int_0^1 p_k (1 - p_k)^{n-2}\, dp_k$$

$$= (n-1)\, B(2, n-1)$$

where $B(2, n-1)$ is the beta function with parameters 2 and $n-1$. Hence

$$E(p_k) = \frac{1}{n},$$

$$E(p_k^2) = (n-1) \int_0^1 p_k^2 (1-p_k)^{n-2} \, dp_k$$

$$= (n-1) B(3, n-1)$$

$$= \frac{2!}{n(n+1)},$$

$$E(p_k^3) = (n-3) B(4, n-1)$$

$$= \frac{3!}{n(n+1)(n+2)},$$

$$E(p_k^4) = (n-1) B(5, n-1)$$

$$= \frac{4!}{n(n+1)(n+2)(n+3)}.$$

Applying the relation:

$$E\left\{\sum_i x_i\right\} = \sum_i (E\{x_i\})$$

we obtain:

$$E\left(\sum_{i=1}^n p_i^2\right) = \frac{2}{n+1}$$

$$E\left(\sum_{i=1}^n p_i^3\right) = \frac{6}{(n+1)(n+2)} \tag{19}$$

$$E\left(\sum_{i=1}^n p_i^4\right) = \frac{24}{(n+1)(n+2)(n+3)}.$$

In order to calculate $E\left\{\left(\sum_{i=1}^n p_i^2\right)^2\right\}$, which we shall write as F_n, we note that

$$\left(\sum_{i=1}^n p_i^2\right)^2 = \left(\sum_{i=1}^{n-1} p_i^2\right)^2 + 2p_n^2 \sum_{i=1}^{n-1} p_i^2 + p_n^4$$

which leads to:

$$F_n = (n-1) \int_0^1 \left[E\left\{\left(\sum_{i=1}^{n-1} p_i^2\right)^2\right\} + 2p_n^2 \, E\left(\sum_{i=1}^{n-1} p_i^2\right) + p_n^4 \right] (1-p_n)^{n-2} \, dp_n.$$

Making the change of variable:

$$p_i = M_i (1 - p_n)$$

the condition $\sum\limits_{i=1}^{n-1} p_i = 1 - p_n$ leads to $\sum\limits_{i=1}^{n-1} M_i = 1$. Hence:

$$E\left\{\left(\sum_{i=1}^{n-1} p_i^2\right)^2\right\} = (1-p_n)^4 \, E\left\{\left(\sum_{i=1}^{n-1} M_i^2\right)^2\right\}$$

and similarly:

$$E\left(\sum_{i=1}^{n-1} p_i^2\right) = (1-p_n)^2 \, E\left(\sum_{i=1}^{n-1} M_i^2\right)$$

$$= (1-p_n)^2 \times \frac{2}{n}.$$

Hence:

$$F_n = (n-1)\int_0^1 \left[(1-p_n)^{n+2}\,F_{n-1} + \frac{4}{n}\,p_n^2(1-p_n)^n + p_n^4(1-p_n)^{n-2}\right]dp_n$$

$$= \frac{n-1}{n+3}\,F_{n-1} + [8(n-1)+24]\,\frac{(n-1)!}{(n+3)!}.$$

This difference equation can also be written:

$$\frac{(n+3)!}{(n-1)!}\,F_n = \frac{(n+2)!}{(n-2)!}\,F_{n-1} + 8n + 16$$

or, going backwards to successively smaller values of n:

$$\frac{(n+3)!}{(n-1)!}\,F_n = 4!\,F_1 + 8[n+(n-1)+\cdots+2] + 16(n-1)$$

$$= 4n^2 + 20n.$$

Thus, finally:

$$F_n = \frac{4(n+5)}{(n+1)(n+2)(n+3)}.\quad^1 \tag{20}$$

Eqs. (19) and (20) can be used to find the values of the expressions required for calculating the genotypic distance between individuals (Eq. (17)), for a system with any number of alleles. Notice that in this case, as for the equal-frequency model, as n increases all the terms tend towards zero, but the changes are much slower for this model than if all the allele frequencies are assumed equal.

(ii) *The second Bayesian model.* Now let us consider the following model. The various alleles at a given locus were discovered in a certain historical order, say $A_1, A_2, \ldots, A_n$, and the number n of known alleles at a locus must be regarded as provisional, since new alleles can always be

---

[1] Result due to T. Leviandier (personal communication).

discovered, controlling new phenotypes of this locus; for instance, at the locus for the AB0 blood groups, the allele for blood-group A has been found to be separable into two distinct alleles, A_1 and A_2. We may expect that the alleles which are discovered first are common, and that the later discoveries will be of low-frequency alleles. We can therefore make the following model: the frequency p_1 of allele A_1 is a random variable with uniform probability density in the interval $0-1$; if n is greater than 2, the frequency p_2 is another such random variable, with uniform probability density between 0 and $1-p_1$. Similarly for all the alleles, up to and including the last but one allele; the last allele, however, is not a random variable, but is determined by the formula:

$$p_n = 1 - p_1 - \cdots - p_{n-1}.$$

In order to evaluate $E\left(\sum_i p_i^2\right)$, we note that:

$$E_n\left(\sum_1^n p^2\right) = \int_0^1 E_{n|p_1 = x} dx$$

$$= \int_0^1 [U_{n-1}(1-x) + x^2] dx \qquad (21)$$

where $E_{n|p_1 = x}$ is the expectation of the sum $\sum_1^n p_i^2$ when the variable p_1 takes the value x, and $U_{n-1}(1-x)$ is the expectation of the sum of squares of the last $n-1$ allele frequencies, when $p_2 + \cdots + p_n = 1 - x$.

There is a difference equation relating U_n, U_{n-1} etc., and this enables us to calculate $E\left(\sum_i p_i^2\right)$. This difference equation is as follows:

$$U_{n-1}(x) = \int_0^x [U_{n-2}(x-y) + y^2] dy$$

$$= \int_0^x U_{n-2}(u) du + \frac{x^3}{3}$$

which leads to the results:

$$U_1(x) = x^2$$

$$U_2(x) = \int_0^x u^2 du + \frac{x^3}{3} = \frac{4x^3}{3!}$$

$$U_3(x) = \frac{4x^4}{4!} + \frac{2x^3}{3!}$$

$$\cdots\cdots\cdots\cdots\cdots\cdots\cdots\cdots\cdots$$

$$U_{n-1}(x) = \frac{4x^n}{n!} + 2\sum_{i=3}^{n-1} \frac{x^i}{i!}$$

and Eq. (21) gives:

$$E_2 = \int_0^1 [x + x^2]\, dx = \tfrac{2}{3}$$

$$E_3 = \int_0^1 \left[4\frac{x^3}{3!} + x^2 \right] dx = \tfrac{1}{2}$$

. . . .

$$E_n = 2 \left[\frac{2}{(n+1)!} + \sum_{i=3}^{n} \frac{1}{i!} \right].$$

As n increases, the term $\dfrac{2}{(n+1)!}$ rapidly becomes negligible and E_n tends towards $2\sum_3^n \dfrac{1}{i!} = 2(e-2.5) = 0.436$. We can therefore write $E_n \approx$ 0.436, to a good approximation, when n is greater than 5. (The exact figures are: $E_4 = 0.450$, $E_5 = 0.439$, $E_6 = 0.437$.)

We can obtain the expectations of the other sums in a similar way. These are:

$$E\left(\sum_1^n p_i^3\right) = 3! \left[\frac{2}{(n+2)!} + \sum_4^{n+1} \frac{1}{i!} \right] \approx 0.312 \qquad \text{for } n > 5$$

$$E\left(\sum_1^n p_i^4\right) = 4! \left[\frac{2}{(n+3)!} + \sum_5^{n+2} \frac{1}{i!} \right] \approx 0.240 \qquad \text{for } n > 5$$

$$E\left(\sum_1^n p_i^2\right)^2 = 8 \left[\frac{3}{(n+3)!} + \frac{1}{4!} - \sum_n^{n+2} \frac{1}{i!} \right] \approx 0.254 \qquad \text{for } n > 5.$$

Finally, therefore, the expectation of the distance between two individuals can be expressed as a function of the coefficients of identity appropriate for the relationship they have to one another and of the number of alleles at the locus in question. The formula is as follows:

$$E[D(AB)] = \alpha_n(\Delta_2 + \Delta_3 + \Delta_5 + \Delta_8) + \beta_n(\Delta_4 + \Delta_6) + \gamma_n\,\Delta_9 \qquad (22)$$

where the coefficients α_n, β_n and γ_n can be calculated using different models for the distribution of allele frequencies.

The computed values of the coefficients show that identity modes $S'2$, $S'3$, $S'5$ and $S'8$ have the least weight in this measure of distance. As might be expected, mode $S'9$, which corresponds to the absence of all relationship between the individuals in question, carries the greatest weight, except in the two-allele case, when $S'4$ and $S'6$ have a slightly greater weight.

The Δ_i coefficients evaluated in Chapter 6 can be used to give the expected genotypic distances for some of the most frequently studied types of relationship (Table 14.4).

Table 14.4. Genotypic distances (D) under different assumptions about the distribution of allele frequencies

Hypothesis	Equal frequencies			First Bayesian model			Second Bayesian model		
Number of alleles	2	3	∞	2	3	∞	2	3	∞
Relationship									
Unrelated	0.611	0.815	1	0.468	0.668	1	0.466	0.666	0.732
Parent-offspring	0.500	0.666	1	0.333	0.500	1	0.333	0.500	0.564
Half-sibs	0.555	0.741	1	0.400	0.583	1	0.400	0.583	0.748
Full sibs	0.403	0.538	0.750	0.283	0.417	0.750	0.283	0.417	0.465
First-cousins	0.583	0.777	1	0.434	0.625	1	0.434	0.625	0.690
Double first-cousins	0.511	0.681	0.937	0.364	0.532	0.937	0.364	0.532	0.592
Second cousins	0.604	0.806	1	0.460	0.657	1	0.460	0.657	0.721
Sibs from consanguineous matings									
Sib-mating	0.382	0.502	0.718	0.259	0.383	0.718	0.259	0.379	0.424
Matings between first-cousins	0.397	0.527	0.734	0.277	0.408	0.734	0.277	0.408	0.453

Table 14.4 shows, for example, that a pair of sibs are closer to one another than a father/son pair, and that this difference increases with the number of alleles at the locus.

Some of the results in Table 14.4 may seem odd. For example, when the number of alleles is high, the second Bayesian model gives the distance between unrelated individuals as about $\frac{3}{4}$; in other words, the probability that two unrelated individuals will have the same genotype is about $\frac{1}{4}$. It might, however, seem probable that, when the number of alleles is high, the probability that two randomly-chosen individuals will have the same genotype should be very small, since the number of possible genotypes is so high.

In fact, the result we have obtained with the genotypic measure of distance results from our assumption that the allele frequencies can be specified by the "Bayesian" model set up above. This model represents a very different situation from the model with all the allele frequencies the same; it is far more similar to a model of a large number of loci, each with one or two common, normal alleles, and many rare mutant alleles. In this case, the large number of alleles, and thus of possible genotypes, does not lead to a rapid decrease in the probability that two randomly-chosen individuals will have the same genotype. As an example, consider a locus with two "normal" alleles, each at a frequency of 0.45, and 20 rare alleles, with frequencies 0.005; the probability that two individuals have the same genotype is $2\left(\sum_i p_i^2\right)^2 - \sum_i p_i^4 = 0.24$, which would not be affected if we were subsequently to discover ten further very rare alleles.

2.3. Other Measures of Distance between Relatives

2.3.1. Genic distance. We have discussed a measure of distance based on the probability that the two individuals A and B we are studying have the same genotype. One could make the objection to this measure that it treats a pair of individuals whose genotypes are $(A_i A_j)$ and $(A_i A_k)$ and a pair whose genotypes are $(A_i A_j)$ and $(A_k A_l)$ as equally distant, despite the fact that the first pair differ less than the second. If we do not wish this to be so, we may prefer to use a different measure of distance based on the similarity between the genes of the two individuals in question, rather than of their genotypes.

The following definition, for example, defines such a measure of distance:

The four genes carried by A and B may all be the same allelic type, or they may include two, three or four different allelic types. There are seven possible cases, as shown in Table 14.5. If we choose at random one of the allelic types represented among these four genes we can ask: what is the probability that this type is not represented in one of the individuals A or B. The probability of choosing such an allele, which is not present in one of the individuals A or B is called the genic distance $d(AB)$. Table 14.5 gives the values of this measure, for each of the seven possible cases.

Table 14.5 shows that the measure d is better than D in cases when there is partial resemblance between the individuals, i.e. when they have a gene in common, but do not have the same genotype.

d is a distance because, if the two individuals have one gene which is not the same allele in both, then $d \geq \frac{1}{2}$, while, by definition, $d \leq 1$.

In order to relate this measure of distance to the coefficients of identity measuring the relationship between A and B, and to the allele

Table 14.5. Genic and genotypic distances between pairs of individuals of defined genotype

Number of allelic types	Genotypes of the individuals		Genic distance $d(AB)$	Genotypic distance $D(AB)$
	A	B		
1	$(A_i A_i)$	$(A_i A_i)$	0	0
2	$(A_i A_i)$	$(A_j A_j)$	1	1
	$(A_i A_j)$	$(A_i A_j)$	0	0
	$(A_i A_i)$	$(A_i A_j)$	$\frac{1}{2}$	1
3	$(A_i A_i)$	$(A_j A_k)$	1	1
	$(A_i A_j)$	$(A_i A_k)$	$\frac{2}{3}$	1
4	$(A_i A_j)$	$(A_k A_l)$	1	1

frequencies at the locus in question, we proceed as before by examining each of the different modes of identity which are possible between A's and B's genes. Gathering up terms with the same coefficients, we obtain, finally:

$$d(AB) = \left(\Delta_2 + \frac{\Delta_3}{2} + \Delta_4 + \frac{\Delta_5}{2} + \Delta_6\right)\left(1 - \sum_i p_i^2\right) + \Delta_8\left(\tfrac{2}{3} - \sum_i p_i^2 + \tfrac{1}{3}\sum_i p_i^3\right)$$
$$+ \Delta_9\left[1 - \tfrac{4}{3}\sum_i p_i^2 + \tfrac{2}{3}\sum_i p_i^3 + \tfrac{1}{3}\sum_i p_i^4 - \tfrac{2}{3}\left(\sum_i p_i^2\right)^2\right]. \tag{23}$$

This formula shows how one can finish up with very laborious calculations, even starting from the simplest criteria.

It is, of course, possible to calculate the expectation of this distance, given some assumption about the allele frequencies, and to determine the distances of different types of relatives, just as in the case of the genotype measure of distance.

2.3.2. Genealogical distance. The two measures of distance which we have just been discussing depend on the relationship between the individuals, and also on the genic structure of the population they belong to. Moreover, these measures of distance will differ for different loci, and this may be an undesirable property, for certain problems. We might therefore wish to define an "absolute" distance, in terms of the relatedness of the individuals alone.

We saw earlier that we cannot use Φ_{AB}, the coefficient of kinship, for this purpose. Once again, therefore, we have to find a more sensitive measure, based on the coefficients of identity.

If we wish to take the sex of the individuals into account (for example if we assume that the distance between a pair of half-brothers with the same father might be different from that of a pair of half-brothers with the same mother), we would obviously need to use the 15 full identity coefficients δ_i of Chapter 6. If, on the other hand, we assume that sex does not affect the distance between individuals, the nine condensed identity coefficients Δ_i will be sufficient to solve the problem. We shall make this assumption in what follows.

It is intuitively obvious that our definition of distance should be such that:

1. With identity modes $S'1$ or $S'7$, when A and B necessarily have the same genotype at the locus in question, the distance must be zero.

2. With identity modes $S'2$, $S'4$, $S'6$ and $S'9$, where the lines of descent are such that A and B do not have any identical genes with one another, the distance should be the maximum possible, which we may give the value 1.

3. With identity modes $S'3$, $S'5$ and $S'8$, the distance should be between 0 and 1 in value.

The problem is to choose values of d for these three situations, so that they express our feelings about the closeness of A and B.

The choice of these values is obviously arbitrary, but they must satisfy the triangle inequality, if we are to use them as a measure of distance.

By analogy with the genic definition of distance given above, we may make the following definition:

The four genes carried by A and B may all be identical, or may fall into two, three or four classes, within which all the alleles are identical by descent. The number of such "identity classes" depends on the identity mode for the two individuals. If we choose one of the identity classes represented among A and B's four genes we can ask: what is the probability that one of the two individuals has a gene which does not belong to this class. The probability of choosing such an identity class defines a genealogical distance $G(AB)$.

From the definitions of the identity modes, in Chapter 6, we see immediately that $G=0$ for modes $S'1$ and $S'7$, $G=1$ for modes $S'2$, $S'4$, $S'6$ and $S'9$, $G=\frac{1}{2}$ for $S'3$ and $S'5$, and that $G=\frac{2}{3}$ for mode $S'8$.

$G(AB)$ is therefore a linear function of the coefficients of identity Δ_i, and is defined by:

$$G(AB) = \Delta_2 + \Delta_4 + \Delta_6 + \Delta_9 + \tfrac{1}{2}(\Delta_3 + \Delta_5) + \tfrac{2}{3}\Delta_8$$

or, equivalently:

$$G(AB) = 1 - \left(\Delta_1 + \Delta_7 + \frac{\Delta_3 + \Delta_5}{2} + \frac{\Delta_8}{3}\right) \tag{24}$$

which, finally, is equivalent to:

$$G(AB) = 1 - (\Phi + \tfrac{1}{2}\Delta_7 + \tfrac{1}{12}\Delta_8).$$

This last formula shows how the coefficient of kinship, Φ, must be modified in order to represent a distance.

2.3.3. The probability of graft acceptance. It is known that graft rejection is due to the presence of antigens in the donor material which are not present in the recipient individual (and against which the recipient will therefore produce antibodies). The probability that two individuals can accept grafts from one another, for a given histocompatibility system, is obviously dependent on the two following factors:

1. The number of alleles in the system, and their frequencies.
2. The relationship between the individuals.

Many workers have studied this problem, and have tried to find out how these two factors affect the probability of graft acceptance or rejection. For example, Elandt-Johnson (1968), Kilpatrick and Gamble (1968), Lunghi (1965) and Serra and O'Mathuna (1966) and others have studied the effect of relationship, but have restricted their attention to the simplest cases. Feingold *et al.* (1970) gave formulae which are valid in more complex cases, based on detailed measures of relatedness, such as the identity coefficients Δ_i. It is clear that the probability of graft-rejection is not really a "distance", since the fundamental property $d_{ij}=d_{ji}$ does not hold. It is, however, a measure of dissimilarity, and so it is appropriate to discuss it in the present chapter.

By finding the probability that a graft with A as donor will carry genes which B does not have, for each in turn of the nine identity modes, it is easy to show that the probability of non-compatibility is given by:

$$P=(\Delta_2+\Delta_3)\left(1-\sum_i p_i^2\right)+\Delta_4\left(1-\sum_i p_i^3\right)+(\Delta_6+\Delta_8)\left(1-2\sum_i p_i^2+\sum_i p_i^3\right)$$
$$+\Delta_9\left[1-2\sum_i p_i^3+3\sum_i p_i^4-2\left(\sum_i p_i^2\right)^2\right]$$

where the p_i are the frequencies of the different co-dominant alleles of the particular histocompatibility system under study.

At present, the genetics of histocompatibility systems is not fully worked out; the numbers (n) of alleles, and their frequencies p_i are unknown, but we can probably assume that n is quite large, of the order of ten or more.

If we assume equal allele frequencies, as is usually done, we obtain:

$$P=(\Delta_2+\Delta_3)\left(1-\frac{1}{n}\right)+\Delta_4\left(1-\frac{1}{n^2}\right)+(\Delta_6+\Delta_8)\left(1-\frac{2}{n}+\frac{1}{n^2}\right)$$
$$+\Delta_9\left(1-\frac{4}{n^2}+\frac{3}{n^3}\right)$$

or, if n is sufficiently large for $1/n$ to be negligible:

$$P=\Delta_2+\Delta_3+\Delta_4+\Delta_6+\Delta_8+\Delta_9$$
$$=1-\Delta_1-\Delta_5-\Delta_7.$$

If, on the other hand, we adopt the second "Bayesian model" of Section 2.2, the summation terms have the following expectations (provided that n is greater than 5):

$$E\left(\sum_i p_i^2\right)=0.436$$

$$E\left(\sum_i p_i^3\right)=0.312$$

$$E\left(\sum_i p_i^4\right)=0.240$$

$$E\left[\left(\sum_i p_i^2\right)^2\right]=0.254.$$

The probability of rejection then becomes:

$$P = 0.564\,(\Delta_2 + \Delta_3) + 0.688\,\Delta_4 + 0.440\,(\Delta_6 + \Delta_8) + 0.588\,\Delta_9$$

and this leads to very different results from those obtained on the assumption of equal allele frequencies. Table 14.6 gives the probabilities of graft rejection for various types of relatives as donor and recipient, for the two models.

Table 14.6. Probabilities of graft rejection with different assumptions about the distribution of allele frequencies

Relationship between donor and recipient	Probability of graft rejection	
	Assuming equal allele frequencies	Second Bayesian model
Unrelated	$1 - \dfrac{4}{n^2} + \dfrac{3}{n^3} \approx 1$	0.59
Parent-offspring	$1 - \dfrac{2}{n} + \dfrac{1}{n^2} \approx 1$	0.44
Half-sibs	$1 - \dfrac{1}{n} - \dfrac{3}{2\,n^2} + \dfrac{3}{2\,n^3} \approx 1$	0.52
Full sibs	$\dfrac{3}{4} - \dfrac{1}{n} - \dfrac{1}{2\,n^2} + \dfrac{3}{4\,n^3} \approx 0.75$	0.37
First-cousins	$1 - \dfrac{7}{2\,n^2} - \dfrac{11}{4\,n^2} + \dfrac{9}{4\,n^3} \approx 1$	0.55
Double first-cousins	$\dfrac{15}{16} - \dfrac{3}{4\,n} - \dfrac{15}{8\,n^2} + \dfrac{27}{16\,n^3} \approx 0.94$	0.49

Table 14.6 shows that the probability of rejection depends strongly on the model, but there are some similarities. In both cases, the relations who have the greatest chance of accepting a graft from one another are a pair of sibs; the "Bayesian" model gives them a considerably lower advantage over the other types of relative than the model with equal allele frequencies. With the "Bayesian" model, the probability of graft acceptance between unrelated individuals is 41 %.

In certain cases, for example marrow grafting, both donor and recipient must have the same genotype, in order for the graft to be successful. Assuming that histocompatibility is controlled by only one locus, the probability of graft acceptance in this case is given by $1 - D$, where D is the genotypic distance defined in Section 2.2.1.

3. Distances between Populations

The choice of a definition of the distance between two populations depends on what information we have about them. We may, for example, know about the origins of the populations (in other words, we might have information that would allow us to draw up the genealogies of the members of the populations), or we might have information about their present state, in terms of the genetic constitution of the individual population members. We shall examine each of these two cases.

3.1. Distance between the Genetic Structures of Populations

When we discussed the problem of comparing two individuals, we set the distance between them equal to zero when they have the same genotype at a particular locus, and 1 when this is impossible; this enabled us to define a genotypic measure of distance $D(AB)$, between two individuals, A and B.

When we come to consider distances between populations, we obviously cannot talk about the genotype at a locus. We can, however, characterise the populations by the frequencies of the various genotypes among the population members, in other words by the populations' genotypic structures S.

We could, for example, call two populations identical, and the distance between them zero, if the distribution of numbers of individuals among the various genotypes is the same in both, in other words if they have the same genotypic structures.

This definition of distance gives a number which condenses the information contained in the set, Δ_{ij}, of the differences between the genotype frequencies in the two populations P_i and P_j:

$$\Delta_{ij} = S_i - S_j.$$

Similarly, if we are interested in the frequencies of the various genes at a particular locus, we could set in the distance between two populations equal to zero when their genic structures are equal. This measure of distance between two populations, P_i and P_j, is a number which condenses the information from the vector δ_{ij} of the differences in the allele frequencies in the populations:

$$\delta_{ij} = s_i - s_j.$$

In either case, for genotype or gene frequencies, the choice of the number is, as we have seen, an arbitrary one. Several measures of distances have been proposed, each with its own particular advantages.

3.1.1. χ^2 measures of distance. We may note that the considerations of Section 1.2.2, which led to the definition of distance as a χ^2, are particularly relevant to the problem of defining measures of distance based on the genetic structures of populations. We may recall the arguments, in genetical terms, as follows:

1. A difference between the frequencies $P_{rs}^{(i)}$ and $P_{rs}^{(j)}$ of the genotype $(A_r A_s)$, or between the frequencies $p_r^{(i)}$ and $p_r^{(j)}$ of the allele A_r, must receive a greater weight, the rarer this genotype or gene is in the population.

2. If, in all the populations, two genotypes, or genes, occur in frequencies that are proportional to one another, we should be able to condense the information from them, without altering the distance between the populations.

To satisfy these two conditions, we may choose a definition of the distance between two populations, based on χ^2:

$$D_{ij}^2 = \sum_{rs} \frac{(P_{rs}^{(i)} - P_{rs}^{(j)})^2}{P_{rs}}$$

where the summation is over all the genotypes $(A_r A_s)$ at the locus; there are therefore $\dfrac{n(n+1)}{2}$ terms if the number of alleles at the locus is n. Also, P_{rs} is the mean frequency of the genotype $(A_r A_s)$ in the whole set of populations.

We can similarly define the "genic distance" between two populations by:

$$d_{ij}^2 = \sum_r \frac{(p_{ir} - p_{jr})^2}{p_r}$$

where the summation is over all the alleles at the locus.

Here, we may note the following important point. When all the populations being considered are such that the Hardy-Weinberg law is obeyed, their genotypic structures are determined fully by their genic structures. We would therefore expect that distances calculated on the basis of either genotype frequencies or of gene frequencies should lead to descriptions of the ensemble of populations which agree with one another. With the definitions of distance which we have adopted this is, however, not so; a single set of populations may have different topographies, depending on which of the two measures of distance we use.

In order to show this, we can consider the example of three populations, whose genic and genotypic structures for a given locus with two alleles are as follows:

Population A: $s_A = \frac{1}{2} (1 \quad 1)$

$$S_A = \frac{1}{4} \begin{vmatrix} 1 & \\ 2 & 1 \end{vmatrix}$$

Population B: $s_B = \frac{1}{6} (1 \quad 5)$

$$S_B = \frac{1}{36} \begin{vmatrix} 1 & \\ 10 & 25 \end{vmatrix}$$

Population C: $s_C = \frac{1}{6} (5 \quad 1)$

$$S_C = \frac{1}{36} \begin{vmatrix} 25 & \\ 10 & 1 \end{vmatrix}$$

We shall also assume that these three populations belong to a larger group of populations, whose mean genetic structures are:

$$\bar{s} = \frac{1}{4} (1 \quad 3), \qquad \bar{S} = \frac{1}{6} \begin{vmatrix} 1 & \\ 1 & 4 \end{vmatrix}$$

The mean genotypic structure of the whole group of populations will obviously not obey the Hardy-Weinberg law, since the mean of a set of panmictic populations always includes an excess of homozygotes.

Applying the definitions of genic and genotypic distance given above, we have, for these populations:

Genic distances: $d_{AB} = 0.77$
                          $d_{BC} = 1.54$
                          $d_{CA} = 0.77$

Genotypic distances: $D_{AB} = 0.94$
                          $D_{BC} = 1.82$
                          $D_{CA} = 1.56$

If we use these values to represent the three populations in a two-dimensional space, the first measure of distance gives a straight line, with population A equidistant from populations B and C:

B _____ A _____ C

The genotypic measure of distance, however, gives a triangle, with population A closer to B than to C:

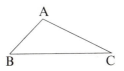

The picture we obtain of the ensemble of populations under study can thus differ very greatly, depending on whether we characterise the populations in terms of the gene frequencies or of the genotype frequencies, despite the fact that the genotype frequencies are fully determined by the gene frequencies, according to the Hardy-Weinberg formula.

For this reason, we may be led to reject the use of measures of distance based on χ^2, in spite of the advantages of these measures, which we discussed above.

Of course, the example which we have chosen to illustrate this problem is an extreme case; the mean population is very different from the populations we wish to compare, and its genotypic structure is different from the corresponding Hardy-Weinberg structure (see Section 2.1 of Chapter 8). In practice, agreement between the genic and genotypic measures of distance would often be much better than in this example. It is necessary, however, to check the agreement in every case, so as to avoid unjustifiable interpretations of the results. It would therefore seem to be preferable to find a measure of distance which does not suffer from this disadvantage.

3.1.2. The arc-cosine measure of distance. The measure of distance proposed by Cavalli-Sforza and Edwards (1967), called the arc-cosine measure of distance, proves to have the desirable property of correspondence between genic and genotypic distances, at least in cases where Hardy-Weinberg genotype frequencies can be assumed to hold.

The definition of this measure of distance between two populations is given by:

$$\Theta\,(AB) = \text{arc}\cos \sum_r \sqrt{\mu_{ir}\,\mu_{jr}}$$

where μ_{ir} is the frequency of the r-th character in the i-th population, and the summation is over all possible characters (so that $\sum_r \mu_{ir} = 1$). Θ varies between 0 and 1.

When the characters we are considering are the frequencies p_r of different alleles at a locus, the μ_{ir} represent the allele frequencies p_{ir} in the i-th population. The distance between two populations then becomes:

$$g_{ij} = \text{arc}\cos \sum_r \sqrt{p_{ir}\,p_{jr}}\,.$$

When the characters are the genotype frequencies at a locus, the μ_{ir} are the frequencies $P_{rs}^{(i)}$ of the genotype $(A_r A_s)$ in the i-th population; if the population is in Hardy-Weinberg equilibrium, these are related to the p_r values for the same population, by the formulae:

$$P_{rr} = p_r^2, \qquad \text{for all } r$$

$$P_{rs} = 2\,p_r\,p_s, \qquad \text{for } r \neq s.$$

The genotypic measure of distance is therefore given by:

$$G_{ij} = \text{arc cos} \sum_r \sqrt{P_{rs}^{(i)} P_{rs}^{(j)}}$$

where the summation is over all the $\dfrac{n(n+1)}{2}$ genotypes. This definition leads to:

$$G_{ij} = \text{arc cos} \left[\sum_r \sqrt{p_{ir}^2 p_{jr}^2} + \sum_{r \neq s} \sqrt{4 p_{ir} p_{jr} p_{is} p_{js}} \right]$$

where the second summation is over the $\dfrac{n(n-1)}{2}$ heterozygotes.

But the term in the square brackets is equal to the expansion of:

$$\left[\sum_r \sqrt{p_{ir} p_{jr}} \right]^2$$

so that the measures of distance, g and G are related by:

$$\cos G_{ij} = (\cos g_{ij})^2. \tag{25}$$

The arc cosine measure of distance thus guarantees correspondence between the two measures of genetic distance.

3.2. Distance between Populations of Known Ancestry

By comparing the genic or genotypic structures of populations, we obtain information about the extent of their similarity or difference with respect to the genes at a locus. These methods do not, however, take into account information we may have about the common origins of part or all of the populations, or about exchanges of members between them, though these will obviously result in the populations being to some extent similar to one another. Resemblance in terms of the genes or genotypes might be the result of similar selective pressures acting on both the populations, or even simply the result of chance, and may not be due to any true relatedness between the populations. We might therefore come to quite different conclusions about the distances between populations, depending on which locus we study. These methods are based entirely on the present states of the populations, and any attempt to draw conclusions about their past history must make gratuitous assumptions, which may well be wrong.

From the point of view of genetics, the history of a population is fully contained in the genealogies of the individual population members. If we know these, we can apply measures of the distances between populations, in terms of the genetic relationships between them. In our discussion of measures of distance between individuals, we remarked on the problems that arise if we use the coefficient of kinship Φ as a distance. To measure distances between populations, we can use a measure based on the same fundamental reality, the "probability of origin" of genes from given sources.

3.2.1. Probabilities of origin of genes from the founding members of a population.

If we choose a gene at random out of the two genes carried by an individual, the probability that it came from his father is $\frac{1}{2}$, and the probability that it came from his mother is $\frac{1}{2}$; the probability that it came from one particular, specified grandparent is $\frac{1}{4}$, provided that he had four distinct grandparents. Going back in the genealogy of an individual, in this way, we finally come to the most distant known ancestors, whose parents we do not know. We shall call these the "founders" of the group. The concept of founders thus covers all the ancestors where the genealogical information stops short; some of these will be true members of the initial group who actually founded the population, while others will be immigrants at later times. Thus the "founders" will belong to more than one generation. They represent the source of the genes at present found in the population, and our knowledge of the ancestry of the population enables us to trace their transmission to the present-day population members.

If we can follow the ancestry of an individual's parents sufficiently far back, we will obviously find that they have ancestors in common; a given common ancestor may be linked to these descendants by one or more lines of descent. In order to calculate the probabilities of origin of genes from such an ancestor, we simply add the probabilities for each of the lines of descent of which he is a common ancestor (see Chapter 6). This type of calculation is simple, but laborious; it can, however, be done very easily by computer.

We can thus find, for each individual I_j, the probabilities, $\omega_{j1}, \ldots, \omega_{jn}$, that a gene taken at random from I_j is derived from founder number 1, 2 etc. The mean values of ω_{ji} over all the members of a population A constitute a vector:

$$\Omega_A \equiv \{\omega_{A1}, \ldots, \omega_{Ai}, \ldots, \omega_{An}\}$$

which characterises the origin of the genes of this population with regard to the founders of the population. We shall call this vector the population's "origin vector".

3.2.2. Distance between populations.

To compare two populations, on the basis of their "origin vectors", we are therefore faced with a very similar problem to the one we had before, when we wished to compare populations characterised by the frequencies of the various alleles at a locus. We wish to define a measure of distance by choosing a single number d which condenses "as well as possible" the set of n numbers we obtain by taking the differences between the origin vectors of the two populations from their combined founders:

$$\boldsymbol{\Omega}_A - \boldsymbol{\Omega}_B = (\omega_{A1} - \omega_{B1}), \ldots, (\omega_{An} - \omega_{Bn}).$$

We may note here that:

1. If, in each population, the probabilities of origin of genes from two founders are equal (i.e. if $\omega_{Ai} = \omega_{Aj}$, $\omega_{Bi} = \omega_{Bj}$ etc.), which is in particular the case for two founders who were monogamous mates of one another, we can replace these two founders by the mating which they constitute, without changing the distances between the populations.

2. Suppose that, in the group of populations we wish to compare, one founder F_j contributes a small fraction of the gene-pool. Then a difference $\omega_{Aj} - \omega_{Bj}$ between two populations should weigh more in the distance $(A-B)$ than the same absolute difference with respect to a founder who contributes a large fraction of the present-day gene-pool.

We have seen that a χ^2 measure of distance has these properties, and it would therefore seem reasonable to use this type of measure for this case. The distance between two populations would thus be given by:

$$Z^2(AB) = \sum_i \frac{(\omega_{Ai} - \omega_{Bi})^2}{\omega_i} \tag{26}$$

where the summation is over all the founders of the populations being compared, and ω_i is the mean of the probabilities of origin of genes from founder i, over all the populations. This measure of the distance between populations therefore measures the differences between the contributions of the different founders to the genetic constitutions of the different contemporary populations.

3.2.3. The evolution of a population.

We can apply the measure of distance Z that we have just developed, to measure the distance between two generations in the history of the same population. This can be considered as a measure of the extent of evolutionary change over a number of generations, within the same group. We simply classify the members of the population's genealogy (which is assumed to be fully recorded), according to which generation they belong to, and use formula (26) to calculate the distances between different generations.

The only problem which arises is to determine to which generation each individual belongs. Since marriages occur between individuals of different "generations", the word "generation" loses any precise meaning; for example, if an uncle-niece marriage takes place, will the children belong to the generation after that of their father (i.e. the same generation as their mother) or to the generation following that of their mother? The answer to this question is arbitrary. In order to avoid ambiguity, we must lay down a rule, and assume, for example, that an individual belongs to generation $n+1$, if his mother belongs to generation n, and *vice versa*. This rule raises no problems over a short time-span, but, in the long-term, it will lead to individuals who lived at different times being assigned to the same generation.

Another method would be to assign to each individual a generation number which is the mean of those of his parents. This gives non-integer numbers, but we can group the individuals into classes with the nearest whole number.

Yet another method is to ignore generations and to consider the individuals living at a given period, in other words a "horizontal section" of the history of the population.

Whatever method we choose, we end up with a set of probabilities of origin of the genes, for each generation:

$$\Omega_g = \{\omega_{1g}, \dots, \omega_{ng}\}.$$

Changes in Ω_g are indicative of changes in the genetic make-up of the population, in other words of the population's evolution. The measure $Z(g, g+1)$ of the distance between successive generations is therefore a measure of the rate of the population's evolution.

3.3. Biometrical Estimation of the Relatedness of Two Populations

We have seen that measures of the distance between populations, based on their gene or genotype frequencies has the disadvantage that the distance will depend on which locus we study; it measures the end result of the process of gradual differentiation of the populations, without taking account of the important biological reality of the relationships between them due to genes of common origin present in both populations.

A full description of this biological reality can be given only in terms of the exact lines of descent followed by the genes in the populations, starting from the initial group of founders of the populations. Unfortunately, complete genealogies are rarely available. They can only be compiled in certain exceptional cases, and with a great deal of work.

Morton *et al.* (1971 b) have proposed a method of treating this problem, which they call "bioassay of kinship", and which overcomes the inherent difficulty of drawing conclusions about the biological processes which have taken place, from data about the present state of a population.

The aim of this method is to use data on present-day gene frequencies to evaluate the parameters which measure the relationship of the populations. In order to do this, we obviously have to have a model which relates the known gene frequencies to the parameters which we wish to estimate. We shall next describe this model, laying particular stress on the assumptions involved.

3.3.1. The evaluation of the coefficient of kinship of two populations.

Consider a set of p populations, all assumed to be descended from the same original population. Under the effects of selection and genetic drift, these populations will have become differentiated from one another, but will nevertheless, because of migration, maintain certain similarities.

To measure the similarity of the genetic constitutions of two populations i and j, we can use the coefficient of kinship Φ_{ij}, defined as the probability that two genes taken at random, one from population i, and one from population j will prove to be identical by descent, i.e. to be descended from the same ancestral gene.

We can relate Φ_{ij} to the frequencies q_{ir} and q_{jr} of the allele A_r in the two populations, by means of the following argument:

Consider an "F_1 hybrid" population consisting of the offspring of matings each of which is between a member of population P_i and a member of population P_j. Then the following properties hold:

1. The probability that one of these "F_1" individuals will be of the genotype $(A_r A_r)$ is equal to $q_{ir} \times q_{jr}$.

2. The probability that these two genes are copies of one ancestral gene, i.e. are identical by descent, is equal to Φ_{ij}. If this is the case, the probability that the individual's genotype is $(A_r A_r)$ is q_r^*, the frequency of allele A_r among the founders of the populations P_i and P_j, which have descendants among the present members of these populations.

3. The probability that these two genes are non-identical is $1 - \Phi_{ij}$. In this case, the probability that the individual's genotype is $(A_r A_r)$ is $\hat{q}_r^2$, where $\hat{q}_r$ is the frequency of the A_r allele among all the founders of the two populations.

Finally, therefore, we can write:

$$q_{ir} q_{jr} = \Phi_{ij} q_r^* + (1 - \Phi_{ij}) \hat{q}_r^2. \tag{27}$$

This allows us to evaluate Φ_{ij}, assuming that we know q_r^* and $\hat{q}_r$.

In order to obtain values for these parameters, we make the following assumptions:

a) The founders of the two populations P_i and P_j are a random sample of all the possible founders of these populations.

b) In the global population consisting of all the populations which we are considering, there has been no genetic drift, and no natural selection (these assumptions are obviously unrealistic). We can then write:

$$\hat{q}_r = Q_r = \frac{\sum_i N_i p_{ir}}{\sum_i N_i}$$

where N_i is the size of the i-th population, and Q_r is the frequency of allele A_r, calculated over all the populations.

We shall also assume that:

c) The founders of the populations who are the sources of identity between the genes carried by the members of populations P_i and P_j, are a random sample of all the founders. This is equivalent to assuming that the allele A_r which we are considering has not been involved in whatever selective processes led to certain of the founders having more descendants that others. With this assumption, we can write:

$$q_r^* = \hat{q}_r = Q_r.$$

Under these three assumptions, Eq. (27) becomes:

$$\Phi_{ij} = \frac{q_{ir}q_{jr} - Q_r^2}{Q_r - Q_r^2}. \tag{28}$$

Finally, we assume that:

d) The three above assumptions are valid for all the alleles at the locus in question, so that:

$$\Phi_{ij} = \frac{\sum_r q_{ir}q_{jr} - \sum_r Q_r^2}{1 - \sum_r Q_r^2}. \tag{29}$$

This equation enables us to estimate Φ_{ij} from the frequencies of the alleles.

Since the measure Φ_{ij} is based on the concept of the identity of genes, it has a clear biological meaning, and to this extent is superior to the other measures of distances between populations which we have considered. We must, however, be cautious in our interpretation of Φ_{ij}, since it characterises not only the relationship between the two populations P_i and P_j, but also their relations to the overall, global population. This is clear from formulae (28) and (29); if the two populations we are comparing have the same frequency of an allele, and this frequency is close to the general frequency of the allele, Φ_{ij} will be zero, despite the fact that the two populations have identical genetic compositions, and ought therefore to be close to one another.

In general, we can say that the limits 0 and 1 of the values of Φ_{ij} are of uncertain significance, for the following reasons:

1. For Φ_{ij} to be equal to 0, we must have $q_{ir}q_{jr} = Q_r^2$. This can occur with $q_{ir} = q_{jr} = Q_r$ (similar populations), or with $q_{ir} = Q_r^2/q_{jr}$ (e.g. with $Q_r = \frac{1}{2}$, $q_{ir} = \frac{1}{4}$, $q_{jr} = 1$; i.e. with very dissimilar populations).

2. For Φ_{ij} to be equal to 1, we must have $q_{ir}q_{jr} = Q_r$. This can occur with similar or dissimilar populations.

Finally, it can also be shown that $\Phi_{ij} > \Phi_{kl}$ means that populations P_i and P_j have both become different from the global population, to a greater extent than P_k and P_l; but it does not necessarily mean that P_i and P_j are closer to one another than P_k and P_l — these might, for example, be genetically identical, and similar to the global structure.

3.3.2. The coefficient of hybridity. The coefficient Φ_{ij} has, as we have seen, many disadvantages; Morton *et al.* (1971 b) have therefore proposed another measure of the distance between populations, the "coefficient of hybridity", which we shall discuss in the next few paragraphs.

The coefficient Φ_{ij} is equivalent to the mean inbreeding coefficient α of the imaginary F_1 populations (Φ_{ij} is defined as the probability that the two genes at a locus carried by an individual chosen at random from the F_1 population, are identical by descent).

We saw in Chapter 8 that we can also define the "mean coefficient of kinship", β, of a population, as the probability that two genes at the same locus taken at random from two separate individuals in the population are identical. We also saw that, if the population is panmictic, the coefficients α and β of successive generations are related by the obvious relation:

$$\alpha_g = \beta_{g-1}.$$

In order to characterise the population's deviation from panmixia, we can calculate a number δ, which depends on the difference in the values of β_{g-1} and α_g:

$$\delta = \frac{\beta_{g-1} - \alpha_g}{1 - \beta_{g-1}}. \tag{30}$$

Now let us return to the problem of comparing two populations, P_i and P_j. We shall assume that they are of equal size. They can be considered to be equivalent to a single population, P_{i+j}, whose mean coefficient of kinship β_{g-1}, before the imaginary fusion, is given by:

$$\beta_{g-1} = \tfrac{1}{4}(\Phi_{ii} + \Phi_{jj} + 2\,\Phi_{ij}).$$

We can see this as follows. If we take two genes at random from the ensemble P_{i+j}, their possible origins are:

1. Both from individuals belonging to population P_i. The probability that this is the case is $\tfrac{1}{4}$. In this case, the probability that they are identical by descent is Φ_{ii}, the coefficient of kinship of population P_i.

2. Both from population P_j. In this case the probability that they are identical is Φ_{jj}.

3. One from population P_i and one from P_j. The probability that this is the case is $\tfrac{1}{2}$, and in this case the probability that the two genes are identical is Φ_{ij}.

If in Eq. (30) we substitute Φ_{ij} for α_g and the value just derived for β_{g-1}, we find that the combined population P_{i+j} deviates from panmixia by:

$$\delta = \frac{\Phi_{ii} + \Phi_{jj} - 2\Phi_{ij}}{4 - \Phi_{ii} - \Phi_{jj} - 2\Phi_{ij}}. \tag{31}$$

This parameter is larger, the greater the difference between the two populations. Morton *et al.* propose the term "coefficient of hybridity" for this measure, and employ the symbol Θ_{ij} for it.

It is easy to express δ, or Θ_{ij}, in terms of gene frequencies, as can be shown by the following argument.

1. The probability that an individual taken at random from population P_i has the genotype $(A_r A_r)$ is equal to q_{ir}^2. But the two genes this individual carries are both descended from genes present in the founders of the population. The probability that they are identical is therefore Φ_{ii}.

We thus have:

$$q_{ir}^2 = \Phi_{ii}\,\hat{q}_r + (1 - \Phi_{ii})\,\hat{q}_r^2$$

where $\hat{q}_r$ is the frequency of allele A_r among the founders of population P_i.

If we make the first three assumptions a), b) and c) of Section 3.3.1, we can write $\hat{q}_r = Q_r$, which leads to the following formula for estimating Φ_{ii}:

$$\Phi_{ii} = \frac{q_{ir}^2 - Q_r^2}{Q_r - Q_r}.$$

If we make assumption d) as well, and consider all the alleles at the locus, we have:

$$\Phi_{ii} = \frac{\sum\limits_r q_{ir}^2 - \sum\limits_r Q_r^2}{1 - \sum\limits_r Q_r^2}.$$

Similarly, we have:

$$\Phi_{jj} = \frac{\sum\limits_r q_{jr}^2 - \sum\limits_r Q_r^2}{1 - \sum\limits_r Q_r^2}.$$

Combining the equations thus derived, we finally obtain the formula:

$$\delta \equiv \Theta_{ij} = \frac{\sum\limits_r q_{ir}^2 + \sum\limits_r q_{jr}^2 - 2\sum\limits_r q_{ir} q_{jr}}{4 - \sum\limits_r q_{ir}^2 - \sum\limits_r q_{jr}^2 - 2\sum\limits_r q_{ir} q_{jr}}$$

$$= \frac{\sum\limits_r \left(\dfrac{q_{ir} - q_{jr}}{2}\right)^2}{1 - \sum\limits_r \left(\dfrac{q_{ir} + q_{jr}}{2}\right)^2}. \tag{32}$$

This equation shows that Θ_{ij}, which measures the relation between two populations P_i and P_j, can be estimated from the allele frequencies in the populations alone, without reference to Q_r, the general frequencies in the ensemble of populations.

This measure is therefore easier to interpret than Φ_{ij}, which, as we have seen, depends on the frequencies Q_r. However, if this parameter is to be considered to have a biological meaning and to be related to the modes of identity of the genes in the two populations, we have to be able to assume that all four assumptions a)–d) of Section 3.3.1 are valid. The terms containing Q_r in Eq. (32) then disappear, simply because of the algebraic structure of the formulae.

To put it another way, the biological interpretation of Θ_{ij} is a result of its definition in terms of Φ_{ii}, Φ_{jj}, and Φ_{ij}, given in Eq. (31). In order to estimate it from the values of the q_{ir} and q_{jr}, according to Eq. (32) all the assumptions that were made in deriving this equation must be valid. This measure is therefore somewhat unrealistic for use in any real population.

Note. If we choose to use gene and genotype frequencies, rather than probabilities of identity, we can proceed as follows. In the "F_1", the frequency of allele A_r is $\dfrac{q_{ir}+q_{jr}}{2}$, and the frequency of the genotype $(A_r A_r)$ is $q_{ir} q_{jr}$. The difference between this frequency and the corresponding frequency under random mating enables us to define the inbreeding coefficient, f, by the classical formula:

$$q_{ir} q_{jr} = \frac{(q_{ir}+q_{jr})^2}{2} - f\frac{(q_{ir}+q_{jr})}{2}\left\{1 - \frac{(q_{ir}+q_{jr})}{2}\right\}.$$

Hence, assuming that f is the same for all alleles:

$$f = \frac{\displaystyle\sum_r \frac{(q_{ir}-q_{jr})^2}{2}}{1 - \displaystyle\sum_r \frac{(q_{ir}+q_{jr})^2}{2}}$$

which is the same as the expression derived above for the estimation of Θ_{ij}.

3.4. Conclusion

It is obvious that populations are such complex entities that we cannot hope to condense information about the many ways in which two populations differ from one another, into a single number, the "distance" between them. Nevertheless, there are certain problems in biology for whose solution we need to be able to provide answers to such questions as: "is this group of individuals more similar to this second group, or to a third?". The present chapter has given a discussion of some of the methods which have been suggested for use in this type of study. We have seen that there are many methods to choose from, some offering particular advantages for certain types of problem. We have also studied the limitations of some of the methods.

The methods which we have discussed illustrate two rather different aims of scientists working in this area. First, they wish to measure the differences between populations, whatever their causes. Second, they wish to find out the causes of these differences, and to discover the nature of the biological connections between the populations.

These two points of view are both important in the next phase of this type of investigation, when we try to establish a "topography" of the group of populations being studied, on the basis of the distances we have calculated between them. In the next chapter, we shall study methods for doing this.

Further Reading

Yasuda, N., Morton, N. E.: Studies on human population structure. Proc. III Int. Congress Hum. Genet., p. 249–265. Baltimore: Johns Hopkins Press, 1966.

Chapter 15

Genetic Distance.
II. The Representation of Sets of Objects

When we have chosen a measure of distance between the objects we wish to study, whether these are individuals or populations, we will next want to represent the group of objects in some way that displays their distances from one another, and which enables us to detect subgroups of objects which lie close to one another; this will lead to the detection of classes of closely-related objects among the original group, and thus to increased understanding of the set of objects we are studying. Here, we are simply trying to describe a set of objects; we want to find the clearest possible representation of the real relations between them.

This is a very difficult task for the human mind, which finds it hard to think in terms of more than three (or, in many cases, two) dimensions. If we are considering a set of n objects, defined in terms of p characteristics, we will represent them in a space of p dimensions, E^p. When there are only a small number of objects, it may be possible to represent them in a space of lower dimension than p (for example, two points are situated on a straight line, and can be represented in a one-dimensional space, three points can be represented in two dimensions, and so on—n points can be represented in $(n-1)$-dimensional space). When $n-1$ is less than p, we can therefore represent the n objects in E^{n-1}.

The problem which we have to solve is to find the "best" representation of an $(n-1)$-dimensional, or p-dimensional, space, in two or three dimensions. This is similar to the problem geographers face when they want to represent the relief of an area on a two-dimensional map. Let us first consider what we mean by the "best" representation.

To make the problem concrete, let us consider the simplest case of reducing the number of dimensions; consider a set of points in two dimensions, which we wish to represent on a straight line (see Fig. 15.1). Fig. 15.1 shows that if we perform the reduction of dimensions by projecting the points onto the abscissa, for example, the form of the set of points will be considerably changed: two points A and B which are distant from one another can end up close in the new representation. A projection onto a line such as Δ, which is oriented in the same general

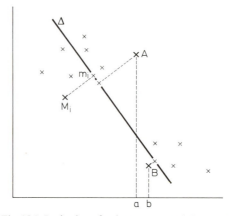

Fig. 15.1. Projection of points onto a straight line, Δ

direction as the scatter of the points themselves, preserves the distances between the objects (and the groupings into which they fall) much better.

If we replace each point M_i by its projection m_i onto the line Δ, we lose the information about the distance, $M_i m_i$ from M_i to Δ. Since we want to alter the set of distances between the objects as little as possible, we must choose the straight line Δ in such a way that we lose as little as possible of the information we possess. In practice, the line which leads to the least loss of information is taken to be the line for which the sum of the squares of the distances $M_i m_i$ is a minimum; this is called a "regression line". Writing $\overline{M_i m_i}$ for the vector representing the straight line joining M_i and m_i, we know that the square of the length of this line is equal to the scalar product:

$$\overline{M_i m_i} \cdot \overline{M_i m_i}$$

which we can also write as $(\overline{M_i m_i})^2$. Thus the quantity which we have to minimise in order to determine the equation of the regression line can be written in vector notation, as:

$$\sum_{i=1}^{n} (\overline{M_i m_i})^2 .$$

Note. The method of minimising the sum of squares is not the only possible one; an equally valid alternative is to choose a method which conserves as well as possible the order of the set of distances; yet another method might be to choose to maximise the correlation coefficient between the distances in the original space and those calculated after performing the projection. The choice of methods has been discussed by Lalouel (1973 a, 1973 b) who has shown that one can obtain very different representations of a set of objects, depending on the criterion chosen for the "best" projection.

1. Principal Components Analysis

First let us consider the case when each point M_i represents one of n objects which are each defined according to p co-ordinates, $x_{i1}, \ldots, x_{ip}$, in a system of orthogonal axes defined by the unit vectors $e_1, \ldots, e_p$. Distances between points are assumed to behave in the classical Euclidean manner.

If O is the origin of the axes, we have, using vector notation:

$$x_{ir} = \overline{OM_i} \cdot e_r$$

and the distance between two points is given by the formula:

$$d_{ij}^2 = \sum_r (x_{ir} - x_{jr})^2.$$

1.1. The First Principal Axis

We wish to find a straight line Δ, such that the sum of the squared distances:

$$\sum_{i=1}^{n} \overline{M_i m_i}^2$$

is a minimum, where m_i is the projection of M_i on Δ. We will call this line the "first principal axis" of the swarm of points.

We proceed by finding:
1. a point on Δ, and
2. a vector which has the same direction as Δ.

Theorem: The first principal axis passes through the centre of gravity of the points.

We can regard the points M_i representing the n objects in p-dimensional space, as points with unit mass. The co-ordinates of the centre of gravity, G, of the points are therefore equal to the means of the co-ordinates of the points in question:

$$\bar{x}_r = \frac{1}{n} \sum_{i=1}^{n} x_{ir}.$$

This can be written in vector notation:

$$\overline{OG} = \frac{1}{n} \sum_i \overline{OM_i}.$$

This result is also equivalent to:

$$\sum_i \overline{GM_i} = \text{the zero vector.} \tag{1}$$

The expression $\sum_i M_i m_i^2$, which we have to minimise, can now be seen to be equal to the moment of inertia of the scatter of points about the straight line Δ.

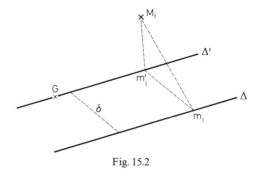

Fig. 15.2

Suppose that Δ does not pass through G. Draw the line Δ' parallel to Δ, and passing through G (see Fig. 15.2). Let the perpendicular distance between Δ and Δ' be δ, and let m_i' be the projection of M_i on Δ' (Fig. 15.2). We then have:

$$\sum_i (\overline{M_i m_i})^2 = \sum_i (\overline{M_i m_i'} + \overline{m_i' m_i})^2$$

$$= \sum_i (\overline{M_i m_i'})^2 + 2 \sum_i \overline{M_i m_i'} \cdot \overline{m_i' m_i} + \sum_i (\overline{m_i' m_i})^2$$

$$= \sum_i (\overline{M_i m_i'})^2 + n \delta^2. \tag{2}$$

This follows from the result:

$$\sum_i \overline{M_i m_i'} \cdot \overline{m_i' m_i} = \sum_i \overline{M_i G} \cdot \overline{m_i' m_i} + \sum_i \overline{G m_i'} \cdot \overline{m_i' m_i}.$$

Now, from Eq. (1) the first term is equal to zero, and the second is also zero, because $\overline{G m_i'}$ and $\overline{m_i' m_i}$ are orthogonal. Therefore the whole expression is equal to zero.

Eq. (2) is equivalent to the classical theorem of Huyghens:

"The moment of inertia of a solid about a straight line Δ is equal to the sum of the moment of inertia about a line Δ', through the centre of gravity of the object, and parallel to Δ, and the product of the mass of the solid multiplied by the square of the perpendicular distance between the two lines."[1]

Eq. (2) shows that $\sum_i \overline{M_i m_i}^2$ can be minimised only with respect to a line passing through G, since it is a function of δ, and has a minimum value when $\delta = 0$.

Theorem: The directional vector of the first principal axis is the eigenvector of the inertial matrix which corresponds to the largest eigenvalue of this matrix.

In order to define Δ, we have to find a vector with the same direction, in other words the unit vector y along the direction of Δ; the property of being a unit vector is equivalent to the statement that the sum of squares of its p elements is equal to 1, in any system of orthonormal axes. This can be expressed in vector notation as:

$$y \cdot y = 1. \tag{3}$$

Since $\overline{M_i m_i}$ and Δ are orthogonal, Pythagoras' theorem leads to (see Fig. 15.3):

$$\overline{M_i m_i}^2 = \overline{GM_i}^2 - \overline{G m_i}^2.$$

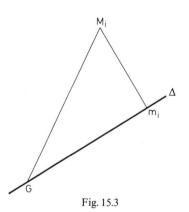

Fig. 15.3

---

[1] This result can also be expressed in non-mechanical terms. We note that the expression $\dfrac{1}{n} \sum_i \overline{m_i M_i}^2$ is the second moment of the distribution of the distances of the points M_i from Δ; $\dfrac{1}{n} \sum_i \overline{m_i' M_i}^2$ is the variance of this distribution. Thus, Eq. (2) corresponds to the classical expression:

$$\text{Variance} = 2\text{nd moment} - (\text{mean})^2.$$

Now $\sum_i \overline{\mathrm{GM}_i}^2$ is a constant (it is equal to the moment of inertia of the points about their centre of gravity). Therefore, in order to minimise $\sum_i \overline{\mathrm{M}_i\,\mathrm{m}_i}^2$, we have to maximise $\sum_i \overline{\mathrm{G}\,\mathrm{m}_i}^2$ (the moment of inertia of the set of points with respect to the plane orthogonal to Δ and passing through G). We shall designate this quantity to be maximised by I. It corresponds to the part of the total moment of inertia which remains, after we substitute the projections on Δ, m_i, for the points M_i.

In order to simplify the notation, we shall now take G as the origin of vectors, so that we can write M_i and m_i for the vectors $\overline{\mathrm{GM}_i}$ and $\overline{\mathrm{G}\,\mathrm{m}_i}$, respectively.

m_i is a vector in the same direction as the unit vector y which we have to find; its length is $y \cdot M_i$, or $M_i \cdot y$. We therefore have:

$$m_i = (M_i \cdot y)\,y,$$

$$m_i = y\,(y \cdot M_i).$$

Hence:

$$m_i^2 = m_i \cdot m_i$$

$$= (y \cdot M_i)(M_i \cdot y)(y \cdot y).$$

Shifting to matrix notation, and writing y' and M_i' for the transposes of y and M_i respectively, this is equal to the product:

$$y'\,(M_i\,M_i')\,y, \quad \text{since } y'y = 1.$$

Thus, finally:

$$I = \sum_i m_i^2 = y'\left[\sum_i (M_i\,M_i')\right]y.$$

The product $M_i M_i'$ is a matrix of order $p \times p$, whose general term is equal to the product:

$$(x_{ir} - \bar{x}_r)(x_{is} - \bar{x}_s)$$

of the r-th and s-th elements of the vector M_i.

The expression $\sum_i (M_i\,M_i')$ is also a $(p \times p)$ matrix, whose r, s-th term is:

$$\sum_i (x_{ir} - \bar{x}_r)(x_{is} - \bar{x}_s).$$

It is obvious that this matrix is symmetric. It is called the "inertial matrix" of the set of points. The same matrix, with each element divided by $1/n$ is called the "variance-covariance matrix", V, of the p characters:

$$V = \frac{1}{n}\sum_i M_i\,M_i'. \tag{4}$$

Finally, we can write:

$$I = n\,y'\,V\,y.$$

The problem now consists of finding a vector y such that:

$$y' V y$$

is maximised, subject to the condition $y' y = 1$.

It is shown in Appendix B, Section 4.2, that any real, symmetric matrix S of order p can be written in the form:

$$S = \lambda_1 v_1 v_1' + \cdots + \lambda_p v_p v_p'$$

where v_i is the eigenvector of the matrix S that is associated with the i-th eigenvalue. This is known as the "spectral analysis" of S. In this equation, the λ_r are the eigenvalues of S, and it can be shown (see Appendix B) that they are all real; the v_r are the corresponding eigenvectors, i.e. the vectors such that:

$$S v_r = \lambda_r v_r.$$

It can be shown that they are orthogonal to one another. We shall assume that the eigenvectors have been normalised.

Let λ_r and v_r be the eigenvalues and eigenvectors of the variance-covariance matrix V. By definition, we have:

$$V v_r = \lambda_r v_r$$

$$v_r' v_s = 0 \quad (r \neq s)$$

(5)

and, assuming that the eigenvectors have been normalised:

$$v_r' v_r = 1.$$

We can therefore write:

$$V = \sum_i \lambda_r v_r v_r'.$$

Hence:

$$y' V y = \sum_r \lambda_r y' v_r v_r' y.$$

But $y' v_r$ and $v_r' y$ represent the length ξ_r of the projection of the vector y which we wish to find, onto the eigenvector v_r. Hence:

$$y' V y = \sum_r \lambda_r \xi_r^2.$$

We have to maximise this expression, under the condition:

$$\sum_r \xi_r^2 = 1$$

which follows from the properties that y is a unit vector, and the eigenvectors v_r are orthonormal.

The classical method for finding the values of a set of variables $z_1, \ldots, z_p$ which maximise the function $f(z_1, \ldots, z_p)$, under the condition

$g(z_1, \ldots, z_p) = 0$, is to use a "Lagrange multiplier" K. We then set the p derivatives of the expression:

$$f(z_1, \ldots, z_p) - K g(z_1, \ldots, z_p)$$

equal to zero. This gives a set of $p+1$ equations in $p+1$ unknowns:

$$\frac{\partial f}{\partial z_i} - K \frac{\partial g}{\partial z_i} = 0 \qquad \text{for } i = 1 \ldots p$$

$$g(z_1, \ldots, z_p) = 0$$

In the present problem, this system of equations is:

$$\frac{\partial}{\partial \xi_1} \left[\sum_r \lambda_r \xi_r^2 - K \sum_r \xi_r^2 \right] = 2(\lambda_1 - K) \xi_1 = 0$$

$$\cdots\cdots\cdots\cdots\cdots\cdots\cdots\cdots\cdots\cdots\cdots\cdots\cdots\cdots\cdots$$

$$\frac{\partial}{\partial \xi_p} \left[\sum_r \lambda_r \xi_r^2 - K \sum_r \xi_r^2 \right] = 2(\lambda_p - K) \xi_p = 0$$

$$\sum_r \xi_r^2 = 1.$$

If K were different from all the eigenvalues λ_r, all the ξ_r would be zero, which is inconsistent with the condition:

$$\sum_r \xi_r^2 = 1.$$

K must therefore be equal to one of the eigenvalues, λ_t, say. Then all the ξ_r values other than ξ_t are zero, and:

$$\sum_r \xi_r^2 = \xi_t^2 = 1.$$

The corresponding value of I is:

$$I = \sum_r \lambda_r \xi_r^2 = \lambda_t.$$

The maximum value of I is therefore attained when K is equal to the largest eigenvalue, λ_1. The direction of the line Δ is then given by the direction of the corresponding eigenvector v_1; the moment of inertia of the set of points with respect to the plane perpendicular to Δ, and passing through G, is equal to λ_1.

1.2. The First Principal Surface

If we replace the original set of points M_i by their projection onto the first principal axis, we have seen that the loss of the information corresponding to the distances $\overline{M_i\,m_i}$ is a minimum; this loss of information may, however, be larger than we are prepared to accept, so that we would next try to represent the objects on a two-dimensional plane Γ, which we again choose so as to minimise the expression

$$\sum_i \overline{M_i\,m_i}^2$$

where m_i is now the projection of M_i on Γ (see Fig. 15.4).

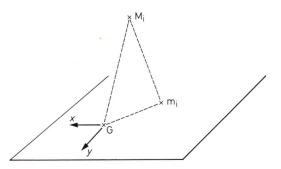

Fig. 15.4. Projection of a point onto a plane Γ

In order to define the plane Γ, it is sufficient to determine:
1. a point on Γ, and
2. two vectors y_1 and y_2 in the plane Γ. We shall assume that these are unit vectors, and that they are orthogonal to one another, so that:

$$y_1'\,y_1 = 1, \quad y_2'\,y_2 = 1, \quad y_1'\,y_2 = 1, \quad y_2'\,y_1 = 0, \tag{6}$$

From the appropriate version of Huyghens' theorem for two dimensions, rather than for straight lines, it follows that the plane Γ contains the centre of gravity, G, of the points.

Using Pythagoras' theorem, we can write:

$$\overline{M_1\,m_1}^2 = \overline{G\,M_i}^2 - \overline{G\,m_i}^2.$$

In order to minimise $\sum_i \overline{M_i\,m_i}^2$, we must maximise $\sum_i \overline{G\,m_i}^2$, since $\sum_i \overline{GM_i}^2$ is a constant. We shall designate $\sum_i \overline{G\,m_i}^2$ by J. It corresponds to the part of the total moment of inertia which remains after we have performed the projection onto the plane Γ, from the initial space.

As before, we shall take G as the origin of vectors. The elements of the vector $\overline{G\,m_i}$, in the system of axes defined by y_1 and y_2 are, respectively:

$$y_1'\,M_i \quad (\text{or } M_i'\,y_1)$$

and
$$y_2' \, M_i \quad \text{(or } M_i' \, y_2).$$

We therefore have:
$$m_i = (y_1' \, M_i) \, y_1 + (y_2' \, M_i) \, y_2$$
$$= (M_i' \, y_1) \, y_1 + (M_i' \, y_2) \, y_2.$$

Hence:
$$m_i^2 = m_i' \, m_i = [(y_1' \, M_i) \, y_1' + (y_2' \, M_i) \, y_2] \, [(M_i' \, y_1) \, y_1 + (M_i' \, y_2) \, y_2].$$

From Eqs. (6), this is equal to:
$$y_1' (M_i \, M_i') \, y_1 + y_2' (M_i \, M_i') \, y_2.$$

Finally, therefore:
$$J = \sum_i m_i^2$$
$$= y_1' \left(\sum_i M_i \, M_i' \right) y_1 + y_2' \left(\sum_i M_i \, M_i' \right) y_2.$$

The general term of the matrix $\sum_i M_i \, M_i'$ is equal to:
$$\sum_i (x_{ir} - \bar{x}_r)(x_{is} - \bar{x}_s).$$

This matrix is the inertial matrix $n \, V$, as defined above. Writing the matrix V in the form of its spectral analysis:
$$V = \sum_r \lambda_r \, v_r \, v_r'$$

we obtain:
$$J = \sum_r \lambda_r (y_1' \, v_r \, v_r' \, y_1 + y_2' \, v_r \, v_r' \, y_2).$$

But $y_1' \, v_r$ (or $v_r' \, y_1$) represents the length ξ_r of the projection of the vector y_1, which we have to find, on the eigenvector v_r; similarly, $y_2' \, v_r$ (or $v_r' \, y$) is the length η_r of the projection of the vector y_2 on this eigenvector. We thus have:
$$J = \sum_r \lambda_r (\xi_r^2 + \eta_r^2).$$

Our problem is to determine the values of ξ_r and η_r which maximise this expression, under the two conditions:
$$\sum_r \xi_r^2 = 1$$

and
$$\sum_r \eta_r^2 = 1.$$

This can be done by using two Lagrange multipliers, K_1 and K_2, and writing out the system of $2p+2$ equations in $2p+2$ unknowns (the p

elements ξ_r of y_1, the p elements η_r of y_2, and K_1 and K_2). The system of equations is as follows:

$$\frac{\partial}{\partial \xi_t}\left[\sum_r \lambda_r(\xi_r^2+\eta_r^2)-K_1\sum_r \xi_r^2-K_2\sum_r \eta_r^2\right]=0, \quad \text{for all } t$$

$$\frac{\partial}{\partial \eta_t}\left[\sum_r \lambda_r(\xi_r^2+\eta_r^2)-K_1\sum_r \xi_r^2-K_2\sum_r \eta_r^2\right]=0, \quad \text{for all } t$$

$$\sum_r \xi_r^2 = 1$$

$$\sum_r \eta_r^2 = 1$$

or

$$(\lambda_r-K_1)\,\xi_r=0, \quad \text{for all } r$$

$$(\lambda_r-K_2)\,\eta_r=0, \quad \text{for all } r$$

$$\sum_r \xi_r^2 = 1$$

$$\sum_r \eta_r^2 = 1.$$

To solve this system, we set K_1 equal to one of the eigenvalues, λ_t, which leads to $\xi_r=0$ if $r\neq t$ and $\xi_t=1$; K_2 is set equal to another eigenvalue, λ_s, so that $\eta_r=0$ when $r\neq s$, and $\eta_s=1$.

We then have:

$$J=\sum_r \lambda_r(\xi_r^2+\eta_r^2)$$

$$=\lambda_t+\lambda_s.$$

The maximum value of J is thus attained when the two eigenvalues λ_t and λ_s are the two largest eigenvalues λ_1 and λ_2. The plane is therefore defined by two vectors y_1 and y_2 which have the same directions as the two eigenvectors v_1 and v_2 of the inertial matrix which correspond to the two highest eigenvalues. It follows that the plane Γ contains the first principal axis Δ.

1.3. Generalisation

We have seen how to transform a set of points in multi-dimensional space into a set of points in one-dimensional space (the first principal axis, Δ, in the same direction as the eigenvector v_1), or into a set of points in two-dimensional space (the plane Γ defined by the eigenvectors v_1 and v_2). In a similar way, we could transform to a space of any dimensionality s, where $3\leq s\leq p$, in such a way as to lose the minimum of information; such a transformation distorts the distances between the objects as little as possible, and thus can be said to be the "best" representation of the set of points, in the sense in which we have defined this concept.

The projection onto a space of s dimensions is defined by s orthogonal vectors with the centre of gravity, G, of the points as their origin; the directions of these vectors are given by the eigenvectors of the variance-covariance matrix V, that correspond to the s largest eigenvalues of this matrix.

The moment of inertia which remains after this projection has been carried out is proportional to the sum of the s largest eigenvalues.

When the space onto which the points are projected is itself of p dimensions, it is identical with the original space. The inertia which is conserved after this projection is equal to n times the sum of the p eigenvalues of the matrix V, and this is equal to the total inertia.

We therefore see that the part of the total inertia which is conserved after projection onto a space of s dimensions (i.e. the fraction of the original information which has been retained) depends on the ratio:

$$\frac{\lambda_1 + \cdots + \lambda_s}{\lambda_1 + \cdots + \lambda_p}$$

where the $\lambda_1, \ldots, \lambda_p$ are the eigenvalues of the covariance matrix V, arranged in order of decreasing magnitude.

In p-dimensional space, when we have found the p "principal axes", we can use this to substitute new axes for the initial system of unit vectors $e_1, \ldots, e_p$; the new axes will be based on the eigenvectors $v_1, \ldots, v_p$ of the matrix V. The original measures $x_{i1}, \ldots, x_{ir}, \ldots, x_{ip}$ of the character will be replaced by new measures $y_{i1}, \ldots, y_{is}, \ldots, y_{ip}$, which are defined by $y_{is} = M'_i \cdot v_s$; these are called the "principal components" of the objects O_i (see Fig. 15.5).

The projection of the points in p-dimensional space onto a s-dimensional space is thus equivalent to considering the point O_i in terms of the first s components $y_{i1}, y_{i2}, \ldots, y_{is}$, and neglecting the others.

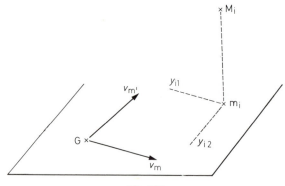

Fig. 15.5

We have seen that the fraction of the inertia of the original set of points that is conserved by this procedure is:

$$\frac{\lambda_1 + \lambda_2 + \cdots + \lambda_s}{\sum_k \lambda_k}.$$

This, however, is only a mean value. A given point may be affected much more or much less strongly than this. It is important to remember this, when attempting to interpret data by these methods.

1.4. Normalisation of Measures

We have so far assumed nothing about the measures x_{ir}, which we are using to characterise the objects we are studying. x_{i1} might, for example, be a weight, x_{i2} a length, x_{13} the intensity of some factor, x_{14} might be a code-number, and so on. The co-ordinates of the point M_i representing object i are obviously dependent on what units we choose for these measures.

The variance-covariance matrix V, and hence its eigenvalues, will also depend on these units. We can see this by means of an example. Consider a set of objects whose weights (in kg) and lengths (in cm) are known. Suppose the variance-covariance matrix is:

$$V = \begin{bmatrix} 8 & 2 \\ 2 & 5 \end{bmatrix}.$$

The eigenvalues of this matrix are the roots of the equation:

$$\lambda^2 - 13\lambda + 36 = 0.$$

Their values are $\lambda_1 = 9$, $\lambda_2 = 4$. The portion of the total inertia which is explained by the first principal axis is thus 9/13.

Now suppose that the weights were expressed in decigrams, and the lengths in millimetres. It is obvious that these changes do not affect the problem under study at all. In these units, the variance-covariance matrix becomes:

$$V = \begin{bmatrix} 8 \times 10^4 & 2 \times 10^3 \\ 2 \times 10^3 & 5 \times 10^2 \end{bmatrix}.$$

and the eigenvalues of this matrix are $\lambda_1 = 80050$, $\lambda_2 = 450$, so that the first principal axis explains 99.3 % of the total inertia.

In order to avoid this problem of the result of the principal components analysis depending on the units chosen, it is usual to "normalise" the units.

This is done by choosing as units the standard deviations of each of the characters studied (i.e. we replace x_{ir} by the ratio x_{ir}/σ_r, where

$\sigma_r^2 = 1/n \sum_i (x_{ir} - \bar{x}_r)^2$. In this system of units, the variance of each of the characters is equal to 1, and the variance-covariance matrix V becomes equal to the "correlation matrix":

$$\mathcal{V} = \begin{bmatrix} 1 & \cdots & r_{r1} & \cdots & r_{s1} & \cdots & r_{p1} \\ \vdots & & \vdots & & \vdots & & \vdots \\ r_{r1} & \cdots & 1 & \cdots & r_{rs} & \cdots & r_{pr} \\ \vdots & & \vdots & & \vdots & & \vdots \\ r_{s1} & \cdots & r_{rs} & \cdots & 1 & \cdots & r_{ps} \\ \vdots & & \vdots & & \vdots & & \vdots \\ r_{p1} & \cdots & r_{pr} & \cdots & r_{ps} & \cdots & 1 \end{bmatrix}.$$

1.5. Interpretation of the Projections Obtained. Representation of Characters

To summarise the procedure described above, we have n objects characterised in terms of their measures x_{ir}, with respect to p characters. The steps to be carried out are as follows:

1. Calculate the means $\bar{x}_r = \dfrac{1}{n} \sum_i x_{ir}$ of the various characters. These give the co-ordinates of the centre of gravity, G, of the set of objects.

2. Calculate the standard deviations:

$$\sigma_r = \left[\frac{1}{n} \sum_i (x_{ir} - \bar{x}_r)^2 \right]^{\frac{1}{2}}.$$

3. Find the correlation matrix $\mathcal{V}$; the r, s-th element of this matrix is equal to:

$$\frac{1}{n \sigma_r \sigma_s} \sum_i (x_{ir} - \bar{x}_r)(x_{is} - \bar{x}_s).$$

4. Calculate the eigenvalues $\lambda_1, \dots, \lambda_p$ of $\mathcal{V}$, in order of decreasing magnitude.

5. Calculate the normalised eigenvectors $v_1, \dots, v_r, \dots, v_p$ corresponding to the eigenvalues. The elements of the eigenvectors are $v_{r1}, \dots, v_{rp}$; in terms of the original system of axes, we can write:

$$v_{rt} = v_r' e_t.$$

6. Finally, calculate the co-ordinates $y_{i1}, \dots, y_{ip}$ of the n objects in the system of axes defined by the eigenvectors. These are calculated from the formula:

$$y_{ir} = \sum_s x_{is} v_{rs} = M_i' v_r.$$

In practice, we will usually be concerned with a two-dimensional representation, on the plane Γ defined by the centre of gravity G and by the vectors v_1 and v_2; the objects are represented by the points m_i, whose co-ordinates are y_{i1} and y_{i2}. We have seen that these points constitute the "best" representation in two dimensions of the original set of points in p dimensions. The distances between the points are, in general, as near as is possible to the actual distances between the objects; to put this precisely, we have:

$$\sum_i \overline{Gm_i}^2 = \frac{\lambda_1 + \lambda_2}{\sum_r \lambda_r} \sum_i \overline{GM_i}^2.$$

The ratio $\dfrac{\lambda_1 + \lambda_2}{\sum_r \lambda_r}$ is a measure of the quality of the reproduction

of the set of objects, in two dimensions; if this ratio appears to us to be large, the grouping of the points in the two-dimensional representation will enable us to class the objects into sets of "close" objects, with reasonable accuracy. We can thus use the arrangement of the set of projected points m_i to give us an idea of the structure of the original set of n objects.

If, however, we consider that this ratio is low, the plane representation of the objects is insufficiently informative. The next step is to try a projection into three-dimensional space, defined by the plane Γ together with the vector v_3. In some cases, a four-dimensional representation may be necessary.

When we come to try to interpret the set of points m_i, it is helpful to understand the relation between the co-ordinates of these points, i.e. the principal components y_{i1} and y_{i2}, and the original measures x_{ir}, which we shall assume to have been normalised. In order to help us in seeing these relationships, we shall calculate the correlation r_{rs} between the principal component y_s and the original character x_r.

By definition:

$$r_{rs} = \frac{1}{n\sigma_r\sigma_s} \sum_i (y_{ir} - \bar{y}_r)(x_{is} - \bar{x}_s)$$

$$= \frac{1}{n\sigma_r\sigma_s} \sum_i (v'_s M_i)(M'_i e_r).$$

Now the x_{ir} are normalised measures, so we have $\sigma_r = 1$.

$$\sigma_s^2 = \frac{1}{n} \sum_i (y_{is} - \bar{y}_s)^2$$

$$= \frac{1}{n} \sum_i (v'_s M_i)(M'_i v_s).$$

From Eq. (4) we have:

$$V = \frac{1}{n} \sum_i M_i M_i'.$$

Hence:

$$\sigma_s^2 = v_s' V v_s$$
$$= \lambda_s v_s' v_s$$
$$= \lambda_s, \quad \text{from Eq. (5).}$$

Consequently we find:

$$r_{rs} = \frac{1}{\sqrt{\lambda_s}} v_s' \left(\sum_i M_i M_i'\right) e_r$$

$$= \frac{1}{\sqrt{\lambda_s}} v_s' V e_r$$

$$= \frac{1}{\sqrt{\lambda_s}} v_s' \left(\sum_t \lambda_t v_t v_t'\right) e_r.$$

From point (5) on p.476, this is equal to:

$$\sqrt{\lambda_s} \, v_s' e_r = \sqrt{\lambda_s} \, v_{sr}.$$

The correlation coefficient between the s-th principal component and the r-th character is thus equal to the r-th element of the s-th eigenvector v_s, multiplied by the square root of the s-th eigenvalue λ_s.

It is helpful to mark the points γ_r, whose co-ordinates are $(\sqrt{\lambda_1} \, v_{1r}, \sqrt{\lambda_2} \, v_{2r})$, on the plane diagram on which the projected points m_i are plotted. Each point γ_r can be considered as a representation of the r-th character, just as the points m_i represent the objects O_i. If the r-th character is strongly related to the first principal component, the co-ordinate of γ_r with respect to the first axis, will be large. The γ_r therefore help us to interpret the principal components, in terms of the original characters.

Notice that the distance $G\gamma_r$ from the origin of the axes to the point γ_r is such that:

$$(\overline{G\gamma_r})^2 = r_{1r}^2 + r_{2r}^2.$$

In other words it is equal to the partial correlation coefficient between character C_r and the two first principal components. This distance takes values between 0 and 1; it enables us to assess the extent to which the r-th character plays a part in the plane representation of the objects (see Fig. 15.6).

In summary, the interpretation of the projection of a set of points onto the plane Γ involves the consideration of the following distances:

1. Between the points m_i, which represent the objects.
2. Between the points m_i and the principal axes.

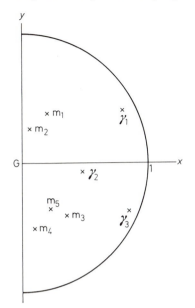

Fig. 15.6. Representation of a set of objects on the first principal plane. The character points, γ_r, are also shown

3. Between the principal axes and the points representing the characters, γ_r.

4. Between the points γ_r and the circle of radius 1 centred at the origin of the axes.

The distances between the object-points m_i and the character-points γ_r have however, no meaning. We shall see below that this is not the case in a special type of principal components analysis, called principal components analysis of contingency tables.

2. Principal Components Analysis of Contingency Tables[2]

Consider two systems for characterising objects: I, with n discrete classes $1, \ldots, i, \ldots, n$, and R, with p classes $1, \ldots, r, \ldots, p$. Suppose we make N observations and draw up a "contingency table" of the numbers N_{ir} of objects that fall into the i-th class of classification-system I, and the r-th class of system R (Table 15.1).

---

[2] The method described below was developed by Benzecri (1970), who called it "analyse factorielle des correspondances". A fuller treatment is given by Morlat et al. (1971).

For example, suppose that the classes of system I are populations, and those of system R the various alleles at a particular locus. In this example, the numbers N_{ir} are the numbers of A_r alleles found in the i-th population.

Table 15.1. A contingency table

		R					
		1	...	r	...	p	Totals
I	1	N_{11}	...	N_{1r}	...	N_{1p}	.
	$\vdots$	$\vdots$		$\vdots$		$\vdots$	$\vdots$
	i	N_{i1}	...	N_{ir}	...	N_{ip}	N_i
	$\vdots$	$\vdots$		$\vdots$		$\vdots$	$\vdots$
	n	N_{n1}	...	N_{nr}	...	N_{np}	.
	Totals	.	...	N_r	...	.	N

It sometimes seems natural to consider one of the systems of classification, say I, as defining objects, while the other, R, can be considered to define a set of discrete characters. There is, however, no fundamental difference between "objects" and "characters", since "characters" can be defined in terms of the objects which manifest them, just as objects can be defined in terms of the characters they manifest. For example, we could study resemblances between alleles in terms of the populations in which they are found, or resemblances between populations in terms of the alleles which are found in them.

To simplify the notation, we shall use the indices i, j, etc. for the "object" classes, and r, s etc. for the "character" classes. It is important to remember that this distinction is completely arbitrary.

2.1. The χ^2 Metric

Consider a table of N observations, giving the $n \times p$ numbers N_{ir} of observations of the r-th character in the i-th type of object (see Table 15.1). The table shows, in addition to the N_{ir} values, the marginal totals:

$$N_i = \sum_r N_{ir} \quad \text{and} \quad N_r = \sum_i N_{ir}.$$

It is obvious that:

$$N = \sum_{ir} N_{ir} = \sum_i N_i = \sum_r N_r.$$

In what follows, we shall be concerned, not with the numbers N_{ir} themselves, but with the frequencies:

$$p_{ir} = \frac{N_{ir}}{N}$$

$$p_i = \frac{N_i}{N}$$

$$p_r = \frac{N_r}{N}.$$

The i-th type of object is characterised, not by the numbers N_{ir}, or the proportions p_{ir}, but by the relative numbers of that type of object which fall into the various possible character classes, in other words by the conditional frequencies:

$$\frac{N_{ir}}{N_i} = \frac{p_{ir}}{p_i}.$$

We shall use the notation $p_{r|i}$ for these conditional frequencies. The i-th and j-th types of objects are identical, from the point of view of the p characters studied, if $p_{r|i} = p_{r|j}$, for all the p characters. The more different are these values, the more dissimilar are the objects. We can therefore measure the dissimilarity of two types of object by using their $p_{r|i}$ values.

The most appropriate measure of distance in the present case would seem to be χ^2, in view of the properties of this measure, which we discussed in Section 1 of Chapter 14. We can therefore write the distance between the i-th and j-th types of object as:

$$D^2(i, j) = \sum_r \frac{1}{p_r} (p_{r|i} - p_{r|j})^2. \tag{7}$$

By symmetry, the distance between the r-th and s-th characters is defined as:

$$D^2(r, s) = \sum_i \frac{1}{p_i} (p_{i|r} - p_{i|s})^2. \tag{8}$$

We now make the change of variable:

$$x_{ir} = \frac{p_{r|i}}{\sqrt{p_r}} = \frac{p_{ir}}{p_i \sqrt{p_r}} = \frac{N_{ir} \sqrt{N}}{N_i \sqrt{N_r}}.$$

Then we have:

$$D(i, j) = \sum_r (x_{ir} - x_{jr})^2.$$

This shows that this measure of distance, D, is equivalent to classical Euclidean distance in p-dimensional space, in which points are defined by the co-ordinates $x_{i1}, \ldots, x_{ip}$.

By another change of variable:

$$\xi_{ir} = \frac{p_{ir}}{p_r \sqrt{p_i}}$$

we see that:

$$D^2(r, s) = \sum_i (\xi_{ir} - \xi_{is})^2.$$

2.2. The Projection of the Object-Points Onto the Principal Plane

The objects O_i are defined by their co-ordinates x_{ij} in p-dimensional space, and we have just seen that this space has the properties of classical Euclidean space. We can now carry out a principal components analysis, by the steps outlined in Section 1 of this chapter. In other words, we calculate the following parameters:

1. The means:

$$\bar{x}_r = \sum_i p_i x_{ir}$$

$$= \sum_i \frac{p_{ir}}{\sqrt{p_r}}$$

$$= \sqrt{p_r}.$$

2. The variances:

$$V_r = \sum_i p_i x_{ir}^2 - \bar{x}_r^2$$

$$= \sum_i \frac{p_{ir}^2}{p_i p_r} - p_r. \tag{9}$$

3. The covariances:

$$\text{Cov}(r, s) = \sum_i (p_i x_{ir} x_{is}) - \bar{x}_r \bar{x}_s$$

$$= \sum_i \frac{p_{ir} p_{is}}{p_i \sqrt{p_r p_s}} - \sqrt{p_r p_s}. \tag{10}$$

4. The variance-covariance matrix, V_c, for the characters. The above formulae (Eqs. (9) and (10)) give the elements of this matrix. If we write T for the matrix of order $n \times p$ whose i, r-th element is:

$$t_{ir} = \frac{p_{ir}}{\sqrt{p_i p_r}}$$

and r for the vector of p elements $r_r = \sqrt{p_r}$, it is clear from the formulae giving the elements of V_c that:

$$V_c = T' T - rr'.$$

5. The eigenvalues λ_k and the eigenvectors v_k of V_c. These are such that:

$$V_c v_k = \lambda_k v_k$$

or

$$T' T v_k - r r' v_k = \lambda_k v_k. \tag{11}$$

Inspection of Eq. (11) shows that the vector r is itself one of the eigenvectors of V_c; it corresponds to the eigenvalue $\lambda = 0$. We can see this as follows. $T' Tr$ is a vector with p elements, whose r-th element is:

$$\sum_s \sum_l \frac{p_{lr} p_{ls}}{p_l \sqrt{p_r p_s}} \sqrt{p_s} = \frac{1}{\sqrt{p_r}} \sum_l \frac{p_{lr} p_l}{p_l} = \sqrt{p_r}.$$

Hence:

$$T' Tr = r. \tag{12}$$

Now $r r' r$ is also equal to r, since $r' r = \sum_r p_r = 1$. We therefore have:

$$T' Tr - r r' r = 0 \times r.$$

The moment of inertia of the set of object-points with respect to the hyperplane passing through the centre of gravity and orthogonal to r is, as we have seen, equal to the eigenvalue with which r is associated. Now we have just seen that this eigenvalue is zero. It follows that all the points lie on this hyperplane. (This corresponds to the property, which follows directly from the definition of x_{ir}, that:

$$\sum_r \sqrt{p_r}(x_{ir} - \sqrt{p_r}) = 0.)$$

The other eigenvectors v_k which we have to find are orthogonal to r; this is equivalent to the statement:

$$r' v_k = 0.$$

Eq. (11) thus becomes:

$$T' T v_k = \lambda_k v_k. \tag{13}$$

The eigenvectors of V_c are therefore eigenvectors of $T' T$. From Eq. (12), however, it is obvious that Eq. (13) has the solution $v = r$ and $\lambda = 1$, which corresponds to the zero-valued eigenvalue of V. We are thus only concerned with the other eigenvectors of $T' T$ which define the "principal components" of the set of object-points.

To summarise, therefore, the plane we wish to find is defined by:

1. The point G, whose co-ordinates are given by $\sqrt{p_r}$.

2. The two eigenvectors v_1 and v_2 of the matrix $T'T$ corresponding to the two largest eigenvalues other than 1. T is the matrix of order $n \times p$ whose general term is:

$$t_{ir} = \frac{p_{ir}}{\sqrt{p_i p_r}}.$$

2.3. The Principal Plane of the Character-Points

The measure of distance between two characters, as defined by Eq. (8), shows that we can replace each of the original characters by a vector with n elements $\xi_{1r}, \ldots, \xi_{ir}, \ldots, \xi_{nr}$, where

$$\xi_{ir} = \frac{p_{ir}}{p_r \sqrt{p_i}}$$

and where distances are measured in n-dimensional Euclidean space.

The set of character-points can obviously be analysed by the type of principal components analysis which we have just described, with reference to the object-points. The calculations are similar to the earlier ones, except that summation over i is replaced by summation over r, and N_i is replaced by N_r, and *vice versa*.

It is easy to see that the variance-covariance matrix for the objects, V_o, has the general term:

$$\mathrm{Cov}\,(i,j) = \sum_r (p_r \xi_{ir} \xi_{jr}) - \bar{\xi}_i \bar{\xi}_j$$

$$= \sum_r \frac{p_{ir} p_{jr}}{p_r \sqrt{p_i p_j}} - \sqrt{p_i} \sqrt{p_j}.$$

If we write s for the vector of n elements $s_i = \sqrt{p_i}$, we can write the matrix V_o as:

$$V_o = TT' - ss'$$

where T is the same matrix as in the last section.

It is possible to show that s is an eigenvector of the matrix V_o, and that it corresponds to the eigenvalue zero. The other eigenvectors of V_o are orthogonal to s, and are therefore such that:

$$TT'w_l = \lambda_l w_l. \tag{14}$$

This equation for the eigenvectors of V_o shows that they are simply related to the eigenvectors v_k of V_c; the corresponding equation for the v_k is:

$$T'Tv_k = \lambda_k v_k. \tag{15}$$

Multiplying on the left by T, we have:

$$TT'(Tv_k) = \lambda_k(Tv_k).$$

Comparing this with Eq. (14), we see that Tv_k is an eigenvector, and λ_k an eigenvalue, of V_o. By a similar argument, we can show that every eigenvalue of V_o is an eigenvalue of V_c; the two variance-covariance matrices thus have the same eigenvalues, and the normalised eigenvectors v_k and w_k corresponding to a given eigenvalue λ_k are related according to:

$$w_k = KTv_k$$

where K is a number that normalises w_k, i.e. such that the following equation is valid:

$$w'_k w_k = K^2 v'_k T' Tv_k = 1.$$

From Eq. (15), we can write this in the form:

$$K^2 v'_k \lambda_k v_k = 1$$

and, since v_k is normalised by definition:

$$K = \frac{1}{\sqrt{\lambda_k}}.$$

In summary, the eigenvectors w_k which define the principal axes of the set of character-points are related to those for the object-points by:

$$w_k = \frac{1}{\sqrt{\lambda_k}} Tv_k \tag{16}$$

and the symmetrical equation:

$$v_k = \frac{1}{\sqrt{\lambda_k}} T' w_k. \tag{17}$$

2.4. Interpretation of the Simultaneous Representation of Objects and Characters

Each of the first two axes explains a part of the total inertia of the system; this part is equal to $\lambda_1/\sum \lambda$ for the first axis, and $\lambda_2/\sum \lambda$ for the second. This is true for the analysis applied to the objects, or to the characters. It is therefore possible to project both sets of points onto the same plane. On this plane, the co-ordinates of the i-th object-point are given by:

$$y_{i1} = \sum_r v_{1r} x_{ir} = \sum_r v_{1r} \frac{p_{ir}}{p_i \sqrt{p_r}} \tag{18}$$

$$y_{i2} = \sum_r v_{2r} \frac{p_{ir}}{p_i \sqrt{p_r}}$$

and the co-ordinates of the r-th character-point are given by:

$$y_{1r} = \sum_i w_{1i} \xi_{ir} = \sum_i w_{1i} \frac{p_{ir}}{p_r \sqrt{p_i}}$$

$$y_{2r} = \sum_i w_{2r} \frac{p_{ir}}{p_r \sqrt{p_i}}.$$

Eq. (17) shows that:

$$v_{1r} = \frac{1}{\sqrt{\lambda_1}} \sum_k \frac{p_{kr}}{\sqrt{p_k p_r}} w_{1k}.$$

Hence, substituting into Eq. (18):

$$y_{i1} = \frac{1}{\sqrt{\lambda_1}} \sum_r \sum_k \frac{p_{ir} p_{kr}}{p_i p_r \sqrt{p_k}} w_{1k}$$

$$= \frac{1}{\sqrt{\lambda_1}} \sum_r p_{r|i} y_{1r} \tag{19}$$

$$y_{i2} = \frac{1}{\sqrt{\lambda_2}} \sum_r p_{r|i} y_{2r}.$$

The i-th object-point is therefore situated at a point whose co-ordinates are those of the centre of gravity of the character-points, after weighting by the frequencies $p_{r|i}$ of the r-th character in the i-th type of object, and multiplying by $1/\sqrt{\lambda_1}$, and $1/\sqrt{\lambda_2}$.

Similarly, we can show that:

$$y_{1r} = \frac{1}{\sqrt{\lambda_1}} \sum_i p_{i|r} y_{i1}$$

$$y_{2r} = \frac{1}{\sqrt{\lambda_2}} \sum_i p_{i|r} y_{i2}$$

which shows that the point representing the r-th character lies at a position whose co-ordinates are those of the centre of gravity of the object-points, weighted by the frequencies with which the different types of object are found among objects manifesting the r-th character, and multiplied by the coefficients $1/\sqrt{\lambda_1}$ and $1/\sqrt{\lambda_2}$.

These properties lead to a set of simple rules for interpreting the sets of points on a plane diagram, which we obtain from this type of analysis. We no longer have to try to interpret the principal axes in terms of the original characters, but need only note that the i-th type of object is represented by a point m_i, which is close to the point, γ_r, representing the r-th character, if this character occurs with high frequency in this type of object. The clusters of object-points which we can notice in the plane representation are therefore easy to characterise in terms of the character-points γ_r which lie close to them. We shall study some examples which illustrate this later on.

3. Cluster Analysis

Principal components analysis allows us to obtain the best possible description of a set of objects, given that we cannot easily take into account more than two or three dimensions. We must, however, realise that the clusters of objects on the projection plane may be spurious; it is possible for the projection m_i of an object M_i to fall into a given cluster of points, although the object in question is really not close to the objects represented by the cluster. The method we have described ensures that the risk of this type of error is, in general, the lowest that is possible, but this does not give any assurance with respect to a particular object. It is therefore advisable to subject the data to an analysis which takes all the characteristics of the objects into account, and which does not involve a decrease in the number of dimensions of the space in which the objects are defined.

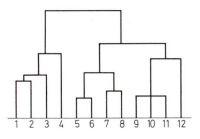

Fig. 15.7. A tree diagram

In this type of analysis, we are no longer in the position of a geographer trying to make a plane "map" of the set of objects under study; we try to group the objects into classes of similar objects, and then to group these classes into higher-order classes, and so on. This type of analysis can be represented in the form of a tree-diagram (see Fig. 15.7). Object 8, for example, is close to object 7, and these two objects can be considered to form a class of similar objects; in the next stage of classification into higher-order groupings, the pair of objects 7 and 8 might be associated with the pair 5 and 6, and these, at a higher level yet, with objects, 9, 10, 11 and 12, and so on. The number of classes that will be defined will depend on what criterion we adopt to decide whether two objects are sufficiently close to be assigned to the same class, or not (in other words on the sensitivity of the system of classification we use).

Many methods of making such hierarchical classifications have been proposed. Some, for example the algorithm of Williams and Lambert (1960), divide up the set of objects by means of "cuts" into successively narrower classes; this class of methods is referred to as "descending" classification. The "ascending" methods of classification group the objects into successively broader classes. Referring to Fig. 15.7, descending methods start by dividing the objects into two groups (objects 1–4 and 5–12), then by dividing the first group into two (objects 1–3 and 4) and so on. The ascending methods would start by forming the class of the closest elements (in this example, the pair 5 and 6), then the next closest (7 and 8), then these four would be classed together, and so on.

Whatever method is used, it is important to realise the extent to which the classification is arbitrary. When we define an algorithm to perform a classification, we will choose one that is compatible with the aim of the study we are carrying out; the results we obtain will depend on this choice, and could be quite different had another algorithm been used. In what follows, we shall confine ourselves to a study of an "ascending" method of analysis, which is related in a simple fashion to the χ^2 measure of distance between objects. We shall consider a set of n objects, each defined in terms of p characters. We will thus have a table of numbers, such as Table 15.1.

3.1. Information and Variance

What does a table such as Table 15.1 tell us? It gives the numbers N_{ir} (or the proportions p_{ir}) of the associations of the i-th type of object with the r-th character. We may compare the information contained in such a table with the information we would have if we knew only the marginal totals N_i and N_r (or the marginal proportions p_i and p_r).

The ratio of the information contained in the full table to that contained in the marginal totals is given by the formula (Shannon, 1948):

$$I = \sum_{ir} p_{ir} \log_e \frac{p_{ir}}{p_i p_r}. \tag{20}$$

Suppose that p_{ir} is not very different from the product $p_i p_r$; more precisely, suppose that a_{ir} is defined by:

$$p_{ir} = (1 + a_{ir}) p_i p_r$$

and that a_{ir} is sufficiently small for its square to be negligible compared with itself. Expanding Eq. (20), and neglecting terms of the order of a_{ir}^2, compared with a_{ir}, we obtain:

$$I = \sum_{ir} p_i p_r (1 + a_{ir}) \log_e (1 + a_{ir})$$

$$\approx \sum_{ir} p_i p_r (1 + a_{ir}) \left(a_{ir} - \frac{a_{ir}^2}{2} \right)$$

$$\approx \sum_{ir} p_i p_r \frac{a_{ir}^2}{2}$$

since:

$$\sum_{ir} p_i p_r a_{ir} = \sum_{ir} p_{ir} - \sum_{ir} p_i p_r = 0.$$

Hence:

$$I \approx \frac{1}{2} \sum_{ir} p_i p_r \left(\frac{p_{ir} - p_i p_r}{p_i p_r} \right)^2$$

$$= \frac{1}{2} \left(\sum_{ir} \frac{p_{ir}^2}{p_i p_r} - 1 \right) \tag{21}$$

$$= \frac{1}{2} \sum_r V_r$$

where V_r is the variance of the r-th character, on the scale:

$$x_{ir} = \frac{p_{ir}}{p_i \sqrt{p_r}}$$

defined in the previous section. The total amount of information contained in the table of N_{ir} values is therefore proportional to the total variance of the set of object-points.

When we group two objects into a single "class", we would like to lose as little as possible of the information which we have; from Eq. (21) we see that this is equivalent to the requirement that the variance should be diminished by as small a quantity as possible.

3.2. Aggregation of Two Objects

Consider two objects i and j; when we aggregate them into a class of similar objects, we are replacing the two original objects by a single, imaginary object C, the class to which they belong. We shall assign a weight to the class, according to the frequencies of the two types of object; the weight $p_i + p_j$ will be used. The proportion of objects of class C which show character r is:

$$p_{r|C} = \frac{p_{ir} + p_{jr}}{p_i + p_j}.$$

The general mean of character r remains the same, $\sqrt{p_r}$, as before the aggregation, but the variance of this character is decreased; by definition, this variance is:

$$V_r = \sum_i \frac{p_{ir}^2}{p_i\, p_r} - p_r$$

so that, writing V_r' for the variance after the aggregation, we have:

$$V_r - V_r' = \frac{p_{ir}^2}{p_i\, p_r} + \frac{p_{jr}^2}{p_j\, p_r} - \frac{(p_{ir} + p_{jr})^2}{(p_i + p_j)\, p_r}$$

$$= \frac{p_{ir}^2\, p_j^2 + p_{jr}^2\, p_i^2 - 2\, p_{ir}\, p_{jr}\, p_i\, p_j}{p_r\, p_i\, p_j (p_i + p_j)}$$

$$= \frac{p_i\, p_j}{p_i + p_j} \left[\frac{1}{p_r} \left\{ \frac{p_{ir}}{p_i} - \frac{p_{jr}}{p_j} \right\}^2 \right].$$

The aggregation of the i-th and j-th objects therefore causes the least decrease in the amount of information, when $\sum_r V_r'$ is as close as possible to $\sum_r V_r$, in other words when the expression:

$$\Delta_{ij}^2 = \frac{p_i\, p_j}{p_i + p_j} \sum_r \frac{1}{p_r} \left\{ \frac{p_{ir}}{p_i} - \frac{p_{jr}}{p_j} \right\}^2$$

is a minimum. But, by the definition of the χ^2 measure of distance D_{ij} between two objects, this is equal to:

$$\frac{p_i\, p_j}{p_i + p_j}\, D_{ij}^2. \tag{22}$$

The method of carrying out the first aggregation step is therefore as follows:

1. Calculate the value of Δ^2 for each of the $\dfrac{n(n-1)}{2}$ pairs which can be formed among the objects.

2. Find the pair, i and j, for which Δ^2 is the smallest.

3. Replace objects i and j by the imaginary object C, to which we assign the weight $p_i + p_j$, and which has a profile with respect to the p characters which is defined by:

$$p_{r|C} = \frac{p_{ir} + p_{jr}}{p_i + p_j}.$$

At the end of this procedure, we have a set of $n-1$ objects, which can then be subjected to the same procedure, and reduced to $n-2$ objects, and so on. The sequence stops when we reach a single "object", which is, of course, the set of the n original objects.

3.3. Interpretation of the Decrease in Variance:
The Diameter of a Class

We have seen that the aggregation of two objects i and j decreases the overall variance of the set of characters, $\sum_r V_r$, by an amount Δ_{ij}^2, according to Eq. (22); equivalently, Δ_{ij}^2 can be given a concrete meaning, by the following analogy. We can consider the points representing the objects as having masses p_i and as situated at points in Euclidean space which are defined by the co-ordinates:

$$x_{ir} = \frac{p_{ir}}{p_i \sqrt{p_r}}.$$

The co-ordinates of the centre of gravity g of two points P_i and P_j are given by:

$$x_{gr} = (p_i x_{ir} + p_j x_{jr})/(p_i + p_j)$$

$$= \frac{p_{ir} + p_{jr}}{(p_i + p_j)\sqrt{p_r}}.$$

The moment of inertia of these two points about their centre of gravity is:

$$I_g = p_i \sum_r (x_{ir} - x_{gr})^2 + p_j \sum_r (x_{jr} - x_{gr})^2.$$

Replacing the x_{ir} by the expression given above, we have:

$$I_g = \frac{p_i p_j}{p_i + p_j} \sum_r \frac{1}{p_r} \left\{ \frac{p_{ir}}{p_i} - \frac{p_{jr}}{p_j} \right\}^2$$

$$= \Delta_{ij}^2.$$

This is the classical result which we encountered earlier (Section 1.1): the moment of inertia of a set of points about a point 0 is equal to their moment of inertia about their centre of gravity, plus the moment of inertia of the centre of gravity, with respect to 0.

Thus, when we replace two objects i and j by the class C, we substitute the centre of gravity g for the two original masses, so that the total inertia I of the set of points is decreased by an amount equal to the moment of inertia I_g of the two points with respect to g. I_g is called the "diameter" of the class C. Similarly, if we aggregate the objects into successively fewer and fewer classes, the aggregation of two classes C_1 and C_2 into a single class C_3 is equivalent to the substitution of two masses g_1 and g_2 by their centre of gravity, g_3; the total inertia lost by all the aggregation steps before the aggregation into the point g_3 is called the "diameter" of the class C_3.

It is useful to be able to display the diameters of the various classes on the tree-diagram representing the set of objects. This can be done if we assign co-ordinates to the nodes corresponding to the aggregation steps, such that these co-ordinates are equal to the diameters of the classes formed (see Fig. 15.8).

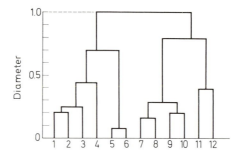

Fig. 15.8. The results of a cluster analysis, displayed as a tree diagram

When the diameter of a class of two objects is small, the objects are "close" to one another. Thus a tree-diagram showing the diameters of the classes gives an excellent representation of the distances between the objects. In Fig. 15.8, for example, objects 5 and 6 are seen to be very close, since their aggregation into a single class involves a loss of information of less than 10 %; objects 5 and 8, on the other hand, are very distant from one another, and cannot be grouped together without the loss of all the information obtained originally.

One can also use such a tree-diagram to choose the number of classes which conserves the maximum amount of information, consistent with the sensitivity of the classification that is desired. For example, for the objects represented in Fig. 15.6, if we group the 12 objects into four classes, we lose only 45 % of the original information, whereas, if we group into three classes, 70 % of the information is lost. In this example, provided that this is compatible with the aim of the study, we would stop grouping when four groups had been formed.

3.4. Phylogenetic Trees

The type of tree-diagram which we have been discussing must not be confused with the well-known method of representing evolutionary relationships by a "phylogenetic tree". In Section 3.3, we simply wished to group the objects under study (which might, for example, be populations) into classes of defined homogeneity. The aim of a phylogenetic tree is the far more ambitious one of displaying the paths of descent which lead to present-day populations; the nodes in such a diagram are intended to represent real populations, which once existed. This is naturally a fascinating field, but the interpretation of data represented in a phylogenetic tree is often of doubtful accuracy, because of the large number of assumptions that are necessary. For a review of this field, see Cavalli-Sforza and Bodmer (1971).

In the next chapter, we shall discuss some applications of the descriptive methods which we have studied in this and the preceding chapter.

Chapter 16

Some Studies of Human Populations

In this chapter, we shall discuss some of the results obtained by the French school of human population geneticists. Some of these studies are still unfinished, since studies of human populations are necessarily long-term projects; some of the results are provisional, but we feel that they may be of interest to the English-speaking reader, since accounts of this work have not until now been available in English.[1]

The studies which we shall describe concern populations which seem likely to constitute "genetic isolates". This type of population appears to us to be of particular interest, for the following reasons. First, isolates may give us the opportunity to study the effects of different environmental factors on human populations, and thus to make some of the kinds of observations which would be possible if we were working with an animal with which we could do experiments. It is obvious that experiments on human populations are impossible: it would be immoral to perform them, and it would take too long. But we may sometimes have the good fortune to be able to make observations on human groups who are living in circumstances that allow us to study the effects of the particular factors in which we are interested.

Second, many human isolates live in extreme conditions, and they are frequently populations of limited size. Both these factors may promote relatively fast changes in their genetic structures and this makes them particularly favourable material for the study of human evolution.

Thirdly, the effects of evolutionary factors, such as natural selection and genetic drift, are often quite small, whereas migration can have an important effect on populations. If, therefore, we wish to study factors other than migration, we will try to choose material for study where migration is known to be absent, or at least very slight.

---

[1] Other studies which make use of the technique of collecting population pedigrees, include the following: Gessain's (1970) study of the Ammassalik eskimos of the Greenland coast; Robert's (1970) study of the Ammassalik colony at Scoresbysund; Jaeger's (1973) work on the village of Sara, Central African Republic; Djemai's (1972) work on a Berber village in southern Tunisia. Sutter (1968) has also studied a number of Breton villages, using this method.

Fourthly, it is only in a small, completely defined population, in other words an isolate, that we can hope to describe in detail the changes in genetic composition of the group, to understand fully the structure of the group itself, and to place it in its true relation to the structure of the set of populations it belongs to.

1. The Jicaque Indians of the Montaña de la Flor, Honduras

1.1. History of the Group

In the year 1870, eight Jicaque Indians escaped from their Spanish and half-caste masters, and took refuge in the mountains, where they lived in complete isolation (in the region of the Montaña de la Flor, about 100 km north of the capital of Honduras, Tegucigalpa). The founding members of the group were a man called Pedro, his wife and daughter (Petrona and Juana), his sister Polinaria and her husband Juan, Petrona's brother Francisco and his wife Caciana, and finally Juana's fiancé, Leon. The group lives under very harsh conditions, but nevertheless has maintained its system of beliefs and myths, and a distinctive way of life, and has grown in number to a present population size of over 300 individuals.

During the century since this group was founded, it has remained almost completely cut off from the outside world. This isolation has been deliberate on the part of the Jicaque, at first because they were in revolt against their masters, and later because they hoped to avoid the illnesses and the vices that outsiders would bring. They blocked off the two points of entry into their territory, in order to ensure themselves isolation. The group therefore appears to be a text-book example of an ethnological and genetic isolate.

The Jicaque are clearly distinguishable from the surrounding population, both with respect to their physical characteristics, and also by their language (they still speak an Indian dialect, and only a few of them can speak Spanish). They also differ with respect to their religion (they rejected Christianity and now have no religious ceremonies) and in their dress (most men wear a poncho-like garment made of blue cotton, rather than trousers and shorts, and the women wear long cotton skirts and wide-sleeved blouses).

Until recently, the only way to reach the Jicaque was by mule. This, however, was not the main reason for the continued isolation of the group. As mentioned above, they had long noticed the danger to which they were exposed if a white or half-caste person were to infect them with influenza or the common cold; an Indian in apparently perfect health can die in a few days, of a "cold". In order to avoid contagion, they forbade outsiders to enter their territory, except with the express permission of their chiefs. Contact with the outside world has also been avoided as a result of the Jicaques' attitude to alcohol, which is strictly forbidden among them. They have maintained this attitude since the earliest years of the group's foundation, and it has been supported not only by the two chiefs but also by the entire population.

Whatever the reasons for the isolation of the group, this has led to its maintaining its own specific "personality". Other people of Jicaque descent living in the province of Yoro have been gradually assimilated into the half-caste population, so that the group living on the Montaña de la Flor is now the last remaining Jicaque community.

One result of the isolation which this group has maintained is an increase in the level of inbreeding, which may have harmful effects. Also, the small size of the group means that genetic drift must have occurred, and the genetic make-up of the present group may differ considerably from that of their ancestors.

1.2. Inbreeding among the Jicaque Indians

Chapman (Chapman and Jacquard, 1971) has made a thorough ethnological study of the Jicaque Indians, and she has been able to compile a complete genealogy of the group, going back to the original founders. The only individuals who were left out were children who died too young to be remembered. In 1965, the genealogy comprised 565 individuals, including the founders. Among the more recent individuals were included 81 children who died too young for their sex to be recorded; the sex-ratio among the survivors was 267 males to 217 females.

During the century since the group was founded, there have been five generations. There have, however, been marriages between individuals from different generations, so that we cannot assign generation numbers unambiguously. In order to overcome this difficulty, we can adopt the convention that an individual will be assigned to generation $n + 1$, if his father belongs to generation n. Thus, at the time of the founding of the group, Juana would be assigned to generation 2, since her parents were founders of the group; her two children would also belong to generation 2, since their father, Leon, was also a founder of the original group. This is obviously an arbitrary procedure, but it is satisfactory over so short a period as a hundred years. The distribution of the 565 individuals according to generation is shown in Table 16.1.

Despite the restricted choice of mates in such a small population, the Jicaque Indians have maintained a strict avoidance of brother-sister matings. They could not, however, escape consanguineous marriages altogether, since the children of Pedro and Petrona were first cousins of

Table 16.1. The distribution of individuals according to generations in the Jicaque population

Generation	Males	Females	Sex unknown	Total
1 (Founders)	4	3	0	7
2	19	15	0	34
3	96	80	0	176
4	110	93	44	247
5	38	26	37	101
Total	267	217	81	565

Juan and Polinaria's children, and of Francisco and Caciana's children, and Leon and Juana's children were their nephews and nieces. In the second and later generations, therefore, most marriages must have been between relatives. Most of the lines of descent of pairs of individuals therefore contain many common ancestors, and these are usually connected to the individuals in question by many different paths.

The lines of descent naturally become more and more complex, with the passage of the generations. By the 4th generation they are in some cases almost inextricably entangled (see, for example the pedigree of Julio and his wife Mencha − Fig. 6.6).

Julio and Mencha are "one-and-a-half cousins", since Julio's mother is the sister of Mencha's mother, and his father is the half-brother of Mencha's father. If this were the only relationship between them their coefficient of kinship would be $\frac{3}{32} = 0.09375$. However, they also have other, more remote, common ancestors, so that we have to use the full pedigree, and calculate their coefficient of kinship by the formula:

$$\Phi = \sum_i \left(\tfrac{1}{2}\right)^{n_i + p_i + 1}$$

(see Chapter 6) where the summation is over all the 28 possible paths of descent by which Julio and Mencha could have received a gene from a common ancestor; the result of this calculation is $\Phi = 0.1836$. This is a 98 % increase in the value of the coefficient of kinship, compared with the value we obtained when we ignored the more distant ancestors.

Φ has been calculated for each of the 150 married couples in the genealogy of the Jicaque, during their 100-year history.

We saw in Chapter 8 that the coefficient of kinship of a mating couple is also the inbreeding coefficient of their offspring, i.e. the probability that at a given locus the two genes such an individual carries are identical by descent. Weighting the coefficients of kinship of the matings in a given generation by the number of offspring they produced, we can thus obtain estimates of the mean inbreeding coefficients of the different generations.

This mean is called Bernstein's α coefficient; it is the probability that the two genes at a locus which are carried by an individual taken at random from the population, are identical. If the genes are indeed identical, the individual cannot be heterozygous since he received from his parents two copies of the same gene, rather than two distinct genes. α is therefore a measure of the decrease in genetic variability from generation to generation (see Chapter 8).

Table 16.2. Mean coefficients of kinship in the Jicaque population

Generation (g)	Number of individuals	$\alpha_g \times 10^2$
1	7	0
2	34	0
3	176	6.87
4	247	9.22
5	101	6.56

Table 16.2 shows the mean inbreeding coefficients, α_g, for the five generations. The value for generation 4 is extremely high.

As a basis for comparison, the following are two of the highest values of the mean inbreeding coefficient which have been reported to date for human populations:

1. Andhra Pradesh (East coast of India): $\alpha = 2280 \times 10^{-5}$ (Dronamraju, 1964).

2. The island of Hosojima, Japan: over five generations, $\alpha \doteq 3341 \times 10^{-5}$ (Ishinuki et al., 1960).

The mean inbreeding coefficients found among the Jicaque are thus two to three times higher than even the very high values reported for these populations.

We can also compare these values with the values we would expect if mating were at random; we saw in Chapter 8 that, under random mating, the coefficients α_g of successive generations are related by the difference equation:

$$\alpha_g = \frac{1}{2 N_e^{(g-2)}} + \left(1 - \frac{1}{N_e^{(g-2)}}\right) \alpha_{g-1} + \frac{1}{2 N_e^{(g-2)}} \alpha_{g-2}$$

where $N_e^{(g-2)}$ is the effective population size in generation $g-2$ (see Chapter 8).

Applying this formula to the Jicaque population, we obtain the results shown in Table 16.3. In generations 3 and 4 the values obtained agree quite well with the observed values of α; the observed value in

Table 16.3. Expected mean coefficients of kinship for the Jicaque population, under random mating

Generation	N_m	N_f	N_e	$\alpha \times 10^2$
1	4	3	6.86	0
2	19	15	33.52	0
3	96	80	174.54	7.29
4	—	—	—	8.56
5	—	—	—	8.82

generation 3 is lower than the expected value under panmixia, but this would be expected in a population which forbids brother-sister marriages. The value for generation 4, however, is higher than the observed value, although the same restriction on sib-mating must have been in force then too.

In generation 5, the value of α decreases, which is contrary to what theory would suggest should happen. The explanation for this is, however, a simple one—the population did not remain closed, but a few young men married girls from other communities. Even this slight amount of immigration had a detectable effect on the mean inbreeding coefficient.

A study of the mean inbreeding coefficient α thus allows us to detect some of the major features of the history of the group, though this obviously uses only a small part of the information contained in the complete genealogies. We find, for example, that the probability that an individual taken at random from generation 4 is heterozygous is less by 9 % than the same probability in the initial generation. But other changes have also taken place; in particular, some couples have had more offspring than others, so that genes descended from them have become more widespread in the population than genes from other, less fecund couples. Thus, in addition to a decreased probability of heterozygosity, the population will also change in terms of the alleles which are present in it. We can study this effect by calculating the probabilities of origin of genes from the various founders of the population, as discussed in Chapter 14.

1.3. Changes in the Genetic Composition of the Group

For each of the five generations, we can calculate the probability that a gene taken at random from among the genes carried by the individuals in the generation is derived from each of the original seven founders, or from one of the later migrants who joined the population. Table 16.4 shows the results of these calculations. It is obvious that this table contains

Table 16.4. Probabilities of origin of genes from the different founders of the Jicaque population

Generation	Population	Founders							Total	Immigrants	
		Leon	Francisco	Caciana	Juan	Polinaria	Pedro	Petrona		Indians	Half-castes
1	7	0.143	0.143	0.143	0.143	0.143	0.143	0.143	1.000	0	0
2	34	0.088	0.074	0.074	0.191	0.176	0.199	0.199	1.000	0	0
3	176	0.066	0.078	0.078	0.167	0.151	0.208	0.208	0.957	0.043	0
4	247	0.054	0.068	0.068	0.162	0.146	0.195	0.195	0.888	0.081	0.031
5	101	0.024	0.031	0.031	0.132	0.126	0.129	0.129	0.602	0.186	0.212

more information than a simple list of the values of α for the five generations.

Table 16.4 shows that the variation in fecundity between the couples has caused considerable changes in the genetic make-up of the group since its foundation. For example, Leon and Petrona, who contribute equally in generation 1, soon come to have very different contributions to the gene pool of the population. In generation 2, genes descended from Petrona are 2.26 times more frequent than genes from Leon. This disparity continues to increase with time, and by generation 5 genes descended from Petrona are 5.4 times commoner than genes from Leon.

Table 16.4 also shows the strong effect, starting in generation 3, of immigration to the group. The genetic effect of the immigration is much stronger than might be expected, given the number of immigrants, which was only 17 in total (these were mainly women, of whom seven were of Indian descent, while ten were half-caste). On the basis of numbers alone, we would have estimated the immigrants' contribution to the population's gene pool as only 4 %.

Such a small number of immigrants appears insignificant, both to the population which receives them, and to the ethnologist who studies the population, and yet in this case immigration has had a profound effect on the genetic make-up of the population. In generation 5, 40 % of the genes in the population were descended from immigrant genes. The explanation for this lies in the superior mean fecundity of the couples with one immigrant member.

Paula Flores, who joined the group in generation 2, is an example of the process; she had six children, barely more than the mean number at that period. The total number of children produced in this generation was 176, so that Paula Flores' genes form a fraction $\frac{6}{176} \times \frac{1}{2} = 1.7\%$ of the gene pool in generation 3. But all her six children in turn had children of their own, to a total of 35 altogether; genes descended from Paula Flores thus came to represent $\frac{35}{247} \times \frac{1}{4} = 3.5\%$ of all the genes in generation 4.

The fraction of genes descended from this woman thus doubled in one generation. In generation 5 they formed 4.0 % of the genes, so that, although she was not one of the original founders of the group, Paula Flores already contributes more to the group, in terms of her genes, than the two true founders, Francesco or Caciana.

This example shows how rapidly the genetic make-up of a small population can change, and also that we cannot estimate the significance of immigration for a population simply in terms of the number of immigrant individuals; the effect which the immigrants will have on the population depends on the numbers of descendants they have.

If we know, for each generation, the probabilities of origin ω_{gr} of the genes in the population from the various founders F_r, we can calculate the "distance" between two generations g and i, by the method described in Chapter 14. The measure of distance Z, which we shall use, is based on χ^2, and is calculated by the formula:

$$Z_{gi}^2 = \sum_r \frac{(\omega_{gr} - \omega_{ir})^2}{\omega_r}.$$

Table 16.5. Approximate distances between the generations in the Jicaque population

Generation	2	3	4	5
1	16	18	21	44
2		6	11	39
3			7	36
4				30

The results for the Jicaque Indians are given in Table 16.5. The values of the distances show that the population changed relatively slowly between generations 2 and 4, when its size was increasing rapidly, but that it changed very fast between generations 4 and 5, because of the contributions of the immigrants. We shall study another example later (the Kel Kummer Tuareg) which shows that this does not always happen.

2. The Bedik of Eastern Senegal

2.1. History and Ecology

During his ethnological studies in the province of Kedougou in Senegal, Gessain (1963) was informed by the Rev. Paravy of the existence of a small group which was apparently very different from the neighbouring groups of people. These people have been given a variety of names, but are now generally called the Bedik tribe. They live in an area of about 300 km² in the basaltic hills near the bend of the river Gambia, to the east of Niobpolo-Koba.

This tribe has been studied by workers in several fields, and has been described in detail by Gomila (1971). The Bedik language has been discussed by Ferry (1967), the genetic structure of the tribe by Langaney and Le Bras (1972) and haematological studies of the group are described by Langaney *et. al.* (1972).

The 1500 present-day members of the Bedik tribe belong to the Tenda ethno-linguistic group. According to their traditions, they once occupied a large territory, but were forced to take refuge in the hills where they now live, as a result of an unsuccessful war with the Fula tribesmen of the Fouta-Djalon uplands, which took place at the end of the 19th century, and which probably caused a drastic decrease in the size of the group.

The Bedik have retained their animist beliefs, while most of the surrounding tribes have become converted to Islam. Their language differs somewhat from that certain of their neighbours, who speak the related Bassari and Conagui languages. Their whole way of life, with its complex set of social structures, is also distinct from that of other tribes of the neighbourhood. The environment in which the Bedik live, however, is very similar to that of the other tribes of the area. The climate is typical for this part of Africa, with a rainy season from May to September, and then a long, dry season. Infectious and parasitic diseases are very common, despite the efforts of the health authorities to improve the health of the people: malaria causes heavy mortality, especially among the children; filariasis and related diseases are found in up to 80% of the population of certain villages, and onchocerca infestations, which may cause blindness, are particularly serious; leprosy, bilharzia and rickettsial diseases are also common.

2.2. Marriages among the Bedik

In order to study the transmission of genes from generation to generation in a group such as the Bedik, we must study the pattern of marriages which take place. Such a study among the Bedik shows that, although the group is genetically isolated from other tribes, it is by no means homogeneous. The population consists of two sub-groups, which differ with respect to certain details of their language and culture: these are the Banapas (about 400 individuals), and the Biwol (1100). The two groups live in separate villages (there are two Banapas villages, and four Biwol villages).

Table 16.6. Movement between the Bedik villages. The table shows numbers of women according to their villages of origin and the villages in which they were found to be living

Village of residence		Village of origin						Other	Total
		Banapas		Biwol					
		1	2	3	4	5	6		
Banapas	1	23	1	4	1	5	1	1	38
	2	8	57	4	0	11	2	6	88
Biwol	3	1	2	78	28	11	5	1	126
	4	0	0	40	47	2	10	0	99
	5	6	1	22	1	17	1	1	49
	6	0	2	3	12	1	64	4	86
Total		38	63	151	89	47	83	13	486

The barriers separating the six villages and the two subgroups from one another, and the group as a whole from nearby groups, can be characterised in terms of the exchange of mates between the different groups. Gomila (1971) enumerates the origins of couples alive in 1964, and of some of their parents; on the basis of these data, Langaney and Gomila (1973) drew up tables showing the origins of the males and females living in each of the villages (Tables 16.6 and 16.7).

Tables 16.6 and 16.7 show that the Bedik are a rather self-contained group. There were no male immigrants to the tribe, and only 13 out of 486 (2.7%) of the women came from other groups. Table 16.7, for the men, shows that nearly all the men stay in their own village when they marry.

Table 16.7. Movement between the Bedik villages. The table shows numbers of men according to their villages of origin and the villages in which they were found to be living

Village of residence		Village of origin						Other	Total
		Banapas		Biwol					
		1	2	3	4	5	6		
Banapas	1	22	0	0	0	0	0	0	22
	2	2	48	1	1	5	1	0	58
Biwol	3	0	0	66	2	1	0	0	69
	4	0	0	0	53	0	3	0	56
	5	0	0	1	0	21	0	0	22
	6	0	1	0	0	0	45	0	46
Total		24	49	68	56	27	49	0	273

Table 16.8. Gene frequencies in the Bedik villages. Where only one allele is shown

Vil-lage	$A_1/A_2/B/0$				Lewis L/l	Inv $1+2+/$ $1-2-$	M/N	P+/ P−	PGM 1/2	6PGD A/B
	A_1^*	A_2	B*	0*	L*	$1+2+$	M*	P+*	1*	B
1	0.185	0.021	0.172	0.622	0.837	0.391	0.403	0.565	0.721	0.054
2	0.170	0.015	0.226	0.589	0.836	0.547	0.500	0.668	0.777	0.032
3	0.146	0.049	0.224	0.581	0.829	0.419	0.477	0.733	0.857	0.065
4	0.162	0.048	0.258	0.532	0.763	0.383	0.534	0.737	0.835	0.046
5	0.224	0.066	0.229	0.481	0.844	0.333	0.524	0.760	0.766	0.039
6	0.143	0.038	0.261	0.588	0.815	0.240	0.427	0.748	0.897	0.038

The women also tend to remain within their own sub-group, Banapas or Biwol, as the case may be: 91.8% of the women married men of their own group, whereas if there was random-mating among the Bedik in general this percentage would be about 64%. This tendency to marry within one's own specific group is even more clearly shown by the figures for marriages between people from the different villages. Out of 286 women, 60.4% remained in their home villlage when they married, compared with an expected figure of 9.4% if mating were at random.

This rigidity in the choice of marriage-partners must clearly have an effect on the genetic structure of the group. These effects have been studied in terms of both haematological and anthropometric characters.

2.3. Haematological Characters − Distances between Villages

Two expeditions to the Bedik villages have been made, and 798 blood-samples were collected (Bouloux et al., 1972). The Centre d'Hémo-typologie, Toulouse, typed the samples for nine red-blood-cell antigens, six erythrocyte iso-enzyme systems, two serum isozymes, the immuno-globulin Gm and Inv types, and the haptoglobin, transferrin and haemo-globin types. Table 16.8, which is taken from Langaney and Le Bras (1972), shows the results for the most informative systems.

We may also mention the finding of a high haemoglobin S frequency: out of 784 haemoglobin types determined, 226 AS heterozygotes were found, but no SS homozygotes. The frequency of the S allele is therefore 0.144; this high value may be connected with the high frequency of malaria in the area.

Using the data on these haematological systems, and the frequencies of the alleles at the various loci involved, we can calculate genetic distances between the villages.

for a system, the other allele frequency is obtained by subtraction from 1.0

MNSsS^u						Acid phos-phatase B/A	Rhesus				
MN	Ms*	NS	Ns*	MS^u	NS^u	B*	r*	R_0^*	R_2^*	r'	R_z
0.000	0.434	0.019	0.321	0.039	0.187	0.641	0.252	0.624	0.112	0.006	0.006
0.013	0.499	0.035	0.281	0.058	0.115	0.736	0.308	0.597	0.092	0.003	0.000
0.024	0.313	0.069	0.351	0.148	0.095	0.769	0.217	0.681	0.094	0.003	0.005
0.059	0.326	0.147	0.376	0.079	0.013	0.799	0.204	0.618	0.140	0.036	0.002
0.094	0.449	0.090	0.242	0.010	0.115	0.756	0.251	0.666	0.050	0.033	0.000
0.000	0.491	0.070	0.199	0.103	0.137	0.707	0.291	0.594	0.067	0.048	0.000

In Chapter 14, we discussed the choice of measure of distance, and we saw that this choice is often arbitrary. In the present instance, the choice is made difficult by the differences in the information provided by the different haematological systems. We can get over this difficulty in the following way. If we calculate the gene frequencies by the method of maximum likelihood, we can also calculate the amount of information which each system provides, and we can use these values to assign weights to the systems, in the calculation of the distances between the villages. We can also calculate several different measures of the distances; this helps

Table 16.9. Distances between the Bedik villages

Pair of villages	χ^2 distance		Arc cosine distance		Θ
	Unweighted	Weighted	Unweighted	Weighted	
1, 2	1	1	1	1	1
1, 3	1.92	2.53	1.45	1.92	2.26
1, 4	3.22	4.47	2.65	3.76	2.94
1, 5	2.48	3.26	2.07	2.82	2.17
1, 6	2.16	2.62	1.93	2.28	3.01
2, 3	1.23	1.62	1.97	1.12	1.00
2, 4	2.10	2.94	1.72	2.24	1.23
2, 5	1.82	2.13	1.18	1.34	1.51
2, 6	1.92	1.59	1.58	1.53	2.44
3, 4	1.05	1.70	0.69	1.07	0.12
3, 5	1.95	2.97	1.24	1.89	0.89
3, 6	1.44	2.00	1.16	1.64	0.93
4, 5	1.47	2.26	1.06	1.70	0.30
4, 6	2.11	3.15	1.63	2.46	0.98
5, 6	1.84	2.56	1.37	1.87	1.07

to overcome the difficulty that the results may simply depend on the measure of distance we have adopted. Table 16.9 shows the results of five different calculations of the distances between all possible pairs of villages; the measures used were χ^2 (unweighted and weighted), the arc-cosine method of Cavalli-Sforza and Edwards (unweighted and weighted), and Morton *et al's.* (1971 b) coefficient of hybridity, Θ. In order to make comparisons of the distances easier, the distance between villages 1 and 2 has arbitrarily been assigned the value 1.

Table 16.10. Correlations between different measures of distance

Measure of distance		χ^2 Unweighted	Weighted	Arc cosine Unweighted	Weighted	Θ	Anthropometric
χ^2	Unweighted	1.0	0.913	0.956	0.947	0.730	−0.100
	Weighted		1.0	0.824	0.953	0.443	−0.118
Arc cosine	Unweighted			1.0	0.937	0.809	−0.126
	Weighted				1.0	0.601	−0.122
Θ						1.0	−0.165

The distances given in Table 16.9 agree reasonably well with one another. To measure the extent of the agreement or disagreement between the different measures, we can calculate the correlation coefficients between the measures, for the 15 pairs of villages. Table 16.10 shows the results of these calculations, together with the correlations of the various distances with the distance based on anthropometric data (these consisted of 16 lengths of body parts measured on 280 men between the ages of 22 and 51 (Gomila, 1971); the 16 most informative measures were used by Langaney and Le Bras (1972) to calculate the distances between the villages).

Table 16.10 shows that there is a high correlation between the χ^2 and arc cosine measures of distance, but that distances based on the coefficient of hybridity do not agree so well. The anthropometric distances are uncorrelated with the distances based on the haematological characters.

2.4. Representation of the Structure of the Population

The set of distances between the populations can be studied by the methods described in Chapter 15. Having calculated the set of 15 distances

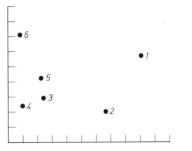

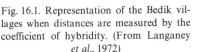

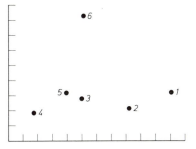

Fig. 16.1. Representation of the Bedik villages when distances are measured by the
coefficient of hybridity. (From Langaney
et al., 1972)

Fig. 16.2. Representation of the Bedik villages when distances are in terms of the arc
cosine measure of distance. (From Langaney
et al., 1972)

between the six villages, we can represent any given village by a vector
of five elements – the distances of this village from the other five. The
points representing the six villages can be projected onto a plane, by the
method of principal components analysis discussed in Chapter 15.

Figs. 16.1 and 16.2 show the plane representations of the "topography"
of the set of villages, when distances are measured by the unweighted
arc cosine measure of distance, and by the coefficient of hybridity,
respectively. The first representation preserves 68 % of the information,

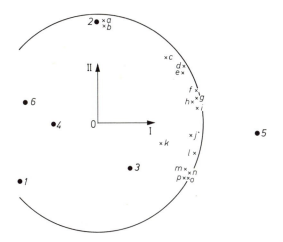

Fig. 16.3. Distances between the Bedik villages based on the anthropometric data. a: Fore-
arm length; b: Forehead height; c: Arm length; d: Biacromial width; e: Height of tibiale;
f: Foot width; g: Waist measurement (sitting); h: Morphological face height; i: Height of
the iliac spine; j: Head length; k: Biiliocristal diameter; l: Stature; m: Bigoniac width;
n: Head breadth; o: Minimum frontal diameter; p: Bizygomatic width. (From Langaney
et al., 1972)

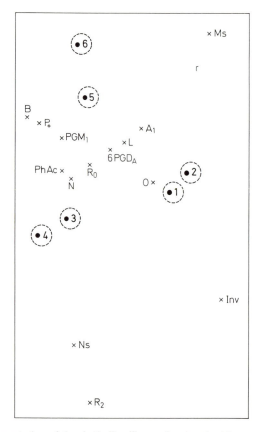

Fig. 16.4. Representation of the six Bedik villages, showing the 15 gene-points, using the unweighted χ^2 measure of distance. (From Langaney *et al.*, 1972)

while the second preserves 80 %. Fig. 16.3 shows the representation of the anthropometric distances between the villages, on the first principal plane; the "character-points" for the 16 characters that were used, are also shown (we saw in Chapter 15 that a character which has a high correlation with either of the first two principal axes will be represented by a point close to the circle of radius 1, centred at the origin, 0).

Fig. 16.4 shows the results of a principal components analysis of the set of distances based on χ^2, when the distances were calculated from the gene frequencies for the various haematological systems, in the different populations.

In order to minimise sampling error, we have included only the 15 alleles whose frequencies are greater than 0.05; these are marked with an asterisk in Table 16.8.

Figs. 16.1, 16.2 and 16.4 show that the χ^2, chord and coefficient of hybridity measures of distance based on the haematological data all give broadly similar representations of the set of Bedik villages. The two Banapas villages, 1 and 2, can be distinguished from the four Biwol villages, thus confirming the conclusion which the ethnological workers had drawn from their studies of these people's language and customs. The cultural barrier to exchange of mates between the two sub-groups is therefore sufficiently strong to have permitted a certain amount of genetic differentiation. We also find that, among the Biwol sub-group, there are clear differences between village 6 and the others. This village lies at some distance from the others and is therefore more endogamous, at least at present; this village is also subjected to harsher conditions than the others, from the epidemiological and nutritional points of view.

Fig. 16.4 is perhaps superior to the other representations in that it displays clearly which genes play the greatest part in differentiating the villages.

If we now consider the distances based on anthropometric data (Fig. 16.3) we see that the topography of the villages is quite different. Village 5 is the most distinct from the others, and this is due entirely to separation in the direction of the first principal axis. Now, 14 out of the 16 measures used show high correlations with the first principal component, which can therefore be regarded as an "intensity" axis, such as is often encountered in this type of analysis. Gomila (1971) notes that in village 5, which is smaller and more exogamous than the others, the men tended to have larger values of the physical dimensions that were measured, than men from the other villages. We must not jump to the conclusion that this is due to heterosis, but it is nevertheless clear that some factor or factors have caused a distinct difference in the anthropometric characters of this village, which is not manifest when we study the haematological data.

3. The Kel Kummer Tuareg of Mali

3.1. History, Ecology and Social Organisation

At the end of the 17th century, the mountainous region of Adrar des Iforas, in the southern Sahara close to the elbow of the river Niger, was under the domination of a powerful tribe of Tuareg, the Kel Tademaket. At this time, quarrels arose over the method of determining the succession of the chieftaincy, and a group seceded, led by the great warrior Kari-Denna. This group gradually came to dominate the area, and at present their descendants, the Kel Kummer, are the chief tribe of all the Aulimidan Tuareg. The Kel Kummer Tuareg consist of three sub-groups, the Kel Tagiwelt, the Kel Telataye and the Kel Ahara, corresponding to the inheritances of the three grandsons of Kari-Denna.

The Kel Kummer have moved slowly southwards, along the rich pastures near the river. They have remained powerful, and control the numerous nomadic groups who live in the area.

At the time of the French conquest of Mali at the end of the 19th century, the Kel Kummer resisted the colonial authorities and were driven towards the east. Their resistance to the colonists ended tragically: in the rising of 1916 hundreds of men were killed in battle, and many women and children were taken prisoner after the defeat and died; above all, the humiliation of their defeat led to a drastic, voluntary curtailment of the number of births, which has amounted almost to racial suicide. Although they still continue to dominate the other tribes of the area, the Kel Kummer are therefore threatened in their very existence; in 1970 there were only 367 members of the tribe left (171 men and 196 women).

They are a nomadic tribe and range over an area of 175000 km^2 in the region to the north and east of the town of Menaka (Republic of Mali). The relief of the area is slight, and is most marked around three wadis which run from north to south and which die out south of the latitude of Menaka in a monotonous land of sterile sand-dunes covered with alumuz, a species of graminaceous plant with thorny flowers.

The climate is chiefly determined by the amount of rain. This varies considerably from year to year; in 1958 there was more than 400 mm at Menaka, but less than 150 mm in 1965 and 1967. On the other hand, the rain falls at very regularly-determined periods in the year; the masses of warm, humid air from the gulf of Benin do not reach Menaka until July or, more usually, August, and half of the total annual precipitation occurs during this month.

The Kel Kummer are one of a set of distinct nomadic tribes which each play a particular role with respect to the whole group. Each group has a complex set of clearly-defined rights and obligations, and these are rigorously adhered to.

The highest position in this hierarchy belongs to the Kel Kummer themselves, who constitute the caste of "Imajeghen" or nobles. In Tuareg society, nobility is equated with being a warrior, even today when the people live a pastoral life. The superior position of the Kel Kummer is reflected in many aspects of their everyday life, and the group maintains its isolation and its traditions by its matrimonial system: a noblewoman may marry a man of lower rank, but their children will be members of their father's group, and will not be allowed to marry a noble. It is rare for a man of noble rank to marry beneath his rank. When this occurs, usually because of limited choice of young women available, this is not regarded as a scandal, but the children of the marriage must either remain unmarried, or else leave the tribe and marry a member of another group.

The Kel Kummer therefore form an extremely endogamous group. A study of the actual lines of descent in the group is, however, necessary before we can be certain that the social isolation of this group does indeed result in genetic isolation. We saw in the case of the Jicaque Indians that a good deal of genetic exchange had occurred between this apparently isolated population and its neighbours, and this might also be the case for the Kel Kummer, despite their distinctive mode of life, language differences, and their attitude to other groups.

3.2. The Genealogy of the Kel Kummer People

In an attempt to describe in a complete form the history and matrimonial system of the tribe, the ethnologist Chaventré succeeded in finding out the complete genealogy of the Kel Kummer group. The task of listing all the 2420 members of the tribe since its foundation by Kari-Denna at the end of the 17th century, together with the father and mother of each individual, took him 15 years. We therefore have at our disposal a full description of the paths of descent in the tribe, in other words we know the exact pathways by which the genes were transmitted from one generation to the next, through a period of three centuries. Fig. 16.5 shows part of this complete genealogy; it shows all the ancestors of a present-day sibship, going back to the parents of Kari-Denna, together with several collaterals (Chaventré, 1972).

We must, of course, realise that these genealogies may not be entirely reliable; actual and official paternities are not always the same.

However, ethnological studies of this group show that the social organisation of the Tuareg excludes illegitimate children as a practical possibility; haemotological studies also confirm this. An expedition in 1971 obtained blood samples from 285 individuals, which represents nearly the whole of the present-day tribe, except for very young children. Blood-groups, serum proteins (for all the samples) and leukocyte antigens (for 116 samples) were studied, and in no case could the stated paternity be excluded. Furthermore, when rare genes were discovered they could be traced back in a consistent fashion to a few individuals many generations ago; it would not be possible to do this, if the pedigrees contained many errors. As a further safeguard, data from different, independent sources could sometimes be compared; for example data based on the traditions accepted by the group could in certain instances be compared with documents in the local archives, or with the records of the French army, or with the accounts of travellers in the area. All the information we have points to the conclusion that Chaventré's genealogies are reliable.

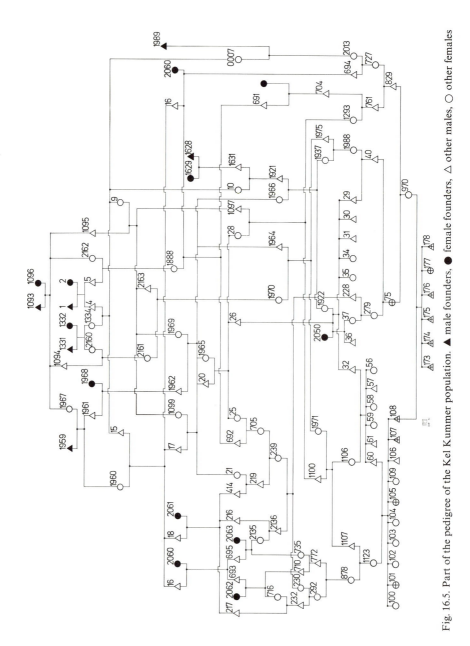

Fig. 16.5. Part of the pedigree of the Kel Kummer population. ▲ male founders, ● female founders, △ other males, ○ other females

The task which Chaventré had to undertake was to some extent simplified by the fact that no two individuals are ever given the same name. (It is a custom in the tribe that no-one may ever speak or hear the name of any of his dead ancestors.) The list of 2 420 Tuareg members of the tribe is thus a list of 2 420 different names. In order to simplify the handling of this data, each individual in this list was assigned a number, and was referred to by a set of three numbers – his own identification number, and the numbers of his father and his mother, in order. This enables us to calculate the parameters measuring the relationship between any two individuals, or to characterise the overall changes in the genetic composition of the group.

All the lines of descent go back, often by several different pathways, to the founder of the tribe, Kari-Denna, but also to other "founders" in the sense of the term given in Chapter 14, i.e. individuals whose ancestors we do not know. For the whole group, there are 156 "founders" of the Kel-Kummer, if we use the data from the entire 300 years' history of the group.

If we consider the fact that the 2 420 individuals studied can be traced to only 156 "founders" it becomes obvious that the pedigrees of the present-day members of the group, and of many of their ancestors, must be extremely involved. The 367 Kel Kummer Tuareg who were alive in 1970 were all related to one another, and they all have at least 15 ancestors in common; each of these common ancestors is related to most of the present-day tribesmen by multiple paths of descent.

In such a complex genealogy, where relationships are so close, the coefficient of kinship is not an informative enough measure of the changes in the population's genetic make-up. We must therefore, as in the case of the Jicaque Indians, turn to the study of the probabilities of the origin of genes from the various ancestors.

3.3. Changes in the Genetic Make-Up of the Population

The members of the Kel Kummer tribe who were included in the genealogy were grouped into "generations" according to the rule that an individual was assigned to generation $n + 1$, if his mother was a member of generation n. This is an arbitrary procedure, and leads to some inconsistencies over the long period of time for which we have data on this group (16 "generations"). However, it provides a way of describing changes in the make-up of the tribe, over the period studied.

Table 16.11 summarises the results of the calculations of the probabilities of origin of the population's genes from the different founders. In this table, the generations have been grouped so as to give larger numbers for comparison; this should also lessen any biases due to the arbitrary

Table 16.11. Origins of genes in the Kel Kummer population

Founders	Generation					
	1–5	6–7	8–9	10–11	12–13	14–16
	Population size					
	126	371	601	653	268	245
1093 and 1096	0.198 ⎫	0.142 ⎫	0.158 ⎫	0.146 ⎫	0.155 ⎫	0.186 ⎫
1 and 2	0.199 ⎬ 0.571	0.110 ⎬ 0.350	0.116 ⎬ 0.380	0.105 ⎬ 0.357	0.105 ⎬ 0.361	0.116 ⎬ 0.426
1331 and 1332	0.174 ⎭	0.098 ⎭	0.106 ⎭	0.106 ⎭	0.101 ⎭	0.124 ⎭
1919	0.030	0.066	0.076	0.064	0.071	0.107
1959	0.047	0.055	0.070	0.060	0.063	0.071
2060	0.016	0.060	0.048	0.033	0.037	0.033
2067	0.016	0.021	0.049	0.045	0.043	0.046
1968	0.029	0.035	0.039	0.034	0.040	0.053
2062	0	0.044	0.025	0.024	0.025	0.027
1628 and 1629	0.038	0.020	0.019	0.015	0.019	0.024
2063	0	0.023	0.023	0.015	0.022	0.020
15 principal founders	0.747	0.674	0.729	0.647	0.681	0.807
10 next most important	0.103	0.112	0.102	0.111	0.101	0.078
Set of 131 other founders	0.150	0.214	0.169	0.242	0.218	0.115

definition of the generation. The founders have been separated into three groups: the 15 founders who contributed most to the present-day population are listed singly; the ten founders of next lower importance are not listed singly, but their group contribution is given; the contribution of the 131 least important founders is also shown, for each of the groups of generations.

Table 16.11 shows that the different founders contribute very unequally to the population. The two parents of Kari-Denna (founders numbers 1 and 2) contribute 11.5 % of the genes in the present-day population. At the time of the secession of the group, their contribution was 20 %; this value declined to 11 % by generation 6, due to the introduction of genes from new "founders"; after this time, their contribution has remained practically constant.

Despite their historically important place in the tribe, Kari-Denna and his family have not proved to be the most important founders of the population, in terms of their genetic contributions. The couple numbered 1093 and 1096[2] can be shown to be the origin of 18.6 % of the genes of the present-day population, which is approximately 30 % higher than

---

[2] The grandparents of the wife of Kari-Denna's only son.

the contribution of Kari-Denna's parents. In the original group, however, their contributions were equal; the predominance of the couple 1093–1096 became established during the first six generations, and has been maintained since then. Their predominance has been less affected by the influx of new "founders" than has the contribution of the couple 1–2.

Genes descended from couple 1331–1332 (Kari-Denna's wife's parents) originally formed a smaller part of the total genetic endowment of the population (17.4 %) than those from couple 1–2, but their contribution gradually increased to equal the contribution of couple 1–2, and eventually even exceeded it.

The three couples we have just mentioned contribute 42.6 % of all the genes of the present population. In other words, *the probability that a gene chosen at random from a present-day member of the Kel Kummer group is descended from one of these six ancestors is greater than 40 %*.

Continuing with the founders, in order of decreasing importance, we find that each of the next nine individuals contributes between 2 and 11% of the genes in the present-day population, the next ten individuals each contributes between 0.5 % and 2 % and the 131 least important individuals each contributes less than 0.5 %.

In summary, we find that the group of the 15 most important founders contribute 80 % of the present-day population's genes, the next ten individuals contribute 8 %, and the other 131 founders contribute less than 12 % in total. These 131 founders have thus had only a very slight effect on the genetic history of the population; these individuals themselves belonged to the group, but either left no descendants, or else these were excluded from the marriageable group of individuals in the next generation. Eighty-two of these 131 founders have no descendants in the three last generations, so that their genetic contribution to the tribe was only a temporary one.

The biological meaning of these detailed calculations is simply that *the genes in the Kel Kummer tribe are descended, not from 156 founders, but essentially from only 25 individuals, whom we may regard as the true "founders" of the population;* these 25 individuals have maintained their important place in the genetic history of the group, despite immigration into the population, over the years. We can therefore claim that the Kel Kummer represent a true genetic isolate.

Another approach to the study of changes in the genetic structure in this group is to calculate the "distance" Z_{ij} between two generations, i and j. The method for doing this was described in Chapter 14. Table 16.12 shows the distances between the different generations (2–16) of the Kel Kummer.

These distances tell us not only the *rates* of change of the population, but also show the *directions* of the changes. For instance, the distance

Table 16.12. Distances between the generations for the Kel Kummer population, using the χ^2 measure of distance. (Generation 1 has been omitted because the number of individuals was small)

	3	4	5	6	7	8	9	10	11	12	13	14	15	16
2	92	156	193	170	142	139	134	145	156	156	218	168	126	142
3		157	195	174	144	141	138	150	159	160	219	171	129	145
4			162	158	159	160	159	168	177	179	233	192	155	165
5				124	149	152	154	162	174	175	230	191	155	163
6					94	111	117	123	138	140	202	161	118	129
7						65	69	85	100	106	183	130	70	87
8							61	81	98	99	178	121	60	92
9								66	92	98	178	122	46	77
10									95	110	186	134	75	97
11										110	194	143	93	103
12											190	144	94	114
13												194	177	189
14													112	134
15														67

between generations 4 and 5 is 162, while that between generations 5 and 6 is 124. The population thus changed more slowly between generations 5 and 6 than between generations 4 and 5. The distance between generations 4 and 6 is 158, so that the population moved back again in generation 6 to a state closer to generation 4 than to generation 5. Similarly, we may note that in generations 15 and 16 the population resembled generations 7 to 11 more than generations 12 to 14. The sequence of changes can be represented as in Fig. 16.6. This figure is based on a principal components analysis of the set of distances between the generations, as described in Section 2.4, on the Bedik villages.

These results confirm the conclusion we drew earlier, that this tribe is a genetic isolate. Contributions of genes from outside individuals may sometimes significantly change the composition of the population (for instance in generation 5, or 13), but the customs of the group ensure that the immigrants' contributions to the genes in the population are reduced, in the succeeding generations, and that the original composition is restored.

The mechanism by which the genetic composition of the population is maintained constant can also be studied by finding the numbers of individuals, in the various generations, who are descended from particular "founders" of the population. Using the complete genealogy of the tribe,

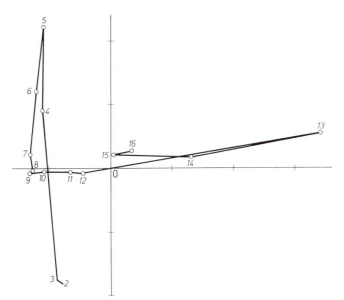

Fig. 16.6. Projection of the different generations of the Kel Kummer population onto the first principal plane

it is easy to calculate the numbers, z_{rg}, of individuals in generation g who are descended from the r-th founder. Table 16.13 shows some data of this kind. When the number z_{rg} is equal to the total population size for generation g, this is indicated by a heavy type. Table 16.13 shows that, by generation 9, founders numbers 1 and 2 (Kari-Denna's parents) were ancestors of the entire population. By generation 12, this is also true

Table 16.13. Numbers of descendants of some of the founders of the Kel Kummer population. (Heavy type indicates that the number is the total population size)

Founders	Generation								
	2	3	8	9	10	11	12	15	16
	Population size								
	11	19	298	303	340	313	204	118	61
1 and 2	4	13	297	**303**	**340**	**313**	**204**	**118**	**61**
1093 and 1096	4	7	285	298	337	312	**204**	**118**	**61**
1331 and 1332	3	12	293	299	332	312	**204**	**118**	**61**
1628	0	2	93	140	213	219	152	112	**61**
1777	0	0	0	0	2	12	0	0	0
1788	0	0	0	0	5	0	0	0	0
2050	0	0	13	21	29	72	84	34	22

for founders numbers 1093, 1096, 1331 and 1332. By generation 16, all of the 15 "principal founders" are ancestors of all the population members.

In contrast, "founders" of later date have only contributed temporarily to the population. For instance, founder number 1777 had two descendants in generation 10, 12 in generation 11, but none thereafter. This is the fate of nearly all the "foreign" genes that have been introduced into the population: immigrant individuals themselves are accepted into the group and apparently integrated into the tribe, but the tribe's marriage customs ensure that their descendants must seek mates from another tribe, so that the immigrant genes only remain temporarily in the population.

3.4. Haematological Studies of the Kel Kummer Population

Thanks to the co-operative attitude of the Kel Kummer people, and of the authorities of Mali, it was possible to obtain samples from 285 of the 367 present-day population. These were all typed for several red-blood-cell and serum markers, and the HL-A types of 116 individuals were also determined.

The results of the study of these systems are described by Degos (1973), Lefèvre-Witier (1973) and Colombani et al. (1972).

3.4.1. The HL-A system. The effect of small initial population size and of genetic drift due to the population's isolation is particularly clearly demonstrated by the genetic structures of the genes govering the HL-A system of leukocyte antigens.

The HL-A system is known to be controlled by two tightly-linked loci (recombination fraction less than 1 %). Fourteen co-dominant alleles have so far (1972) been distinguished at the first locus (locus LA or SD1), and 15 at the second (locus 4 or SD2); there is evidence for the existence of other alleles, for which specific antisera have yet to be discovered; these alleles are referred to collectively as "blank". Among the gametes there are therefore $(14+1) \times (15+1) = 240$ possible combinations of the alleles at these two loci; these combinations are called "haplotypes". Each individual has two haplotypes, and these determine his HL-A genotype; the number of possible genotypes (if we distinguish between coupling and repulsion types) is therefore $\dfrac{240 \times 241}{2} = 28\,920$, and there are

$$\left(\frac{14 \times 15}{2} + 1 \right) \times \left(\frac{15 \times 16}{2} + 1 \right) = 12\,826$$

different possible phenotypes. This system is therefore extremely useful for comparing populations, or for grouping the individual members of a single population.

The samples from the Kel Kummer population were subjected to testing for nine of the alleles at the LA locus, and 14 of the locus 4 alleles. For five of these 23 alleles, no positive reactions were found; these antigens were HL-A10 and HL-A13 and W14 and W15 which are widespread in all the populations that have been tested from Europe, the Middle East, and Africa, and W22 which is chiefly found in Oceania. It is probable that the absence of the first four of these alleles is due to genetic drift and, more specifically, to the "founder effect", which must have been very important for this population, since the majority of its genes are descended from a small number of individuals.

If we consider the haplotype frequencies, this effect is even more marked: among the individuals whose haplotypes could be determined, only 21 different associations of the alleles at the two loci were found; of these, only two were common (the association HL-A1, W21 formed 19% of all the haplotypes studied, and the association W28, HL-A7 represented 17%); more than a third of all the gametes among the Kel Kummer population were thus of one or other of these two types.

The high degree of uniformity in this population makes it especially suitable for testing whether the genotypic structure differs from that predicted by the Hardy-Weinberg law. Indeed, Degos *et al.* (1973) found that the frequency of homozygotes was considerably lower than the expected value: out of 106 individuals both of whose haplotypes could be determined, only one is a possible homozygote (this individual has two "blank" alleles, which may or may not both be the same). Now the expected proportion of homozygotes, in the absence of selection, is $\sum_i p_i^2$, where p_i is the frequency of the i-th haplotype; in the present case, this is equal to 0.084, or, in a sample of 106, to 9 individuals.

We reach the same conclusion if we take account of the population's inbreeding, which can be done by using the pedigree data referred to above. The mean inbreeding coefficient of the 106 individuals whose haplotypes are known is 0.1108. This means that the expected number of individuals both of whose haplotypes are copies of one of the haplotypes in an ancestral individual is 11.7.

The observed absence, or at least deficit, of homozygotes could be explained by selection either acting on the zygotes themselves, or else at the time of conception, perhaps by preventing the fertilisation of an egg by a spermatozoon carrying the same HL-A genes. Such a process might, however, be governed by a gene in the same region of the chromosome as the HL-A loci, rather than by the HL-A system itself.

Notice that this kind of observation could not be made in a population containing a large number of HL-A alleles, each at low frequency, as is usually the case in populations that are not isolated from others, like the Kel Kummer population.

3.4.2. Classification of the individual members of the population.
We can characterise individual population members in several different ways, as follows:

1. By their ancestors; we can determine, from the family trees, the probabilities of origin ω_{ir} of a gene carried by any individual I_i, from each of the founders F_r of the group.

2. By their HL-A genotypes.

3. By the alleles which they carry at loci governing erythrocyte and serum systems.

We can then calculate the distances between the individuals, based on these different criteria. This analysis has been done for 69 individuals, each from a different sibship; distances were calculated using the following formulae:

Genealogical distance: $d_1(i, j) = \sum_r \frac{(\omega_{ir} - \omega_{jr})^2}{\omega_r}$;

Distance based on HL-A: $d_2(i, j) = \frac{n(i, \bar{j}) + n(\bar{i}, j)}{20}$

where $n(i, \bar{j})$ is the number of HL-A genes which individual I_i carries, but individual I_j does not. A distance, $d_3(i, j)$, based on 40 markers determined using the red blood cell and serum systems, was calculated by an analogous formula.

These calculations give three sets of $\frac{69 \times 68}{2} = 2\,346$ distances between pairs of individuals. The correlation coefficients between the different measures of distance were calculated, and the results were as follows:

$$r(d_1, d_2) = 0.104,$$
$$r(d_2, d_3) = 0.024,$$
$$r(d_3, d_1) = 0.058.$$

Although these correlation coefficients are positive, their values are very low. This results from the fact that the complement of genes which a new individual receives at the moment of conception is chosen at random out of a huge number of possible combinations; if, therefore, two individuals are descended from precisely the same set of ancestors, who each contribute the same proportions of the genes the individuals possess, their genotypes can still be quite different, at many loci. This shows that this technique is of limited usefulness when applied to the relations between individuals. We shall next consider its use in the study of the similarities and differences between populations.

3.4.3. Comparisons with other populations. Lefèvre-Witier (1973) has carried out a biological and anthropological study of a village (Ideles) of Ahaggar in Algeria. He made comparisons between the genetic structures of the members of the various groups of people living in this village, and those of some other African and Mediterranean population samples, based on the AB0, Rhesus, MN, P-Tj(a) and Kell-Cellano blood systems (these types were determined at the Centre d'Hémotypologie, Toulouse, and at the Serological Population Genetics Laboratory, London). A total of 27 populations were compared, and these furnished two series of $\dfrac{27 \times 26}{2} = 351$ distances, one series based on the "chord" measure of distance, and one on χ^2. The correlation coefficient between

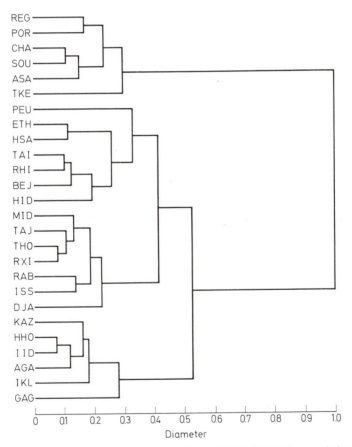

Fig. 16.7. Cluster analysis of the set of populations of Table 16.14. The populations are designated by three letters, as in Table 16.14. (From Lefèvre-Witier, 1973)

Table 16.14. Distances between 26 populations. χ² distances based on gene frequencies (AB0, MN, Rhesus, P and K systems). The populations are designated as follows: 01: Regueibat; 02: Chaamba; 03: Soufi; 04: Ahl Azzi; 05: Tuareg (Ajjer); 06: Tuareg (Ahaggar); 07: Tuareg (Air); 08: Tuareg (Kel Kummer); 09: Beja; 10: Ethiopia; 11: Wadi Rhir and Mya; 12: Djanti; 13: Harratines (Saoura); 14: Harratines (Ahaggar); 15: Agadesians; 16: Peul (Bororo); 17: Iklan (Algeria);18: M'Rabtines (Ideles);19: Isseqamarene Tuareg (Ideles); 20: H'Rabtines Issequam (Ideles);21: Harratines (Ideles);22: Iklan (Ideles); 23: Recent half-castes (Ideles); 24: Eastern Pyrenees (France); 25: Saudi Arabia; 26: Gagou (Ivory Coast)

	02	03	04	05	06	07	08	09	10	11	12	13	14	15	16	17	18	19	20	21	22	23	24	25	26
01-REG	962	816	1373	1218	1521	1227	1158	1106	1445	1028	1512	1423	1567	1493	1292	1473	1262	1155	1550	1347	1651	1446	799	998	1798
02-CHA		509	1231	691	959	699	1164	622	1014	668	1176	928	1169	1108	1327	962	696	677	990	702	1213	1073	779	810	1351
03-SOU			1280	914	1103	840	1179	811	932	601	1166	1016	1257	1137	1341	1103	867	760	1157	1043	1393	1163	771	553	1419
04-KAZ				942	1049	831	1905	1140	1081	815	998	983	518	725	963	898	1216	1080	1268	1137	767	723	1645	1487	872
05-TAJ					496	609	1295	792	870	783	964	789	828	933	1244	863	605	640	514	693	907	630	1264	1088	964
06-THO						830	1528	1056	789	938	929	697	811	1019	1484	1035	737	706	427	1036	843	533	1572	111	843
07-TAI							1403	558	855	462	815	710	651	569	1001	635	756	681	876	745	794	676	1293	1182	920
08-TKE								1243	1666	1469	1519	1660	1883	1837	1885	1746	1273	1275	1380	1470	1888	1583	1176	1271	2000
09-BEJ									990	686	1213	871	1018	982	1214	1002	758	829	1093	1077	999	898	1070	1095	1228
10-ETH										706	1147	625	805	910	1399	1049	1076	1008	1032	1047	672	823	1438	1042	766
11-RHI											896	675	796	651	1004	748	901	709	1088	892	823	836	1131	940	1019
12-DJA												1021	939	1006	1423	1081	1107	898	999	1301	1068	777	1648	1317	1142
13-HSA													722	814	1244	1098	940	842	942	1019	750	753	1457	1130	874
14-HHO														486	1081	726	1097	994	1046	1025	368	548	1725	1463	480
15-AGA															987	669	1144	963	1176	1037	648	790	1641	1473	782
16-PEU																1182	1332	1429	1529	1366	1253	1128	1547	1698	1277
17-JKL																	1095	949	1109	890	803	919	1557	1431	951
18-RAB																		695	651	973	1213	799	1232	960	1209
19-ISS																			765	1035	1109	810	1318	856	1263
20-RXI																				1108	1121	702	1602	1199	1074
21-HID																					1012	1113	1125	1393	1164
22-IID																						710	1732	1486	440
23-MID																							1629	1284	681
24-POR																								1023	1866
25-ASA																									1587
26-GAG																									

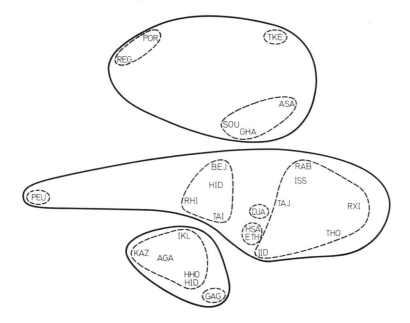

Fig. 16.8. The projection onto the first principal plane of the populations of Table 16.14. Solid lines are drawn around the three major clusters from Fig. 16.7, and broken lines around the minor clusters. (From Lefèvre-Witier, 1973)

the two measures of distance was 0.96, which shows that the two measures agree almost perfectly in the distances between populations to which they lead.

Table 16.14 shows the χ^2 distances calculated from these data. The data have also been analysed by principal components analysis and by the method of cluster analysis, using the algorithm described in Chapter 15, and the results are shown in Figs. 16.7 and 16.8. The tree diagram (Fig. 16.7) shows that the 27 populations fall into three major groups:

1. A "Mediterranean" group, including the Arabs of Saudi Arabia, populations of the Eastern Pyrenees (France), and the Regueibat of Algeria.

2. A "black African" group, which includes the Gagou of the Ivory Coast, the Iklan (descendants of African slaves deported to the Sahara), and the "Harratine" groups.

3. An "intermediate" group.

It is remarkable that the Kel Kummer Tuareg fall into the first group, whereas the populations classified as "Tuareg" in the Ahaggar, Ajjer, Aïr or Ideles areas all belong to the "intermediate" group of populations.

The projection onto the principal plane (Fig. 16.8) confirms the results of the cluster analysis. Fig. 16.8 shows the individual population points, and also the groupings into the three major clusters and into the ten classes of lower rank (corresponding to a diameter of 0.2), derived from the cluster analysis.

This analysis is, of course, based on a small number of populations. We must therefore not attempt to draw firm conclusions from such incomplete information. Nevertheless, this study illustrates the usefulness of the techniques described in Chapter 15, when we wish to study the biological similarity and differences of populations.

4. Classification of Populations Using the HL-A Systems

4.1. Data and Methods of Calculation

At a Workshop held in 1972, data on HL-A types from about 50 populations was presented. In order to make this mass of data comprehensible, we must condense it down to a small number of parameters, if possible in such a way that the results bear a reasonably close relation to the original data, so that the information which has been collected is not wasted. It is obviously not possible to meet these requirements simultaneously, and certain compromises have to be made.

First of all, we would like to estimate the distances between the populations. The simplicity of the phrase: "the distances between populations" must not lead us to forget that the definition of the measures of distance is arbitrary (see Chapter 14), and that the final results can differ greatly, depending on which method we choose; moreover, real populations are not at all the same as our simple mathematical model of a population, consisting of a totally isolated group of individuals. It is an abuse of language which we must not allow to confuse us, that we use the same words for both, though they do not have the same properties at all. We must also be aware that in any study of real populations we are only able to study samples of their members.

The results discussed below must thus be regarded simply as an attempt to apply the descriptive methods which were described in Chapters 14 and 15, and not as a detailed or accurate description of the actual state of affairs; this will always be too complex to understand fully with the crude tools which we have at our disposal.

The data consist of the frequencies of 29 alleles at the two HL-A loci, 13 at the LA locus and 16 at locus 4, in 50 populations. We can use these frequencies to calculate the distances between each of the $\dfrac{50 \times 49}{2} = 1\,225$

pairs of populations. The calculations have been done for the two loci separately, and two measures of distance, the arc cosine measure $(d_{ij} = \text{arc} \cos \sum_r \sqrt{p_{ir} p_{jr}})$ and the χ^2 measure $\left(D_{ij}^2 = \sum_r \left[\dfrac{p_{ir} - p_{jr}}{p_r}\right]^2\right)$ were used. There are therefore four distances for each pair of populations.

4.2. Results

The results of these calculations are described by Degos *et al.* (1972). We may summarise them, as follows:

1. In order to find out whether the measure of distance chosen had a large effect on the set of distances between the populations, the correlation coefficients between the measures were determined. For the LA locus, the value obtained was 0.90, and for locus 4 it was 0.94. These two measures of distance thus result in essentially the same topography of the set of populations.

2. Although the two HL-A loci control related functions, changes in allele frequencies in one are not necessarily accompanied by changes in the other. It is quite possible that the action of natural selection or genetic drift may have had different effects on the two loci. If this has happened, the topography of the set of populations will be completely different, depending on which locus we study. In order to see if this is so, we can calculate the correlation coefficient between the distances based on the two loci; that for the arc cosine measure of distance comes out at 0.68, and for the χ^2 distance at 0.66. The two loci therefore lead to similar representations of the set of populations: we will find that in general populations which are similar on the basis of one of the loci also tend to be similar when we study the other.

3. We can next draw up a contingency table for each locus, giving the frequencies of the different alleles in the various populations. Using the method of analysis of contingency tables described in Chapter 15, we can find the representations on the same plane of the set of populations and the set of alleles.

For the LA locus, the proportions of the total variance that are removed by the first, second and third axes are 21 %, 18 % and 13 %, respectively; the values for locus 4 are 27 %, 17 % and 12 %. The projection onto the first principal plane therefore preserves sufficient information for us to have some hope of obtaining a meaningful interpretation of the points. These projections are shown in Figs. 16.9 and 16.10.

These figures show that a number of groups of points, clearly separate from one another, can be seen. For instance, taking locus LA, the Western European populations of Britain (SC, OX, EN), Italy (FE, SG, ST, SA, SR) and France (F, BA) fall into a cluster at the opposite end of the diagram

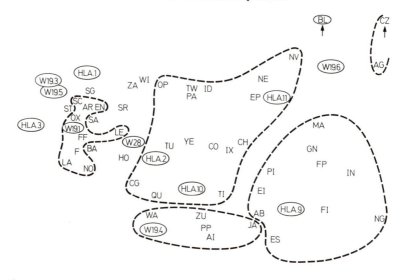

Fig. 16.9. Analysis of the locus LA allele frequencies: representation of 50 populations on the first principal plane. The dotted lines enclose the five major clusters from Fig. 16.11. (From Degos *et al.*, 1972)

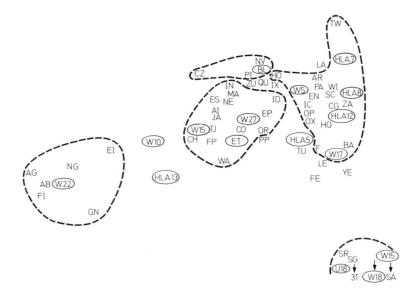

Fig. 16.10. Analysis of the locus 4 allele frequencies: representation of 50 populations on the first principal plane. The dotted lines enclose the five major clusters from Fig. 16.12. (From Degos *et al.*, 1972)

from the populations of Oceania — New Guinea (NG, GN), Australian Aborigines (AG, AB), Indonesia (IN), the Fiji Islands (FI). Similarly for locus 4, the populations AG, FI, AB, NG, GN, and EI (Easter Island) cluster together at one extremity of the diagram, while the points representing the Sardinian populations (SR, SG, ST, SA) are grouped at the other extremity. Since the points representing the characters are also shown on the same diagram, we can see, for instance, that W 22 is the allele which is most markedly characteristic of the first group, and that W 18, U 18 and W 15 are characteristic of the Sardinian populations.

We must, however, always bear in mind that the groupings may be artifacts of the projection onto the first principal plane, and may not necessarily represent true clusters of similar populations.

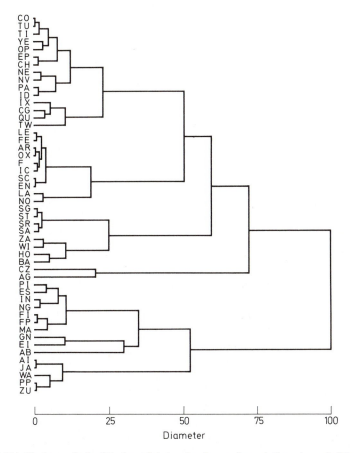

Fig. 16.11. Cluster analysis of the locus LA data for the set of populations shown in Fig. 16.9. (From Degos *et al.*, 1972)

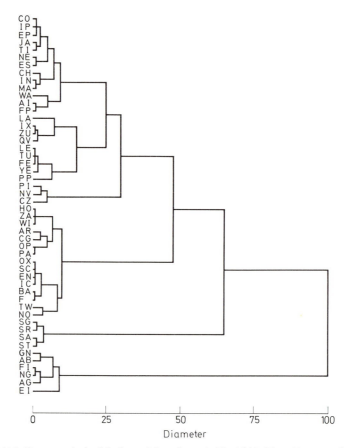

Fig. 16.12. Cluster analysis of the locus 4 data shown in Fig. 16.10. (From Degos *et al.*, 1972)

4. In order to study this question, the data were also subjected to a cluster analysis, based on the algorithm developed in Section 3 of Chapter 15; the tree diagrams shown in Figs. 16.11 and 16.12 were obtained. We can compare the groups of populations revealed by this technique with those which we found before. To illustrate the procedure, we can use Figs. 16.11 and 16.12 to group the populations into six groups; this involves a loss of 35 % of the information, for locus LA, and 20 % for locus 4. Populations which fall into the same group have been ringed around in Figs. 16.9 and 16.10 for the five groups for which this is possible. This shows that the two methods give results which are broadly in agreement, though we may note that, for locus 4, populations LE, YE, FE, TU, ZU, QU, IX, LA and PP, which all belong to the same "branch" of the tree, are not clustered together on the first principal plane.

The fact that the two methods of grouping the populations show a good measure of agreement must not induce too confident an attitude to the classification thus obtained. The methods used were developed simply in order to make the "best" use of the information contained in the data. The problem is to know what is really the "best" method. In the example we have just studied, the method of cluster analysis allowed us to make use of nearly all the information about the distances between the populations, but involved an arbitrary choice of the method of grouping them, while the method of principal components analysis of contingency tables involves a partial loss of the information, so that the resulting representation is to some extent a distortion of the true topography.

Above all, when we use these methods, we must remember that all they can do is to provide evidence of the similarity of populations, with respect to certain characters (loci LA and 4, in this instance); they cannot reveal the true, biological connections between the populations in question. Figs. 16.11 and 16.12 are therefore not to be thought of as phylogenetic trees, and they do not tell us anything about the history or relationships of the groups studied.

Conclusion

The fundamental aim of population genetics is to understand the mechanisms by which living matter evolves, and to develop methods by which we can control the natural, spontaneous processes involved. As regards the study of human evolution in particular, we want to find out what caused the divergence of the branch of the evolutionary tree that led to *Homo sapiens*, to ascertain how the various sub-branches of this main branch may be classified into distinct races, to discover the means by which the present diversity of the human species is preserved, and to learn how we can guide human evolution into favourable directions. The task which we set ourselves in this book was to show how population genetics has contributed to the achievement of these aims. In conclusion, we would like to lay stress on the inadequacy of our present knowledge about the fundamental process which we have been trying to understand.

The first truly "living" organism was the first entity which had the capacity to maintain its organisation in a stable state in a changing environment, and to transmit the principles of this organisation to its "descendants". Since the beginning of life, innumerable different species have evolved, nearly all of which are now extinct. The main stages in the history of life are shown in the following table:

Formation of the earth	5×10^9 years ago
Beginning of life	$2-3 \times 10^9$ years ago
First fossils	$0.5-1 \times 10^9$ years ago
First mammals	13×10^7 years ago
First "humans" (*Australopithecus*)	2×10^6 years ago
"Modern" man (Cro Magnon)	3×10^4 years ago

To make these time-periods more concrete, suppose that the time since the origin of life is represented by a book of 500 pages, with the beginning of life on the first page. Then the first fossils would appear on p. 350, the first mammals on p. 470, the first men on one of the last lines of the final page, and Cro Magnon man would appear in the middle of the last line of the whole book.

During this unimaginably large span of time, the only source of evolutionary innovation has been the process of mutation: mutation alone can produce new genes, and thus lead to the synthesis of new protein molecules; usually the mutant individuals died as a result of these changes, but, in a tiny proportion of cases, the new mutation conferred an advantage, perhaps because some part of the organism became better adapted to its function, or maybe due to the creation of a new, useful metabolic reaction.

From all the new forms created by chance, most were doomed to be unsuccessful in their struggle to survive in a hostile environment; only a few better adapted forms survived, and these have gradually evolved into ever more complex forms, have given rise to the higher animals, and to the final, marvellous product of the slow, evolutionary process: Man.

This qualitative description leaves many questions unanswered. At least one of these is the concern of population genetics: if the changes in living organisms brought about by mutation are purely random, has there been sufficient time, in the period of somewhat more than 10^9 years since the first appearance of life on earth, for all the unsuccessful forms that must have been necessary in the evolution of man from unicellular organisms, to have been developed, perhaps elaborated, and then abandoned?

The results which we have studied in this book, relating to the rates of mutation, genetic drift, and selective change in populations, provide a partial answer to this question. But if we turn back and examine the many simplifying assumptions which we had to make in order to obtain these results, we see immediately that the methods we have used, despite their forbidding mathematical complexity, are powerless to answer a question of this magnitude.

We see that our efforts have led to only a tiny advance; we have learnt a little about the mechanisms which control changes in the living world; we have succeeded in drawing certain limits to our present knowledge, thereby decreasing the risk of accepting hasty, over-simplified conclusions. But we cannot answer the fundamental question: "Is there an underlying 'force', which directs evolution?" Science has no reply to the anguished cry for knowledge, expressed so forcefully by André Gide in his diary[1]:

---

[1] «La science il est vrai, ne progresse qu'en remplaçant partout le *pourquoi* par le *comment*; mais si reculé qu'il soit, un point reste toujours où les deux interrogations se rejoignent et se confondent. Obtenir l'homme...des milliards de siècles n'y sauraient suffire, par la seule contribution du hasard. Si antifinaliste que l'on soit, que l'on puisse être, on se heurte là à de l'inadmissible, à de l'impensable, et l'esprit ne peut s'en tirer qu'il n'admette une propension, une pente, qui favorise le tâtonnant, confus et inconscient acheminement de la matière vers la vie, vers la conscience; puis à travers l'homme, vers Dieu.»

«Mais comme Dieu se fait attendre!...»

André Gide.

"It is true that science progresses only by replacing *why* by *how*; but there is always a point, however distant it may be, where the two questions meet and merge. To produce man by chance alone, millions of centuries would not suffice. However strongly we reject finalism, however strongly we can do so, we are here up against something inadmissible, un-thinkable, from which we can escape only by assuming that there is some tendency, some gradient, which helps on its way the groping, confused, mindless path of matter towards sentient life; then towards man, and finally towards God."

Journal. 8 June, 1942.

"But how long God keeps us waiting! ..."

Journal. 9 June, 1942.

Appendix A

Linear Difference Equations

1. Definitions

Let us consider a function, u_n, of the integer variate n which takes the values $0, 1, \ldots, \infty$. The values of u_n for $n = 0, 1, \ldots, \infty$ therefore constitute a sequence, which we may denote by $\{u_n\}$. $\{u_n\}$ might, for example, represent the values of the inbreeding coefficient of a population over successive generations. It is often the case that, from our knowledge of the biological situation under consideration, we can derive a relation connecting u_n with the values of u in a certain array of preceding generations:

$$u_n = f(u_{n-1}, u_{n-2}, \ldots, u_{n-m}). \tag{A.1}$$

Here, m is a fixed number, characteristic of the system under study.

Equations such as (A.1) are known as difference equations, and arise frequently in population genetics. Eq. (A.1) is referred to as a difference equation of the "m-th order".

A type of difference equation which is encountered very frequently is that known as a "linear difference equation"; such an equation is of the form:

$$u_n = a_1(n) u_{n-1} + a_2(n) u_{n-2} + \cdots + a_m(n) u_{n-m} + b_n. \tag{A.2}$$

$a_1(n), \ldots, a_m(n)$ are known functions of n, and the $b_1, \ldots, b_n$ constitute a sequence, the values of whose elements are given.

If $b_n = 0$ for all n, Eq. (A.2) is called a "homogeneous linear difference equation".

If the functions $a_1(n), \ldots, a_m(n)$ are independent of n, Eq. (A.2) is known as a "linear difference equation with constant coefficients".

In what follows, we shall confine our attention to equations of the form (A.2). The problem is to express the values of the elements u_n in the sequence $\{u_n\}$ as functions of n alone, rather than in terms of the values of u earlier in the sequence. If we can do this, then we have obtained the solution of Eq. (A.2).

2. The Solution of Linear Difference Equations

In this Appendix, we shall discuss the solution of some types of difference equation which are used in this book. We start by considering some general properties of homogeneous linear difference equations.

2.1. The General Solution of a Homogeneous Equation of Order m

In an equation of this type, the b_n of Eq. (A.2) are zero for all n. Moreover, we take $u_0, u_1, \ldots, u_{m-1}$ to be given as the initial conditions of the system. Eq. (A.2) is therefore applicable only for $n \geq m$; we thus have:

$$u_n = a_1(n) u_{n-1} + a_2(n) u_{n-2} + \cdots + a_m(n) u_{n-m} \quad (n \geq m). \quad \text{(A.3)}$$

First consider the equation of the first order, i.e. with $m = 1$. The system is determined by u_0 and the relation

$$u_n = a_1(n) u_{n-1} \quad (n \geq 1). \quad \text{(A.4)}$$

Suppose that we find a function $f(n)$ which satisfies Eq. (A.4). Then we can write:

$$f(n) = a_1(n) f(n-1).$$

It is obvious that the product $Af(n)$, where A is an arbitrary constant, will also satisfy this equation. If, however, we are to express u_n in terms of $f(n)$, for all values of n (i.e. if we are to obtain the solution of the difference equation), we must have:

$$Af(0) = u_0.$$

The constant A is therefore determined uniquely by $f(0)$ and u_0.

The solution $Af(n)$, where A is found in this way, is unique; if there is another function $g(n)$ which satisfies both Eq. (A.4) and the condition $g(0) = u_0$, then it is easy to see that $g(n) = Af(n)$, for all n.

With an equation of the second order, the system is determined by the initial values u_0 and u_1, and the equation:

$$u_n = a_1(n) u_{n-1} + a_2(n) u_{n-2} \quad (n \geq 2). \quad \text{(A.5)}$$

If we can find two functions $f_0(n)$ and $f_1(n)$ which satisfy Eq. (A.5), such that $f_0(n) \neq f_1(n)$, the function $A_0 f_0(n) + A_1 f_1(n)$ will also satisfy (A.5). The coefficients A_0 and A_1 are uniquely determined by the equations:

$$u_0 = A_0 f_0(0) + A_1 f_1(0)$$
$$u_1 = A_0 f_0(1) + A_1 f_1(1)$$

provided that $f_0(0) f_1(1) \neq f_0(1) f_1(0)$. By an argument similar to that for the first order equation, it can be seen that this solution is unique.

We can extend these considerations to the general case. The general solution of a homogeneous linear difference equation of order m is:

$$u_n = \sum_{i=0}^{m-1} A_i f_i(n) \tag{A.6}$$

where the $f_i(n)$ are independent functions, each of which satisfies Eq. (A.3), and where the constants $A_0, A_1, \ldots, A_m$ are determined by the m linear constraints:

$$u_k = \sum_{i=0}^{m-1} A_i f_i(k) \quad (k=0, 1, \ldots, m-1). \tag{A.7}$$

2.2. The General Solution of an Inhomogeneous Equation of Order m

Suppose now that we are given the m initial values of the system, $u_0, \ldots, u_{m-1}$, and that we have the inhomogeneous equation:

$$u_n = a_1(n) u_{n-1} + \cdots + a_m(n) u_{n-m} + b_n \quad (n \geq m). \tag{A.8}$$

Suppose we can obtain a solution of (A.8), say $\bar{f}(n)$. It is easy to see that $\bar{f}(n) + \sum_i A_i f_i(n)$ will also satisfy Eq. (A.8) together with the set of initial conditions.

2.3. The Solution of the Homogeneous Equation of Order m, with Constant Coefficients

In this case, the system is determined by the m initial values $u_0, \ldots, u_m$ and the equation:

$$u_n = a_1 u_{n-1} + a_2 u_{n-2} + \cdots + a_m u_{n-m} \quad (n \geq m) \tag{A.9}$$

where $a_1, \ldots, a_m$ are now independent of n.

The general solution of this equation has a particularly simple form, which can be determined as follows. Let us substitute into Eq. (A.9) the trial solution $f(n) = \lambda^n$, where λ is a real or complex number. We have, from Eq. (A.9):

$$\lambda^n = a_1 \lambda^{n-1} + a_2 \lambda^{n-2} + \cdots + a_m \lambda^{n-m}$$

i.e.

$$1 = a_1 \lambda^{-1} + a_2 \lambda^{-2} + \cdots + a_m \lambda^{-m}. \tag{A.10}$$

Eq. (A.10) is called the "auxiliary equation", or the "characteristic equation" for Eq. (A.9). It can be rewritten as a polynomial of degree m:

$$\lambda^m - a_1 \lambda^{m-1} - a_2 \lambda^{m-2} - \cdots - a_m = 0. \tag{A.11}$$

Such a polynomial equation has m real or complex roots. In the case when all the roots are distinct (which is commonly encountered in practice), Eq. (A.11) therefore yields m distinct solutions to Eq. (A.9); these are of the form λ_i^n, where λ_i is one of the m distinct roots of Eq. (A.11). In this case, therefore, we can write the general solution of (A.9) as:

$$u_n = \sum_{i=0}^{m-1} A_i \lambda_i^n \tag{A.12}$$

where the λ_i are the roots of (A.11), and the A_i are determined by the m linear equations:

$$u_k = \sum_{i=0}^{m-1} A_i \lambda_i^k \quad (k=0, 1, \ldots, m). \tag{A.13}$$

When there is a repeated root, λ_j say, with multiplicity k, it can be shown by an extension of the above method that the functions λ_j^n, $n\lambda_j^n, \ldots, n^{k-1}\lambda_j^n$ each satisfy Eq. (A.9). The general solution in this case becomes:

$$u_n = \sum_{i \neq j} A_i \lambda_i^n + \sum_{r=0}^{k-1} A_j^{(r)} n^r \lambda_j^n. \tag{A.14}$$

This can easily be extended to the case when several roots are multiple.

For large values of n, Eq. (A.13) can be approximated as follows. Any real or complex number λ_k can be expressed in the form:

$$\lambda_r = r_k e^{i \Theta_k}$$

where r_k is the modulus of λ_k. If one of the roots of Eq. (A.13), say λ_0, has a larger modulus than the others, then for $k \neq 1$:

$$\lim_{n \to \infty} \left\{ \frac{\lambda_k}{\lambda_0} \right\}^n = \lim_{n \to \infty} \left\{ \frac{r_k}{r_0} \right\}^n e^{i n (\Theta_k - \Theta_0)} = 0.$$

Eq. (A.13) therefore tends, with increasing n, towards:

$$u_n = A_0 \lambda_0^n. \tag{A.15}$$

2.4. The Renewal Equation of Order m

In Chapter 7, we dealt with the problem of evaluating the course of change of the number of female births per cycle, in a population with overlapping generations. We found that, for a time greater than or equal to ω, the age of last female reproduction, we could express the number of female births $B_f(t)$ as a linear function of the numbers of births in preceding cycles. If we equate $B_f(t)$ with u_n in Eq. (A.9), and $l(1) f(1)$ with a_1, $l(2) f(2)$ with a_2, etc., we have the equation giving $B_f(t)$ for $n \geq m$ (i.e. for $t \geq \omega$, in the terminology of Chapter 7):

$$u_n = a_1 u_{n-1} + a_2 u_{n-2} + \cdots + a_m u_{n-m}. \tag{A.16}$$

For $n<m$, however, individuals who were alive before cycle 0 will contribute some of the births that take place in the population. To take this contribution into account, it is easy to see that Eq. (A.16) must be supplemented by the equation:

$$u_n = b_n + a_1 u_{n-1} + \cdots + a_n u_0 \qquad (n<m). \tag{A.17}$$

The terms $b_0, \ldots, b_{m-1}$ measure the contributions of individuals who were alive before cycle 0 to the number of births in the cycles $0, 1, \ldots, m-1$.

A pair of equations such as (A.16) and (A.17) is known as a "renewal equation". The solution of the renewal equation can be obtained by treating it as a homogeneous linear difference equation with constant coefficients, regarding the set of values $u_0, \ldots, u_{m-1}$ as given. These values can easily be calculated from the initial conditions, together with the parameters of the process we are studying (e. g. the mortality and fecundity parameters of a population). The solution of (A.16) and (A.17) is then given in terms of the roots of the characteristic Eq. (A.10), in precisely the same way as for Eq. (A.9). For large n, the solution of the renewal equation is therefore approximated by Eq. (A.15).

Appendix B

Some Definitions and Results in Matrix Algebra

It is obviously impossible to give a complete or rigorous treatment of matrix methods in a small appendix such as this. However, it would seem useful to recall some definitions concerning matrices, and explain a few of the properties of matrices which are used frequently in population genetics.

1. Definitions

1.1. Types of Matrix

A matrix of order $m \times n$ is a set of $m \times n$ elements arranged in a rectangle consisting of m rows and n columns. The element at the intersection of the i-th row and the j-th column is denoted by a_{ij}. Thus a matrix A of order $m \times n$ is the set:

$$A \equiv \begin{bmatrix} a_{11} \dots a_{1j} \dots a_{1n} \\ \dots\dots\dots\dots\dots \\ a_{i1} \dots a_{ij} \dots a_{in} \\ \dots\dots\dots\dots\dots \\ a_{m1} \dots a_{mj} \dots a_{mn} \end{bmatrix}.$$

A row vector with n elements, $R \equiv (r_1, \dots, r_n)$, can be considered as a matrix of order $1 \times n$; a column vector with n elements

$$C \equiv \begin{bmatrix} c_1 \\ \vdots \\ c_n \end{bmatrix}$$

can be considered as a matrix of order $n \times 1$.

A matrix is called a "square matrix" if the number of rows is the same as the number of columns: $m = n$.

A square matrix is said to be "diagonal" if all its elements are zero, except for the elements a_{ii} which lie on the principal diagonal, in other words, if it is of the form:

$$D \equiv \begin{bmatrix} a_{11} \dots & 0 & \dots & 0 \\ \hdotsfor{4} \\ 0 & \dots a_{ii} & \dots & 0 \\ \hdotsfor{4} \\ 0 & \dots & 0 & \dots a_{nn} \end{bmatrix}.$$

A "unit matrix" is a diagonal matrix with all the elements on the principal diagonal equal to 1:

$$I \equiv \begin{bmatrix} 1 \dots 0 \dots 0 \\ \hdotsfor{1} \\ 0 \dots 1 \dots 0 \\ \hdotsfor{1} \\ 0 \dots 0 \dots 1 \end{bmatrix}.$$

A matrix B is called the "transpose" of the matrix A if the rows of B are equal to the columns of A, in other words, if:

$$b_{ij} = a_{ji}.$$

The notation A' is used to designate the transpose of A. The transpose of a matrix of order $m \times n$ is thus of order $n \times m$; in particular, the transpose of a row vector is a column vector, and *vice versa*.

A square matrix S is said to be "symmetric", if it is equal to its transpose, i.e. if, for all values of i and j:

$$s_{ij} = s_{ji}.$$

1.2. The Determinant of a Matrix

If A is a square matrix of order n, the *determinant* of A, which we shall denote det A, is a polynomial of the n-th degree obtained by multiplying together all possible sets of n elements taken one from each row and one from each column and giving the products plus and minus signs in a specific way. A determinant is usually represented as:

$$\det A = \begin{vmatrix} a_{11} \dots & a_{1n} \\ \vdots & \vdots \\ a_{n1} \dots & a_{nn} \end{vmatrix}.$$

The determinant of a 2×2 matrix is simply:

$$\det A = \begin{vmatrix} a_{11} & a_{12} \\ a_{21} & a_{22} \end{vmatrix} = a_{11} a_{22} - a_{12} a_{21}.$$

The determinant of a higher order matrix can be evaluated as follows. Let us call the determinant of order $n-1$ formed by deleting the i-th row and the j-th column of det A the *minor* of a_{ij}; we shall write this as det A_{ij}. The signed minor:

$$C_{ij} = (-1)^{i+j} \det A_{ij}$$

is called the *cofactor* of a_{ij}. It is possible to show that:

$$\det A = \sum_j a_{ij} C_{ij} = \sum_i a_{ij} C_{ij}.$$

By repeatedly expanding the cofactors in terms of their cofactors until we get down to determinants of order 2, we can therefore evaluate a determinant of order n entirely in terms of products of its elements.

Note: det $A = 0$ if and only if the columns of A are linearly dependent, i.e. the elements of each row satisfy a relation $\sum_j \mu_j a_{ij} = 0$.

1.3. Matrix Addition and Multiplication

Consider two matrices A and B which are of the same order. The sum of A and B is a matrix C whose elements are given by:

$$c_{ij} = a_{ij} + b_{ij}.$$

The *product* of two matrices is defined only when the number of columns of the first matrix is the same as the number of rows of the second.

If U is a matrix of order $m \times n$ and V is a matrix of order $n \times p$, the product UV is a matrix W of order $m \times p$ whose elements w_{ij} are found from the elements of U and V by the formula:

$$w_{ij} = \sum_{k=1}^{k=n} u_{ik} v_{kj}. \tag{B.1}$$

Notice that the product VU is defined only when $m = p$.

Consider a matrix M of order $m \times n$, a row vector R with m elements and a column vector C with n elements. The row and column vector are matrices of order $1 \times m$ and $n \times 1$, respectively, Hence:

1. The product RM is defined. It is a matrix of order $1 \times n$, in other words a row vector with n elements.

2. The product MC is defined. It is of order $m \times 1$, i.e. it is a column vector with m elements.

3. The products MR and CM are not defined.

4. If $m = n$, the product RC is a matrix of order 1×1, i.e. a single number. This is called the scalar product of two vectors. In particular, the product RR' of a row vector with its transpose is equal to $\sum_i r_i^2$, and similarly for the product $C'C$ of a column vector with itself.

5. The product CR is a matrix of order $m \times n$, in other words a set of $m \times n$ numbers such that:

$$m_{ij} = c_i r_j.$$

The product CC' is a matrix whose general term is equal to the product $c_i c_j$.

The scalar products of vectors can also be written in the notation $R \cdot C$, which is equivalent to the expression RC; the product of a vector with its transpose can thus be written as $R \cdot R$ (or even as R^2), or as RR'.

Taking account of the definition summarised by formula (B.1), it is easy to show that matrix multiplication is an associative operation: if U is a matrix of order $m \times n$, V of order $n \times p$, and W of order $p \times q$, we have:

$$(UV)W = U(VW).$$

However, matrix multiplication is not a commutative operation, and VU may be undefined, even though UV is defined. If both the products are defined, they may be different from one another, for example:

$$\begin{bmatrix} 1 & 1 \\ 1 & 1 \end{bmatrix} \begin{bmatrix} 1 & 0 \\ 1 & 1 \end{bmatrix} = \begin{bmatrix} 2 & 1 \\ 2 & 1 \end{bmatrix}$$

whereas

$$\begin{bmatrix} 1 & 0 \\ 1 & 1 \end{bmatrix} \begin{bmatrix} 1 & 1 \\ 1 & 1 \end{bmatrix} = \begin{bmatrix} 1 & 1 \\ 2 & 2 \end{bmatrix}.$$

Matrix expressions cannot therefore be handled in the same way as ordinary algebraic expressions.

It follows directly from the definition of the product of two matrices that the unit matrix is the neutral element in matrix multiplication, in other words that:

$$UI = U \quad \text{and} \quad IU = U.$$

A square matrix U is said to be "non-singular" if a matrix V exists, such that:

$$UV = VU = I.$$

The inverse of U, if one exists, is designated by U^{-1}. By definition, therefore, we have:

$$UU^{-1} = U^{-1}U = I.$$

It can be shown that a matrix is non-singular if its determinant is not equal to zero.

Let A and B be two non-singular matrices of the same order. It can then be shown that:

$$(AB)^{-1} = B^{-1}A^{-1}. \tag{B.2}$$

The proof is as follows. By definition, we have:

$$(AB)(AB)^{-1} = I.$$

Multiplying each side of this equality on the left by first A^{-1} and then by B^{-1}, and using the associative property of matrix multiplication, we obtain the relations:

$$A^{-1} = A^{-1} A B (AB)^{-1} = B (AB)^{-1}$$

and

$$B^{-1} A^{-1} = B^{-1} B (AB)^{-1} = (AB)^{-1}.$$

It is also simple to show that:

$$(AB)' = B' A' \qquad\qquad (B.3)$$

and that:

$$(A')^{-1} = (A^{-1})'. \qquad\qquad (B.4)$$

2. Diagonalisation of a Square Matrix

2.1. The Powers of a Matrix

The square $A^2 = A \times A$ of a square matrix of order $n \times n$ is also a square matrix of order $n \times n$. It can, therefore, itself be multiplied by A. By successive multiplications, we can therefore define the successive powers $A^2, A^3, \ldots, A^g$ of a square matrix. Many problems require us to use such powers of matrices. For example, in population genetics, we often want to define the passage from an initial state to a final state which is reached by a series of intervening steps. In Chapter 11, when we were studying the effect of recurrent mutation, we used just such a formulation in Eq. (7), which gave us the genic structure in any generation in terms of the initial genic structure and the matrix of mutation rates from one allele to another:

$$s_g = s_0 V^g.$$

It would be very laborious to calculate A^g by finding all the successive powers of A; also, this would not help us to find the behaviour of A^g as g is increased indefinitely. It is therefore useful to have direct methods for handling the higher powers of matrices.

We note that the problem is solved immediately if the matrix is a diagonal matrix, i.e. if it is of the form:

$$D \equiv \begin{bmatrix} \lambda_1 \ldots 0 \ldots 0 \\ \cdots\cdots\cdots\cdots \\ 0 \ldots \lambda_i \ldots 0 \\ \cdots\cdots\cdots\cdots \\ 0 \ldots 0 \ldots \lambda_n \end{bmatrix}.$$

In this case, the rule for forming a matrix product (Eq. (B.1)) gives:

$$D^2 \equiv \begin{bmatrix} \lambda_1^2 \ldots 0 \ldots 0 \\ \cdots\cdots\cdots\cdots \\ 0 \ldots \lambda_i^2 \ldots 0 \\ \cdots\cdots\cdots\cdots \\ 0 \ldots 0 \ldots \lambda_n^2 \end{bmatrix}.$$

Hence, by successive multiplications,

$$D^g \equiv \begin{bmatrix} \lambda_1^g \ldots 0 \ldots 0 \\ \cdots\cdots\cdots\cdots \\ 0 \ldots \lambda_i^g \ldots 0 \\ \cdots\cdots\cdots\cdots \\ 0 \ldots 0 \ldots \lambda_n^g \end{bmatrix}.$$

The calculation of the powers of A would therefore be very much simplified if we could write A as the product of three matrices:

$$A = CDC^{-1} \tag{B.5}$$

where D is a diagonal matrix and C is a non-singular matrix. The square of A would then be:

$$A^2 = (CDC^{-1})(CDC^{-1}) = CD(C^{-1}C)DC^{-1}$$
$$= CD^2 C^{-1}$$

and, by successive multiplications:

$$A^g = CD^g C^{-1}. \tag{B.6}$$

This would enable us to calculate the g-th power of A without having to find all the intervening powers. The problem, therefore, is to find out whether the matrices C and D exist, and if so, to determine them.

2.2. The Eigenvalues of a Matrix

Eq. (B.5) can be written, multiplying both sides by C, as:

$$AC = CDC^{-1}C = CD. \tag{B.7}$$

Let the elements of C be c_{ij}. Now consider any column of the matrix product AC. For the i-th column, Eq. (B.7) gives:

$$\sum_k a_{jk} c_{ki} = \lambda_i c_{ji} \quad \text{for all } j. \tag{B.8}$$

This is equivalent to a set of n homogeneous equations with n unknowns:

$$c_{1i}, \ldots, c_{ki}, \ldots, c_{ni}.$$

Writing these equations out in full, we have:

$$(a_{11} - \lambda_i) c_{1i} + \cdots + a_{1k} c_{ki} + \cdots + a_{1n} c_{ni} = 0$$
$$\cdots\cdots\cdots\cdots\cdots\cdots\cdots\cdots\cdots\cdots\cdots\cdots\cdots\cdots\cdots$$
$$a_{k1} c_{1i} + \cdots + (a_{kk} - \lambda_i) c_{ki} + \cdots + a_{kn} c_{ni} = 0 \tag{B.9}$$
$$\cdots\cdots\cdots\cdots\cdots\cdots\cdots\cdots\cdots\cdots\cdots\cdots\cdots\cdots\cdots$$
$$a_{n1} c_{1i} + \cdots + a_{nk} c_{ki} + \cdots + (a_{nn} - \lambda_i) c_{ni} = 0.$$

It is known that such a set of homogeneous equations has a non-zero solution only if the determinant of the coefficients is zero, i.e. if:

$$\begin{vmatrix} (a_{11} - \lambda_i) & \cdots & a_{1k} & \cdots & a_{1n} \\ a_{k1} & \cdots & (a_{kk} - \lambda_i) & \cdots & a_{kn} \\ a_{n1} & \cdots & a_{nk} & \cdots & (a_{nn} - \lambda_i) \end{vmatrix} = 0. \tag{B.10}$$

When this determinant is expanded, it takes the form of a polynomial in λ_i, of degree n. The equation symbolised by (B.10) is therefore a polynomial of degree n. It is called the "*characteristic equation*" of the matrix A.

The n roots $\lambda_1, \ldots, \lambda_i, \ldots, \lambda_n$ of this equation (which may be real or complex numbers, distinct or coincident) are called the "eigenvalues" of the matrix A.

If we then solve the system of Eqs. (B.9) with one of the eigenvalues, λ_i say, we obtain the relative values of the n elements $c_{1i}, \ldots, c_{ni}$ of the i-th column of C. These n elements define the "column eigenvector" V_i associated with the eigenvalue λ_i. Eq. (B.8) shows that this vector is such that:

$$A V_i = \lambda_i V_i. \tag{B.11}$$

The set of the n column eigenvectors constitute the matrix C which we required.

Similarly, we could multiply the terms of Eq. (B.5) on the left by C^{-1}, instead of on the right by C^{-1}. We then obtain:

$$C^{-1} A = C^{-1} CDC^{-1} = DC^{-1}. \tag{B.12}$$

Let γ_{ij} be the elements of the matrix C^{-1}. Now consider any row of the product $C^{-1} A$. Eq. (B.12) shows that, for the i-th row:

$$\sum_k \gamma_{ik} a_{kj} = \lambda_i \gamma_{ij} \quad \text{for all } j \tag{B.13}$$

which is equivalent to the set of n homogeneous equations in n unknowns:

$$(a_{11} - \lambda_i) \gamma_{i1} + \cdots + a_{k1} \gamma_{ik} + \cdots + a_{n1} \gamma_{in} = 0$$
$$a_{1k} \gamma_{i1} + \cdots + (a_{kk} - \lambda_i) \gamma_{ik} + \cdots + a_{nk} \gamma_{in} = 0 \tag{B.14}$$
$$a_{1n} \gamma_{i1} + \cdots + a_{kn} \gamma_{ik} + \cdots + (a_{nn} - \lambda_i) \gamma_{in} = 0.$$

This set of equations has a non-zero solution if the determinant of the coefficients is zero, i.e. if:

$$\begin{vmatrix} (a_{11}-\lambda_i) & \dots & a_{k1} & \dots & a_{n1} \\ a_{1k} & \dots & (a_{kk}-\lambda_i) & \dots & a_{nk} \\ a_{1n} & \dots & a_{kn} & \dots & (a_{nn}-\lambda_i) \end{vmatrix} = 0. \tag{B.15}$$

This determinant is the same as the one in Eq. (B.10), but with the rows and columns interchanged; the two Eqs. (B.10) and (B.15) therefore represent the same polynomial equation in λ_i, of degree n.

Solving the set of Eqs. (B.14) with one of the eigenvalues λ_i, we obtain the relative values of the n elements $\gamma_{i1}, \dots, \gamma_{in}$ of the i-th row of the matrix C^{-1}. These constitute the "row eigenvector" W_i associated with the eigenvalue λ_i. Eq. (B.13) shows that this vector is such that:

$$W_i A = \lambda_i W_i. \tag{B.16}$$

To summarise, therefore, we can decompose the matrix A into a product of three matrices, as follows:

1. We find the n eigenvalues λ_i, which are the roots of the equation:

$$\det (A - \lambda I) = 0.$$

2. We solve the sets of Eqs. (B.9) and (B.14) with each of the eigenvalues λ_i, to give the column and row eigenvectors V_i and W_i associated with each value of λ.

3. Finally, we can write:

$$A = (V_1, \dots, V_k, \dots, V_n) \times \begin{bmatrix} \lambda_1 & & & & 0 \\ & \ddots & & & \\ & & \lambda_k & & \\ & & & \ddots & \\ 0 & & & & \lambda_n \end{bmatrix} \times \begin{bmatrix} W_1 \\ \vdots \\ W_k \\ \vdots \\ W_n \end{bmatrix}.$$

From Eq. (B.5) it is obvious that the problem is not solved unless the matrix $(V_1, \dots, V_k, \dots, V_n)$ is the inverse of $\begin{bmatrix} W_1 \\ \vdots \\ W_k \\ \vdots \\ W_n \end{bmatrix}$. This can be shown to be the case when the sets of vectors $W_1, \dots, W_n$ and $V_1, \dots, V_n$ are linearly independent sets. It can be shown that this is, in particular, the case when all the eigenvalues are distinct.

3. The Spectral Analysis of a Matrix

Let us now consider a square matrix A whose eigenvalues are all distinct. We have just seen that the eigenvectors of such a matrix are independent and therefore C is non-singular. We can therefore write:

$$A = CDC^{-1}.$$

D is a diagonal matrix, and can therefore be written in the form:

$$D = \lambda_1 I_1 + \cdots + \lambda_i I_i + \cdots + \lambda_n I_n$$

where I_i is a square matrix with all its elements zero except for the term on the principal diagonal at the intersection of the i-th row and the i-th column, which is equal to 1.

We then have:

$$A = C(\lambda_1 I_1 + \cdots + \lambda_n I_n) C^{-1}$$
$$= \lambda_1 CI_1 C^{-1} + \cdots + \lambda_n CI_n C^{-1}.$$

Writing $P_i = CI_i C^{-1}$, this becomes:

$$A = \lambda_1 P_1 + \cdots + \lambda_n P_n. \tag{B.17}$$

The right-hand side of Eq. (B.17) is called the "spectral analysis" of the matrix A. The n matrices P_i are called the "projections" of A. They have the following properties:

a) $\quad P_1 + \cdots + P_n = CI_1 C^{-1} + \cdots + CI_n C^{-1}$
$$= C(I_1 + \cdots + I_n) C^{-1}$$
$$= CIC^{-1} = I,$$

b) $\quad P_i \times P_i = (CI_i C^{-1})(CI_i C^{-1}) = CI_i C^{-1} = P_i,$

c) $\quad P_i \times P_j = (CI_i C^{-1})(CI_j C^{-1}) = CI_i I_j C^{-1} = 0.$

It follows immediately from the last two properties that if we raise both sides of Eq. (B.17) to the power g, we have:

$$A^g = \lambda_1^g P_1 + \cdots + \lambda_n^g P_n. \tag{B.18}$$

Finally, notice that:

$$P_i = CI_i C^{-1} = \begin{bmatrix} 0 & \cdots & c_{1i} & \cdots & 0 \\ & & \cdots\cdots & & \\ 0 & \cdots & c_{ki} & \cdots & 0 \\ & & \cdots\cdots & & \\ 0 & \cdots & c_{ni} & \cdots & 0 \end{bmatrix} \times \begin{bmatrix} \gamma_{11} & \cdots & \gamma_{1n} \\ \gamma_{k1} & \cdots & \gamma_{kn} \\ \gamma_{n1} & \cdots & \gamma_{nn} \end{bmatrix} \tag{B.19}$$

$$= \begin{bmatrix} c_{1i}\gamma_{i1} & \cdots & c_{1i}\gamma_{in} \\ c_{ki}\gamma_{i1} & \cdots & c_{ki}\gamma_{in} \\ c_{ni}\gamma_{i1} & \cdots & c_{ni}\gamma_{in} \end{bmatrix} = V_i W_i.$$

The behaviour of the successive powers of A can therefore be studied easily if we know the eigenvalues λ_i and the projections P_i of A.

If one of the eigenvalues, λ_m say, is larger in absolute magnitude than the other eigenvalues, Eq. (B.18) shows that A^g is more and more exclusively determined by λ_m^g as g increases. In other words, we can make the approximation:

$$A^g \approx \lambda_m^g \, V_m \, W_m. \tag{B.20}$$

This formula can be generalised to the case when r of the eigenvalues have the same value λ_m, provided that r of the eigenvectors V_m and W_m associated with λ_m are independent, which is usually the case in practice. In this case, we can write:

$$A^g \approx \lambda_m^g (V_m^{(1)} \, W_m^{(1)} + \cdots + V_m^{(r)} \, W_m^{(r)}).$$

4. Real Symmetric Matrices

A real symmetric matrix S is one in which the elements s_{ij} are real numbers such that $s_{ij} = s_{ji}$, in other words $S' = S$.

4.1. The Eigenvalues of a Real Symmetric Matrix are all Real

For any column vectors X and Y we can write:

$$X' \, SY = (X' \, SY)'$$

since the product is a scalar. But $S' = S$. Hence this expression is also equal to:

$$Y' SX. \tag{B.21}$$

Now suppose that the vector X is the eigenvector V_i of the matrix S corresponding to the eigenvalue λ_i, i.e. such that:

$$SV_i = \lambda_i \, V_i$$

and that Y is the complex conjugate $\overline{V}_i$ of V_i (i.e. the real parts of the elements of $\overline{V}_i$ are the same as those of V_i, but the complex parts are of opposite sign). Then Eq. (B.21) becomes:

$$V_i' \, S\overline{V}_i = \overline{V}_i' \, SV_i = \overline{V}_i' \, \lambda_i \, V_i = \lambda_i \, \overline{V}_i' \, V_i.$$

In this identity, the product $\overline{V}_i' \, V_i$ is a real number. The first expression, $V_i' \, S\overline{V}_i$ is also a real number, since it is equal to its complex conjugate. It follows that λ_i is a real number.

4.2. The Eigenvectors of a Real Symmetric Matrix Corresponding to Distinct Eigenvalues are Orthogonal

Let the eigenvectors corresponding to the eigenvalues λ_i and λ_j be V_i and V_j. Then, since $SV_i = \lambda_i V_i$, we obtain:

$$V_j' S V_i = \lambda_i V_j' V_i \tag{B.22}$$

and since $SV_j = \lambda_j V_j$, we have:

$$V_i' S V_j = \lambda_j V_i' V_j. \tag{B.23}$$

But $V_j' V_i = (V_j' V_i)' = V_i' V_j$, since the product is a scalar, and also:

$$V_j' S V_i = (V_j' S V_i)' = V_i' S V_j$$

since the matrix S is symmetric. Subtracting (B.23) from (B.22), we therefore have:

$$0 = (\lambda_i - \lambda_j) V_i' V_j.$$

But $\lambda_i \neq \lambda_j$ by hypothesis, therefore this result implies that $V_i' V_j = 0$; in other words the eigenvectors are orthogonal.

When the solutions of the characteristic equation are all distinct, all the eigenvectors are therefore orthogonal to one another. It can be shown that when there are multiple roots, each root of order k has k associated vectors, which are each orthogonal with the eigenvectors associated with the other eigenvalues.

We have seen that we could determine the relative values of the eigenvectors, but not their absolute values. We can therefore choose the eigenvectors such that they are normalised, in other words such that:

$$V_i' V_i = 1. \tag{B.24}$$

Now suppose that C is a matrix whose columns are the n normalised eigenvectors V_i. Eq. (B.7) gives:

$$SC = CD.$$

Taking the transposes of each side of this expression, and remembering that S and D are both symmetric matrices, we have:

$$C'S = DC'.$$

Hence $CC'S = CDC'$. But, from Eqs. (B.23) and (B.24), $CC' = I$. Hence:

$$S = CDC'. \tag{B.25}$$

We can therefore write the spectral analysis of S in the form:

$$S = \lambda_1 V_1 V_1' + \cdots + \lambda_n V_n V_n'.$$

5. Stochastic Matrices

A matrix A is called a "stochastic matrix" when all its elements are positive or zero, and when the sum of the elements of each row is equal to 1, in other words if:

for every value of i and j: $a_{ij} \geq 0$ and
for every value of i: $\sum_j a_{ij} = 1$.

Matrices of this sort are encountered when, for example, the elements a_{ij} represent the probabilities of passage during a given stage of a process from a state i to a state j (and where the terms a_{ii} on the diagonal represent the probabilities of not changing states); the matrix, V, of mutation rates which we used in Chapter 11 is an example of a stochastic matrix.

Let $p_i^{(g)}$ be the probability that the state i is observed at the g-th stage of the process. The set of $p_i^{(g)}$ values constitute a vector E_g. It is obvious that the probability $p_i^{(g)}$ is related to the probabilities $p_j^{(g-1)}$ for the previous stage of the process, by:

$$p_i^{(g)} = p_1^{(g-1)} a_{1i} + \cdots + p_j^{(g-1)} a_{ji} + \cdots + p_n^{(g-1)} a_{ni}$$

which can be written in the condensed notation of matrix multiplication as:

$$E_g = E_{g-1} \times A.$$

This equation is of the same form as the one in Chapter 11 for the change in genic structure due to mutation.

5.1. The Eigenvalues of Stochastic Matrices

The eigenvalues of stochastic matrices have several special properties. We shall mention a few which are particularly useful in population genetics.

a) *The eigenvalues of a stochastic matrix are of modulus equal to or less than* 1. Let λ_i be an eigenvalue of a stochastic matrix, and let $c_{1i}, \ldots, c_{ni}$ be the elements of the column eigenvector associated with this eigenvalue. Suppose that c_{ki} is the element with the largest modulus. The k-th equation out of the set of Eqs. (B.9) is:

$$\lambda_i c_{ki} = \sum_j a_{kj} c_{ji}.$$

Hence:

$$|\lambda_i c_{ki}| = |\lambda_i| \, |c_{ki}| \leq \sum_j |a_{kj} c_{ji}| \leq \sum_j a_{kj} |c_{ki}| = |c_{ki}|$$

which shows that $|\lambda_i| \leq 1$.

b) *Every stochastic matrix has the eigenvalue* $\lambda=1$. The proof is as follows. The eigenvalues must be such that $\det(A-\lambda I)=0$. But $\det(A-I)=0$, since all the rows of the matrix $(A-I)$ consist of elements whose sum is zero and this implies linear dependence of the columns of $A-I$ (cf. Section 1.2). Therefore $\lambda=1$ is a root of the characteristic equation of a stochastic matrix.

c) *If all the elements of the principal diagonal of a stochastic matrix are non-zero, then* $\lambda=1$ *is the only eigenvalue of modulus* 1. Let $\lambda_r=e^{i\Theta}$ be an eigenvalue with modulus 1, and let c_{jr} be the element of the corresponding eigenvector which has the largest modulus, M_j. Thus $c_{jr}=M_j e^{i\Theta_j}$. Let c_{kr} be any element of the corresponding eigenvector; $c_{kr}=M_k e^{i\Theta_k}$. The j-th equation in the set of Eqs. (B.9) then becomes:

$$\lambda_r c_{jr} = M_j e^{i\Theta} e^{i\Theta_j} = M_j e^{i(\Theta+\Theta_j)}$$
$$= \sum_k a_{jk} c_{kr} = \sum_k a_{jk} M_k e^{i\Theta_k}. \tag{B.26}$$

Hence:

$$M_j = \sum_k a_{jk} M_k \frac{e^{i\Theta_k}}{e^{i(\Theta+\Theta_j)}}$$
$$= \sum_{k\neq j} a_{jk} M_k e^{i(\Theta_k-\Theta_j-\Theta)} + a_{jj} M_j e^{-i\Theta}.$$

If $a_{jj}\neq0$, this last term must be taken into account. We can write the last equation in the form:

$$M_j = \sum_{k\neq j} a_{jk} M_k [\cos(\Theta_k-\Theta_j-\Theta)+i\sin(\Theta_k-\Theta_j-\Theta)]$$
$$+ a_{jj} M_j [\cos(-\Theta)+i\sin(-\Theta)].$$

Equating real parts of both sides of this equation, we have:

$$M_j = \sum_{k\neq j} a_{jk} M_k \cos(\Theta_k-\Theta_j-\Theta) + a_{jj} M_j \cos(-\Theta)$$
$$\leq \sum_{k\neq j} a_{jk} M_j + a_{jj} M_j \cos(-\Theta)$$
$$\leq M_j \sum_{k\neq j} a_{jk} + a_{jj} M_j \cos(-\Theta).$$

But, since $\sum_k a_{jk}=1$, we know that:

$$M_j = M_j \sum_k a_{jk}.$$

Hence, the above equation cannot be an inequality; assuming that $a_{jj}\neq0$, this equation therefore proves that $\cos(-\Theta)=1$, $\Theta=0$. Therefore the largest eigenvalue λ_r is equal to $e^0=1$. This completes the proof of proposition c).

When the elements of A represent probabilities of passing from a state i to a state j during one stage of a process, the condition that, for all values of i, $a_{ii} > 0$ implies that, for each state, the probability of remaining in the same state is non-zero. In the case of a matrix of mutation rates, this means that no allele is certain to mutate each generation. We can certainly accept this assumption in this case, without restricting the generality of the results. A matrix of mutation rates will therefore not have any complex eigenvalues with absolute value 1.

5.2. The Spectral Analysis of a Stochastic Matrix

Finally, a stochastic matrix whose eigenvalues are all distinct can be broken down into a spectral analysis according to the equation:

$$A = P_1 + \lambda_2 P_2 + \cdots + \lambda_n P_n$$

where $|\lambda_i| \leq 1$ for all values of i, and $|\lambda_i| < 1$ for all values of $i > 1$, when none of the elements a_{ii} is equal to zero.

Applying Eq. (B.18) we therefore have:

$$A^g = P_1 + \lambda_2^g P_2 + \cdots + \lambda_n^g P_n.$$

If all the values λ_i are less than 1, all the terms other than P_1 will tend towards zero as g increases, so that $A^g \to P_1$. The speed with which A^g approaches P_1, i.e. the "rate of convergence" towards the final state, is given by the value λ_m of the eigenvalue with the largest modulus, other than $\lambda = 1$.

Finally, we note that the projection P_1 corresponding to the eigenvalue $\lambda = 1$, which, as we have seen, is unique, has a particular form. This follows from the fact that if λ is set equal to 1 and $\sum_j a_{kj} = 1$, the set of Eqs. (B.9) has the obvious solution:

$$c_{11} = c_{21} = \cdots = c_{n1}.$$

Let the value of all the elements c_{i1} of the column eigenvector V_1 associated with the eigenvalue $\lambda = 1$ be 1. Then, from Eq. (B.19):

$$P_1 = V_1 W_1 = \begin{bmatrix} 1 \\ \vdots \\ 1 \\ \vdots \\ 1 \end{bmatrix} \times W_1 = \begin{bmatrix} W_1 \\ \vdots \\ W_1 \\ \vdots \\ W_1 \end{bmatrix}.$$

Therefore P_1 is a matrix with every row identical to the row eigenvector W_1 associated with the eigenvalue $\lambda = 1$.

To summarise, if the stochastic matrix A represents the set of probabilities of passage from a state i to a state j we have seen that the vectors E representing the successive stages are related according to:

$$E_g = E_{g-1} A.$$

This gives:

$$E_g = E_0 A^g.$$

As g increases, we therefore have:

$$E_g \rightarrow E_0 P_1 = E_0 \begin{bmatrix} W_1 \\ \vdots \\ W_1 \\ \vdots \\ W_1 \end{bmatrix}. \tag{B.27}$$

But the elements of E_0 are the frequencies or the probabilities of the various possible states before the first stage of the process has taken place; their sum is equal to 1. Eq. (B.27) therefore implies that $E_g \rightarrow W_1$. Thus, finally:

If a system undergoes a process of transformation defined by a matrix of probabilities of passage from one state to another, and this matrix has just one eigenvalue equal to 1, *the final structure of the system is independent of its initial structure; the final structure is represented by the row eigenvector associated with the eigenvalue whose value is* 1.

References

Allison, A.C.: Protection afforded by sickle-cell trait against subtertian malarial infection. Brit. Med. J. **1954 I**, 290–294.

Allison, A.C.: Genetic factors in resistance to malaria. Ann. N.Y. Acad. Sci. **91**, 710–729 (1961).

Allison, A.C.: Polymorphism and natural selection in human populations. Cold Spr. Harb. Symp. Quant. Biol. **29**, 137–149 (1964).

Bajema, C.: Estimation of the direction and intensity of natural selection in relation to human intelligence by means of the intrinsic rates of natural increase. Eugen. Quart. **10**, 175–187 (1963).

Beet, E.A.: Sickle cell disease in the Balovale district of Northern Rhodesia. E. Afr. Med. J. **32**, 75–86 (1946).

Bemiss, S.M.: Report on influence of marriages of consanguinity upon offspring. Trans. Amer. Med. Ass. **11**, 319–425 (1858).

Benzecri, J.P.: Distance distributionelle et métrique du chi-deux en analyse factorielle des correspondances. Paris: Laboratoire de Statistique Mathématique 1970.

Bernstein, F.: Fortgesetzte Untersuchungen aus der Theorie der Blutgruppen. Z. indukt. Abstamm.- u. Vererb.-L. **56**, 233–273 (1930).

Bodmer, W.F., Cavalli-Sforza, L.L.: A matrix model for the study of random genetic drift. Genetics **59**, 565–592 (1968).

Bouloux, C., Gomila, J., Langaney, A.: Hemotypology of the Bedik. Hum. Biol. **44**, 289–302 (1972).

Bourgeois-Pichat, J.: La mesure de la mortalité infantile. Population **6**, 233–248 (1951).

Cantrelle, P.: L'endogamie des populations du Fouta-Sénégalais. Population **15**, 665–676 (1960).

Cavalli-Sforza, L.L.: The distribution of migration distances: Models and applications to genetics. In: Les déplacements humains. Entretiens de Monaco en sciences humaines. Paris: Hachette 1962.

Cavalli-Sforza, L.L., Edwards, A.F.W.: Phylogenetic analysis: models and estimation procedures. Amer. J. Hum. Genet. **19**, 233–257 (1967).

Cavalli-Sforza, L.L., Kimura, M., Barrai, I.: The probability of consanguineous marriages. Genetics **54**, 37–60 (1966).

Cavalli-Sforza, L.L., Zei, G.: Experiments with an artificial population. Proc. III Int. Cong. Hum. Genet., p. 473–478. Baltimore: Johns Hopkins Press 1966.

Chapman, A.M., Jacquard, A.: Un isolat d'Amérique Centrale-Les Indiens Jicaques du Honduras. In: Génétique et populations. I.N.E.D. Paris: Presses Universitaires de France 1971.

Chaventré, A.: Les Kel Kummer, isolat social. Population **27**, 771–783 (1972).

Choisy, A.: Généalogies Genevoises. Familles admises à la Bourgeoisie avant la Réformation. Genève: Imprimerie Albert Kundig 1947.

Clarke, C.A.: Correlation of ABO blood group with peptic ulcer and other diseases. Amer. J. Hum. Genet. **11**, 400–404 (1959).

Clayton, G.A., Morris, J.A., Robertson, A.: An experimental check on quantitative genetical theory. 1. Short-term responses to selection. J. Genet. **55**, 131–151 (1957).

Colombani, J., Degos, L., Petrignani, C., Chaventré, A., Lefèvre-Witier, P., Jacquard, A.: HL-A gene structure of the Kel Kummer Tuareg. In: Histocompatibility testing. Copenhagen: Munksgaard 1972.

Cotterman, C. W.: A calculus for statistico-genetics. Unpubl. thesis, Ohio State University, Columbus, Ohio 1940.

Courgeau, D.: Les champs migratoires en France. Doctoral thesis, Faculté des Lettres et Sciences Humaines, Paris 1969.

Courgeau, D.: Migration continue dans le temps et structure génique. In: Génétique et populations. I.N.E.D. Paris: Presses Universitaves de France 1971.

Coursol, J.: Recherches sur la distribution des fréquences d'une géne dans les populations d'effectif limité. Thesis, Faculté des Sciences, Paris 1969.

Crow, J.F.: Some possibilities for measuring selection intensities in man. Hum. Biol. **30**, 1–13 (1958).

Crow, J.F., Kimura, M.: The effective number of a population with overlapping generations: a correction and further discussion. Amer. J. Hum. Genet. **24**, 1–10 (1972).

Dahlberg, G.: Mathematical methods for population genetics. Basel: Karger 1948.

Davenport, G.C., Davenport, C.B.: Heredity of skin pigmentation in man. Amer. Naturalist **44**, 641–672 (1910).

Davenport, C.B.: Heredity of skin color in negro-white crosses. Carnegie Inst. Wash. Publ. **188** (1913).

De Finetti, B.: Considerazioni matematiche sul l'ereditarieta Mendeliana. Metron **6**, 1–41 (1926).

Degos, L.: Le système HL-A des Kel Kummer. Population **28**, in press (1973).

Degos, L., Chaventré, A., Jacquard, A.: Le polymorphisme HL-A est-il maintenu par un déavantage des homozygotes. C.R. Acad. Sci. (Paris), in press (1973).

Degos, L., Jacquard, A., Landre, M.F., Salmon, D., Valat, M.T.: Study of distances between populations from the variation of HL-A gene frequency. In: Histocompatibility testing. Copenhagen: Munksgaard 1972.

Denniston, C.: Probability and genetic relationship. Thesis, University of Wisconsin 1967.

Depoid, P.: Reproduction nette en France depuis l'origine de l'Etat-Civil. Paris: Imprimerie Nationale 1941.

Dice, L.R.: Measures of the amount of ecologic association between species. Ecology **23**, 297–302 (1945).

Djemaï, H.: Les Berbères de Taoudjout, un isolat au Sud de la Tunisie. Mém. Inst. Démographie, Université de Paris I, 1972.

Dronamraju, K.R.: Le système des castes et les marriages consangiuns en Andhra Pradesh (Inde). Population **19**, 291–308 (1964).

Eaton, J., Mayer, A.: Man's capacity to reproduce. Hum. Biol. **25**, 1–58 (1954).

Ehrman, L., Petit, C.: Genotype frequency and mating sucess in the *Willistoni* species group of *Drosophila*. Evolution **22**, 649–658 (1968).

Elandt-Johnson, R.C.: General purpose probability models in histocompatibility testing. I. Ann. Hum. Genet. **31**, 293–308 (1968).

Feingold, N., Feingold, J., Frézal, J., Dausset, J.: Probability of graft compatibility. Ann. Hum. Genet. **33**, 285–290 (1970).

Felsenstein, J.: Inbreeding and variance effective number in populations with overlapping generations. Genetics **68**, 581–597 (1971).

Ferry, M.P.: Pour une histoire des Bedik. Cahier du C.R.A. No. 7. Bull. et Mém. Soc. Anthrop., Paris, 13th series, Vol. 2 (1967).

Fisher, R.A.: The correlation between relatives on the supposition of Mendelian inheritance. Trans. Roy. Soc. Edinb. **52**, 399–433 (1918).

Fisher, R.A.: Average excess and average effect of a gene substitution. Ann. Eugen. (Lond.) **11**, 53–63 (1941).

Franklin, I., Lewontin, R.C.: Is the gene the unit of selection? Genetics **65**, 707–734 (1970).

Fraser, A.S., Burnell, D.: Computer models in genetics. London: McGraw-Hill 1970.

Freire-Maia, N.: Inbreeding in Brazil. Amer. J. Hum. Genet. **9**, 284–298 (1957).

Galton, F.: Hereditary Genuis. London: Macmillan 1869. Reprinted by Horizon Press, New York, 1952.

Gantmacher, F.R.: The theory of matrices. New York: Chelsea Publishing Co.1959.

Générmont, J.: Paramètres caractéristiques d'une population homogame, Population **24**, 735–756 (1969).

Georges, A.: Méthodes démographiques en génétique des populations. Doctoral thesis, University of Nancy 1969.

Georges, A., Jacquard, A.: Effects de la consanguinité sur la mortalité infantile. Resultats d'une observation dans le départment des Vosges. Population **23**, 1055–1064 (1968).

Gessain, R.: Étude du Sénégal Oriental. Bull. et Mém. Soc. Anthrop., Paris. Cahiers du C.R.A. 11th series, **5**, 5–85 (1963).

Gessain, R.: Ammassalik ou la civilisation obligatoire. Paris: Flammarion 1970.

Gilbert, J.P., Hammel, E.A.: Computer simulation and analysis of problems in kinship and social structure. Amer. Anthrop. **68**, 71–93 (1966).

Gillois, M.: La relation d'identité génétique. Thesis, Faculté des Sciences, Paris 1964.

Glass, B.: On the unlikelihood of significant admixture of genes from the North American Indians in the present composition of the Negroes of the United States. Amer. J.Hum. Genet. **7**, 368–385 (1955).

Glass, B., Li, C.C.: The dynamics of racial admixture. An analysis based on the American Negro. Amer. J. Hum. Genet. **5**, 1–20 (1953).

Gomila, J.: Les Bedik, barrières culturelles et hétérogénéité biologique. Montreal. Les Presses de l'Université de Montréal 1971.

Hägerstrand, T.: Migration and area. Lund studies in geography, No. **13**, 27–158 (1957).

Hajnal, J.: Random mating and the frequency of consanguineous marriages. Proc. Roy. Soc. B **159**, 125–177 (1963).

Haldane, J.B.S.: A mathematical theory of natural and artificial selection, I. Proc. Camb. Philos. Soc. **23**, 195–217 (1924).

Haldane, J.B.S.: The rate of spontaneous mutation of a human gene. J. Genet. **31**, 317–326 (1935).

Haldane, J.B.S.: Selection against heterozygosis in man. Ann. Eugen. (Lond.) **11**, 333–340 (1942).

Haldane, J.B.S.: The mutation rate of the gene for haemophilia and its segregation ratio in males and females. Ann. Eugen. (Lond.) **13**, 262–271 (1947).

Haldane, J.B.S.: Distance and evolution. Ricerca Sci. (Suppl. 1) 3–10 (1949).

Haldane, J.B.S.: The cost of natural selection. J. Genet. **55**, 511–524 (1957).

Harris, H., Hopkinson, D.A.: Average heterozygosity in man: an estimate based on the incidence of enzyme polymorphisms. Ann. Hum. Genet. **36**, 9–20 (1972).

Harris, P.L.: Genotypic covariance between inbred relatives. Genetics **50**, 1319–1348 (1964).

Henry, L.: Fécondité des marriages. Paris: I.N.E.D. 1953.

Henry, L.: Anciennes familles genevoises. Études démographiques, XVIe-XXe sièdes. Paris: Presses Universitaires de France 1956.

Henry, L.: Schémas de nuptialité: déséquilibre des sexes et célibat. Population **24**, 458–486 (1969).

Hiraizumi, Y.: Prezygotic selection as a factor in the maintenance of variability. Cold Spr. Harb. Symp. Quant. Biol. **29**, 51–60 (1964).

Hiraizumi, Y.: Are the MN blood groups maintained by heterosis? Amer. J. Hum. Genet. **16**, 375–379 (1964a).

Holt, S.: Dermatoglyphic patterns. In: Genetical variation in human populations, ed. G.A. Harrison. New York: Pergamon Press 1961.

Ishinuki, N., Nemoto, H., Neel, J. V., Drew, A. L., Yanase, T., Matsumoto, Y. S.: Hosojima. Amer. J. Hum. Genet. **12**, 67–75 (1960).

Jaccard, P.: Nouvelles recherches sur la distribution florale. Bull. Soc. Vaud. Sci. Nat. **44**, 223–270 (1908).

Jacquard, A.: Logique du calcul des coefficients d'identité entre deux individus. Population **21**, 751–776 (1966).

Jaeger, G.: Étude anthropo-sociale et généalogique de la population Sara Kaba Ndindjo d'un village centrafricain. Population **28**, 361–382 (1973).

Karlin, S., Feldman, M. W.: Analysis of models with homozygote × heterozygote matings. Genetics **59**, 105–116 (1968a).

Karlin, S., Feldman, M. W.: Further analysis of negative assortative mating. Genetics **59**, 117–136 (1968b).

Kendall, M. G., Stewart, A.: The advanced theory of statistics, 3rd ed. London: Griffin 1960.

Kidd, K. K., Sgaramella-Zonta, L. A.: Phylogenetic analysis; concepts and methods. Amer. J. Hum. Genet. **32**, 235–252 (1971).

Kilpatrick, S. S. J., Gamble, E. L.: A model for incompatiblity in kidney transplants. Ann. Hum. Genet. **31**, 309–317 (1968).

Kimura, M.: On the probability of fixation of mutant genes in a population. Genetics **47**, 713–719 (1962).

Kimura, M., Crow, J. F.: The number of alleles that can be maintained in a finite population. Genetics **49**, 725–738 (1964).

Kimura, M., Ohta, T.: The average number of generations until fixation of a mutant gene in a finite population. Genetics **61**, 763–771 (1969a).

Kimura, M., Ohta, T.: The average number of generations until extinction of an individual mutant gene in a finite population. Genetics **63**, 701–709 (1969b).

King, J. L.: Continuously distributed factors affecting fitness. Genetics **55**, 483–492 (1967).

King, J. L., Jukes, T. H.: Non-Darwinian evolution: random fixation of selectively neutral mutations. Science **164**, 788–798 (1969).

Kojima, K.: Is there a constant fitness value for a given genotype? No! Evolution **25**, 281–285 (1971).

Kojima, K., Kelleher, T. M.: Changes of mean fitness in random-mating populations when epistasis and linkage are present. Genetics **46**, 527–540 (1961).

Kojima, K., Lewontin, R. C.: Evolutionary significance of linkage and epistasis. In: Mathematical topics in population genetics, ed. K. Kojima. Berlin-Heidelberg-New York: Springer 1970.

Kulczynski, S.: Die Pflanzenassoziationen der Peininen. Bull. Int. Acad. Pol. Sci. (Sci. Nat.) Suppl. **2**, 57–203 (1927).

Laberge, C.: La consanguinité des canadiens français. Population **22**, 861–896 (1967).

Lalouel, J. M.: Topology of population structure. Proc. Int. Congress on population structure. Hawaii: University of Honolulu Press 1973a.

Lalouel, J. M.: Reduction of a multidimensional configuration: metric versus non-metric methods. In preparation (1973b).

Lamotte, M.: Recherches sur la structure génétique des populations naturelles de *Cepaea nemoralis*. Bull. biol. France Belg., Suppl. **35**, 1–239 (1951).

Langaney, A.: Panmixie, "pangamie" et systèmes de croisement. Population **24**, 301–308 (1969).

Langaney, A., Gomila, J.: Bedik and Niokholonko, intra- and inter-ethnic migration. Hum. Biol. **45**, 137–150 (1973).

Langaney, A., Gomila, J., Bouloux, C. J.: Bedik bioassay of kinship. Hum. Biol. **44**, 475–488 (1972).

Langaney, A., Le Bras, H.: Description de la structure génétique transversale d'une population. Application aux Bedik. Population **27**, 83–116 (1972).

Lefèvre-Witier, P.: L'hémotypologie des Kel Kummer. In preparation (1973).

Lejeune, J., Turpin, R., Gautier, M.: Le mongolisme, premier example d'aberration auto-somique humaine. Ann. Génét. Humaine 1, 41–49 (1959).

Levin, B. R.: The effect of reproductive compensation on the long term maintenance of the Rh polymorphism: the Rh crossroad revisited. Amer. J. Hum. Genet. 19, 288–302 (1967).

Lewontin, R. C.: A general method for investigating the equilibrium of gene frequency in a population. Genetics 43, 421–433 (1958).

Lewontin, R. C.: The genetics of complex systems. Proc. V Berkley Symp. Math. Stat. and Prob. 1967.

Lewontin, R. C., Dunn, L. C.: The evolutionary dynamics of a polymorphism in the house mouse. Genetics 45, 705–722 (1960).

Lewontin, R. C., Krakauer, J.: Distribution of gene frequency as a test of the theory of the selective neutrality of polymorphism. Genetics 74, 175–195 (1973).

L'Heritier, P.: Traité de génétique. Paris: Presses Universitaires de France 1959.

Li, C. C.: Fundamental theorem of natural selection. Nature (Lond.) 214, 505–506 (1967).

Lotka, A. J.: The stability of the normal age-distribution. Proc. Nat. Acad. Sci. (Wash.) 8, 339–345 (1922).

Lunghi, G.: Teoria della frequenz d'incompatibilita de transpianto. Atti Ass. Genet. Ital. 10, 311–319 (1965).

MacCluer, J. W.: Monte Carlo methods in human population genetics: a computer model incorporating age-specific birth and death rates. Amer. J. Hum. Genet. 19, 303–312 (1967).

MacCluer, J. W., Schull, W. J.: Estimating the effective size of human populations. Amer. J. Hum. Genet. 22, 176–183 (1970).

Mahalanobis, P. C.: On tests and measures of group divergence. J. and Proc. Asiat. Soc. Bengal 26, 541–588 (1930).

Mahalanobis, P. C.: On the generalised distance in statistics. Proc. Nat. Inst. Sci. India B, 2, 49–55 (1936).

Malécot, G.: Probabilités et heredité. Cahier No. 47, Travaux et documents I. N. E. D. Paris: Presses Universitaires de France 1966.

Matsunaga, E.: Selection in AB0 polymorphism in Japanese populations. Amer. J. Hum. Genet. 11, 405–413 (1959).

Matsunaga, E., Itoh, S.: Blood groups and fertility in a Japanese population, with special reference to intra-uterine selection due to maternal-foetal incompatibility. Ann. Hum. Genet. 22, 111–131 (1958).

Milkman, R. D.: Heterosis as a major cause of heterozygosity in nature. Genetics 55, 493–495 (1967).

Mørch, E. T.: Chondrodystrophic dwarfs in Denmark. Copenhagen: Munksgaard 1941.

Morlat, G., Caillez, F., Mailles, J. P., Nakache, J. P., Pages, J. P.: Analyse de donnés multi-dimensionelles. Paris: C. E. E. E. 1971.

Moroni, A.: Evoluzione della frequenza de matrimoni consanguiniti in Italia negli ultima cinquant'anni. Atti Ass. Genet. Italia 9, 207–223 (1964).

Morton, N. E.: Sequential tests for the detection of linkage. Amer. J. Hum. Genet. 7, 277–318 (1955).

Morton, N. E.: Detection and estimation of linkage between the genes for elliptocytosis and the Rh blood type. Amer. J. Hum. Genet. 8, 80–95 (1956).

Morton, N. E.: Further scoring types in sequential linkage tests, with a critical review of autosomal and partial sex linkage in man. Amer. J. Hum. Genet. 9, 55–75 (1957).

Morton, N. E.: Empirical risk in consanguineous marriages: birth weight, gestation time and measurements of infants. Amer. J. Hum. Genet. 10, 344–349 (1958).

Morton N. E., Chung, C. S.: Are the MN blood groups maintained by selection? Amer. J. Hum. Genet. 11, 237–251 (1959).

Morton, N. E., Crow, J. F., Muller, H. J.: An estimate of the mutational damage in man from data on consanguineous marriages. Proc. Nat. Acad. Sci. (Wash.) 42, 855–863 (1956).

Morton, N. E., Krieger, H., Mi, M. P.: Natural selection on polymorphisms in northeastern Brazil. Amer. J. Hum. Genet. **18**, 153–171 (1966).

Morton, N. E., Yee, S., Harris, D. E., Lew, R.: Bioassay of kinship. Theoret. Pop. Biol. **2**, 507–524 (1971).

Nadot, R., Vayssein, G.: Apparentement et identité. Algorithme du calcul des coefficients d'identité. Biometrics **29**, 347–359 (1973).

Neel, J. V.: Mutations in the human population. In: Methodology in human genetics, ed. W. J. Burdette. San Francisco: Holden-Day 1962.

Neel, J. V., Falls, H. F.: The rate of mutation of the gene responsible for retinoblastoma in man. Science **114**, 419–422 (1952).

Ohta, T., Kimura, M.: Linkage disequilibrium between two segregating nucleotide sites under the steady flux of mutation in a finite population. Genetics **68**, 571–580 (1971).

Olsson, G.: Distance and human interaction: a review and bibliography. Regional Sci. Res. Inst. Bibliography Series No. 2, Philadelphia, 1965 a.

Olsson, G.: Distance and human interaction: a migration study. Geografiska Annaler **47**, 3–43 (1965 b).

Pearson, K.: On the coefficient of racial likeness. Biometrika **18**, 105–117 (1926).

Penrose, L. S.: Detection of autosomal linkage. Ann. Eugen. (Lond.) **6**, 133–138 (1935).

Penrose, L. S.: A further note on the sib-pair linkage method. Ann. Eugen. (Lond.) **13**, 25–29 (1946).

Penrose, L. S.: The general-purpose sib-pair linkage test. Ann. Eugen. (Lond.) **18**, 120–124 (1953).

Penrose, L. S.: Distance, size and shape. Ann. Eugen. (Lond.) **18**, 337–343 (1954).

Perrin, E. B., Sheps, M. C.: Human reproduction: a stochastic process. Biometrics **20**, 28–45 (1964).

Pressat, R.: Cours d'analyse démographique de l'I.D.U.P. Paris: I.N.E.D. 1966.

Race, R. R., Sanger, R.: Blood groups in man. 4th edition. Oxford: Blackwell 1962.

Reed, T. E., Chandler, J. H.: Huntington's chorea in Michigan. Amer. J. Hum. Genet. **10**, 201–221 (1958).

Reed, T. E., Gershowitz, H., Soni, A., Napier, J.: A search for natural selection in six blood group systems and ABH secretion. Amer. J. Hum. Genet. **16**, 161–179 (1964).

Reed, T. E., Neel, J. V.: Huntington's chorea in Michigan. 2. Selection and mutation. Amer. J. Hum. Genet. **11**, 107–136 (1959).

Renwick, J. H.: The mapping of human chromosomes. Ann. Rev. Genet. **5**, 81–120 (1971).

Robert, J.: Anthropologie démographique et socio-économique de la population du Scoresbysund. Thesis, Faculté des Lettres, Paris (1970).

Robert, J. M.: La génétique de la vie. Cah. Medicaux Lyonnais **42**, 123–224 (1966).

Roberts, D. F., Hiorns, R. W.: The dynamics of racial intermixture. Amer. J. Hum. Genet. **14**, 261–277 (1962).

Robertson, A.: A theory of limits in artificial selection with many linked loci. In: Mathematical topics in population genetics, ed. K. Kojima. Berlin-Heidelberg-New York: Springer 1970.

Rucknagel, D. L., Neel, J. V.: The haemoglobinopathies. Proc. Med. Genet. **1**, 158–260 (1961).

Scheuer, P. A. G., Mandel, S. P. H.: An inequality in population genetics. Heredity **13**, 519–524 (1959).

Schull, W. J.: Empirical risks in consanguineous marriages: sex ratio, malformation and viability. Amer. J. Hum. Genet. **10**, 294–343 (1958).

Schull, W. J., Neel, J. V.: Sex linkage, inbreeding and growth in childhood. Amer. J. Hum. Genet. **15**, 106–114 (1963).

Schull, W. J., Neel, J. V.: The effects of inbreeding on Japanese children. New York: Harper and Row 1965.

Serra, A., O'Mathuna, D.: A theoretical approach to the study of genetic parameters of histocompatibility in man. Ann. Hum. Genet. **30**, 97–118 (1966).

Serra, A., Soini, A.: La consanguinite d'une population. Application a trois provinces de l'Italie du nord. Population **14**, 47–69 (1959).

Shannon, C.F.: A mathematical theory of comunication. Bell Syst. Tech. J. 379–423 and 623–656 (1948).

Shields, J.: Monozygotic twins brought up apart and brought up together. London: Oxford University Press 1962.

Siniscalco, M., Bernini, L., Latte, B., Motulsky, A.: Favism and thalassaemia in Sardinia and their relationship to malaria. Nature (Lond.) **190**, 1179–1180 (1961).

Smith, C.A.B.: The detection of linkage in human genetics. J. Roy. Stat. Soc. **15**, 153–192 (1953).

Smith, C.A.B.: Some comments on the statistical methods used in linkage investigations. Amer. J. Hum. Genet. **11**, 289–304 (1959).

Smith, S.M., Penrose, L.S.: Monozygotic and dizygotic twin diagnosis. Ann. Hum. Genet. **19**, 273–289 (1955).

Smith, S.M., Penrose, L.S., Smith, C.A.B.: Mathematical tables for research workers in human genetics. London: Churchill 1961.

Snyder, L.H.: The inheritance of taste deficiency in man. Ohio J. Sci. **32**, 436–440 (1932).

Sokal, R.R., Sneath, P.H.A.: Principles of numerical taxonomy. San Francisco: Freeman 1963.

Stern, C.: Model estimates of the frequency of white and near white genes in the American negro. Acta Genet. (Basel) **4**, 281–297 (1953).

Stouffer, S.: A theory relating mobility and distance. Amer. Sociol. Rev. **5**, 845–867 (1940).

Sutter, J.: Evolution de la distance séparant le domicile des futurs époux. Population **13**, 227–258 (1958).

Sutter, J.: L'atteinte des incisives latérales supérieures. I.N.E.D. Paris: Presses Universitaires de France 1966.

Sutter, J.: Fréquence de l'endogamie et ses facteurs au XIX siècle. Population **23**, 303–324 (1968).

Sutter, J.: La luxation congénitale de la hanche. I.N.E.D. Paris: Presses Universitaires de France 1972.

Sutter, J., Goux, J.M.: Evolution de la consanguinité en France de 1926 à 1958 avec des donnés récentes détaillées. Population **17**, 683–702 (1962).

Sutter, J.: Fréquence de l'endogamie et ses facteurs au XIX century. Population **23**, 303–324 (1968).

Sutter, J., Tabah, L.: Fréquence et répartition des mariages consanguins en France. Population **3**, 607–630 (1948).

Sutter, J., Tabah, L.: Effets des mariages consanguins sur la déscendance. Population **6**, 59–82 (1951).

Sutter, J., Tabah, L.: Effets de la consanguinité et de l'endogamie. Population **7**, 249–266 (1952).

Sutter, J., Thanh, L.M.: Contribution à l'étude de la répartition des distances séparant les domiciles des époux dans un départment français. Influence de la consanguinité. In: Les déplacements humains. Entretiens de Monaco en sciences humaines. Paris: Hachette 1962.

Sved, J.A., Reed, T.E., Bodmer, W.J.: The number of balanced polymorphisms that can be maintained in a natural population. Genetics **55**, 469–481 (1967).

Thoma, A.: Evolution et différenciation du polymorphisme AB0. In: Génétique et populations, Cahier No. 60, I.N.E.D. Paris: Presses Universitaires de France 1971.

Tjio, J.H., Levan, A.: The chromosome number of man. Hereditas (Lund.) **42**, 1–6 (1956).

Twiesselmann, F., Moureau, P., François, J.: Evolution du taux de consanguinité en Belgique de 1918 à 1959. Population **17**, 241–266 (1962).

Wahlund, S.: Zusammensetzung von Populationen und Korrelationserscheinungen vom Standpunkt der Vererbungslehre aus betrachtet. Hereditas (Lund.) **11**, 65–106 (1928).

Williams, W. T., Lambert, J. M.: Multivariate models in plant ecology. II. The use of an electronic digital computer for association analysis. J. Ecol. **48**, 689–710 (1960).

Wold, H. O. A.: Introduction to La technique des modèles dans les sciences humaines, ed. H. O. A. Wold. Monaco: Editions Sciences Humaines 1964.

Workman, P. L.: The maintenance of heterozygosity by partial negative assortative mating. Genetics **50**, 1369–1382 (1964).

Workman, P. L., Blumberg, B. S., Cooper, A. J.: Selection, gene migration and polymorphic stability in a U.S. white and Negro population. Amer. J. Hum. Genet. **15**, 429–437 (1963).

Wright, S.: Systems of mating. Genetics **6**, 111–178 (1921a).

Wright, S.: Correlation and causation. J. Agric. Res. **20**, 557–585 (1921b).

Wright, S.: Coefficients of inbreeding and relationship. Amer. Naturalist **56**, 330–338 (1922).

Wright, S.: Evolution in Mendelian populations. Genetics **16**, 97–159 (1931).

Wright, S.: The distribution of gene frequencies in populations. Proc. Nat. Acad. Sci. (Wash.) **32**, 307–320 (1937).

Wright, S.: Isolation by distance. Genetics **28**, 114–138 (1943).

Wright, S.: The differential equation of the distribution of gene frequencies. Proc. Nat. Acad. Sci. (Wash.) **31**, 382–389 (1945).

Wright, S.: Isolation by distance under diverse systems of mating. Genetics **31**, 39–59 (1946).

Wright, S.: Adaptation and selection. In: Genetics, paleontology and evolution, eds. G. L. Jepsen, E. Mayr, and G. G. Simpson. Princeton: Univ. Press 1949.

Wright, S.: The genetical structure of populations. Ann. Eugen. (Lond.) **15**, 323–354 (1951).

Subject Index

Springer-Verlag
Berlin · Heidelberg · New York
München · Johannesburg · London · Madrid · New Delhi
Paris · Rio de Janeiro · Sydney · Tokyo · Utrecht · Wien

Biomathematics

Edited by K. Krickeberg, R. C. Lewontin, J. Neyman,
and M. Schreiber

Vol. 1

With 55 figures
IX, 400 pages. 1970
Cloth DM 68,–
ISBN 3-540-05054-X

Mathematical Topics in Population Genetics

Edited by **Ken-ichi Kojima,** Department of Zoology,
University of Texas, Austin, Tex.

This book is unique in bringing together in one volume many,
if not most, of the mathematical theories of population genetics presented in the past which are still valid and some of
the current mathematical investigations.

Vol. 2

With 200 figures
XIV, 495 pages. 1971
Cloth DM 49,–
ISBN 3-540-05522-3

Distribution rights for India:
The Universal Book Stall
(UBS), New Delhi

Introduction to Mathematics for Life Scientists

By **Edward Batschelet,** Ph. D., Professor of Mathematics,
University of Zurich, Switzerland

This book introduces the student of biology and medicine to
such topics as sets, real and complex numbers, elementary
functions, differential and integral calculus, differential equations, probability, matrices and vectors.

■ **Prospectus on request**

Biomathematics   Edited by K. Krickeberg, R. C. Lewontin, J. Neyman, and M. Schreiber

Stochastic Processes and Applications in Biology and Medicine

Part I: **Theory**
331 pages. 1973
(Biomathematics, Vol. 3)
Cloth DM 53,—
ISBN 3-540-06270-X

Part II: **Models**
337 pages. 1973
(Biomathematics, Vol. 4)
Cloth DM 53,—
ISBN 3-540-06271-8

Distribution rights for the Socialist
countries: Editura Academiei,
Bucharest

By **M. Iosifescu** and **P. Tăutu,** Academy of the
Socialist Republic of Romania,
Centre of Mathematical Statistics, Bucharest

Intended to introduce mathematicians and biologists with strong mathematical background to the study of stochastic processes and their applications in medicine and biology, this book is a considerably expanded and improved version of the authors' book which appeared in Rumanian in 1968.

The first two chapters (Part I) treat those parts of the theory of stochastic processes which are the most relevant to biological applications.

Chapter 1 deals with discrete-parameter stochastic processes with emphasis on denumerable Markov chains. Chapter 2, dealing with continuous-parameter processes, treats in particular processes with independent increments, Markov (jump and diffusion) processes, and semi-Markov processes.

Chapter 3 (Part II) surveys the most important stochastic models in demography, genetics, epidemics, chronic diseases, physiology and pharmacology, with special reference to recent developments. Each model is described from both biological and probabilistic points of view.

Prospectus on request

Springer-Verlag
Berlin Heidelberg New York
München Johannesburg London Madrid New Delhi
Paris Rio de Janeiro Sydney Tokyo Utrecht Wien